Physical constants

Quantity	Symbol	Value
Speed of light in a vacuum	c	$2.997\,925 \times 10^{8}\ \mathrm{m\,s^{-1}}$
Avogadro constant	L	$6.0221 \times 10^{23}\ \mathrm{mol^{-1}}$
Permeability of a vacuum	μ_0	$4\pi \times 10^{-7}\ \mathrm{H\,m^{-1}}$
Permittivity of a vacuum	$\varepsilon_0 = \mu_0^{-1} c^{-2}$	$8.854\,18 \times 10^{-2}\ \mathrm{F\,m^{-1}}$
Charge of proton (charge on the electron is $-e$)	e	$1.602\,19 \times 10^{-19}\ \mathrm{C}$
Faraday constant	$F = Le$	$9.6487 \times 10^{4}\ \mathrm{C\,mol^{-1}}$
Boltzmann constant	k_B	$1.3806 \times 10^{-23}\ \mathrm{J\,K^{-1}}$
Planck constant	h	$6.626\,196 \times 10^{-34}\ \mathrm{J\,s}$
Rydberg constant	$\mathcal{R}$	$2.179\,908 \times 10^{-23}\ \mathrm{J}$
		$1.097\,373 \times 10^{5}\ \mathrm{cm^{-1}}$
Gas constant	$R = Lk_B$	$8.3143\ \mathrm{J\,K^{-1}\,mol^{-1}}$
		$1.9872\ \mathrm{cal\,K^{-1}\,mol^{-1}}$
		$8.2056 \times 10^{-2}\ \mathrm{dm^{3}\,atm\,K^{-1}\,mol^{-1}}$
Atomic mass unit	u	$1.660\,53 \times 10^{-27}\ \mathrm{kg}$

Defined constants

Quantity	Symbol	Value
Standard acceleration of gravity	g	$9.80665\ \mathrm{m\,s^{-2}}$
Standard atmosphere	$p^{\ominus}$	$101.325\ \mathrm{kPa}$
		$1.013\,25 \times 10^{5}\ \mathrm{N\,m^{-2}}$
		$1.013\,25 \times 10^{6}\ \mathrm{dyne\,cm^{-2}}$
		1 atm
		760 mmHg
		760 Torr
Temperature of ice point		273.15 K

Symbols

See p. xiii

Conversion factors and relations

1 eV	$1.602\,189 \times 10^{-19}$ J molecule^{-1}
	96.485 kJ mol^{-1}
	8065.5 cm^{-1}
1 cm^{-1}	1.986×10^{-23} J molecule^{-1}
	11.96 J mol^{-1}
	1.240×10^{-4} eV
1 debye	$3.335\,64 \times 10^{-30}$ C m
1 inch	2.540 cm
1 pound	453.59 g
1 fluid ounce	29.5727 cm^3
1 °F	0.5556 °C
1 litre	1000.028 cm^3
1 Ångstrom	10^{-8} cm
	0.1 nm
1 bar	10^6 dyne cm^{-2}
	100 kPa
	0.986 92 atm
1 Torr	1 mmHg
	133.3 Pa
	3.24×10^{16} molecule cm^{-3} at 298 K
1 joule	A V s
	10^7 erg
	9.8691 cm^3 atm
	0.23901 cal
ln x	2.302 59 log x
1 newton	J m^{-1}
	10^5 dyn
1 watt	J s^{-1}
1 tesla	V s m^{-2}
1 coulomb	A s

Units

In general, SI units have been used in this book, except in three cases: (i) where apparatus is invariably calibrated in other units, (ii) where the working equations for the experiment require different units, and (iii) where units other than SI units are still invariably used by convention. Examples of these three cases are (i) Pirani pressure gauges which are calibrated in Torr, and mercury manometers read in mmHg, (ii) the use of atm for the unit of pressure in the expression for K_p, and (iii) the use of cm^6 molecule^{-2} s^{-1} for the third-order rate coefficient (Section 8).

Experimental Physical Chemistry

Experimental Physical Chemistry

G. PETER MATTHEWS

Physical Chemistry Laboratory
University of Oxford

CLARENDON PRESS · OXFORD
1985

Oxford University Press, Walton Street, Oxford OX2 6DP

Oxford New York Toronto
Delhi Bombay Calcutta Madras Karachi
Kuala Lumpur Singapore Hong Kong Tokyo
Nairobi Dar es Salaam Cape Town
Melbourne Auckland

and associated companies in
Beirut Berlin Ibadan Nicosia

Oxford is a trade mark of Oxford University Press

Published in the United States
by Oxford University Press, New York

British Library Cataloguing in Publication Data
Matthews, G. Peter
Experimental physical chemistry.
1. Chemistry, Physical and theoretical
I. Title
541.3 QD453.2
ISBN 0-19-855162-2
ISBN 0-19-855212-2 Pbk

Library of Congress Cataloging in Publication Data
Matthews, G. Peter.
Experimental physical chemistry.
Bibliography
Includes index.
1. Chemistry, Physical and theoretical—Laboratory manuals. I. Title.
QD457.M35 1985 541.3'07'8 85-10471
ISBN 0-19-855162-2
ISBN 0-19-855212-2

Set and Printed in the United Kingdom by
The Universities Press, (Belfast), Ltd
Northern Ireland

Preface

There is a danger that the experimental course of any chemistry department becomes its Cinderella—outdated, unloved, and clothed in drab apparel. Any author who has the temerity to produce a book for the prevention of such calamity must explain his philosophy, and how it is translated into print.

Most of my comments will be directed towards staff and students in colleges, polytechnics, and universities. First, however, a word to schoolteachers. One or two of the experiments in this book have been publicised already, and the response from schools has been enthusiastic. Particular interest has been shown in the computer titration experiment (4.6) and the Brownian motion experiment (7.1). A number of other experiments should be of interest to schools, and are labelled with a letter A in the contents list. They may require obvious simplification of their theory.

What, then, of philosophy? Essentially, a laboratory course in physical chemistry should teach a range of practical skills which are generally considered relevant to a modern approach to the subject, and which will be of use to a student who wishes to apply his learning to his career. I have scrapped or ignored many experiments which I feel do not meet this requirement, and have included studies which give practice in the simple preparation and observation of solutions, the use of microscopes, glassblowing, the operation of vacuum lines, the use of spectrometers, and many other skills.

To speak of a modern approach to the subject implies that the experiments should reflect current trends in physical chemistry, and the two most evident developments are in the use of lasers and computers. Lasers present a safety problem, but, by dint of numerous safety interlocks, we have a laser Raman experiment running in the department at Oxford, which I have included (Expt 5.8). With regard to computers, perhaps the most dispiriting aspect of their use in practical courses is as inscrutable machines which magically devour imperfect results and convert them to impressive multi-coloured graphs. Clearly laboratories should have a range of package programs available, and computer programs should be used where the calculation of results by electronic calculator is virtually impossible (Expts 5.4 and 8.5) or very tedious (e.g. Expts 6.2 and 8.6). Apart from this they should be treated with suspicion unless they enhance the experiments and teach new skills. Three experiments have been introduced with this positive aim in mind. The first, Expt 4.6, involves the programming of a microcomputer in BASIC to operate a simple auto-titrator. As the students themselves have pointed out, once the programming has been done, the automation allows them some time to sit back and think about the chemistry involved. In addition there are two experiments which involve computing alone. One is the calculation of conformational energies by extended Hückel theory, in which students have to format the input data, and process the output data to answer several questions about the structure of simple molecules (Expt 5.13). The other (Expt 7.2) is the simulation of phases by molecular dynamics, in which students have to write a simple routine to describe the interactions between molecules. Computer programs for all seven of the experiments mentioned are available from the author.

The idea of modernity leads to my next point, which is that a practical course should relate to the courses which are currently taught as physical chemistry. There would be little point, however, in my repeating experiments from other textbooks on the subject, although the reader will notice that I have not avoided such overlap entirely. To satisfy both these conditions and the criteria mentioned earlier, I have scoured many departments, and here present my own selection of 51 experiments gleaned from polytechnic and university departments of physics, chemistry, and chemical engineering, at thirteen different centres in this country, the United States, and Australia. A further eight experiments have been specially developed with this book in mind.

Nevertheless the selection of experiments presented in this book is open to further improvement. I do not believe that surface chemistry, for example, is well served solely by Expts 2.9 and 2.10. Advanced experiments using specialist, expensive items of apparatus also present a problem. Such experiments tend to be specific to the apparatus in use. Furthermore the existence of the apparatus in the department implies that there is a member of staff with the expertise to use it, who may not be greatly assisted by a text such as this. To ignore advanced experiments entirely, however, would be to betray their importance. Thus I have chosen to include nuclear magnetic resonance, light scattering, laser Raman and mass spectrometry, but have excluded interesting experiments involving rotating disc electrodes, X-ray scattering, neutron activation analysis, electron-spin resonance spectroscopy, and photo-electron spectroscopy.

Finally, much thought has been put into the ordering and structure of the experiment scripts themselves. Students walking with trepidation into a strange laboratory are often discouraged when they pick up instructions which plunge them into unknown areas of their curriculum. A number of measures have been taken to introduce the subject in an interesting way. The most visible is the use of marginal notes, which I hope will enliven the text without detracting from it. Secondly, I have attempted to group the experiments imaginatively. Each of the sections has an introduction to put the experiments into context, and to give a brief idea of the experimental skills taught or required, and the type and level of the theory involved. Each experiment script has an introductory section of its own, shedding further light on these matters, and explicitly stating the purpose of the experiment. The *Theory* section which follows is intended to enable the student to understand the derivation of the working equations, and to relate the experiment to accounts from textbooks, lectures, or classes. It is designed to be read on the laboratory bench, and the approach and notation has therefore been simplified as much as possible. Only when the experimental results demand it do the discussions and equations verge on rigour. The use of Appendices is minimal, since in the laboratory environment they are often ignored. The *Apparatus* section details the instruments and devices that are used to investigate the phenomenon which has been discussed. The *Procedures* are intended to be reasonably general, but asterisks (*) are used to indicate where further information must be sought or supplied. After the *Calculation* is a *Results* section which is intended to give guidance to students, who may then feel less bound to consult a previous 'correct' account, and which is also intended to help instructors when they are setting up and testing the apparatus. The *Technical notes* are also designed to assist instructors and technicians. The apparatus lists which they contain include $ signs to indicate expense, or in the case of $$, considerable expense, and also † signs which indicate that the apparatus can be set up by an enthusiastic amateur, or, in the case of ††, by a skilled technician. The items in square brackets may not be required. Finally there are references which do not for the most part refer to general texts, except where they are particularly apposite, but do indicate the source of the experiment, further and more rigorous theory, and the source of any data.

The reader will notice the absence of techniques experiments and projects. Active consideration has been given to experiments which are purely designed to teach practical techniques, but there are difficulties, as has been found both at Oxford and in other departments. The problems are enshrined in requests of the type: 'Go and play on that obsolete vacuum line until you're confident you won't wreck this one'. Projects are also difficult to encompass in a book of this nature, because they are so often the province of specialist interest.

In the production of this book each experiment has been sought out or developed from scratch, redrafted or rewritten entirely, tried on students, and circulated to experts in the field. Such a task, which at times seemed of a magnitude which threatened to engulf me, would not have been remotely possible but for the able and willing assistance of a great many people. First I must thank Dr W. G. Richards, who proposed the idea of this book, and who has been a source of encouragement throughout its development. I am indebted to Professor J. S. Rowlinson for his permission for me to use my departmental position and the resources of the department as the corner-stone of the book, and I am grateful to both Professor Rowlinson and Professor I. M. Mills for their advice on the book in its entirety. I also thank Dr P. W. Atkins and Dr E. B. Smith for their views on format and the approach to the subject of experimental physical chemistry.

I am grateful to Dr R. Compton, Dr E. B. Smith, and Dr M. J. Pilling for reading entire sections. I also thank the following, whose assistance has ranged from single items of advice to the development of entire sets of experiments: W. J. Albery (Imperial College, London), M. P. Allen (Oxford), R. R. Baldwin (Hull), R. F. Barrow (Oxford), R. J. Bishop (Kingston Polytechnic, London), A. Callear (Cambridge), M. S. Banna (British Columbia, Canada), C. J. Danby (Oxford), J. Eland (Oxford), D. H. Everett (Bristol), R. Freeman (Oxford), R. P. H. Gasser (Oxford), R. Greet (Southampton), G. Hancock (Oxford), A. Hamnett (Inorganic Chemistry, Oxford), T. Hardman (Reading), J. Hatton (Physics, Oxford), D. F. Klemperer (Bristol), J. D. Lambert (Oxford), K. A. McLauchlan (Oxford), G. C. Maitland (Chemical Engineering, Imperial College, London), R. B. Moodie (Exeter), D. H. Napper (Sydney, Australia), A. D. Pethybridge (Reading), R. Popplewell (Oxford), I. Powis (Oxford), W. G. Richards (Oxford), G. Saville (Chemical Engineering, Imperial College, London), D. P. Shoemaker (Oregon, U.S.A.), J. Shorter (Hull), C. J. S. M. Simpson (Oxford), L. E. Sutton (Oxford), F. L. Swinton (Coleraine, N. Ireland), R. K. Thomas (Oxford), D. J. Tildesley (Southampton), C. F. H. Tipper (Liverpool), D. W. Turner (Oxford), B. Vincent (Bristol), R. W. Walker (Hull), R. P. Wayne (Oxford), J. le P. Webb (Sussex), J. W. White (Oxford), and D. A. Young (Chemical Engineering, Imperial College, London). I must also thank the following research students for their very considerable contributions: W. P. Baskett, D. J. F. Chaundy, M. A. Cordiero, A. R. Tindell, and J. R. P. B. Walton.

The production of this book owes much to the artistry of David Koslow, the checking of references by Chris Ilett, the exceptional word-processing expertise of Marjorie Wallace, and the patient cooperation of the staff at Oxford University Press. Despite such a wealth of assistance, I have abrogated no responsibility for the text, and any residual errors of fact or omission, which are so difficult to eradicate in a work of this nature, are mine alone.

My final thanks must go to the scores of undergraduates who have been so keen to test the new experiments, and whose enthusiasm for the project has been my ultimate reward.

Plymouth Polytechnic G. P. M.
September 1984

Contents

Suggested levels of interest
A: School 6th-form projects/Freshmen
B: 1st year and early 2nd year students/Sophomores and early Juniors
C: 2nd and 3rd year students/Juniors and Seniors

List of symbols xiii
Reports xvii
Errors xviii
Graphs xxiv
Safety xxvi

Part I Equilibrium

Section 1 **3**
Gases at equilibrium

1.1 Molar masses by gas density balance B 3
1.2 Gas imperfection by pV measurement B 9
1.3 Joule–Thomson coefficients B 16

Section 2 **21**
Equilibria between phases

2.1 Properties of a system of three partially miscible liquids A, B 22
2.2 Partial molar volume by the buoyancy method A, B 28
2.3 Vapour pressures of liquids A, B 33
2.4 Distillation of ideal and azeotropic liquid mixtures A, B 37
2.5 Molar mass from depression of freezing point A, B 46
2.6 Ideality of solutions from solubility curves A, B 52
2.7 Enthalpy and entropy terms from depression of freezing and transition temperatures A, B 55
2.8 Phase transitions in solids by microcalorimeter B, C 58
2.9 Surface tension of a liquefied gas by the capillary rise method B, C 66
2.10 Adsorption of gas on charcoal by pressure measurements B, C 73

Section 3 **80**
Thermodynamics of chemical reactions

3.1 Enthalpies of combustion by bomb calorimeter B 80
3.2 Enthalpy of dimerization by Dumas' vapour density method B 89
3.3 Enthalpy of dissociation by pressure measurement B 95

3.4 Enthalpy of CO_2 reduction by a flow method B 100
3.5 Enthalpies and entropies of complex formation in a solution calorimeter B, C 104

Section 4 **113**
Ions at equilibrium

4.1 Solubility of ionic salt by titration A, B, C 114
4.2 Standard electrode potentials, solubility product, and water dissociation constant from e.m.f. measurements A, B 120
4.3 Standard electrode potential, activity coefficients, and acid dissociation constant from Harned cell e.m.f.s B, C 137
4.4 Thermodynamics of a chemical reaction from temperature coefficient of a cell e.m.f. A, B 142
4.5 Redox potentials and equivalence points from potentiometric titrations A, B, C 145
4.6 Acid dissociation constants and analysis by microcomputer controlled pH titrations A, B, C 151

Part II Structure

Section 5 **167**
Structure of atoms and simple molecules

5.1 Quantitative and qualitative analysis by mass spectrometry C 170
5.2 Dipole moment from measurements of relative permittivity and refractive index B, C 183
5.3 Ionization potentials and flame temperature from atomic spectra A, B, C 193
5.4 Pure rotation spectra using Michelson interferometer C 206
5.5 Thermodynamics of hydrogen bond formation by infrared spectroscopy B, C 214
5.6 Rotation–vibration spectra of simple diatomic and polyatomic molecules C 220
5.7 Heat capacity of a gas from infrared spectrum and acoustic interferometry C 230
5.8 Vibrational Raman spectroscopy of simple polyatomic molecules C 240
5.9 Formula and stability constant of a complex by spectrophotometry A, B, C 248
5.10 Dissociation energy of iodine from visible absorption spectrum C 253
5.11 Fluorescence spectrum and rate of electron transfer reaction by quenching C 261
5.12 Keto–enol ratios, hydrogen bonding, and functional group analysis by n.m.r. spectroscopy C 269
5.13 Conformational energies by molecular quantum mechanics C 280

Section 6 **288**
Structure of macromolecules

6.1 Conformation of macromolecules by dilute-solution viscometry B, C 289
6.2 Properties of polymer solutions by light scattering photometry C 300
6.3 Properties of a solid polymer by dielectric relaxation C 312

Part III Change

Section 7 **325**
Particles in motion

7.1 Avogadro's constant from Brownian motion A, B, C 326
7.2 Computer simulation of phases by molecular dynamics C 336
7.3 Gas viscosities by capillary flow viscometer B, C 350
7.4 Dissociation constants and solubility from conductance of electrolyte solutions A, B 359
7.5 Temperature variation of conductivity and viscosity B 368
7.6 Transport numbers by moving boundary method B 372
7.7 Liquid junction potentials and transport numbers from cell e.m.f.s B, C 377
7.8 Kinetics of a diffusion-controlled heterogeneous reaction by titration B, C 382
7.9 Kinetics of substituted nitrobenzene reduction at dropping mercury electrode C 386

Section 8 **395**
Reaction kinetics

8.1 Kinetics of propanone iodination by titration A, B 396
8.2 Persulphate–iodide clock reaction by initial rates method A, B 400
8.3 Mutarotation of dextrose by polarimeter B 406
8.4 Kinetics of 2-iodo-2-methylbutane hydrolysis by conductivity B 413
8.5 Kinetics of consecutive and parallel reactions by spectrophotometry B, C 418
8.6 Radiochemical measurements and the kinetics of radioactive decay B, C 425
8.7 Kinetics of ethanal photolysis by pressure measurement and gas chromatography C 441
8.8 Kinetics of atomic oxygen reactions by a flow method C 449
8.9 Kinetics of iodine atom recombination by flash photolysis C 460
8.10 Kinetics of the hydrogen–oxygen reaction from properties of the second explosion limit C 469

Index **479**

List of symbols

Listed below are the most common meanings of the symbols used in this book, which are not necessarily given in the text when the usage is unambiguous. Further specific meanings, limited to single experiments, are defined at their first occurrence.

Symbol	Meaning
a	activity, absorptivity
aq	aqueous solution
A	area, Debye–Hückel constant
b	Langmuir constant
B	second virial coefficient, magnetic field strength, brightness, rotational constant
B'	second virial coefficient of pressure series virial equation
c	concentration, speed of light in vacuum, intercept of a straight line
C	third virial coefficient, number of components, capacitance, count rate
C'	third virial coefficient of pressure series virial equation
C_{p}	heat capacity at constant pressure
C_{V}	heat capacity at constant volume
d	diameter
d	*in calculus* a limitingly small change
D	diffusion coefficient, centrifugal distortion constant
e	electronic charge, emissivity
e	base of natural logarithms ($\simeq$2.7183)
E	potential difference (e.m.f.), electric field intensity, energy, probable error
E_{a}	activation energy
$E^{\ominus}$	standard electrode potential
$E^{\ominus\prime}$	formal redox potential
f	force, function
F	number of degrees of freedom, viscous force
g	gas
g	acceleration due to gravity, gyromagnetic ratio, degeneracy
G	Gibb's free energy, vibrational term value, conductance
h	Planck's constant, height
$\hbar$	$h/2\pi$
H	enthalpy, Hamiltonian operator

i	$\sqrt{-1}$
i	current
I	moment of inertia, intensity of radiation, ionic strength
j	total angular momentum quantum number
J	rotational quantum number, coupling constant, flux
k	rate coefficient
k_B	Boltzmann constant
K	equilibrium constant, rotational quantum number
$K^\ominus$	standard equilibrium constant
K_f	cryoscopic constant
K_p	equilibrium constant in terms of partial pressures
K_{sp}	solubility product
l	liquid
l	length, orbital angular momentum quantum number
ln	natural logarithm
$\log_{10}$	logarithm to base 10
L	Avogadro's constant
m	mass, molality, mean, gradient of a straight line
M	molar mass (molecular weight), molarity
M	mol dm^{-3}
n	number of moles, refractive index, *in spectroscopy* number of molecules
p	pressure, permanent electric dipole moment
pH	$-\log_{10}(a_{H^+})$
P	probability, number of phases, polarization
q	heat absorbed by system, partition function for an isolated molecule, charge
Q	vibrational co-ordinate
r	radius, relative density
R	gas constant, resistance
$\mathcal{R}$	Rydberg constant
s	solid
s	solubility, spin quantum number
S	entropy
t	time, transport number
T	thermodynamic temperature
U	internal energy, potential energy
v	velocity, vibrational quantum number
V	volume, voltage
w	work done on system
W	mass
x	mole fraction
z	number of charges
Z	collision frequency factor
α	degree of dissociation, optical rotation, polarizability, transfer coefficient

γ	activity coefficient, surface tension, C_p/C_V, magnetogyric ratio
$\gamma^{\pm}$	mean ionic activity coefficient
δ	*in calculus* a very small change, chemical shift, film thickness
∂	*in calculus* a partial differential (i.e. a differential with one or more other parameters held constant)
Δ	*in calculus* a finite change, *in thermodynamics* the change in a thermodynamic function from one state to another
ε	molecular energy, dielectric constant, permittivity (Expt 5.2) *or* relative permittivity (Expt. 6.3), decadic extinction coefficient, well depth
ε_0	permittivity of a vacuum
ε_r	relative permittivity
η	viscosity coefficient, overpotential
θ	surface coverage, angle
κ	electric conductivity, Naperian extinction coefficient
λ	wavelength, ionic conductivity, mean free path
Λ	molar conductivity
Λ_0	molar conductivity at infinite dilution
μ	chemical potential, reduced mass, magnetic moment
μ_{JT}	Joule–Thomson coefficient
ν	frequency, *in statistics* degrees of freedom
$\bar{\nu}$	wavenumber
$\bar{\xi}$	average deviation
π	osmotic pressure
$\prod$	product
ρ	density, radius of gyration
σ	standard deviation, charge per unit area, shielding constant, collision diameter
$\sum$	sum
τ	relaxation time, chemical shift, transmittance
ϕ	electric potential, orbital
Φ	quantum yield
χ	atomic orbital
χ^2	distribution of goodness of fit
ψ	wave function
ω	vibrational wavenumber
Ω	omega integral
*	*in instructions* seek or supply further advice; *in calculations* a reduced (dimensionless) quantity
†	can be set up by enthusiastic amateur
††	can be constructed by skilled technician
$	expensive
$$	very expensive
[ ]	*in chemical equations* concentration; *in Apparatus lists* optional item

Reports

Students often regard the proper writing of reports as an unnecessary chore, but entire research projects have been wasted because of poorly recorded results, and exciting new discoveries made (for example in a recent investigation into the nature of quarks) because of good documentation. There are a few cardinal rules which should be followed whichever 'house style' is adopted, and students will ignore them at their peril.

First, experiment reports should, ideally, be written directly into a robust, hard-bound book. If marking arrangements preclude this, the individual pages should at least be *securely* fastened together.

On starting an experiment, it is important that its introduction should be read first. Apart from that, the reading of the experiment script should be determined by the most efficient use of time, subject to a full ultimate understanding of the Theory and the informed completion of the investigation. It is important not to follow the Procedure blindly as one might do a recipe. A little extra effort will greatly enhance the insights gained from the experiment, will make it much more interesting, and may save a good deal of time.

The results of the experiment should be written directly into the book or page as the experiment proceeds. Consequent untidiness is much more acceptable than errors produced by copying results from tatty pieces of paper. If possible, the results should also be plotted on a rough graph during the experiment, so that there is a clear indication of the scatter of the results and whether further experiments are needed.

The calculation should be carried out as soon as the measurements are completed. The reported value of a physical quantity has little meaning unless it is accompanied by a statement of its uncertainty, and therefore at the end of the calculation there should always be an analysis of the precision, repeatability and accuracy of the results, as detailed below. A brief account of the experiment is useful, but there is clearly no point in repeating instructions in detail unless there is no copy of it to take away. The script should be checked to ensure that all questions in the text (shown in italics) have been answered, and that any comments in the margin have been followed up. Finally any additional comments may be added which are felt to be appropriate—what has been learned, the usefulness of the method, how it could be improved, and so on.

Errors

In this section, we discuss briefly the treatment of errors. By errors we do not mean disasters or blunders—supervisors will have their own treatment for students who perpetrate these. Instead we use the term error to describe the unfortunate tendency of any actual experimental result to be different from the true value of the parameter being measured.

Errors have three entirely different properties which should never be confused. These are their precision, repeatability (or reproducibility), and accuracy. The estimation of these properties is illustrated in the first experiment, (p. 7), and it may be helpful to refer to this example as you read the definitions which follow.

Precision is associated with the number of digits which can be meaningfully attributed to a particular measurement. Both calculators and computers have an often irritating capacity for giving out results with unrealistically high precision. A metre rule, for example, is graduated to the nearest millimetre, and even with good eyesight, which is the most direct improver of precision, it can only be read to the nearest half-millimetre. For a measured distance of the order of a metre, this corresponds to a precision of the order of $0.5/1000 = 0.05$ per cent. Suppose that we measure the diameter of a circle with our metre rule and find it to be 675 mm. We indicate the precision by writing the measurement in the form 675 ± 0.5 mm. If we multiply 675 by π to find the circumference, a calculator will give us an answer of, say, 2120.575. However, if we remember the maxim that it is impossible to extract more information from a calculation than we put into it, we realise that the precision of the circumference cannot be greater than the precision of the diameter, and that the circumference should therefore be expressed as 2120.6 ± 1.5 mm.

As computers are used more frequently, so precision is increasingly measured in *bits*, *b*. The maximum precision is $1/2^b$. In Expt 4.6, for example, we employ a convertor chip which changes analogue (continuously variable) information to digital information expressed as a positive integer. If the convertor is 8-bit, its read-out varies from 1 to $2^8 = 256$, and its maximum precision is 1/256, or about 0.4 per cent.

The precision of a single result limits its accuracy, but with repeated measurements statistical methods may be used to gain better accuracy than precision, as described below, and electronic devices can also be employed for this purpose (p. 469).

The repeatability (or reproducibility) of a set of results is the degree to which successive readings agree with one another. The repeatability can be better than accuracy if, for example, a balance repeatedly sticks at a false balance point (p. 8).

The accuracy of a result is its degree of closeness to a true value. Provided that we have resisted the temptation of looking up the answer to an experiment beforehand, there is no direct way in which we can find the difference between our own result and the true value. Instead we employ a well-established mixture of mathematical guidelines and informed guess-work. To be able to do this we need to understand the different types of errors and their causes.

Errors may be divided into two classes—*systematic errors* and *random errors*. Systematic errors arise from errors used in an experiment, in the method by which an experiment is carried out, or, as described below, from errors in the theory on which the experiment is based. They are often difficult to detect, but can be corrected if the source of error is known (p. xx). They may be reduced by carrying out careful experiments with good, well-calibrated apparatus, and by comparing results obtained by different methods and other workers.

Random errors can be caused by lack of precision, either in an instrument or in the way that it is read by an observer. Other causes of random errors are unpredictable fluctuations in environmental conditions, and the fundamental randomness of natural events such as radio-active disintegrations.

Each property and type of error should be dealt with separately and explicitly in the experimental report. It is also necessary to estimate how the uncertainties combine, and we shall describe these calculations after discussing the effects of inaccurate theory.

Inaccurate theory

It may not be immediately obvious why inaccurate theory can lead to errors in experimental results, and such unawareness has often led to entirely false experimental conclusions. We shall therefore discuss two examples which explain this point. The first is the measurement of standard electrode potentials, Expt 4.2. It would be very easy in this experiment to make the approximation that the activity of an ion in solution is the same as its concentration, which is equivalent to assuming that all activity coefficients are unity. Suppose that we measure the e.m.f. between a silver electrode dipping into a 1 molar solution of its ions, and a platinum electrode which has hydrogen gas bubbling around it at a pressure of one atmosphere and which dips into a 1 molar solution of acid. The approximation would lead us to believe that our measurement was the true standard electrode potential of $Ag^+|Ag$, a mistake commonly made in students' essays. However, the actual activity coefficients are considerably less than 1, as shown in Table 4.2.2 on p. 135, and to make a correct determination we must either use literature values of the coefficients, as in Expt 4.2, or find them for ourselves by making a series of determinations and extrapolating to infinite dilution, Expt 4.3.

A more subtle and therefore more pernicious danger exists in Expt 2.3, in which we measure the variation of the vapour pressure p of liquids with absolute temperature T. The interpretation of the result is based on the Clausius–Clapeyron equation

$$\ln p = -\Delta H_{vap}/RT + \text{const.} \tag{0.1}$$

In the discussion of graphs on p. xxiv, we shall see how this equation predicts that if we plot $\ln p$ against $1/T$, we will obtain a straight line graph of slope $-\Delta H_{vap}/R$, i.e. minus the enthalpy of vaporization (evaporation) divided by the gas constant. Well performed experiments do indeed yield excellent straight line graphs, and suggest that the theory behind the equation is justified. However, if correct results do not fit a theory, the theory must be wrong, whereas if they do fit it, the theory *may* be correct but is not necessarily so. In this case, the good straight line graphs conceal the fact that two of the assumptions in the derivation are wrong, namely that ΔH_{vap} is constant with temperature and that the vapour behaves as a perfect gas. A straight line is produced only because these approximations cancel. But if they cancel, why worry? There are two reasons, which are quantified in the second reference quoted in Expt. 2.3 (p. 37). The first is that we do not know whether the enthalpy we are measuring is the true one or an approximation. The second is that we may be deceived into thinking that the approximations cancel under all conditions, whereas if we tried this experiment under conditions near the critical point of the liquid, the theory would break down entirely.

The two experiments we have discussed underline the fact that wherever possible, reports on experiments should include a brief, critical appraisal of the theory of the experiment.

Accumulation of errors

In a typical experiment measurements of several variables are combined to give the desired quantity. We must therefore consider how errors in the individual measurements affect the uncertainty of the final result. We first use calculus to find the expression for the general case, and then apply this to find simple expressions for the majority of the cases which we shall encounter in this book.

General case

Consider the calculations of a quantity X from experimentally observable quantities $A, B, C\ldots$. In mathematical terms, X is a function of $A, B, C\ldots$, or $X = f(A, B, C\ldots)$. The complete differential of X is

$$\mathrm{d}X = \left(\frac{\partial X}{\partial A}\right)_{B,C\ldots} \mathrm{d}A + \left(\frac{\partial X}{\partial B}\right)_{A,C\ldots} \mathrm{d}B + \ldots \tag{0.2}$$

$(\partial X/\partial A)_{B,C\ldots}$ represents the change in X caused by a change in A, when $B, C\ldots$ are constant.

If the changes in A, B, etc. are finite but sufficiently small that the values of the partial derivatives are not appreciably altered, we may write

$$\delta X = \left(\frac{\partial X}{\partial A}\right)_{B,C\ldots} \delta A + \left(\frac{\partial X}{\partial B}\right)_{A,C\ldots} \delta B + \ldots \tag{0.3}$$

In the case of *systematic error*, we can identify δA, δB, etc. with the absolute errors in A, B, etc. These errors can be either positive *or* negative. Equation (0.3) can therefore be used to estimate the total systematic error in X, which we shall call δX, from the errors δA, δB, $\delta C \ldots$ associated with the observables $A, B, C \ldots$.

In the case of *random errors*, on the other hand, the absolute errors in A, B, etc. can now take both positive *and* negative values. We can make no predictions about the sign of a random error, only about its magnitude. Consequently, any equation treating errors must be of a form which is independent of their sign. We can achieve this by squaring eqn (0.3):

$$(\delta X)^2 = \left(\frac{\partial X}{\partial A}\right)^2_{B,C\ldots} (\delta A)^2 + \left(\frac{\partial X}{\partial B}\right)^2_{A,C\ldots} (\delta B)^2 + 2\left(\frac{\partial X}{\partial A}\right)_{B,C\ldots}\left(\frac{\partial X}{\partial B}\right)_{A,C\ldots} \delta A\,\delta B + \ldots \tag{0.4}$$

Where an experiment is repeated a large number of times, it is useful to consider average errors. Equation (0.4) may be averaged, and if it is assumed that the random errors in the variables A, B, etc. are independent, the averages $\overline{\delta A \delta B} \ldots$ of the cross-terms $\delta A \delta B \ldots$ are zero. Our working equation therefore becomes

$$\overline{\delta X^2} = \left(\frac{\partial X}{\partial A}\right)^2_{B,C\ldots} \overline{\delta A^2} + \left(\frac{\partial X}{\partial B}\right)^2_{A,C\ldots} \overline{\delta B^2} + \ldots \tag{0.5}$$

where now $\overline{\delta X^2}$, $\overline{\delta A^2}$, $\overline{\delta B^2} \ldots$ are the mean squared errors in the variables. Because second- and higher-order terms are neglected, equation (0.5) only applies if errors are small ($\delta A/A < 0.15$).

Sums and differences

If $X = (A \pm \delta A) + (B \pm \delta B)$, or $X = (A \pm \delta A) - (B \pm \delta B)$, then

$$\left(\frac{\partial X}{\partial A}\right)_{B,C\ldots} = \left(\frac{\partial X}{\partial B}\right)_{A,C\ldots} = 1.$$

Therefore for systematic errors eqn (0.3) becomes simply

$$\delta X = \delta A + \delta B \quad \text{or} \quad \delta X = \delta A - \delta B, \tag{0.6}$$

and for random errors eqn (0.5) gives

$$\overline{\delta X^2} = \overline{\delta A^2} + \overline{\delta B^2}. \tag{0.7}$$

Products and quotients

If $X = (A \pm \delta A) \cdot (B \pm \delta B)$ or $X = (A \pm \delta A)/(B \pm \delta B)$ then for systematic errors eqn (0.3) gives

$$(\delta X/X) = (\delta A/A) + (\delta B/B), \tag{0.8}$$

and for random errors eqn (0.5) gives

$$(\delta X/X)^2 = (\delta A/A)^2 + (\delta B/B)^2. \tag{0.9}$$

Note that in this case we are dealing with combinations of *relative errors*, $(\delta X/X)$, etc.

Powers

If $X = (A \pm \delta A)^j$, then for systematic errors

$$(\delta X/X) = j \cdot (\delta A/A), \tag{0.10}$$

and for random errors

$$(\delta X/X)^2 = j \cdot (\delta B/B)^2. \tag{0.11}$$

An example of the importance of equations such as these is that the flow times of a gas through a capillary tube are proportional to the fourth power of its radius. It follows that since viscosity is directly proportional to flow time, the error in the viscosity of the gas is four times greater than the error in the measurement of the radius of the capillary tube. The radius of a capillary tube is very small and extremely difficult to measure accurately, and this relation effectively precludes the use of capillary flow viscometry for absolute measurements of gas viscosity (p. 352).

Small samples

Frequently only a small number of repetitions of each measurement are made, and in this situation we are forced to use experience and intuition to obtain reasonable estimates of their precision or uncertainty. Often this is straightforward, as in the example of the metre rule quoted earlier, while on other occasions, p. 7, it requires some knowledge of the devices in use.

Large samples

If six or more measurements of a particular quantity have been made we can, with care, interpret their values by the laws of statistics. When many more measurements have been made, it is essential to use the statistical methods which we now briefly introduce.

Statistical analysis

If the errors in a measurement are truly random, they will, in most cases, follow a *Gaussian* or *Normal distribution*. A particular case when this distribution is not followed is discussed on p. 428. The *true value* of a quantity being measured (assuming no systematic errors) is then given by the arithmetic mean of the individual measurements, provided that the number of these measurements is *very large*. The error of an individual measurement is then taken to be the difference between the measured and mean value.

The Normal distribution is a continuous distribution with probability density

$$P = \frac{\exp(-n^2/2\sigma^2)}{\sigma\sqrt{(2\pi)}}. \tag{0.12}$$

The probability that a random error will lie between n_1 and n_2 is therefore

$$P = \int_{n_1}^{n_2} \frac{\exp(-n^2/2\sigma^2)}{\sigma\sqrt{(2\pi)}}\,\mathrm{d}n. \tag{0.13}$$

The parameter σ is a measure of the width of the Normal error curve, and thus of the uncertainty of the measurements. It is called the *standard deviation* or the *root mean square deviation*, since it may be shown that

$$\sigma = \left(\frac{\sum n^2}{p-1}\right)^{1/2} \tag{0.14}$$

where $\sum n^2$ represents the sum of the squares of all the readings n, and p is the total number of measurements.

In an experiment where a particular measurement has been repeated many times, $p-1 \simeq p$ and the standard deviation is given by

$$\sigma = \left(\frac{\sum n^2}{p}\right)^{1/2}. \tag{0.15}$$

Usually, however, we only have time to make a small number of measurements, and therefore we only obtain *estimates* of the true value of the the result and of the standard deviation of the measurements. When p is small, a good estimate of the standard deviation is given by

$$\sigma = \left(\frac{\sum (n-m)^2}{p-1}\right)^{1/2}. \tag{0.16}$$

This expression is convenient to use, because the difference between a measurement and its mean, $(n-m)$, is smaller and easier to handle than the value of the reading itself. However, great care must be exercised in estimating σ in this way for very small values of p, since it is itself subject to a standard deviation or error of

$$\sigma_\sigma = \left(\frac{2}{p}\right)^{1/2} \sigma. \tag{0.17}$$

Therefore in practice at least six or seven observations are required before the value of σ calculated using equation (0.13) can be relied upon.

Three other measures of uncertainty are in common use.

(i) The *95% error limits* are limits within which there is a probability of 95% that any one result will fall. These are about $m \pm \sigma$ for more than about 7 results and about $m \pm 1.5\sigma$ for between 4 and 7 results. It is difficult to estimate for fewer results.

(ii) The *average deviation* ($\bar{\xi}$) is the arithmetic mean of all the errors without regard to sign:

$$\bar{\xi} = \sigma\sqrt{(2/\pi)} \simeq 0.80\sigma. \tag{0.18}$$

(iii) The *probable error* E is the error such that there is a probability of one half that the error on any single measurement will be greater (or less) than E:

$$E \simeq 0.67\sigma. \tag{0.19}$$

Two graphical methods of analysing randomly distributed results are also available. The first is to use probability graph paper, as explained in for the case of Brownian motion on p. 330. The second, which is convenient for large numbers of measurements as in the radioactivity experiment on p. 430, is to group the readings so that they form a histogram.

Once estimates of σ, $\bar{\xi}$, or E have been found for the various measurements employed in an experiment, the analysis for combination of errors can be applied to these parameters to find the appropriate value for the final result.

Graphs

Experimental results are very often plotted as graphs, showing the variation of a *dependent variable* with change in an *independent variable.* For example in Expt 2.3 we change the temperature of a liquid, and its vapour pressure changes as a result—the former is the independent variable, and the latter the dependent variable. The convention is to plot the independent variable on the horizontal 'x' axis, and the dependent variable on the vertical 'y' axis.

Much the easiest type of graph to plot is one which yields a straight line. To achieve this, we compare the working equation of the experiment with the expression

$$y = mx + c, \tag{0.20}$$

where m is the gradient of the straight line, and c its intercept on the y axis where $x = 0$. Comparing this equation with eqn (0.1), p. xx, we see that $y = \ln p$, that $x = 1/T$ because both the other terms in this part of the equation are constants, and that $m = -\Delta H_{vap}/R$. Thus, as mentioned previously, a plot of $\ln p$ on the vertical axis against $1/T$ on the horizontal axis should yield a straight line of slope $-\Delta H_{vap}/R$.

In cases where it is not possible to arrange a straight line graph, a curve must be drawn through the points to smooth them. It is very important to remember that by drawing a smoothing curve through a set of points we are imposing some functionality on the relationship between the dependent and independent variables. An example of this is in the smoothing of gas imperfection data, Expt 1.2. Some research workers have been tempted to use functions such as the so-called Lennard–Jones 18-6 potential to provide smoothing curves for their experimental data. The trouble is that when theoreticians analyse the virial coefficient data to see what information they can obtain, they themselves get back an 18-6 function, and if they are unaware of the enforced nature of the smoothing, serious errors of interpretation can result. The safest smoothing method is therefore not to use any functional form at all, but to purchase a flexible strip which can be laid along the points and drawn round to produce a smooth curve. The smoothing is then overtly and controllably subjective.

The uncertainties in the experimental results will cause the points on the graph to be scattered, and therefore it may not be immediately obvious where a straight line or curve should be drawn. The standard procedure for fitting a straight line to a set of points is a least squares analysis. This may also be used to position a curve, although it is more normal to carry this out by eye.

Least squares analysis

The criterion used in the least squares analysis is that the best fitting line is that which gives rise to the smallest sum of the squares of the deviations of the experimental points from the line.

Suppose the data consist of a set of p points, with coordinates $(x_1y_1, x_2y_2, \ldots, x_py_p)$. We assume that the values of x are exact, and that all deviations occur in the values of y. The deviation of the point (x_i, y_i) is then $y_i - (mx_i + c)$, Fig. 0.1. The sum of the squares of the deviations is $\sum (y_i - mx_i - c)^2$ summed over all p points, and this quantity must be made a minimum by suitable choice of the values of m and c. The condition for a minimum is obtained algebraically by differentiating the expression with respect to m and c, putting the derivatives equal to zero, and solving

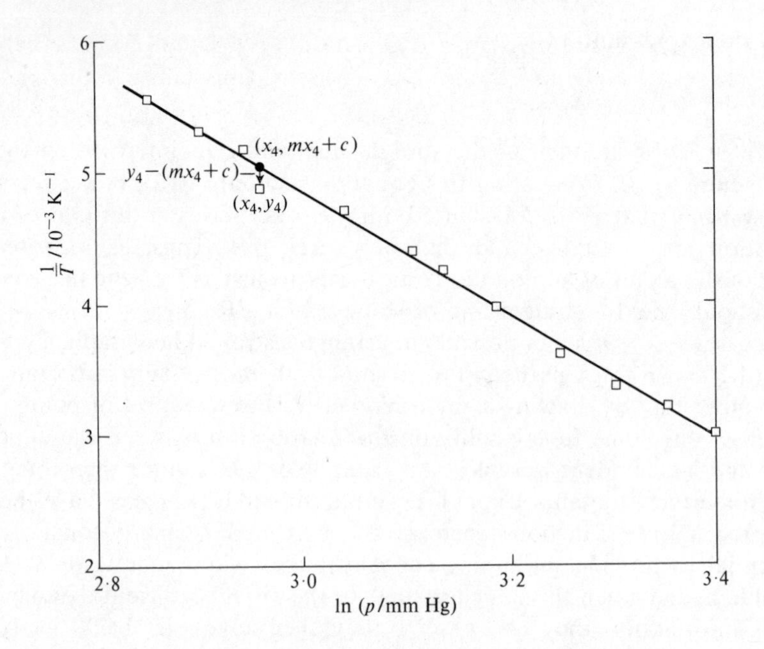

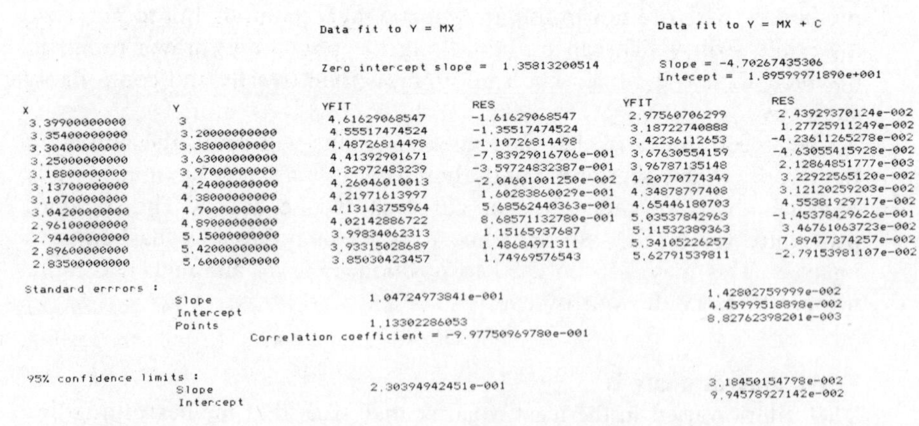

Linear Least Squares Analysis
################################

Data fit to Y = MX

Zero intercept slope = 1,35813200514

Data fit to Y = MX + C

Slope = -4,70267435306
Intecept = 1,89599971890e+001

X	Y	YFIT	RES	YFIT	RES
3,39900000000	3	4,61629068547	-1,61629068547	2,97560706299	2,43929370124e-002
3,35400000000	3,20000000000	4,55517474524	-1,35517474524	3,18722740888	1,27725911249e-002
3,30400000000	3,38000000000	4,48726814498	-1,10726814498	3,42236112653	-4,23611265278e-002
3,25000000000	3,63000000000	4,41392901671	-7,83929016706e-001	3,67630554159	-4,63055415928e-002
3,18800000000	3,97000000000	4,32972483239	-3,59724832387e-001	3,96787135148	2,12864851777e-003
3,13700000000	4,24000000000	4,26046010013	-2,04601001250e-002	4,20770774349	3,22922565120e-002
3,10700000000	4,38000000000	4,21971613997	1,60283860029e-001	4,34878797408	3,12120259203e-002
3,04200000000	4,70000000000	4,13143755964	5,68562440363e-001	4,65446180703	4,55381929717e-002
2,96100000000	4,89000000000	4,02142886722	8,68571132780e-001	5,03537842963	-1,45378429626e-001
2,94400000000	5,15000000000	3,99834062313	1,15165937687	5,11532389363	3,46761063723e-002
2,89600000000	5,42000000000	3,93315028689	1,48684971311	5,34105226257	7,89477374257e-002
2,83500000000	5,60000000000	3,85030423457	1,74969576543	5,62791539811	-2,79153981107e-002

Standard errors :	Y = MX	Y = MX + C
Slope	1,04724973841e-001	1,42802759999e-002
Intercept		4,45999518898e-002
Points	1,13302286053	8,82762398201e-003

Correlation coefficient = -9,97750969780e-001

95% confidence limits :	Y = MX	Y = MX + C
Slope	2,30394942451e-001	3,18450154798e-002
Intercept		9,94578927142e-002

Fig. 0.1. Least squares analysis of a graph of ln p against $1/T$ (Expt 2.3)

the resulting pair of simultaneous equations. The final working formulae for obtaining the best-fit constants are

$$A = p\sum xy - \sum x \sum y \qquad B = p\sum x^2 - \left(\sum x\right)^2$$

$$m = A/B \qquad c = \left(\sum y - m\sum x\right)/p$$

$$\sigma_m^2 = \left(p\sum y^2 - \left(\sum y\right)^2 - mA\right)/(B(p-2))$$

$$\sigma_c^2 = \sigma_m^2\left(\sum x^2/p\right)$$

where σ_m and σ_c are the standard deviations of the slope and intercept of the line respectively, and A and B are intermediate values in the computation. The summations $\sum x$, $\sum y$, $\sum x^2$, $\sum y^2$, $\sum xy$ are easily carried out on a calculator or computer.

Safety

The experiments in this book have been designed to be as safe as possible in terms of the equipment and chemicals they require. Of course, some hazards remain, and specific safety instructions are given in the appropriate Procedure and Technical Notes sections. A more general discussion of safety is beyond the scope of this book, and teachers, supervisors, and students are therefore advised to read the specialist works on this subject, for example those produced by the Royal Society of Chemistry, the American Chemical Society, and the Association for Science Education. They should also be aware of the continual tightening of regulations. Thus, exposed mercury surfaces should be avoided by means of sintered glass discs or ceramic fibre plugs, even though they are shown exposed in the apparatus diagrams for clarity. One should never use open electrical fittings such as screw-type junction blocks, and lasers and their beams should be entirely enclosed.

One should bear in mind that however carefully safety precautions are implemented, unexpected accidents can still occur. One type concerns equipment newly bought from manufacturers, which cannot always be relied upon to be safe. I am mindful, for example, of a new stillhead which shattered when heated by rising vapour, and a discharge-lamp power supply which could, if not very tightly connected, give unpleasant shocks to its users. The lesson is that all new equipment should be carefully tested *in situ*, allowing for student error in following procedures.

Such student error can cause a second type of accident which is difficult to avoid by safety precautions. There is a general suspicion that pupils and students are increasingly lacking in practical skills. In schools, fashion and expense disfavour traditional systematic practical work, and at home constructional hobbies are giving way to more cerebral pursuits such as computer programming. Such trends have given rise to some bizarre accidents, for example a student stabbing himself in the stomach with a pipette while trying to attach a pipette-filler! Supervisors must therefore be aware that students may need extra coaching in basic experimental skills before embarking on the experiments in this book. Finally it is most important that students should never be allowed to work in the laboratory alone.

Part I Equilibrium

Section 1
Gases at equilibrium

We begin our investigation of experimental physical chemistry by examining the properties of gases at equilibrium, the simplest systems to study at an elementary level.

In Expt 1.1, the first of three experiments in this section, the relative molar masses of two gases are calculated from the pressures required to produce the same deflection of a gas density balance. In carrying out the calculation, we assume that the gases obey the perfect gas law.

Experiments 1.2 and 1.3 involve the measurement of the deviations of real gases from perfect behaviour. In Expt 1.2 the deviations are expressed in terms of the second virial coefficient, and in Expt 1.3 in terms of the Joule–Thomson coefficient. These parameters, once determined, can be used to predict the properties of the gases under different conditions, and the temperature dependence of the parameters gives immediate qualitative information about the nature of the gas imperfections. However, to understand how the parameters may be used quantitatively, and how transport processes such as viscosity can be accounted for, we must describe the gas in terms of kinetic theory which includes the effect of the intermolecular forces on the collisions between the gas molecules. This more sophisticated approach is used to interpret viscosity measurements in Expt 7.3.

1.1 Molar masses by gas density balance

One of the most fundamental properties of a gas is the mass of its molecules. In this experiment we set out to measure this quantity by a method which is simple but accurate. The measurements will be relative rather than absolute, i.e. they will relate the mass of the molecules of one gas to the accurately known mass of the molecules of another.

One method would be to fill a flask with a gas at a known pressure, weigh it, and then fill the same flask with the same pressure of a known standard gas and weigh that. However the weight of the gas is small compared to the weight of the container, and errors occur unless very great care is taken.[1]

A better approach is to immerse a balance in the gas. A sealed bulb is attached to one end of the balance beam, and the buoyancy effect on this bulb is measured. The pressure in the balance chamber, measured on a precision manometer, is adjusted until the beam balances. The procedure is then repeated for the standard gas, and the two pressures compared. The complication of surrounding the whole balance by gas is that molecules adsorb onto the beam and the additional weight may affect the readings. To prevent this, a counterpoise is fixed to the other end of the beam. This has the same surface area as the sealed bulb but is much less buoyant—in the present apparatus it takes the form of a smaller hollow sphere with holes through it. The adsorption effects are then very nearly the same at each end of the beam and so almost cancel each other.

For research purposes, several balance points are determined with different weights attached to the beam. The results are then extrapolated to zero gas density. This is known as the *method of limiting gas densities*,[1] and has the great advantage that gas imperfection and adsorption effects are removed entirely.

Theory

When the density balance is stationary and at its balance point, the turning moments on each end of the beam must be equal. The turning moments are caused by the buoyancies of the bulb, balance arm, and counterpoise, and by *Archimedes' principle*, these buoyancies are equal to the weights of gas displaced. The weights are equal to the product of the density (i.e. mass density) of the gas, the local acceleration due to gravity g, and the volumes of the bulb, arm, and counterpoise. Both the volumes and g are constant, and therefore equal buoyancy effects, and balance points, are caused by equal densities of gas.

Since the gas is at low density, the *perfect gas law* can be used to find how density is altered by pressure and temperature:

Symbols not defined in the text are listed on p. xiii

$$pV = nRT. \tag{1.1.1}$$

Therefore

$$\text{density} = nM/V = pM/RT. \tag{1.1.2}$$

Two gases giving rise to the same buoyancy effect must have the same density

$$p_1M_1/RT_1 = p_2M_2/RT_2,$$

and therefore

$$M_1/M_2 = p_2T_1/p_1T_2. \tag{1.1.3}$$

Apparatus A gas density balance is shown in Fig. 1.1.1.[1] The delicate Pyrex glass balance arm and bulbs are suspended on a quartz fibre, and two pointers are arranged to indicate the balance-point. Excessive travel of the arm is prevented by end-stops.

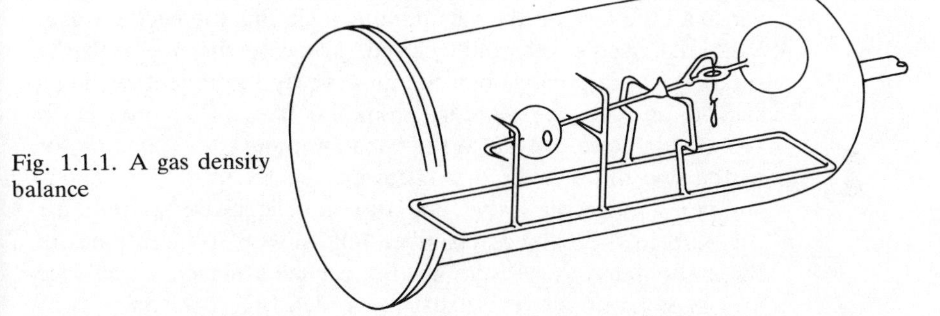

Fig. 1.1.1. A gas density balance

Vibrations are reduced by mounting the balance chamber securely and having a spiral of glass tubing between it and the vacuum line. *What is the disadvantage of the glass spiral*?

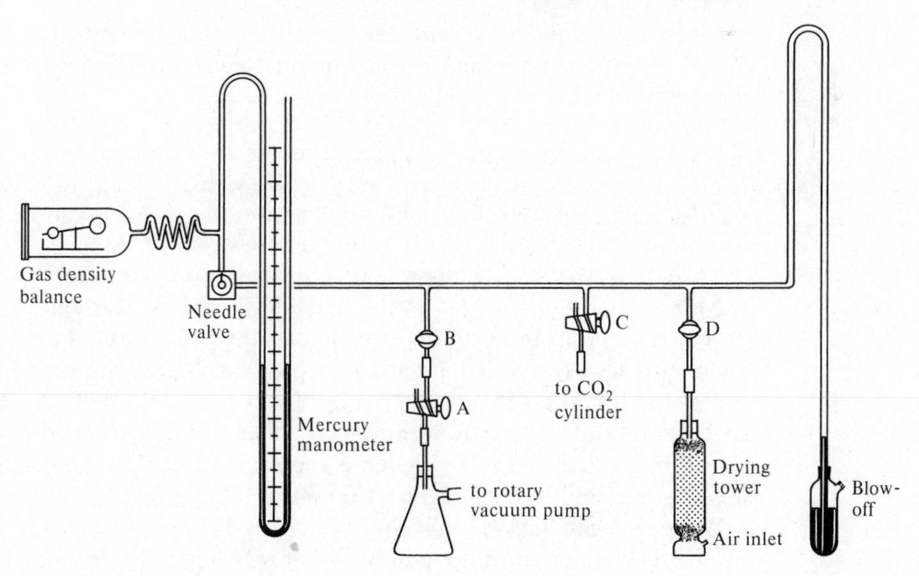

Fig. 1.1.2. The vacuum line

The vacuum line, Fig. 1.1.2, features an open-ended wide-bore mercury manometer with mirror scale, blow-off, rotary vacuum pump with trap (i.e. a filter flask to prevent oil being sucked back into the apparatus), air inlet with drying tower, and gas inlet tap with by-pass.

Procedure

The gas density balance is a delicate instrument which must be treated with great care. Do not touch the balance case. Evacuate and pressurize the balance chamber slowly so that the balance arm does not crash against its end-stops.

(i) Preliminary evacuation

Close all the taps on the vacuum line including the needle valve. With the needle valve still closed, evacuate the line with the rotary vacuum pump; open the taps slowly to prevent air being sucked up with the mercury from the blow-off. If any air is sucked up, close the tap to the vacuum pump (tap B), admit air to the line through the drying tower (tap D) and re-evacuate.

Open the needle valve one turn anticlockwise so that the pressure in the balance chamber falls slowly, as registered on the manometer. A pressure will be reached at which the balance arm gently swings over so that the sealed bulb is resting downwards. Once this has happened, the evacuation may be made more rapid by opening the needle valve a further five turns. Continue pumping until the pressure registered by the manometer remains steady.

Close the needle valve and the tap to the rotary pump (tap B). The manometer reading will remain steady provided that there is no leak.

(ii) Measuring balance-point pressures

By the following procedure, determine balance-point pressures for air, carbon dioxide, and the unknown gas.

With the balance chamber and line evacuated, and needle valve closed, admit the gas to the line. Air is passed through a drying tower (tap D). Carbon dioxide and the unknown gas are connected to a two-way tap with a by-pass tube (tap C); before the gas is admitted to the line, it is blown gently through the by-pass tube for several minutes to flush out air from the connecting tube. The two-way tap is then turned so that the gas fills the line and bubbles *gently* through the blow-off.

With the gas supply still connected to the line, open the needle valve one turn to admit gas slowly into the balance chamber and manometer.

Close the needle valve when the balance beam is just horizontal. If the beam overshoots, close off the gas supply, evacuate the line, and operate the needle valve to bring it to an accurate balance-point. Tap the mercury in the manometer very gently and note the height of the menisci.

Alter the pressure slightly. Repeat the balance-point determination so that you have at least one point approached from too high a pressure and one from too low a pressure.

Use a barometer to find the pressure of the atmosphere, and hence calculate the balance-point pressures. Measure the temperature at the chamber.

Evacuate the apparatus (para (i)).

(iii) Shut down

Bring the balance chamber and line to atmospheric pressure with dry air. Close the needle valve and all taps. Switch off the rotary vacuum pump and *immediately* open the air leak (tap A).

Calculation

Results may also be expressed as the mass of a single molecule of gas by changing the units from (g mol^{-1}) to (u $molecule^{-1}$), where u is the symbol for unified atomic mass unit defined as the mass of $^{12}C/12$.

Using eqn (1.1.3), and given that the effective molar mass of air is 28.96 g mol^{-1}, find the molar masses of cylinder CO_2 and the unknown gas.

Estimate the precision, repeatability, and accuracy of your results, and suggest how the accuracy of the experiment could be improved.

Given that the unknown gas is a perhalogenated alkane, deduce its formula.

Results

Precision

Manometer levels can only be read to nearest 1/2 mm Hg by eye, giving maximum precision of 3 significant figures. This may be increased by observing the meniscus with a *cathetometer*, which comprises a telescope mounted on a vertical vernier scale.

Repeatability

Consecutive measurements on same gas have moderately good repeatability—i.e. same p_{gas} can be obtained to within 1.0 per cent. This does *not* mean the readings are as accurate as this (p. xviii).

Accuracy

Sources of error: (i) Temperature. Chamber not thermostatted. Difference between gas and thermometer temperature could be ~0.5 K = 0.2 per cent at room temperature. (ii) Pressure. Should be accurate to better than 0.1 mm Hg in wide-bore Hg manometer with precision mirror-scale, i.e. error of <0.05 per cent. (iii) Density balance. Incorrect balancing due to imperfect suspension: estimate from difference between balance-points approached from too high a pressure and too low pressure. (iv) Residual gas in line. Should be ~0.2 per cent after chamber has been evacuated and brought to balance pressure. (v) Use of perfect gas equation. Negligible deviation for air, accurate to ~0.3 per cent for CO_2 and perhalogenated alkanes under conditions of experiment.

Thus calculate maximum error and check that this is greater than repeatability of 1.0 per cent.

Identification of unknown. Unknown is a perhalogenated alkane. Therefore formula must be $C_jF_kCl_lBr_mI_n$. *What rules apply to the values of j, k, l, and m?*

Comment

The specimen data emphasize the difference between precision, repeatability, and accuracy of results—three entirely different characteristics which should never be confused. The accuracy of the molar mass of the unknown is sufficient for it to be identified unambiguously if we know that it is a perhalogenated alkane.

Fifty years ago, gas density balance determinations were being used to determine accurate molar and atomic masses to a precision of better than one part in 10 000.[2] Gas density balances are no longer used to determine the formulae of unknown gases, since this can now be performed much more accurately with mass spectrometers (Expt 5.1). However they can still be useful in determining the composition of binary mixtures of pure gases,[3] since mass spectrometers must be specially calibrated before they can be used for quantitative measurements.

Technical notes

Apparatus

Gas density balance,††[1] thermometer mounted near or preferably inside balance chamber, rotary vacuum pump,$ vacuum line and other associated components including $CaCl_2$ drying tower as shown in Fig. 1.1.2,† cylinder of CO_2, small cylinder of readily available Freon or Arcton, barometer.$

Design note

Although it is inefficient to evacuate a large chamber through a needle valve, the absence of a by-pass helps to protect the density balance from sudden changes in pressure.

References

[1] *An advanced treatise on physical chemistry*. J. R. Partington. Longmans, London (1967), vol. 1, sect. VII D.

[2] e.g. *A comparison of the densities of carbon monoxide and oxygen* ..., M. Woodhead and R. Whytlaw-Gray. *J. Chem. Soc.* **846** (1933).
[3] *Intermolecular forces and the gaseous viscosities of argon–xenon mixtures.* I. A. Barr, G. P. Matthews, E. B. Smith, and A. R. Tindell. *J. Phys. Chem.* **85,** 3342 (1981).

1.2 Gas imperfection by *pV* measurement

At low densities, the behaviour of most gases approximates closely to the perfect gas law

$$pV = nRT. \tag{1.2.1}$$

At higher densities, however, the non-zero size of the molecules in a real gas, and the fact there are long-range attractive and short-range repulsive forces between the molecules, causes appreciable deviations from perfect behaviour.

In this experiment we measure these deviations for carbon dioxide by making pV measurements at pressures of up to 20 atmospheres at room temperature.

Theory

(i) The virial equation of state

It may be shown that the second and third virial coefficients B and C in eqn (1.2.2) arise from one- and two-body, and one-, two-, and three-body interactions respectively.

One common method of expressing the non-perfect behaviour of a real gas is to use the *virial equation of state*

$$pV/nRT = 1 + Bn/V + Cn^2/V^2 + \dots \tag{1.2.2}$$

where the coefficients B and C are called the second and third *virial coefficients*. In this experiment it is more convenient to use an expansion in terms of pressure rather than reciprocal molar volume:

$$pV/nRT = 1 + B'p + C'p^2 + \dots \tag{1.2.3}$$

where $B' = B/RT$ and $C' = (C - B^2)/(RT)^2$. The coefficient C', and higher coefficients, are too small to be accurately measured at the pressures used in this experiment, and we therefore ignore them.

The *compression factor* Z of a gas is pV/nRT, and thus from eqn (1.2.3),

$$Z = pV/nRT = 1 + B'p. \tag{1.2.4}$$

Virial coefficients vary with temperature, although not with pressure, and we may interpret the temperature dependence of the second virial coefficient shown in Fig. 1.2.1. The shape of

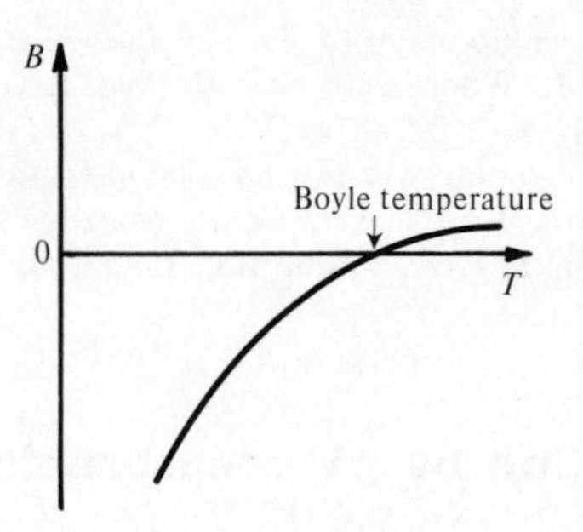

Fig. 1.2.1. Variation of second virial coefficient B with temperature

the curve is similar for any gas, although of course the scaling of the two axes varies. *Why is it not possible to measure the second virial coefficient down to a temperature of absolute zero*? From the figure and eqn (1.2.4) we may deduce that at low temperatures a real gas has a lower pressure than the same number of moles of a perfect gas in an equal volume, whereas at high temperatures its pressure is higher. The basic explanation is that at low temperatures long-range attractive forces dominate the collisions between gas molecules, whereas at high temperatures the gas molecules collide with more energy, and short range repulsive forces dominate. The second virial coefficient passes through zero at the *Boyle temperature*, at which the gas behaves perfectly over a wide range of pressure.

The second virial coefficient B may be determined by plotting the compression factor $Z = pV/nRT$ against p, from which a straight line is obtained of slope $B' = B/RT$ (eqn (1.2.4)).

(*ii*) *The modified Burnett expansion method*[1]

One of the most successful methods for measuring virial coefficients has been Burnett's expansion method. The apparatus comprises a sample vessel connected to an expansion vessel (or several expansion vessels), as shown in Fig. 1.2.2. The sample vessel is filled with the test gas at high pressure, and the expansion vessel is evacuated. The tap between the sample vessel and expansion vessel is then opened so that the pressures equalize. Then the expansion vessel is pumped out and the procedure repeated several times. The second virial coefficient is calculated from measurements of the pressure of the gas at each stage.

In this experiment we use a modification of Burnett's method in that the expansion vessel is larger than the sample vessel, Fig. 1.2.3, and the expansion is carried out in many stages such that the pressure in the expansion vessel is never allowed to be

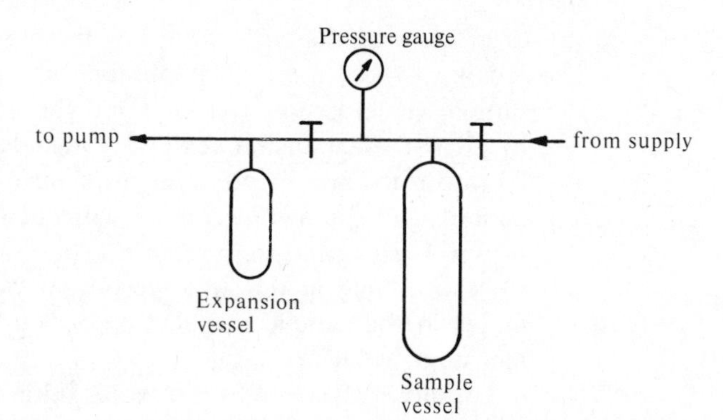

Fig. 1.2.2. Apparatus for the Burnett expansion method

above 1 atmosphere. These alterations allow more measurements to be made for a given starting pressure.

The basis of the present method is to measure the amount of gas contained in the sample vessel of volume V^s at various values of the pressure p. The number of moles of gas n cannot be measured directly, and so we assume that at pressures up to atmospheric deviations from perfect behaviour are negligible, and that we may therefore apply the perfect gas law to each of the small quantities of gas which are allowed to escape from the sample vessel into the expansion vessel.

Suppose that there are initially n_0 moles of gas in the sample vessel at pressure p_0^s. If the volume of the expansion vessel is V^e and the gas from the sample vessel raises its pressure from zero

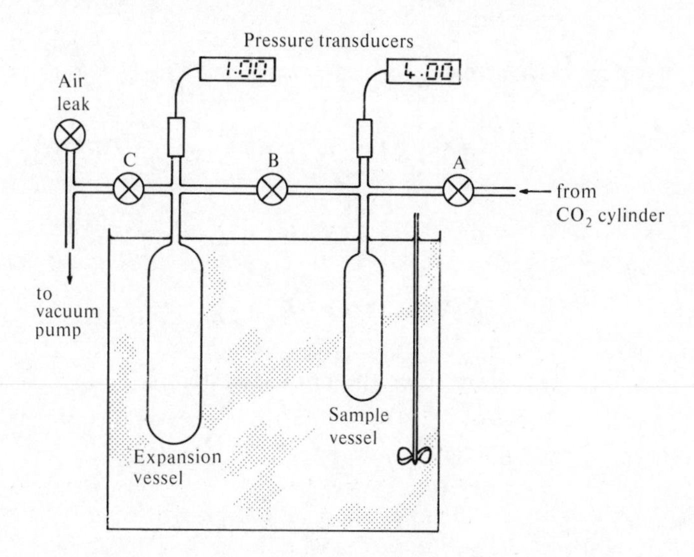

Fig. 1.2.3. Apparatus for the modified Burnett expansion method

to p_1^e, then by the perfect gas law the number of moles transferred is $n_1 = p_1^e V^e/RT$. The number of moles remaining in the sample vessel is now $(n_0 - n_1)$ and the pressure in it is reduced to p_1^s. If the expansion vessel is evacuated again, and more gas let into it to raise its pressure from zero to p_2^e, then the number of moles in the sample vessel will fall to $(n_0 - n_1 - n_2)$, where $n_2 = p_2^e V^e/RT$, and the pressure in the sample vessel will fall to p_2^s. Continuing in this way we obtain values of the number of moles in the sample vessel, $(n_0 - n_1 - n_2 \ldots)$, corresponding to measured values of p^s.

To find n_0, the number of moles initially in the sample vessel, the experiment is continued until it is found that at the mth expansion the pressure in the expansion vessel does not rise above atmospheric, but that with the connection between the two vessels fully open, the pressure in both only reaches the value p_m^e. The number of moles in the sample vessel is now $p_m^e V^s/RT$, and n_0 is found by adding together this quantity and the number of moles from each of the previous expansions into the expansion vessel:

$$n_0 = p_m^e V^s/RT + p_1^e V^e/RT + p_2^e V^e/RT + \ldots + p_m^e V^e/RT$$
$$= p_m^e V^s/RT + (V^e/RT) \sum_{i=1}^{m} p_i^e \tag{1.2.5}$$

where p_i^e represents the pressure rise in the expansion vessel due to the ith expansion.

After r expansions, the number of moles n_r left in the sample vessel is

$$n_r = n_0 - \sum_{i=1}^{r} n_i$$
$$= p_m^e V^s/RT + (V^e/RT) \sum_{i=1}^{m} p_i^e - (V^e/RT) \sum_{i=1}^{r} p_i^e$$
$$= p_m^e V^s/RT + (V^e/RT)\left(\sum_{i=1}^{m} p_i^e - \sum_{i=1}^{r} p_i^e\right)$$
$$= p_m^e V^s/RT + (V^e/RT) \sum_{i=r+1}^{m} p_i^e. \tag{1.2.6}$$

Therefore after the rth expansion,

$$\frac{p_r^s V^s}{n_r RT} = p_r^s V^s \Big/ \left(p_m^e V^s + V^e \sum_{i=r+1}^{m} p_i^e\right)$$
$$= p_r^s \Big/ \left(p_m^e + \frac{V^e}{V^s} \sum_{i=r+1}^{m} p_i^e\right). \tag{1.2.7}$$

The volume ratio of the vessels, V^e/V^s, can be calculated from an expansion at low pressure, and all the pressures in the right-hand term of eqn (1.2.7) may be measured. Thus values of $p_r^s V^s / n_r RT$ may be calculated, and, from eqn (1.2.4), yield a straight line of slope B' when plotted against p_r^s.

Apparatus

The cooling cannot be due to the Joule–Thomson effect (Expt 1.3), because it occurs above as well as below the Joule–Thomson inversion temperature (e.g. 202 K for hydrogen).

The apparatus is shown in Fig. 1.2.3. The pressures are measured by means of calibrated pressure transducers linked to digital read-outs in atmospheres. When gases expand, their temperatures change due to the work of expansion and kinetic effects. The gas vessels must therefore be contained in a thermostatically controlled water-bath. Gas is supplied to a line from a gas cylinder via a regulating valve set at 20 atmospheres maximum.

Procedure

Wear safety glasses. Never open tap A *until you have checked that tap* B *is closed, as the low-pressure transducer is only rated up to* 1 *atm pressure. Never switch off the vacuum pump until you have opened the air leak to the atmosphere, otherwise oil will be sucked back into the apparatus.*

(*i*) *Set up*

Check the temperature of the water bath, and if necessary adjust to 25 °C.

Turn on the pressure transducers. Check that the high-pressure transducer reads less than 1 atm. If it does not, follow the procedure in the paragraph below. Then check the zero of the pressure transducers as follows. Turn on the rotary vacuum pump (and leave it switched on throughout the experiment). Check that the air leak is closed, and then evacuate both vessels (taps B and C open). After 10 minutes the readings on both pressure transducers should be stable at 0.00: if not, then adjust the offset controls until readings of 0.00 are obtained.

If, at the start of the experiment, the pressure p^s is much greater than 1 atm, *do not open tap* B. Do not try to evacuate both vessels. Turn on the pump, and open tap C *only* to evacuate the expansion vessel. Continue the experiment from step (iii), i.e. do not try to check the zero of the pressure transducer.

Leave the pump switched on during the rest of the experiment.

(*ii*) *Filling of sample vessel*

Close off the expansion vessel and pump (taps B and C). Open tap A *gently* and fill the sample vessel with CO_2 to a pressure of not more than 20 atm. (The high-pressure transducer reading tends to go off scale while tap A is open, so close the tap from time to time to read the true pressure, until the right pressure is reached). Shut off the supply by closing tap A *firmly*, and then allow a few minutes for the reading to stabilize. Note this reading (p_0^s).

Note the temperature of the water bath.

(*iii*) *Expansion of sample gas*

Admit gas into the evacuated expansion vessel by *slowly* opening tap B. Continue to admit gas until the pressure reading on the low-pressure transducer is about 1 atm.

Close tap B. Allow a few minutes for the gases to reach thermal equilibrium and for their pressures to stabilize. Then measure the pressure p_1^e in the expansion vessel and the pressure of gas in the sample vessel, p_1^s.

Then re-evacuate the expansion vessel by opening tap C and pumping it out for about 5 min.

Once again, note the pressure p_1^s on the high-pressure transducer. If there is a discrepancy of more than a few hundredths of an atmosphere between this and the previous reading, then not enough time was allowed for thermal equilibrium, or taps A and B may not be firmly closed.

When the expansion vessel has been evacuated, close tap C, and repeat the expansion procedure to obtain p_2^e and p_2^s. Coninue repeating the procedure until, after the mth expansion, the pressure in both vessels with tap B fully open is equal ($p_m^s = p_m^e$) and less than 1 atm.

(*iv*) *Measurement of the volume ratio* $V^e : V^s$

Evacuate both vessels (taps B and C open). Switch off the vacuum pump and immediately open the air leak, so that both vessels fill with air to atmospheric pressure (p_{atm}). Close the air leak and tap B, switch on the pump, and evacuate the expansion vessel. Close tap C. Open tap B and allow the pressure in both vessels to equalize. Use the low-pressure transducer to determine this pressure, (p'). Then

$$p_{atm} V^s = p'(V^s + V^e),$$

so that

$$V^e/V^s = (p_{atm} - p')/p'. \tag{1.2.8}$$

Repeat this determination several times and use the average value of V^e/V^s in your calculation.

(*v*) *Shut down*

On finishing the experiment, open tap C and evacuate both vessels: do not leave air in the apparatus.

Make sure that all three metal taps are closed. Switch off the vacuum pump and immediately open the air leak. Switch off the pressure transducers.

Calculation For each value of p_r^s, calculate $p_r^s V^s/n_r RT$ from eqn (1.2.7). Plot a graph of $p_r^s V^s/n_r RT$ against p_r^s, and find B' (eqn (1.2.4)), and hence B.

Is the Boyle temperature of CO_2 below or above room temperature?

Results See Table 1.3.1, p. 19.

Accuracy

The limiting factor is the accuracy of the pressure measurement. Bell and Howell type pressure transducers are the most suitable instruments, and using these this experiment yields results within 20 per cent of the true virial coefficient. The accuracy may be improved considerably by taking the average value of a series of determinations.

Comment Virial coefficients are notoriously difficult to measure, especially at low temperatures where it becomes increasingly difficult to estimate the effect of adsorption of the gas on to the walls of the vessels. With this in mind, the accuracy of the results from this experiment is quite respectable.

Virial coefficients are an important source of information about intermolecular forces, and, provided they are sufficiently accurately measured, may be 'inverted' to give the shape of the intermolecular pair potential energy function, p. 338.[2]

Technical notes

Apparatus

As shown in Fig. 1.2.3;†$ pressure transducers (e.g. CEC Bell & Howell type 4100-00-01M0 and 4-807-0000-03M0 absolute);$ digital readout for transducers,$ or †† large CO_2 cylinder with regulator.$

Design notes

Sample and expansion vessels are gas cylinders, one about double the volume of the other. Pressure transducers are used to avoid the inaccuracies due to hysterisis effects which occur with Bourdon pressure gauges.

Safety

The line must be entirely of metal or reinforced plastic, rated to 30 atm. Lock the cylinder regulator at 20 atm maximum. Fit a 30 atm bursting disc on the delivery line, and a 2 atm bursting disc on the pumping line. Paint the cylinders with several layers of anti-corrosive and rubberised paint, and add anti-corrosion agent to the water bath.

References

[1] *Measurement of virial coefficients by a modified Burnett expansion method.* W. P. Baskett and G. P. Matthews. *J. Chem. Educ.* (April 1985)

[2] *Intermolecular forces, their origin and determination.* G. C. Maitland, M. Rigby, E. B. Smith, and W. A. Wakeham. Clarendon Press, Oxford (1981), pp. 108 and 136.

1.3 Joule–Thomson coefficients

When a gas expands through a porous plug or nozzle, its pressure decreases. If the gas were perfect, there would be no change in temperature accompanying the pressure drop, but in real gases there is a temperature change, usually a decrease. The extent of the temperature change with pressure is expressed in terms of the *Joule–Thomson coefficient* μ_{JT}, and is a measure of the deviation of the gas from perfect behaviour.

In this experiment we measure the Joule–Thomson coefficient for CO_2 at 25 °C, and compare it with calculations based on gas imperfection data.

Theory

The apparatus to measure the Joule–Thomson coefficient is shown schematically in Fig. 1.3.1. The diagram illustrates the flow of gas through a porous plug, during which it changes from a state (p_1, V_1, T_1) with internal energy U_1 to a state (p_2, V_2, T_2) with internal energy U_2. We consider 1 mole of the gas, and assume that it moves slowly enough for it to be in equilibrium at all times. The work w done on the system by the

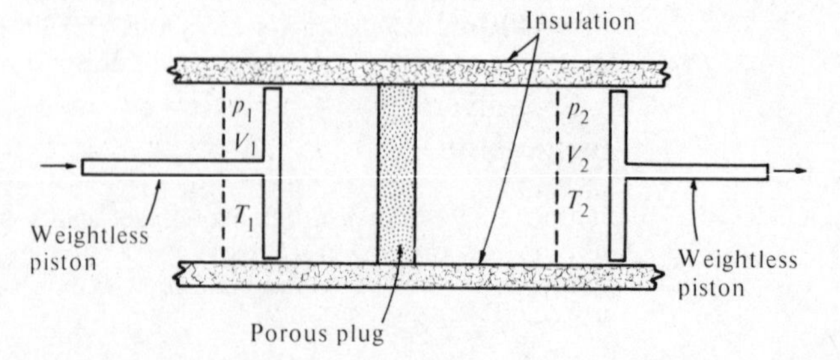

Fig. 1.3.1. Schematic diagram of a Joule–Thomson apparatus

surroundings is given by

$$w = p_1V_1 - p_2V_2. \tag{1.3.1}$$

The system is thermally insulated, and so the heat absorbed by the system, q, is zero. Processes of this type are termed *adiabatic.*

The *First Law of thermodynamics* shows that the change in internal energy is

$$\Delta U = U_2 - U_1 = q + w = w. \tag{1.3.2}$$

Combining eqns (1.3.1) and (1.3.2),

$$U_2 + p_2V_2 = U_1 + p_1V_1, \tag{1.3.3}$$

and therefore

$$H_2 = H_1. \tag{1.3.4}$$

Thus the enthalpy of the gas does not change as it passes through the plug, and the adiabatic process is *isenthalpic.*

The Joule–Thomson coefficient μ_{JT} is defined as the change of temperature with pressure at constant enthalpy,

$$\mu_{JT} = \left(\frac{\partial T}{\partial p}\right)_H. \tag{1.3.5}$$

By standard thermodynamic arguments[1] it may be shown that

Joule–Thomson coefficients are an exception to the general rule that the thermodynamic functions of real gases approach those of a perfect gas as the gas density tends to zero.

$$\mu_{JT} = \left[T\left(\frac{\partial V}{\partial T}\right)_p - V\right] \Big/ C_p, \tag{1.3.6}$$

where C_p is the molar heat capacity at constant pressure, equal to $(\partial H/\partial T)_p$. For a perfect gas, the term in square brackets vanishes, and μ_{JT} is zero.

For a real gas, there will be certain conditions under which the effects of the attractive forces between the molecules will dominate the effects of the repulsive forces. Under these conditions the Joule–Thomson coefficient μ_{JT} will be positive, and the gas will cool as it expands from higher to lower pressure. If the temperature of the gas is then raised, μ_{JT} will diminish until it becomes zero at the Joule–Thomson *inversion temperature.* Above this temperature, μ_{JT} is negative and the gas warms up as it expands.

The inversion temperatures of most gases occur above room temperature—that of nitrogen, for example, is 621 K and that of CO_2 is ~1500 K. In He and H_2, however, the attractive forces are too small to cause a cooling effect on expansion except at low temperatures, and the inversion temperatures are ~40 K and 202 K respectively.

Inspection of Fig. 1.2.1 on p. 10 shows that for a real gas, B and dB/dT are never both zero at the same temperature, and therefore from eqn (1.3.8) we may deduce that the Boyle temperature and Joule–Thomson inversion temperature must inevitably be different; in fact at low pressure, the Joule–Thomson coefficient for simple substances is about twice the Boyle temperature.

The Joule–Thomson coefficient may also be expressed in terms of the *virial equation of state*, (p. 9):

$$pV/nRT = 1 + Bn/V + \dots, \tag{1.3.7}$$

where B is the *second virial coefficient*, and higher terms are ignored. B varies with temperature, but not with pressure. Substituting in eqn (1.3.6) we find that for 1 mole of gas

$$\mu_{JT} \approx (T\,dB/dT - B)/C_p. \tag{1.3.8}$$

Apparatus[2]

The Joule–Thomson apparatus is shown in Fig. 1.3.2. Carbon dioxide is supplied from a cylinder, via a sensitive pressure regulator, and its pressure is measured on a mirror-scale mercury manometer. The gas passes through a 50-turn copper coil in a bath of water, thermostatically controlled at 298 ± 0.01 K. The gas is assumed to have attained this temperature by the time it reaches the porous plug.

The porous plug is a fine sintered glass disc mounted in a glass cell by means of a rubber bung. The disc and rubber bung

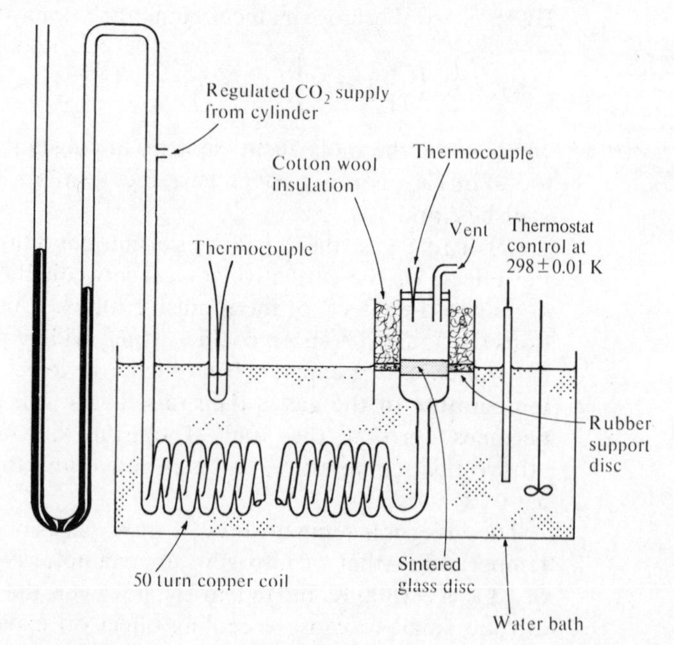

Fig. 1.3.2. The Joule–Thomson apparatus. Note the level of the water in the bath

are at the same level as the water in the tank. The space above the bung is insulated.

After passing through the porous plug the CO_2 is at atmospheric pressure. The temperature difference ΔT between the inlet and outlet gas is measured by a thermocouple. One junction of the thermocouple is immersed in a small amount of oil at the bottom of a test tube in the thermostat, and the other junction is adjacent to the outlet side of the sintered disc. Thus the e.m.f. across the two junctions is a direct measure of ΔT.

Procedure

Take measurements of the e.m.f. of the thermocouple for values of Δp between 350 and 550 mm Hg (0.45–0.7 atm). Measure ΔT periodically until it changes by less than 0.01 K in 5 min. The gas must be admitted to the cell *very* slowly: if the gas pressure is increased too rapidly the coil will not be sufficient to bring the initial surge of gas to bath temperature and the porous plug will be cooled to below its steady state value. This will cause a very slow attainment (over 1 h) of the steady state ΔT. With care in making pressure changes a steady state should be achieved in 30 to 40 min.

Calculation

It is found that the term $(\partial T/\partial p)_H$ is almost constant for the pressure differences Δp of up to 1 atm which are employed in this experiment, and therefore eqn (1.3.5) may be written

$$\mu_{JT} = (\Delta T/\Delta p)_H. \tag{1.3.9}$$

Plot a graph of ΔT against Δp, and draw a straight line through the points which also passes through the origin. Hence obtain a value of the Joule–Thomson coefficient of CO_2 (eqn (1.3.9)) in units of $\mathrm{K\,atm^{-1}}$.

Values of the second virial coefficient of CO_2 are listed in Table 1.3.1. Plot the data on a graph, and draw a smoothing curve through them. Find dB/dT at 298 K by drawing a tangent

Table 1.3.1. Second virial coefficient of carbon dioxide[4]

T/K	$B/\mathrm{cm^3\,mol^{-1}}$
280	−143.3
290	−132.5
298.15	−124.5
300	−122.7
310	−113.9
320	−105.8

to the curve at this temperature. Hence calculate μ_{JT} from eqn (1.3.8), given that $C_p = 0.37\ \text{dm}^3\ \text{atm mol}^{-1}\ \text{K}^{-1}$.[3]

Compare the value of μ_{JT} from your direct measurements with that calculated from second virial coefficients. Which do you think is the more accurate value, and why?

Results

At the maximum pressure difference of 550 mm Hg, ΔT for CO_2 should be about 0.8 K.

Accuracy

The measurements should follow a good straight line and yield a result accurate to 5 per cent.

Comment

Joule–Thomson coefficients should theoretically be a rich source of information about gas imperfections. Unfortunately, however, it has proved very difficult to eliminate heat leaks, and measurements have generally been less accurate than direct measurements of the second virial coefficients, Expt 1.2. The most reliable results have been obtained using an isenthalpic apparatus in which the gas flows through a thimble-shaped plug, and using an isothermal apparatus in which the heat required to bring the gas back to its original temperature is measured.[5]

An important application of the Joule–Thomson effect is in the refrigeration and the liquefaction of gases.[1] The gas or refrigerant is made to pass through a nozzle rather than a porous plug, whereupon it cools and liquefies. Any unliquefied gas is recycled.

Technical notes

Apparatus

Figure 1.3.1 and as described earlier;† potentiometer for thermocouple; thermocouple calibration data.

References

[1] *Heat and thermodynamics.* M. W. Zemansky, McGraw-Hill, New York, 5th edn (1968), p. 335.
[2] *Experiments in physical chemistry*, D. P. Shoemaker, C. W. Garland, J. I. Steinfeld and J. W. Nibler. McGraw-Hill, New York, 4th edn (1981), p. 70.
[3] *CRC handbook of chemistry and physics.* Ed. R. C. Weast, CRC Press Inc., Florida, 64th edn (1983–84), p. D-58.
[4] *The virial coefficients of pure gases and mixtures.* J. H. Dymond and E. B. Smith. Clarendon Press, Oxford (1980), p. 51.
[5] *Intermolecular forces. Their origin and determination*, G. C. Maitland, M. Rigby, E. B. Smith and W. A. Wakeham. Clarendon Press, Oxford (1981), p. 131.

Section 2
Equilibria between phases

The ten experiments in this section are concerned with equilibria between chemical species which are in two or more different phases and which do not react with one another. We study the equilibria either in terms of the quantity of species in each phase, the thermodynamic quantities associated with the position of equilibrium or the phase change, or the properties of the boundary between the phases.

The first experiment introduces this area of physical chemistry with an exposition of the Gibbs phase rule. The phase rule is then applied to a system of three liquids which, when shaken together, either mix completely or separate to form two layers. Experiment 2.2 is concerned with the change in total volume which occurs when two liquids are mixed, and involves the measurement of partial molar quantities.

Experiments 2.3 and 2.4 involve a study of the liquid–vapour phase equilibria of pure components and mixtures respectively. In Expts 2.5 and 2.6, we examine solid–liquid phase equilibria in terms of freezing point depression and solubility, and in Expts 2.7 and 2.8 we also study transitions in the solid phase. All the experiments in this group draw freely on the first and second laws of thermodynamics.

Experiments 2.9 and 2.10 are surface chemistry experiments, the former being concerned with the surface tension of a liquid–vapour interface, and the latter with the adsorption of a gas by a solid surface.

The fact that no chemical reactions are taking place means that the phase equilibria studied in this section are relatively simple to understand. A knowledge of what happens in non-reacting systems is fundamental to a proper understanding of equilibrium reactions, which are studied in the following section, and to the kinetics of chemical reactions which are the subject of Section 8.

2.1 Properties of a system of three partially miscible liquids

In this experiment we investigate the solubility relations of mixtures of water, chloroform and ethanoic (acetic) acid. Ethanoic acid is miscible with both water and chloroform, but water and chloroform do not mix with one another. The behaviour of the system at constant temperature and pressure (in this case 25 °C and 1 atm) is plotted on a *phase diagram*. The *Gibbs phase rule* is then used to account for the behaviour of the system, and analysis of the separated liquid layers allows the *plait point* or *critical solution point* of the system to be determined.

Theory[1]

(*i*) *The Gibbs phase rule*

Let us first define three quantities relating to a chemical system: the number of *phases*, *P*, which is the number of states of matter which are uniform throughout, not only in chemical composition but also in physical state; the number of *components*, *C*, which is the minimum number of chemically distinct constituents necessary to describe the composition of each phase in the system; and the number of *degrees of freedom*, *F*, which is the number of variable factors such as temperature, pressure and concentration, which need to be fixed in order that the system may be completely defined. From thermodynamic considerations, Gibbs derived the *phase rule* connecting these three parameters:

$$F = C - P + 2. \tag{2.1.1}$$

The present system has three chemical constituents, water, chloroform, and ethanoic acid, so that $C = 3$. We study only the condensed system, i.e. the three liquids, and ignore their vapours. Thus when the liquids mix, the number of phases P is 1, and when they separate into two layers, $P = 2$. From the phase rule, the corresponding number of degrees of freedom F is 4 or 3. Since two of these degrees of freedom are taken up by fixing the temperature and pressure, we find that there are two remaining degrees of freedom in the 1-phase region, and only one degree of freedom in the 2-phase region.

The consequence is that in the single phase region, two of the compositions may change independently. However, if we attempt to make up a homogeneous solution with a total composition which lies in the 2-phase region, it separates to give two layers which both have compositions lying on the binodal curve. These layers only have one degree of freedom; if, for example, we know the amount of chloroform in one of the layers, the amount of the other two phases in the layer is fixed in relation to the amount of chloroform by the position of the binodal curve.

(ii) Phase diagram of a condensed three component system

From a theoretical point of view the most convenient method of expressing relative concentrations is in terms of mole fractions x. For example in the present system,

$$x_{H_2O} = \frac{n_{H_2O}}{n_{H_2O} + n_{CHCl_3} + n_{CH_3CO_2H}} \tag{2.1.2}$$

where n is the number of moles of a particular species. It follows that

$$x_{H_2O} + x_{CHCl_3} + x_{CH_3CO_2H} = 1. \tag{2.1.3}$$

However it is experimentally much more convenient to use volume percentages v. For example

$$v_{H_2O} = \frac{(m_{H_2O}n_{H_2O}/\rho_{H_2O}) \times 100 \text{ per cent}}{(m_{H_2O}n_{H_2O}/\rho_{H_2O}) + (m_{CHCl_3}n_{CHCl_3}/\rho_{CHCl_3}) + (m_{CH_3CO_2H}n_{CH_3CO_2H}/\rho_{CH_3CO_2H})} \tag{2.1.4}$$

where m is the molar mass and ρ the density of each species, and

$$v_{H_2O} + v_{CHCl_3} + v_{CH_3CO_2H} = 100 \text{ per cent}. \tag{2.1.5}$$

In the present case where we wish to describe the state of the system in terms of any two of three variables with a constant sum, the state may be conveniently represented on triangular graph paper (Fig. 2.1.1). To plot, or read off, the volume fraction of one of the components, we note that the labelled corner of the triangle represents the pure species, i.e. $v = 100$ per cent, and the side of the triangle opposite the corner represents $v = 0$ per cent. So to plot the composition of a mixture containing 60 per cent by volume of ethanoic acid, we count up six of the horizontal graph lines to reach the line WX, Fig. 2.1.1. If our mixture also contains 30 per cent by volume of water, then we count down three right-sloping lines (/) to the line YZ. This intersects WX at the required point Q. *What must* v_{CHCl_3} *be (eqn (2.1.5))? Does the graph indicate this correctly?*

When labelling the axes on triangular graph paper, add marks round the edge as in Fig. 2.1.1, starting from 0 per cent to discover in which direction to draw them; the marks then indicate the line on the graph to which each number refers, and thus saves much confusion in the plotting of points.

If a whole series of mixtures of the three components were made up and shaken, we would find that some mixed completely, whereas others separated into two layers. After a sufficient number had been tried, we could draw the *binodal curve*

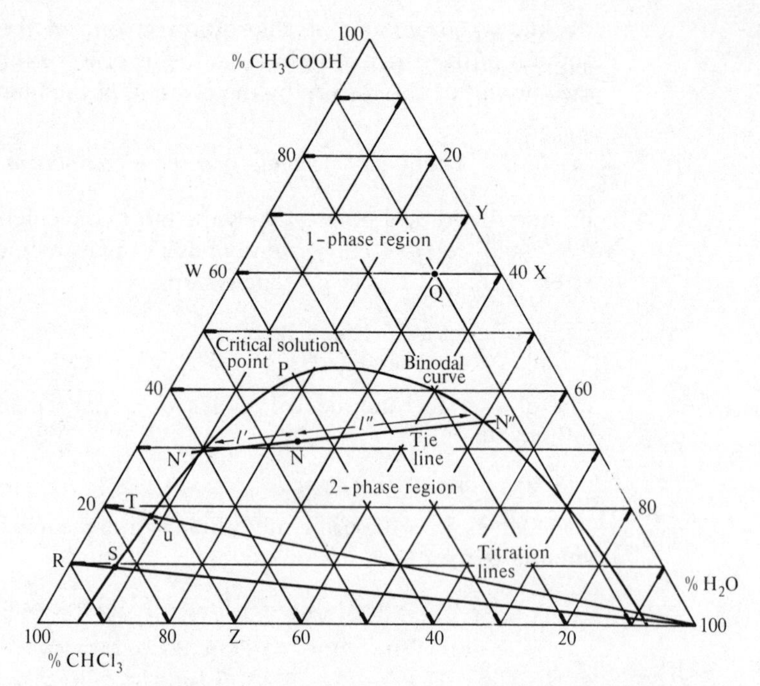

Fig. 2.1.1. Features of the phase diagram of the $H_2O/CHCl_3/CH_3CO_2H$ system. Note that the binodal curve is *not* correctly positioned

which separates the 2-phase from the 1-phase region, Fig. 2.1.1. However a much quicker method of finding the binodal curve is to make up a mixture of two of the components, and titrate with the third. Suppose, for example, we start with a mixture of 18 cm^3 chloroform and 2 cm^3 ethanoic acid, so that $v_{CHCl_3} = 90$ per cent and $v_{CH_3COOH} = 10$ per cent, corresponding to the point R on the diagram. As we titrate with water, the system moves along the line from R to the pure water apex, since this line represents all the mixtures with a constant ratio $v_{CHCl_3} : v_{CH_3COOH}$ of 9:1. When the mixture goes cloudy, its composition must be on the binodal curve at point *S*, which is then marked on the graph. (It is assumed that the amount of the second phase, which causes the cloudiness, is so small that the composition of the first phase corresponds to the amount of the three components added). Suppose that we then add a further 2.5 cm^3 of ethanoic acid to the mixture. The mixture goes clear because we have jumped to the point T on the line of mixtures starting at $v_{CH_3COOH} = (2+2.5)/(20+2.5) \times 100$ per cent $= 20$ per cent and $v_{CHCl_3} = 80$ per cent. Titration to cloudiness then yields the point *U* on the binodal curve.

If the procedure is repeated, and a further series of points determined by titrating ethanoic acid/water mixtures with chloroform, the entire binodal curve may be determined.

(*iii*) *Tie lines and the critical solution point*

If information is required about the compositions of the layers which form in the 2-phase region, an analysis must be carried out. Suppose that we make up a mixture of total composition N, Fig. 2.1.1, which then separates into two layers of composition N′ and N″. Not only will N′ and N″ both lie on the binodal curve, but also N′, N and N″ will all lie on the same straight *tie line*. The position of this tie line may be found by analysing the layers. Note that the tie line is not horizontal because ethanoic acid is not distributed equally between the layers; there is more ethanoic acid in the water-rich phase than in the chloroform-rich phase.

By the *lever rule* (a condition of conservation of mass), the amounts of N′ and N″ are in inverse proportion to their lengths along the tie line from the overall composition, i.e. $N'/N'' = l''/l'$.

If a number of tie lines are determined, they are found to converge on a single *critical solution point* P. At this point, the two layers have the same composition, and become one.

Apparatus

The experiment requires volumetric glassware and a well-stoppered 50 cm^3 conical flask. It may be performed on the bench at room temperature, or using solutions brought to 25 °C in a thermostatically controlled water bath.

Procedure

Glacial ethanoic acid is flammable and can cause severe burns; keep the stock bottle in a fume cupboard at all times. Chloroform vapour is harmful; avoid inhalation, and stopper flasks when not in use. Under no circumstances pipette either glacial ethanoic acid or chloroform by mouth. Any traces of detergent on glassware will drastically affect results; rinse all glassware several times with purified water before use.

Before using pipette fillers, always check that they are entirely clean inside.

(*i*) *Determination of binodal curve*

If a thermostat tank is used, place the liquids and mixtures on a thermostat bath at 25 °C (or as specified*). Leave for at least 10 minutes before carrying out any titrations.

Label a piece of triangular graph paper in the manner described previously.

A light beam, or a black baseboard or tile, can help to highlight the opalescence of the solution. Opalescence before adding any water signifies contamination.

Add 1 cm^3 of glacial ethanoic acid to 25 cm^3 chloroform in a (stoppered) 100 cm^3 conical flask. Add purified water from a burette, while shaking the flask to prevent layers forming, until the first signs of persistent cloudiness appear. Note the composition of the mixture, and plot it on your graph.

Now add 2 cm^3 of ethanoic acid, add water until cloudiness occurs, and note the composition again.

Repeat with further 2 cm^3 additions of ethanoic acid until a total of 21 cm^3 has been added. Then add acid in 4 cm^3 steps until a total of 29 cm^3 has been added, after which add 6 cm^3 portions of acid.

Repeat the procedure, this time titrating with chloroform using a burette with a Teflon (Fluon) tap, and adding the same amounts of ethanoic acid to 25 cm^3 water to make up the initial solutions.

(ii) Determination of tie lines and critical solution point

Rinse all the glassware again with distilled water.

Make up 3 cm^3 of a mixture of the three components with a composition lying well inside the 2-phase region. Shake the mixture well, and separate the layers in a separating funnel. If a thermostat bath is being used, ensure that the mixture reaches thermal equilibrium before separation, and shake the mixture at intervals while it is in the tank. Also, bring the separating funnel to the correct temperature by pouring through liquid which has been in the thermostat bath.

Determine the ethanoic acid in the lower, chloroform layer by titration with sodium hydroxide solution, using phenolphthalein as indicator. When the acid has been completely neutralized, the chloroform will become almost insoluble. It can then be separated and its volume measured. Measure also the volume of the aqueous liquid.

Repeat for the aqueous layer. It is unlikely that any of the chloroform will separate out in this case. However, the position of the mixture on the binodal curve can be estimated by finding the relative volumes of the other two components.

Repeat the entire procedure for two further mixtures of different overall composition.

Calculation

Plot the binodal curve by drawing a line through the compositions at which cloudiness occurs. Label the graph with the temperature at which the experiment was carried out.

Use the density of ethanoic acid determined from Table 2.1.1 to convert the amount of the acid in one of the layers to a

Table 2.1.1. Densities of water, ethanoic acid, and chloroform/kg m^{-3} [2]

T °C	Water	Ethanoic acid	Chloroform
16.0	999.0	1054	1497
18.0	998.6	1052	1493
20.0	998.2	1050	1489
22.0	997.8	1048	1485
25.0	997.1	1044	1480
30.0	995.7	1039	1470
40.0	992.2	1027	1451
50.0	988.1	1016	1431
60.0	983.2	1005	1411
70.0	977.8	993	1391

volume. Add 1 per cent of the volume of the aqueous liquid to the volume of the chloroform to allow for the slight solubility of chloroform in water. Then calculate the composition of the layer in terms of volume percentages. Repeat the calculation for the other layers.

Plot the compositions of the layers on the graph, and draw the tie line between each pair. The tie lines should pass through the corresponding overall compositions.

From the positions of the tie lines, estimate the critical solution point.[3] Does this coincide with the maximum of the binodal curve with respect to percentage volume of ethanoic acid? Are the tie lines parallel?

For one point on the binodal curve calculate the composition in mole fractions.

Briefly discuss the diagram. What do you expect to be the effect of increase of temperature of the system?

Results[4]

Accuracy

Errors due to detergent on glassware must be avoided. Other sources of error, which are all easily recognizable, should be estimated.

Comment

The behaviour of multi-component systems is of considerable importance in the chemical industry. For example, whereas absolute alcohol cannot be obtained from the distillation of ethanol and water, it can be produced by distilling a three-component mixture of ethanol, water and benzene. The mechanism of this process can be explained in terms of phase diagrams of the type produced in this experiment, drawn for different temperatures.[5]

Technical notes

Apparatus

Several 50 cm^3 stoppered conical flasks, two burettes (at least one with a Teflon tap), pipettes, pipette-filler, 5 cm^3 separating funnel; glacial ethanoic acid, chloroform, standard sodium hydroxide (~0.5 M), phenolphthalein; triangular graph paper; [light source or black base or tile]; [thermostat bath].

Effect of temperature

If the experiment is studied over a range of temperatures by different students, the results can be drawn onto a stack of perspex sheets in a frame to give a three-dimensional phase diagram.

References

[1] A. *Findlay's 'The phase rule and its applications'*. A. N. Campbell and N. O. Smith. Dover Publications, New York (1951), ch. 2 and 14.
[2] *International critical tables* (E. W. Washburn, ed.), McGraw-Hill, New York (1928), vol. 3, pp. 24 and 28.
[3] *Tie line correlation and plait point determination.* E. L. Heric. *J. Chem. Educ.* **37,** 144 (1960).
[4] Ref. 2, p. 410.
[5] *The preparation of absolute alcohol from strong spirit.* S. Young. *J. Chem. Soc.* **81,** 707 (1902).

2.2 Partial molar volume by the buoyancy method

When several substances are mixed together, the total volume of the mixture is generally different from the sum of the volumes of the individual components. This expansion or contraction arises from the difference in magnitude of the intermolecular forces between like and unlike molecules, and is very noticeable when both components are liquids. The extent of the volume change is described quantitatively in terms of *partial molar volume.*[1, 2]

In this experiment we measure the partial molar volume of ethanol in water at various concentrations. The partial molar volume is the effective volume occupied by 1 mole of ethanol in the aqueous solution, which is different from the volume of 1 mole of pure ethanol. The most common method is to measure directly the mass of a known volume of solution measured in a calibrated glass vessel known as a pyknometer. Pyknometers are quite difficult to use, however, and so in this experiment we measure the density of known amounts of solutions with a sinker—a weighted quartz sphere, attached to the arm of a chemical balance, which dips into the solution.

Theory

Three concentration scales are used to describe the concentrations of solutions. The most common is *molarity*, i.e. the number of moles of a substance per dm^3 of *solution*. On the *mole fraction* scale, the concentration of a solution is expressed as the number of moles of substance n_e divided by the total number of moles present in the solution $(n_e + n_w)$, where n_w is the number of moles of solvent (in this case water).

The molarity of a solution varies with temperature but the molality does not.

However, for many purposes it is more convenient to keep the mass of the solvent constant. The *molality* of the solution is then defined as the number of moles of substance per unit mass of solvent:

$$m_e = W_e/(M_e W_w). \qquad (2.2.1)$$

The molality scale is used throughout this experiment, and has SI units of $mol\,kg^{-1}$. m_e is the molality of ethanol in an aqueous solution, W_e and W_w are the masses of the ethanol and water in the solution (kg), and M_e is the molar mass of the ethanol ($kg\,mol^{-1}$).

When n_e moles of ethanol are mixed with n_w moles of water at constant temperature and pressure, the resulting volume V is

$$V = n_e \bar{V}_e + n_w \bar{V}_w, \qquad (2.2.2)$$

where $\bar{V}_e$ and $\bar{V}_w$ are the partial molar volumes. For any component i of a mixture, the partial molar volume is defined by the partial derivative $(|\partial V/\partial n_i|)_{n_j,\,T,\,p}$, in which the subscripts indicate that the temperature, pressure, and the number of moles of all the components except i are constant.

According to eqn (2.2.2), the partial molar volume of ethanol, $\bar{V}_e$, could be determined experimentally by measuring the total volume V for different amounts of ethanol, n_e, at constant n_w. If V were than plotted against n_e, the result would be a curve from which $\bar{V}_e$ could be obtained by drawing a tangent at any value of n_e.

In this experiment, however, instead of finding the volume change on mixing we measure the density of the solution ρ. The density is obtained from the density of water, ρ_w, and the density r of the solution relative to water, by the relation

$$\begin{aligned} \rho &= \rho_w \cdot r \\ &= \rho_w \cdot \frac{\text{upthrust of sinker in solution}}{\text{upthrust of sinker in water}} \\ &= \rho_w \cdot \frac{\text{weight of sinker in air} - \text{weight of sinker in solution}}{\text{weight of sinker in air} - \text{weight of sinker in water}} \end{aligned} \qquad (2.2.3)$$

The *excess volume* V^E is the volume of the mixture less the sum of the separate volumes of the components before mixing:

$$V^E = V - V_e - V_w$$
$$= \frac{W_w + W_e}{\rho} - \frac{W_w}{\rho_w} - \frac{W_e}{\rho_e}, \qquad (2.2.4)$$

where W_e is the mass of ethanol and W_w the mass of water in the solution, and ρ_e is the density of the ethanol.

V^E is calculated for a series of solutions all containing a unit mass of water and varying amounts of alcohol, so that

$$V^E = \frac{1 + W'_e}{\rho} - \frac{1}{\rho_w} - \frac{W'_e}{\rho_e}, \qquad (2.2.5)$$

where W'_e is the mass of ethanol in a solution containing a unit mass of water, i.e. $W'_e = W_e \cdot 1/W_w$ (as shown in Table 2.2.1 overleaf).

We then plot a graph of V^E against n_e, which has a slope given by

$$\left(\frac{\partial V^E}{\partial n_e}\right)_{n_w,T,p} = \left(\frac{\partial(V - V_e - V_w)}{\partial n_e}\right)_{n_w,T,p} = \bar{V}_e - V_e^0, \qquad (2.2.6)$$

where V_e^0 is the molar volume of ethanol, equal to $\rho_e M_e$. In practice we plot $(\partial V^E/\partial m_e)$ rather than $(\partial V^E/\partial n_e)$, m_e and n_e being the same numerically.

Apparatus

The sinker used for the density determinations is a sphere, made from fused quartz to minimize the effect of temperature on its volume. It contains enough lead shot to have a near zero weight in the most dense liquid, and is attached by a nylon thread to the arm of a chemical balance so that its weight in each solution may be measured.

Volumes of ethanol are measured out by means of a dispenser which delivers 1 cm^3 of liquid at each depression of the plunger. The solutions reach thermal equilibrium in a water bath at 20 °C.

Procedure

Clean, dry and label the twelve sample bottles provided. Set out a table for the twelve samples similar to that shown in the *Results* section, and record all weighings as they are carried out.

Weigh the first bottle. Add the correct volume of ethanol from the dispenser and weigh the bottle again. Suitable volumes for the twelve samples are 3, 6, 9, 11, 13, 15, 17, 19, 21, 23, 25, and 27 cm^3 ethanol. Quickly, without allowing time for the ethanol to evaporate significantly, make up the liquid to about

60 cm^3 with purified water, and weigh once more. Immediately *mix the solution well* with a glass rod, stopper the bottle and put it in the thermostat. Repeat the procedure for the other bottles. Leave them for at least 30 min to reach thermal equilibrium. Also place samples of water and pure ethanol in the tank.

Check that the sinker is dry and weigh it in air.

Measure the densities of the equilibrated solutions, water and pure ethanol by the buoyancy method as follows. Since the measurements are carried out with the bottles removed from the thermostat, they must be carried out as quickly as possible to avoid undue temperature changes. Measure the weight of the sinker in each of the solutions in turn, and in water and pure ethanol. If there are bubbles on the walls of the bottles on removal from the bath, shake the bottle, holding the stopper in firmly, until they are dispersed. This will prevent their subsequent transfer to the sinker. When making the measurements, take care that the sinker does not touch the walls of the bottle, and that there are no air bubbles sticking to the outside of the sinker. Drain excess liquid from the sinker between measurements, and return the bottles to the bath after use. Inspect the trend of the weighings regularly, and recheck any that seem anomalous.

Calculation

Calculate the molality m_e of each solution from eqn (2.2.1). Calculate the densities of ethanol and the solutions from eqn (2.2.3), given that the density of water ρ_w is as shown in Table 2.1.1, p. 27. Calculate the excess volumes V^E of the solutions by use of eqn (2.2.5). Plot a graph of the excess volume of each solution against its molality, and draw carefully a smooth curve through the points. Find the slope of this curve at ten points by drawing tangents to it. Tabulate the results against molality m_e, and calculate the partial molar volume $\bar{V}_e$ at each point from eqn (2.2.6). Plot the values of $\bar{V}_e$ against m_e on the same graph as the V^E curve, putting in a second, clearly labelled ordinate axis.

The molality scale of concentration has been the most convenient for this experiment. However, as a further exercise, sketch a small freehand graph (*not* a full-scale plot with graduated axes), showing the approximate dependence of the partial molar volume of ethanol on mole fraction, from mole fraction $x_e = 0$ to 1. As you have points only at the low x_e end, guess what the rest of the curve might look like, and show on the sketch which part you have guessed and which comes from your results.

Results

Suppose the weight of the sinker in air = 5.5085 g, in water = 0.7341 g, in ethanol = 1.7323 g. Then the upthrust of the sinker in water = 4.7654×10^{-3} kg. See Table 2.2.1 for specimen results at 20 °C.

Table 2.2.1. Specimen results at 20 °C

Approximate volume ethanol/cm^3	3
Mass empty bottle/g	43.9462
Mass bottle + ethanol/g	46.3347
Mass bottle + ethanol + water/g	100.0786
Mass of sinker in solution/g	0.7858
Upthrust of sinker in solution/kg	4.7227×10^{-3}
Mass of ethanol W_e/kg	2.3885×10^{-3}
Mass of water W_w/kg	5.3744×10^{-2}
Mass of ethanol in solution containing 1 kg of water W'_e/kg	4.4442×10^{-2}
Molality m_e/mol kg^{-1}	0.965
Density of solution ρ/kg m^{-3}	989.3
Excess volume V^E/m^3	-5.570×10^{-6}

Accuracy

Errors can arise if the sinker touches the walls of the bottle, or if air bubbles become attached to the sinker. Consequently one or two points may have to be ignored when drawing the curve of excess volume against molality. Despite such errors and the fact that excess volume is calculated from the small difference between large numbers, this experiment produces results of good accuracy, and values within a few per cent of published data[3] can be obtained with care.

Comment

An understanding of partial molar volume is clearly important in explaining the relations between the scales which can be used to describe the composition of a mixture. The effects we have studied also give some hint as to the role of intermolecular forces in the structure of liquids, examined in detail in Expt 7.2.

Technical notes

Apparatus

Sinker—see *Apparatus* section;[††] chemical balance with bridge and wire to take sinker,[$†] twelve sample bottles, in the form of glass cylinders 3.5 cm diameter by about 10 cm long, marked at the 60 cm^3 level;[†] thermostat tank set at 20 °C; ethanol in dispenser set to 1 cm^3.

References

[1] *Physical chemistry*. P. W. Atkins. Oxford University Press, Oxford, 3rd edn (1985), Sect. 8.1.

[2] *Liquids and liquid mixtures.* J. S. Rowlinson and F. L. Swinton, Butterworths, London, 3rd edn (1982), ch. 4.
[3] *Density of ethyl alcohol and of its mixtures with water.* N. S. Osborne. *Bull. Bur. Stand.* **9,** 424, (1913).

2.3 Vapour pressures of liquids

This experiment involves the measurement of the vapour pressures of ethanol and cyclohexane over a range of temperatures.

The method used in this experiment is a very simple one.[1] A narrow tube, closed at its top end, is dropped into a vessel containing a thermometer and some of the sample liquid, Fig. 2.3.1. The liquid is boiled so that its vapour displaces any air in the narrow tube. Then pressure is applied to equalize the liquid levels inside and outside the tube, whereupon the applied pressure as measured with a mercury manometer, is equal to the vapour pressure.

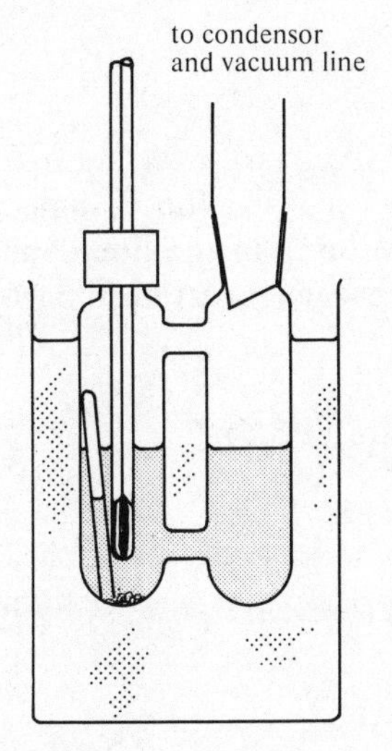

Fig. 2.3.1. Arrangement of the sample vessel

The variation of vapour pressure with temperature yields both the enthalpy of vaporization ΔH_{vap} and the entropy of vaporization ΔS_{vap}. From the magnitude of ΔS_{vap} it may be deduced whether or not the liquid or vapour is associated.

Theory From the definition of free energy, G, we may derive the relation

$$dG = V\,dp - S\,dT. \tag{2.3.1}$$

Consider a system comprising a liquid and its vapour at equilibrium which undergoes a very small reversible change. Equating the changes in free energy in the liquid and gaseous phases

$$V(\mathrm{l})\,\mathrm{d}p - S(\mathrm{l})\,\mathrm{d}T = V(\mathrm{g})\,\mathrm{d}p - S(\mathrm{g})\,\mathrm{d}T. \tag{2.3.2}$$

Therefore

$$\frac{\mathrm{d}p}{\mathrm{d}T} = \frac{S(\mathrm{g}) - S(\mathrm{l})}{V(\mathrm{g}) - V(\mathrm{l})} = \frac{\Delta S_{\mathrm{vap}}}{\Delta V_{\mathrm{vap}}}. \tag{2.3.3.}$$

Since the two phases are in equilibrium

$$\Delta G_{\mathrm{vap}} = \Delta H_{\mathrm{vap}} - T\Delta S_{\mathrm{vap}} = 0, \tag{2.3.4}$$

where T is the temperature of the equilibrium, i.e. the boiling point at the pressure under consideration. So

$$\Delta S_{\mathrm{vap}} = \Delta H_{\mathrm{vap}}/T, \tag{2.3.5}$$

and we obtain the *Clapeyron equation* which is exact for systems at equilibrium:

$$\frac{\mathrm{d}p}{\mathrm{d}T} = \Delta H_{\mathrm{vap}}/T\,\Delta V_{\mathrm{vap}}. \tag{2.3.6}$$

To a good approximation, the term $\Delta V_{\mathrm{vap}} = V(\mathrm{g}) - V(\mathrm{l})$ may be replaced by $V(\mathrm{g})$ since the volume of the gas is so much greater than the volume of the liquid. If the vapour is assumed to obey the perfect gas equation, then for one mole

$$pV(\mathrm{g}) = RT. \tag{2.3.7}$$

Therefore from eqn (2.3.6)

$$\frac{\mathrm{d}p}{\mathrm{d}T} = \Delta H_{\mathrm{vap}}\,p/RT^2, \tag{2.3.8}$$

from which is obtained the *Clausius–Clapeyron equation*

$$\frac{\mathrm{d}(\ln p)}{\mathrm{d}T} = \Delta H_{\mathrm{vap}}/RT^2. \tag{2.3.9}$$

Assuming ΔH_{vap} is independent of temperature, the equation may be integrated to give

Is ΔH_{vap} independent of temperature? What must the value of ΔH_{vap} be at the critical point?[2]

$$\ln p = -\Delta H_{\mathrm{vap}}/RT + \mathrm{const.} \tag{2.3.10}$$

The entropy of vaporization at the boiling point under 1 atm has been found to be approximately the same for all substances not associated in their liquid or vapour phases. For such substances ΔS_{vap} lies in the range 85–90 J K^{-1} mol^{-1}. This observation is known as *Trouton's rule*.

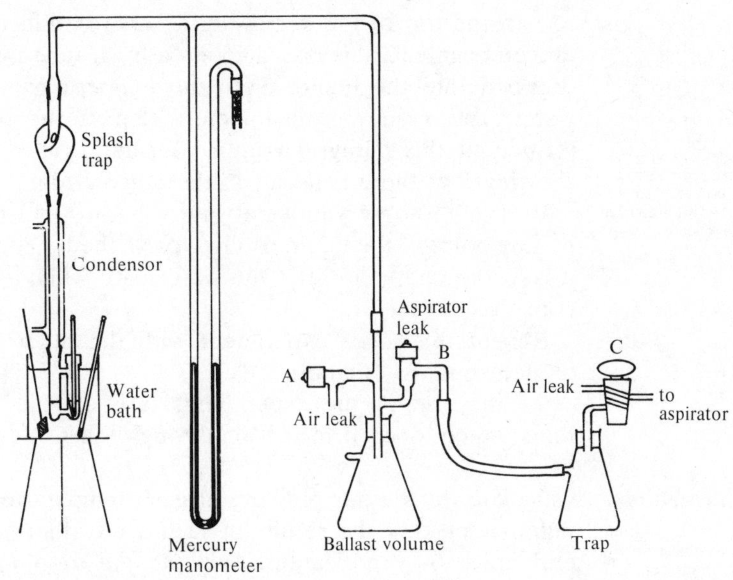

Fig. 2.3.2. The vacuum line

Apparatus The apparatus is shown in Figs. 2.3.1 and 2.3.2. The sample vessel, enclosing the narrow tube, thermometer, and sample liquid, is immersed in a large beaker of water. Into the beaker dips a thermometer and stirrer, and under it is a bunsen burner. The boiling tube is attached via a reflux condenser and splash trap to a ballast volume, adjustable leaks, an aspirator with trap, and mercury manometer.

Procedure Measure the ambient atmospheric pressure p_{atm} with a barometer.

Place one of the narrow tubes supplied for the experiment in the sample vessel, half fill the vessel with the sample liquid, add some anti-bumping granules, and arrange the thermometer, as shown in Fig. 2.3.1. Close the adjustable air leak and aspirator leak (taps A and B).

Turn on the aspirator. Heat the water in the water bath to about 30 °C and stir it.

By fractionally opening the tap between the aspirator and the vacuum line, reduce the pressure in the line until the liquid boils *gently*, thus removing air from the narrow tube. Using the adjustable air leak, increase the pressure in the line *slowly* until the liquid level in the narrow tube rises to the level of the liquid outside it. Maintain the pressure near this equilibrium value for at least 5 min before taking readings. Do not allow the liquid to

A consideration of the relative densities of mercury and the test liquids shows that in practice, sufficient precision will be obtained in this experiment simply by obtaining a stable level of liquid somewhere in the body of the vapour pressure tube.

rise to the top of the narrow tube, as it is difficult to make it descend again. If this does happen a fresh tube may have to be dropped into the liquid. Note the temperature of the stirred water, and read the manometer. Repeat the procedure described in this paragraph until constant results are obtained, showing that there is no air in the narrow tube.

Repeat the observations at intervals of about 6 K up to the boiling point of the liquid. Boiling cools the liquid, so make sure it is at the temperature of the water bath when a measurement is made.

Repeat the entire experiment with the other sample liquid (cyclohexane or ethanol).

At the end of the experiment, turn off the aspirator and immediately open it to the air through tap C.

Calculation

The vapour pressure of water varies in a similar manner; at 500 mmHg, the atmospheric pressure at the summit of Mont Blanc, water boils at 89 °C.

Calculate the vapour pressure at each temperature for the two liquids, and plot the results in such a way that you obtain the enthalpies of vaporization. *Why do the units of pressure not matter?* Calculate the entropies of vaporization of these liquids at their boiling points under a pressure of 760 mmHg (1 atm), which are 78.3 °C and 80.7 °C respectively.

Compare the entropies of vaporization with the value from Trouton's rule. By considering the effect of association of a liquid or vapour on its entropy and hence the entropy change on vaporization, deduce whether ethanol and cyclohexane are associated, and if so in which phase.

Results

Accuracy

The assumptions employed in the theory of this experiment are far from accurate (p. xix), but the errors which they create tend to cancel under conditions far removed from the critical point of the liquid.[2] The apparatus itself is capable of results to ±2 per cent, but errors of 20 per cent may arise if observations are not made carefully.

Comment

There are many other ways of measuring vapour pressure. The boiling point of a liquid under a known applied pressure may be found, or an *isoteniscope* may be used, which incorporates a narrow U-tube on which the balance point between the vapour pressure and applied pressure can be accurately determined.

Technical notes

Apparatus

Narrow tubes† (3 cm × ≥2 mm i.d.) sealed at one end;† sample vessel (Fig. 2.3.1);† vacuum line (Fig. 2.3.2);† accurate barometer.$ Ethanol, cyclohexane, anti-bumping granules.

Other liquids

The apparatus is not suitable for water.

References

[1] *A simple isoteniscope and an improved method of vapor pressure measurement* M. W. Lindauer. *J. Chem. Educ.* **37,** 532 (1960).

[2] *Putting Clapeyron's equation into practical use with no help from Clausius* S. Waldenstrom, K. Stegavik, and K. R. Naqvi. *J. Chem. Educ.* **59,** 30 (1982).

2.4 Distillation of ideal and azeotropic liquid mixtures

This experiment illustrates the effect of temperature on the liquid–vapour phase equilibria of binary mixtures, and is in two parts. In Part A, a mixture of benzene and tetrachloromethane (carbon tetrachloride) is distilled, and the degree of separation of the liquids used to determine the efficiency of a distillation column. In Part B, the characteristics of the *n*-propanol/water system are investigated with the aid of a simple still. Two types of graph are used to analyse the results, and the phase rule (p. 22) confirms that at constant pressure they both describe the systems completely. The first is a plot of liquid composition against vapour composition, and the second a graph of both against temperature.

Theory

A(i) Ideal mixtures

Benzene and tetrachloromethane form an almost *ideal mixture*, because the forces between the molecules of both components of the mixture are almost identical. The vapour pressure of the mixture obeys *Raoult's Law*, changing linearly with mole fraction from that of pure benzene to that of pure tetrachloromethane. Not only do these liquids form a near ideal mixture, but also their boiling points are very similar. They are therefore very difficult to separate by distillation, and can be used as a sensitive test of the efficiency of a distillation column.

A(ii) Distillation columns[1]

Let us consider a bubble-plate fractionating column, Fig. 2.4.1, as used in oil refineries. The more bubble plates a column has and the better the equilibrium between the descending liquid and ascending vapour, the better the separation achieved between the components of a mixture. The vapour rising from a

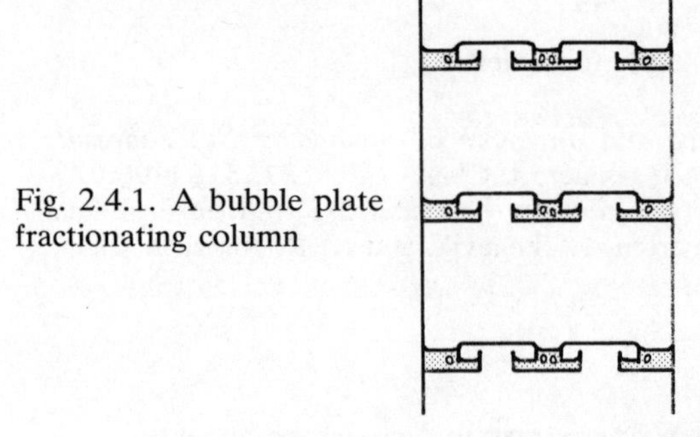

Fig. 2.4.1. A bubble plate fractionating column

perfect bubble plate would be in complete equilibrium with the liquid on the plate, and if this vapour were to be drawn off and totally condensed it would form a liquid of the same composition as the vapour, not as the liquid from which it was formed. However a distillation column need not contain bubble plates—it may be packed or simply be an empty tube. We define a *theoretical plate* as a section of any type of column, of length such that the liquid condensing at the top of the section has the same composition as the vapour in complete equilibrium with the liquid draining from the bottom. The efficiency of a distillation column is directly proportional to the number of these theoretical plates it contains, the number usually being expressed to the nearest integer.

Although it is possible to use an empty tube as a column, it will only correspond to one or two theoretical plates. In a *Vigreux column*, indentations are made in the tube so that fingers of glass point into the path of the vapour, and an efficiency of four or five plates may be achieved. For many laboratory applications this degree of efficiency is acceptable, because the liquid and vapour compositions at each temperature in the column are sufficiently different for a good separation to be achieved.

In columns for gas–liquid chromatography, packed with materials such as fine alumina, efficiencies of several thousand theoretical plates may be achieved; however these columns cannot cope with the high throughput required for fractional distillations.

However, for a useful separation of benzene (b.p. 80.1 °C) and toluene (b.p. 110.8 °C), a column with twelve theoretical plates is needed, and for mixtures with closer boiling points, still more plates are required.[2] To achieve this greater efficiency, the column must be packed. In this experiment, tiny glass helices, known as *Fenske helices*, are used. While achieving good separation because of their high surface area, they also conduct away much of the heat from the vapour. This can cause the vapour to condense in the column and form slugs of liquid which destroy the equilibrium conditions. To minimize heat losses the column we shall use has an evacuated, internally silvered, glass jacket. This does not, of course, prevent the necessary temperature gradient from being set up along it.

A(iii) Calculation of column efficiency

McCabe–Thiele plots are employed by chemical engineers, because by various graphical constructions they can be used to predict the behaviour of a distillation column under industrial operating conditions with different feed rates, reflux ratios, and so on.[4,5]

The graph we shall use to find the efficiency of the distillation column is of mole fraction of tetrachloromethane in the vapour against mole fraction of tetrachloromethane in the liquid, and is known as a McCabe–Thiele plot.[3]

The use of the graph is demonstrated in Fig. 2.4.2. The ordinate (vertical axis) of the curve is of the mole fraction of tetrachloromethane in the vapour which is in equilibrium with the liquid having the composition of the abscissa (horizontal axis). No temperatures are shown on the diagram, although they vary from the boiling point of pure benzene (80.1 °C) to that of pure tetrachloromethane (76.7 °C).

If we start with a mixture of composition a_1 and boil it, the vapour above will be richer in CCl_4, the more volatile (lower b.p.) component. From the nature of the curve, the vapour composition must be a_2. If this vapour is completely condensed, a liquid of composition a_2 is formed. So on passing through one theoretical plate, the liquid will have changed from composition a_1 to a_2. Instead of reading the value a_2 from the ordinate, and then starting again at composition a_2 on the abscissa, a diagonal construction line can be drawn as shown, and the horizontal intercept of the vapour composition on this line then gives the new liquid line. Over the next theoretical plate, the composition

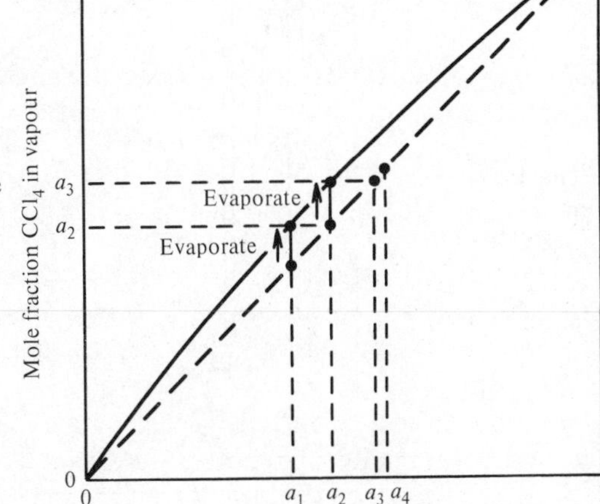

Fig. 2.4.2. McCabe–Thiele plot for benzene/tetrachloromethane. Not to scale

will change from a_2 to a_3. If the distillate has composition a_4, then the efficiency of the distillation is at just over two theoretical plates. However, even without the column, the surface of the liquid and the condensation at the stillhead form one theoretical plate. Therefore

$$\text{efficiency of column} = \text{number of theoretical plates} - 1,$$

and in this case the efficiency is approximately one plate.

B. Azeotropic systems

In part B of this experiment we study the properties of the non-ideal mixture formed by *n*-propanol and water which shows deviations from Raoult's law.

A discussion of the *n*-propanol/water system at this stage would pre-empt, and make worthless, any experimental investigation. We therefore examine an entirely different (and slightly more complicated) system, namely hydrochloric acid and water. In this case the vapour pressures at intermediate compositions are less than predicted. This means that the mixtures are more difficult to boil than expected, and the curve of boiling point against composition goes through a maximum, as illustrated in Fig. 2.4.3. Note the shape of the liquid (boiling point) curve near the maximum.

The diagram shows that if a mixture of hydrochloric acid and water of composition *a* is heated from its initial state a_1, it will

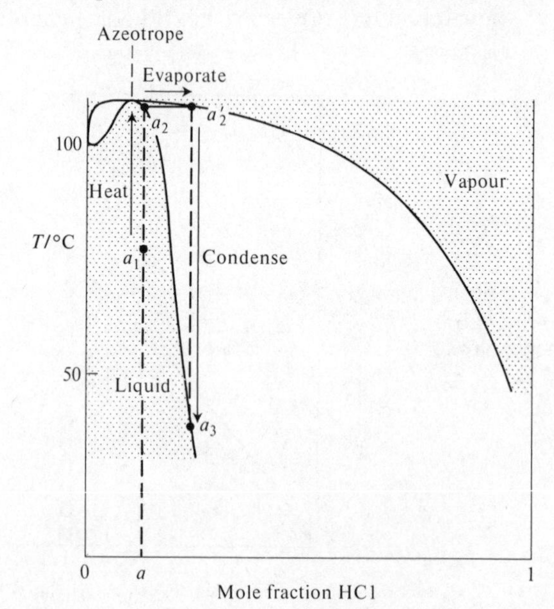

Fig. 2.4.3. The HCl/water system

The HCl/water system is complicated by demixing at temperatures lower than those shown in Fig. 2.4.3.[5,6]

boil at temperature a_2. The composition of the vapour in equilibrium with this liquid will be a_2', and after passing through one theoretical plate the composition of the liquid will reach a_3. Thus in a simple distillation of a mixture of composition a, the vapour will become richer in HCl, and the liquid remaining in the flask will correspondingly become richer in water. This process will continue until the liquid reaches the composition which has the maximum boiling point of 111 °C under a pressure of 1 atm. At this composition (0.11 mole fraction HCl$\approx$20 per cent by weight), the vapour has the same composition as the liquid. As the liquid boils, therefore, its composition and boiling point will stay the same. Constant boiling mixtures of this type are termed *azeotropic*. The azeotropic composition of hydrochloric acid is sufficiently constant at constant pressure for it to be useful in the preparation of hydrochloric acid for volumetric analysis.

Azeotropic systems can exhibit either negative or positive deviations from Raoult's law. The convention is to classify them in terms of their vapour pressure deviation.[5] Hydrochloric acid–water mixtures have *negative azeotropes* because of their negative deviation from Raoult's law, although the boiling point goes through a maximum. In the second part of the experiment which follows, we shall discover whether the n-propanol/water system has a negative or positive azeotrope.

Apparatus The benzene/tetrachloromethane mixture is distilled using a column packed with Fenske helices and surrounded by a silvered vacuum jacket. The column is surmounted by a variable

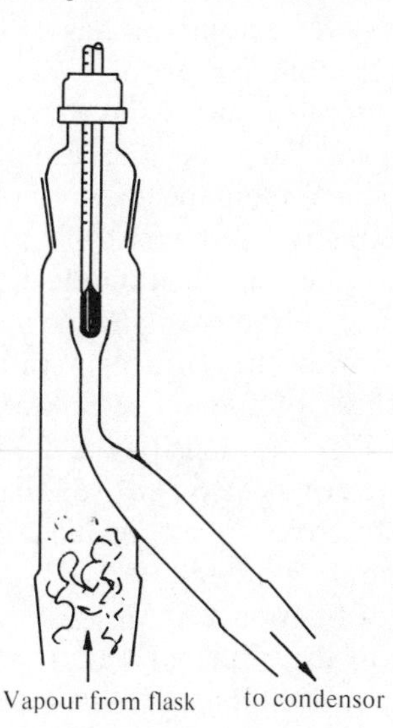

Fig. 2.4.4. Modified still-head

reflux ratio stillhead. The *n*-propanol/water system is distilled with a basic distillation apparatus, although for sufficient accuracy a modified stillhead is needed, as shown in Fig. 2.4.4, or, ideally, an equilibrium still.[7, 8] Both mixtures are analysed with a thermostatted precision Abbé refractometer.

Procedure

A. *Benzene/tetrachloromethane*

The vapours given off from the benzene/tetrachloromethane mixture are toxic. They are therefore contained in a closed system with a vent to outside the laboratory, and you should not expose yourself to them at any time.

Carefully remove the thermometer and its cone from the bottom flask. Add about 6 cm^3 tetrachloromethane. Turn on the water supply to the stillhead. Switch on the heating mantle fully. Carefully adjust the stillhead to give total reflux.* When the vapour reaches the stillhead, adjust the heater control to give reflux at a steady dripping rate but not a trickle. Once set correctly, the still must be left for at least 2 h to equilibrate. It should be checked regularly during this period to make sure that it is not over-heating or flooding.

Meanwhile check that the Abbé refractometer is thermostatted at 20 °C. If the temperature drifts too high, replace some of the water in the tank with ice, and wait for the temperature to stabilize again.

Using pipettes and pipette-fillers, make up three benzene/tetrachloromethane mixtures of varying concentration in small test-tubes.

If the Abbé refractometer has a scale lamp, switch it on. Open the refractometer prism and check that it is clean. Put two or three drops of one of the solutions onto the glass surface. Close the prism. Turn the measurement knob until there is a light/dark boundary on the cross-wires in the eyepiece in the region of refractive index 1.4–1.5. If there is a dispersion control knob, sharpen the boundary and re-set on the cross-wires. Read the refractive index to four decimal places. Open the prism and wipe the surfaces with a soft tissue. Measure the refractive indices of the other mixtures and the pure components. Given that the densities at 20 °C of benzene and tetrachloromethane are 0.8786 and 1.5940 g cm^{-3} respectively, plot a graph of refractive index against mole fraction of tetrachloromethane.

When the distillation column has reached equilibrium, attach a small flask to the stillhead outlet. Move the stillhead control fractionally to give a slow take-off at a reflux ratio of more than

20 : 1, which is effectively total reflux. After a few drops have been collected, replace the collecting flask with another, and discard the first sample. After about 1 cm^3 of liquid has been collected, return the stillhead to total reflux and measure the refractive index of the take-off sample.

Repeat the whole procedure with a reflux ratio of about 1 : 1, discarding the first 10 cm^3.

Obtain a small hypodermic syringe with a long needle. Holding the needle near its tip, push it through the hole at the centre of the seal on the bottom flask. Withdraw a sample of the boiling liquid and measure its refractive index.

Turn off the heating mantle and cooling water.

B. n-Propanol/water

Using burettes, make up at least twelve mixtures of *n*-propanol and water, each of volume 25 cm^3 in a labelled flask. They should cover the whole range of mole fractions; suitable mixtures are listed in Table 2.4.1, based on the densities of *n*-propanol and water at 20 °C of 0.8035 and 0.9982 g cm^{-3} respectively.

Measure the refractive index of each sample and draw a graph of refractive index against mole fraction.

Measure the boiling point of each sample with a thermometer in the liquid. Use the boiling tube device with reflux condenser and add anti-bumping granules to the liquid. Return each

Table 2.4.1. Suggested mole fractions of *n*-propanol and the corresponding volumes of *n*-propanol in the mixtures

Mole fraction propanol	Vol *n*-propanol in 25 cm^3 mixture/cm^3
0.00	0.0
0.08	6.6
0.15	10.6
0.23	13.8
0.30	16.0
0.38	17.9
0.45	19.3
0.50	20.1
0.55	20.9
0.62	21.8
0.70	22.7
0.77	23.3
0.85	24.0
0.92	24.5
1.00	25.0

mixture to its flask. Also measure the boiling point of pure *n*-propanol.

Plot the data on a graph of temperature against mole fraction.

Pour the four mixtures with least *n*-propanol into the 250 cm^3 flask. Add some glass spherules to promote boiling. Set up the simple distillation apparatus with the thermometer bulb enclosed by, but not touching, the central tube of the stillhead. Distil the liquid at the rate of about 1 drop per second. Take a small sample after every 1 K rise in temperature.

Note accurately the distillation temperature of each sample, measure its refractive index and thus find its composition. Then use the points to form a vapour curve on the same graph as the liquid curve.

Repeat the procedure, this time filling the flask with the six mixtures richest in propanol. The surface tension of this composition of mixture is lower than for the previous one, with the result that it is a better wetting agent, does not easily form bubbles on the glass spherules, and may super-heat while boiling. Guard against this by never allowing the distillation rate to rise above one drop per minute, and, if the temperature of the vapour begins to rise steeply at constant composition, turn down the heat at once. Follow the same procedure as before and plot the vapour curve.

Calculation *A*

Plot the data given in Table 2.4.2 on a graph of the type explained previously (Fig. 2.4.2). By drawing steps between the lines from the feed mixture composition in your experiment to

Table 2.4.2. Liquid and vapour composition of benzene/tetrachloromethane mixtures at 1 atm and at temperatures between the boiling points[3,5]

Mole per cent CCl_4 in liquid	Mole per cent CCl_4 in vapour
0.0	0.0
13.64	15.82
21.57	24.15
25.73	28.80
29.44	32.15
36.34	39.15
40.57	43.50
52.69	54.80
62.02	63.80
72.2	73.3
100.0	100.0

the distillate composition, calculate the efficiency of the distillation column. Find whether the efficiency of the column is less at a reflux ratio of 1 : 1 than at total reflux.

B

Find the azeotropic composition of the *n*-propanol/water system, and its boiling point. Describe the fundamental difference between distilling a system such as hydrochloric acid/water and one such as *n*-propanol/water. Using your graph, calculate the approximate composition of the remaining liquid and the 50 cm^3 distillate if 500 cm^3 of a 0.6 mole-fraction mixture of *n*-propanol/water were distilled with the aid of a Vigreux column.

Results

Accuracy

A. The accuracy of the column efficiency measurement is determined by the extent to which the column has achieved equilibrium; complete equilibrium is only attained after periods longer than available for a student experiment.
B. For accurate measurements of liquid–vapour equilibrium compositions, very complicated stills are required.[7,8] With the present still, the *n*-propanol/water points will be accurate to 1 K at best. However, appropriate curves can be drawn close to them to give a good estimate of the correct phase diagram.

Comment

Ethanol/water has a maximum azeotrope at 96 per cent by weight of ethanol which prevents pure ethanol being obtained by fractional distillation; the azeotrope is 'broken' by the addition of benzene.[9,10]

This experiment provides an introduction to distillation, and to some of the methods used by chemical engineers concerned with this crucial industrial process. Very few liquid–vapour systems are ideal, so their properties are usually determined by experiment rather than from theoretical predictions. The measurement and use of temperature–composition diagrams is especially important in the case of azeotropes, which can prevent the efficient separation of liquids by fractional distillation.

Technical notes

Apparatus

A. Distillation column >30 cm × ~2.5 cm i.d. with silvered vacuum jacket;[††] 1 dm^3 flask with three necks for column, thermometer, and septum (Suba-seal); electric heating mantle with variable transformer or other fine control; 1 cm^3 hypodermic syringe with long needle; variable reflux ratio stillhead with thermometer and vent to outside air (Gallenkamp DPR-220-550N with PTFE cone in place of metal spike);[$] 50 cm^3 sample flasks with ground glass joints to stillhead; anti-bumping granules; steps to reach stillhead. Tetrachloromethane; benzene.

A and B. Precision Abbé refractometer (4 decimal places) with thermostat facility (may be shared between two sets of apparatus);[$] thermostat tank and pump.
B. 250 cm^3 flask; heating mantle; stillhead with thermometer, Fig. 2.4.4;[†] condenser; sample tubes. Boiling tube with thermometer and side-arm reflux condenser;[†] anti-bumping granules. Small ground-glass spheres.

Setting up A. Pack column with 3 mm Fenske helices and loose glass wool plugs top and bottom. Fill distillation flask with 100 cm^3 tetrachloromethane, 500 cm^3 benzene and anti-bumping granules. Replace liquids periodically.

References

[1] *Laboratory distillation practice.* E. A. Coulson and E. F. G. Herington. George Newnes, London (1958), ch. 2.
[2] *Experimental physical chemistry.* F. Daniels, J. W. Williams, P. Bender, R. A. Alberty, C. D. Cornwell and J. E. Harriman McGraw-Hill, New York, 7th edn (1970), Expt. 10.
[3] Ref. 1, p. 69.
[4] *Chemical engineers handbook* (R. H. Perry and C. H. Chilton, eds.), McGraw-Hill, New York, 4th and 5th edns (1963 and 1973).
[5] *Liquids and liquid mixtures.* J. S. Rowlinson and F. L. Swinton, Butterworth, 3rd edn (1982), sects. 6.1, 6.2 and 6.5.
[6] *Chemical phase theory.* J. Zernike, Deventer, Antwerp, (1955), pp. 106 and 108.
[7] *Vapour–liquid equilibrium.* E. Hála. J. Pick, V. Fried and O. Vilím, tr. G. Standart, Pergamon, Oxford, 2nd English edn (1967), ch. 5.
[8] Ref 1, p. 139.
[9] Ref. 1, p. 135.
[10] *The preparation of absolute alcohol from strong spirit.* S. Young. *J. Chem. Soc.* **81,** 707 (1902).

2.5 Molar mass from depression of freezing point

Boiling point elevation[1] and osmosis are also colligative effects.

When an involatile solute is added to a solvent, it lowers the solvent's freezing point. This effect is *colligative* in that it depends on the number of molecules of solute in solution, but not on their nature. The colligative nature of freezing point depression can be studied by finding the lowering caused by a known concentration of solute. The result can be used to 'calibrate' the solvent in terms of the freezing point depression constant or *cryoscopic constant*, K_f. This is defined as the depression in freezing point caused by adding 1 mole of solute to 1 kg of solvent to give a *molal* solution. Once the cryoscopic constant is known, the effective molar mass of a test compound can be determined from the freezing point depression it causes in the particular solvent. If the compound associates and dissociates in the solvent, its effective molar mass will differ from its actual molar mass, because there will be fewer or more particles in solution than would be expected from the number of moles added.

The cryoscopic constants of the solvents water, benzene, cyclohexane, and camphor are 1.86, 5.12, 20, and 40 K kg mol^{-1} respectively. Of these the obvious choice is camphor, since the temperature change on adding a given amount of solute is the largest and most easy to measure. This is the basis of Rast's well known method.[1] However, camphor does have the disadvantage that commerical samples are variable, and the cryoscopic constant has to be redetermined before each experiment. In this exercise we use cyclohexane as solvent, and determine the molar mass of an unknown organic compound. We follow a procedure which is designed to cope with *supercooling*—the cooling of a solution below its freezing point with separation of solid.

Theory[2]

The thermodynamic expressions associated with freezing point depression are straightforward if two conditions are met. First, the solution must either be *ideal*, or otherwise so dilute that it behaves ideally, i.e. an *ideal dilute* solution. The second condition is that the solute must dissolve only in the liquid phase of the solvent, and should not form solid solutions with it.

In deriving an expression for the cryoscopic constant, the solid–liquid equilibrium considered is entirely that of the solvent, species 1, *not* that of the solute. At the freezing point, the solid and liquid phases are in equilibrium, and the chemical potential of the solvent in each phase must therefore be equal. Thus

$$\mu_1^{\ominus}(\mathrm{s}) = \mu_1(\mathrm{soln}) = \mu_1^{\ominus}(\mathrm{l}) + RT \ln x_1, \tag{2.5.1}$$

where x_1 is the mole fraction of the solvent in the solution. The difference in the standard chemical potentials of the liquid and solid solvent, $\mu_1^{\ominus}(\mathrm{l}) - \mu_1^{\ominus}(\mathrm{s})$, is equal to the solvent's standard free energy of fusion, $\Delta G^{\ominus}_{\mathrm{fus},1}$. If we differentiate this term with respect to temperature at constant pressure, and employ the *Gibbs–Helmholtz equation*

$$\left[\partial\left(\frac{\Delta G}{T}\right) \Big/ \partial T\right]_p = -\Delta H/T^2, \tag{2.5.2}$$

we obtain

$$\left(\frac{\partial \ln x_1}{\partial T}\right)_p = \frac{\Delta H^{\ominus}_{\mathrm{fus},1}}{RT^2}. \tag{2.5.3}$$

This expression is then integrated from the freezing point of the pure solvent, $T_{fus,1}$ (mole fraction unity), down to the freezing point T of the solution of mole fraction x_1. We obtain:

$$\ln x_1 = \frac{\Delta H^{\ominus}_{fus,1}}{R}\left(\frac{1}{T_{fus,1}} - \frac{1}{T}\right) \tag{2.5.4}$$

$$= -\frac{\Delta H^{\ominus}_{fus,1}}{R}\left(\frac{\Delta T}{T_{fus,1}T}\right). \tag{2.5.5}$$

ΔT is the freezing point depression $(T_{fus,1} - T)$. If we know $\Delta H^{\ominus}_{fus,1}$, the standard enthalpy of fusion of the solvent, then we can use the equation to find the mole fraction x_1 and hence the molar mass of the solute.

The equation is usually simplified by various approximations. As ΔT is usually small, $T_{fus,1}T \simeq T^2_{fus,1}$. If the solution is dilute, x_1 is nearly 1 so $\ln x_1 = \ln(1 - x_2) \simeq -x_2$, where x_2 is the mole fraction of solute. So we obtain

$$x_2 = \Delta H^{\ominus}_{fus,1}\,\Delta T / RT^2_{fus,1}. \tag{2.5.6}$$

For a dilute solution, the number of moles of solvent n_1 is very much greater than the number of moles of solute n_2. Thus $x_2 = n_2/(n_1 + n_2) \simeq n_2/n_1$.

In terms of the molality m of the solution (p. 29), $n_2/n_1 = mM_1$, where M_1 is the solvent's molar mass. Thus

$$mM_1 = \Delta H^{\ominus}_{fus,1}\,\Delta T / RT^2_{fus,1}. \tag{2.5.7}$$

By definition the cryoscopic constant $K_f = \Delta T/m$, and therefore

$$K_f = RT^2_{fus,1}M_1/\Delta H^{\ominus}_{fus,1}. \tag{2.5.8}$$

It also follows from this definition that

$$M_2 = K_f w/\Delta T \tag{2.5.9}$$

where M_2 is the molar mass of the solute and w the mass dissolved in 1 kg of solvent.

Apparatus[1]

Beckmann thermometers do not measure actual temperatures but temperature differences; these they determine, over a moveable range of a few degrees, to 0.01 °C.

For this experiment we use a *Beckmann apparatus*, Fig. 2.5.1. The solution is contained in a 'freezing tube' with a stirrer, and it is convenient if this has a side arm, as shown, for entry of solute pellets. The freezing tube is mounted in a wider glass tube, with an air gap or polystyrene insulation between the two. This is in turn positioned in a Dewar flask, or simply a beaker, which contains a mixture of ice and water and another stirrer. The temperature of the test solution is measured by a Beckmann or cryoscopic thermometer, or a calibrated thermistor

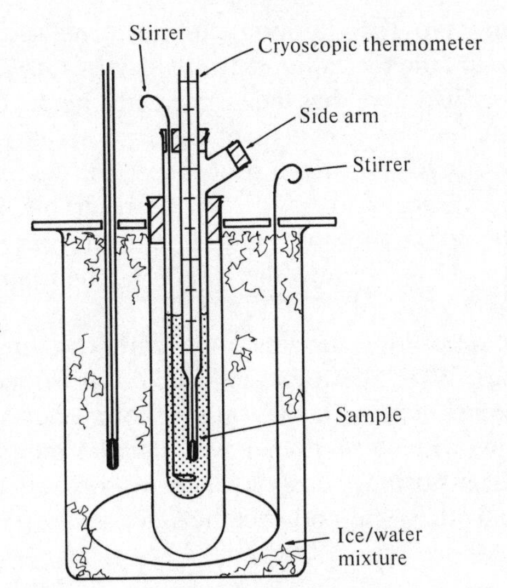

Fig. 2.5.1. Beckmann freezing point depression apparatus [1]

connected to a resistance bridge. A thermometer dipping into the ice/water freezing mixture is also useful.

Procedure *Beckmann thermometers are easily broken*; *take great care if you are using one.*

(*i*) *Setting up*

Check that the freezing tube and inside of the surrounding jacket is perfectly clean and dry, that the complete apparatus assembles properly, and that the stirrers move freely. Fill the outer cooling vessel with cold water and several pieces of ice.

If a thermistor is used, connect it to the resistance bridge and galvanometer.*

If a Beckmann thermometer is used, it must be set up so that the maximum temperature of the experiment, the freezing point of cyclohexane at 6.6 °C, registers on the upper part of the scale. Reduce a beaker of water to about this temperature by adding ice, and insert the thermometer. If the height of the mercury thread is incorrect, carry out the following procedure. Prepare a second beaker of water at a temperature a specified number of degrees above 6.6 °C (between 2 and 16, depending on the thermometer used*). Invert the thermometer and allow the mercury to flow down the column until it joins the mercury in the reservoir. Gently raise the thermometer to the vertical, taking care not to break the thread of mercury, and immerse the bulb in the beaker of water. When the flow of mercury ceases, gently tap the thermometer near the top to break away the mercury in the column from that in the reservoir. Check that the adjustment has been correctly made by dipping the thermometer into a beaker of water at about 6.6 °C. If the reading is too low, repeat the procedure just described to allow a little more mercury to flow from the reservoir. If the reading is too high, warm the bulb in the palm of your hand to allow some mercury to rise into the reservoir, then tap gently to break the thread.

Stopper the freezing tube with one or two plain corks or bungs. Weigh the tube, in a stand if provided or a beaker, to the nearest hundredth of a gram. Pour in cyclohexane to a depth which will more than cover the thermometer bulb when the bulb is positioned about 1 cm above the bottom of the tube. Weigh the tube and cyclohexane.

(*ii*) *Preparation of sample*

Form portions of the sample into five pellets with the pellet press provided.* Each pellet should weigh about 0.2 g. Place each in a weighing bottle and weigh accurately.

(*iii*) *Measurement of temperatures*[1]

The procedure outlined below should be followed carefully to minimize the effects of supercooling. In particular:

(a) Stir the solution uniformly and continuously with an up and down movement about once per second.

(b) If a thermometer is used, tap it *gently* before each reading.

(c) Do not have the temperature of the outer bath more than 4 °C below the freezing point of the sample liquid, otherwise it may cool the liquid faster than it can crystallize.

(d) Never allow the liquid to supercool by more than about 0.5 °C. If the supercooling reaches 0.1 °C, stir more vigorously to encourage crystallization. If this is not successful, the liquid may have to be *seeded* with a tiny crystal of frozen solvent to initiate crystallization. The crystal is best prepared by putting a test-tube of pure cyclohexane into a beaker of ice and a little water. A fine rod or capillary tube is inserted into the cyclohexane, and the test-tube stoppered with glass wool to prevent the entry of moisture. The rod, carrying specks of frozen cyclohexane, is then dipped as gently as possible into the supercooled liquid to seed it.

(e) It is essential that the cyclohexane should separate from the solution in the form of quite fine separate crystals. If a thin transparent coating of solid cyclohexane forms round the wall of the freezing tube the experiment is worthless as there will not be a true thermal equilibrium between the solid and the solution. This is especially likely to happen in the first determination.

(f) The cooling of the freezing tube may be hastened by removing it from the outer tube and immersing it in the water of the cooling vessel. However, particular care must be taken to avoid the formation of solid round the tube as described above. Also, the freezing tube must be well dried before it is replaced in the apparatus.

Bearing these points in mind, find the freezing point of pure cyclohexane on the thermistor or thermometer scale. The freezing point is the maximum temperature reached after supercooling has taken place.

Remove the freezing tube from the apparatus and melt the crystals by resting the tube in the open palm of your hand. Stir the liquid gently. Then replace the tube and repeat the freezing point determination several times. The readings should agree to 0.005 °C.

Drop one of the pellets from its weighing bottle into the freezing tube, through the side-arm if there is one. Accurately weigh the empty bottle to find the amount of sample added. Find the freezing point of the solution, and repeat at least once.

Repeat the whole procedure with another pellet. Continue with as many of the five pellets as time allows.

Calculation

Tabulate your results to give the total depression of freezing point against the total weight of sample added to the solvent to form each solution. Given that the cryoscopic constant of cyclohexane is 20 $K\,kg\,mol^{-1}$, find the molar mass of the sample. How could this experiment be used to find the enthalpy of fusion of a solvent?

Results

Camphor in benzene. Find that average depression = 1.39 K per g solute for 24.17 g solvent. Therefore from eqn (2.5.9), $M_2 = 5.12 \times (1000/24.17)/1.39 = 152\ g\,mol^{-1}$, which agrees closely with actual molar mass of 152.2 g and shows that camphor is not associated in benzene.

Accuracy

The accuracy of the specimen result is probably fortuitous as errors of up to 5 per cent can occur through the effects of precrystallization and the formation of amorphous solid. Further errors may occur if the sample does not form a truly ideal solution in cyclohexane. The precision of the results will decrease with samples of lower molar mass.

Comment Although the measurement of freezing point depression is no longer used as a general investigative technique, this experiment provides a demonstration of the theory of solutions, and of the utilization of colligative properties. The effect is employed qualitatively in determining the purity of a freshly prepared organic compound. Any impurities in the compound will cause a lowering of its melting point below the known melting point of the pure substance.

Technical notes

Apparatus

As in Fig. 2.5.1, suitable sample of high molar mass, pellet press.

References

[1] *Findlay's practical physical chemistry.* Revised by B. P. Levitt. Longman, London, 9th edn (1973), sect. 7B.
[2] *Basic chemical thermodynamics.* E. B. Smith Clarendon Press, Oxford, 3rd edn (1982), sects 6.5, 6.6 and 6.7.

2.6 Ideality of solutions from solubility curves

In this brief experiment we set out to investigate the solubility of solid naphthalene in two different solvents, toluene and hexane. We do this by measuring *liquidus curves,* which are plots of mole fraction of solute against the *liquidus temperature* at which, on cooling, the solid just begins to separate out from the solution. The liquidus curves yield values of the enthalpy of fusion (melting) of the solute, and by comparing these with the known value, we deduce whether or not the solutions we have formed are ideal.

The temperature changes involved are large enough to be measured with an ordinary thermometer in a very simple apparatus.

Theory[1,2] The theory we use is very similar to that of the previous experiment, except that we now consider the chemical potential of the *solid,* rather than the solvent.

If a solid dissolves to form an *ideal solution* then its chemical potential in solution is given by the equation

$$\mu_2(\text{soln}) = \mu_2^\ominus(\text{l}) + RT\ln x_2, \qquad (2.6.1)$$

where x_2 is the mole fraction of solute. The expression relates the chemical potential of the solute in solution, $\mu_2(\text{soln})$, to the standard chemical potential of the pure solute if it were a liquid at the same temperature, $\mu_2^{\ominus}(\text{l})$. The second term sounds paradoxical, but it can be easily calculated.

When a solid solute is added to a liquid solvent it continues to dissolve until a saturated solution is formed. At equilibrium the chemical potential of the solute in solution must be equal to the standard chemical potential of the pure solute:

$$\mu_2(\text{soln}) = \mu_2^{\ominus}(\text{s}). \tag{2.6.2}$$

Thus from eqn (2.6.1),

$$\mu_2^{\ominus}(\text{s}) = \mu_2^{\ominus}(\text{l}) + RT \ln x_2. \tag{2.6.3}$$

x_2 now corresponds to the solubility of the solid in the solution and T is the liquidus temperature at which the solid and liquid are in equilibrium. By definition the standard chemical potentials in eqn (2.6.3) are equal to the standard molar free energies $G^{\ominus}$ of the solute in the respective phases. Therefore

$$RT \ln x_2 = G_2^{\ominus}(\text{s}) - G_2^{\ominus}(\text{l}) = \Delta G_{\text{fus},2}^{\ominus}, \tag{2.6.4}$$

where $\Delta G_{\text{fus},2}^{\ominus}$ is the standard free-energy change when the solute melts at T. If we differentiate this equation with respect to temperature at constant pressure, and incorporate the *Gibbs–Helmholtz equation* (p. 47), we find that

$$\frac{\partial \ln x_2}{\partial T} = \frac{\Delta H_{\text{fus},2}^{\ominus}}{RT^2}, \tag{2.6.5}$$

where $\Delta H_{\text{fus},2}^{\ominus}$ is the standard enthalpy of fusion of the solute.

At its melting point, $T_{\text{fus},2}$, the solute is a liquid and completely miscible with the solvent, so that $x_2 = 1$. The integration of the last equation from this point to temperature T and solubility x_2 gives

$$\ln x_2 = \frac{\Delta H_{\text{fus},2}^{\ominus}}{R}\left[\frac{1}{T_{\text{fus},2}} - \frac{1}{T}\right]. \tag{2.6.6}$$

Note the similarity of this equation to eqn 2.5.4 (p. 48), which relates the mole fraction of *solvent* to its enthalpy of fusion.

Equation (2.6.6) demonstrates that solubility in an ideal solution is independent of the identity of the solvent, although in practice we need to know the solvent's molar volume to calculate the solute mole fraction.

Apparatus

The only apparatus required for this experiment is a wide test-tube fitted with a cork through which passes a 0–100 °C thermometer, and a nickel or stainless-steel wire stirrer. The test-tube is heated in a beaker of water by a bunsen burner, and viewed with a light shining through it.

Various refinements may be made to the apparatus, for example the temperature may be measured with a thermocouple rather than a thermometer, and several samples may be studied simultaneously, provided that constant stirring is maintained in each.

Procedure

Avoid breathing the harmful vapours of toluene and hexane, and avoid contact between the liquids and your skin and eyes. Both liquids are highly inflammable and must be kept well away from the bunsen burner at all times. Always use a water bath—never heat them directly.

Weigh accurately a quantity of naphthalene of about 15 g, and put it into the test-tube provided. Add an accurately measured volume of around 5 cm^3 of toluene from a pipette or burette. Fit the cork, thermometer, and stirrer. Warm the test-tube in a water bath until all the naphthalene is dissolved. Allow it to cool in air. Stir the solution regularly and observe it with the light shining through it. Determine the temperature at which solid first appears.

Repeat the measurement at least six times, adding successive small amounts of toluene, to obtain the liquidus curve.

Repeat the experiment with hexane as solvent. Start with about 4 g naphthalene and 5 cm^3 hexane.

Calculation

Given that the densities of toluene and hexane at 20 °C are 0.867 and 0.659 g cm^{-3} respectively, calculate the mole fraction of naphthalene for each measurement. On a single graph, plot the measured temperatures against mole fraction for the two solutions. Comment on the form of lines joining the points, both overall and near $x = 0$ and $x = 1$.

Given that for naphthalene $\Delta H^{\ominus}_{fus} = 18.8$ kJ mol^{-1} and $T_{fus} = 353.3$ K, draw a straight line on a differently scaled graph which the liquidus temperatures of all ideal solutions of naphthalene should follow. Plot the experimental points on this graph, with error bars indicating their experimental uncertainty. Thus determine whether the solutions are ideal. If they are not, calculate the apparent value of $\Delta H^{\ominus}_{fus}$ and, by drawing error slopes, the uncertainties in your experimental results.

Results

For solutions of naphthalene in benzene, values of $\Delta H^{\ominus}_{fus,naphthalene}$ are obtained in the range 18.7–19.9 kJ mol^{-1}. These agree with the literature value, and we may therefore conclude that the solutions are ideal.

Comment The perfect gas law is well known and much used, but we never encounter an entirely perfect gas in the laboratory. Similarly, truly ideal solutions do not exist, although many are very nearly ideal. Nevertheless, the concept of an ideal solution gives a model against which we can judge the properties of real solutions. The characteristics of an ideal solution can also provide a first approximation to the properties of unknown solutions, and they give a limit towards which all real solutions tend as they become increasingly dilute.

Technical notes

Apparatus

Wide test-tube, two-holed cork with 0–100 °C thermometer and nickel or stainless-steel wire stirrer, beaker, bunsen burner, tripod, and gauze, lamp.

References

[1] *Basic chemical thermodynamics.* E. B. Smith. Clarendon Press, Oxford, 3rd edn (1982), p. 95.
[2] *Thermodynamics.* G. N. Lewis and M. Randall. Revised by K. S. Pitzer and L. Brewer. McGraw-Hill, New York, 2nd edn (1961), ch. 18.

2.7 Enthalpy and entropy terms from depression of freezing and transition temperatures

Carbon tetrabromide solidifies near 92°C and undergoes a *solid state transition* at about 45 °C. The phase transition is not well understood;[1] it involves a change in crystal structure which may be associated with a change in the rotational motion of the molecules. The high temperature phase I forms solid solutions with molecules of approximately spherical symmetry such as camphor, whereas the low temperature phase II does not readily form solid solutions. Thus we find, as indicated in Fig. 2.7.1, that camphor is soluble in liquid CBr_4 and phase I, whereas naphthalene is soluble in the liquid only. So adding naphthalene to CBr_4 depresses the freezing point and allows us to find the enthalpy and entropy of fusion, whereas adding camphor depresses the transition temperature and yields the enthalpy and entropy of transition.

The freezing points and transition temperatures of the various solutions are found by plotting a graph of temperature against time as the mixtures cool. The *cooling curve* levels off while the

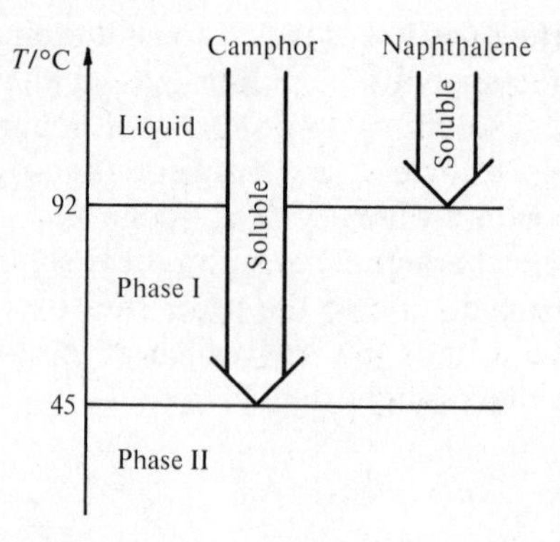

Fig. 2.7.1. The depression of the freezing point and transition temperature of CBr_4 by naphthalene and camphor respectively

enthalpy of freezing or transition is lost to the surroundings. There may be appreciable *supercooling* in each case, i.e. the temperature may drop by several degrees below the freezing or transition temperature before it rises again and equilibrium is attained.

Theory

If we assume that naphthalene forms an ideal solution with CBr_4 in the liquid phase but does not form solid solutions then according to eqn (2.5.6), p. 48,

$$x_2 = \frac{\Delta H_{\mathrm{fus},1}\,\Delta T}{RT^2_{\mathrm{fus},1}}. \tag{2.7.1}$$

x_2 is the mole fraction of solute at equilibrium in an ideal dilute solution at a temperature ΔT below the freezing point $T_{\mathrm{fus},1}$ of the pure solvent. So if we add a small mole fraction x_2 of naphthalene to carbon tetrabromide, the enthalpy of fusion of the solvent can be calculated from the depression in freezing point ΔT.

The entropy of fusion of the solvent, $\Delta S_{\mathrm{fus},1}$, can also be calculated; the system is at equilibrium at constant temperature and pressure, so that $\Delta G_{\mathrm{fus},1} = 0$ and $\Delta S_{\mathrm{fus},1} = \Delta H_{\mathrm{fus},1}/T_{\mathrm{fus},1}$.

Assuming that camphor forms an ideal solution in phase I but is insoluble in phase II, the analogous equation for the depression in the transition temperature is

$$x_2 = \frac{\Delta H_{\mathrm{trans},1}\,\Delta T}{RT^2_{\mathrm{trans},1}}, \tag{2.7.2}$$

from which $\Delta S_{\mathrm{trans},1}$ may also be calculated.

Apparatus

For this experiment, we need an apparatus to measure the cooling curves of solution of known mole fraction. The apparatus can be very simple. However, an electric heating block is useful because it allows controlled heating, which is necessary to prevent the solids subliming. The cooling curve can be measured with a thermometer, or with a thermocouple in a stainless steel sheath linked to a millivoltmeter and digital read-out in °C. As a further refinement, the millivoltmeter may be wired to a chart recorder, or several such devices can be connected to a multiple point plotter so that several cooling curves can be measured at once.

The CBr_4 must be dried and stored in a desiccator.

Procedure

Carbon tetrabromide is harmful by inhalation. Avoid breathing its vapour, and avoid contact with your skin and eyes.

(i) Melting point and transition temperature of pure solvent

Take about 20 g of CBr_4 from the desiccator, and weigh it accurately. Place the sample in a clean, dry test-tube which fits the apparatus being used. Heat the solid to about 95 °C and record the cooling curve.* Note the melting point and transition temperature. Repeat the experiment, or carry out two experiments simultaneously if the apparatus allows.

(ii) Depression of freezing point

Using a weighing bottle, add an accurately known weight of naphthalene (about 0.1 g) to the CBr_4. Melt the mixture and stir it well with the thermometer or thermocouple. Record the cooling curve to find the new freezing point. Also note the transition temperature. Repeat the experiment at least four times with additional weighed portions of naphthalene of about 0.1 g each.

(iii) Depression of transition temperature

Starting with a fresh weighed sample of 20 g CBr_4, carry out similar experiments with camphor instead of naphthalene.

Calculation

Find the quantities $T_{\mathrm{fus},1}$, $T_{\mathrm{trans},1}$, $\Delta H_{\mathrm{fus},1}$, $\Delta S_{\mathrm{fus},1}$, $\Delta H_{\mathrm{trans},1}$ and $\Delta S_{\mathrm{trans},1}$.

Comment on the magnitude of the enthalpy and entropy terms.

Examine the effect of naphthalene on the transition temperature and camphor on the melting point. Do naphthalene and camphor satisfy the criteria necessary for the thermodynamic equations you have used?

Results *Accuracy*

Errors may be caused by the inapplicability of the thermodynamic equations, impure or wet samples, and poor experimental technique. Estimate their relative magnitude.

Comment Although its results may not be accurate, this experiment is an elegant demonstration of the application of thermodynamics to the depression of both freezing points and transition temperatures. It gives a simple introduction to phase transitions in the solid, which are also studied with a much more sophisticated apparatus in the next experiment.

Technical notes

Apparatus

As described previously.

References

[1] *Disorder in crystals*. N. G. Parsonage and L. A. K. Staveley. Clarendon Press, Oxford (1978), p. 606.

2.8 Phase transitions in solids by microcalorimeter

Thermometric determinations can reveal many interesting properties of solids. The experiments involve a measurement of the change in the temperature of a sample while heat is put in or taken from it at a constant rate. The temperature of the sample will remain steady, or nearly so, while a particular transition takes place, and so the temperature of the transition can be found. If a more sophisticated technique is employed, in which both the sample and a reference are heated in a special calorimeter, the enthalpies of transitions may be obtained as well. In this experiment we first calibrate such a calorimeter by measuring the melting point and heat of fusion of indium. We then measure the temperatures of the transitions in ammonium nitrate, and find the eutectic composition of a binary mixture.

Theory (*i*) *Principles of the apparatus*

The temperatures of transitions in a solid may be deduced from the changes in gradient of a heating or cooling curve. If we also wish to determine the enthalpies of the transitions, we may employ one of two methods.

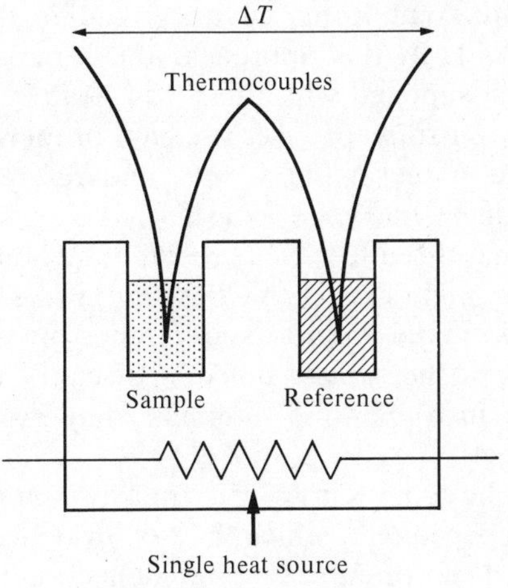

Differential thermal analysis

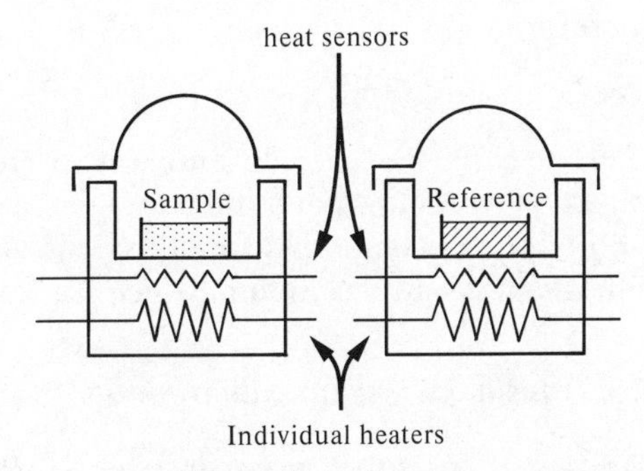

Differential scanning calorimetry

Fig. 2.8.1. Two methods of thermal analysis

The first is differential thermal analysis (DTA). In this technique, the sample and an inert reference are mounted adjacent to one another and heated (or cooled) at the same rate by a single heater (Fig. 2.8.1).[1] The difference in temperature between the sample and reference is measured, and plotted against the sample temperature. Positive or negative peaks occur in this graph at temperatures corresponding to endothermic or exothermic changes, and the areas of these peaks should be proportional to the enthalpy of the transition. However, they are also affected by the characteristics of the instrument and sample. Thus although DTA experiments may be devised to measure transition enthalpies both above[2] and below[3] room temperature, in practice they yield little more than the transition temperatures.

A more informative method (although one which requires

more complicated circuitry) is differential scanning calorimetry (DSC). In this approach, the sample and reference cells are each supplied with their own heaters as shown in Figure 2.8.1. The instrument gives a signal proportional to the difference in heat input to the sample and reference, as required to keep them at the same, constantly rising, average temperature. The signal is fed to a chart recorder. When no transition is occurring one might expect the 'baseline' to be horizontal, but in practice it is curved due to the difference in heat dissipation characteristics of the sample holders. Circuitry is usually incorporated in the instrument to minimize, but by no means eliminate, this effect.

The temperature of a transition is usually assumed to be the temperature at which the first evidence of the change appears in the form of an observable deflection from the baseline. A peak is formed on the trace, and its area is proportional to the integral of $d(\Delta H)/dT$ with respect to temperature, i.e. to ΔH. Therefore

$$\text{peak area} = \Delta H m k \qquad (2.8.1)$$

where ΔH is the specific enthalpy of the transition (i.e. the transition enthalpy per unit mass), m is the mass and k is a calibration constant. k is found by carrying out an experiment with a sample such as indium which has known characteristics.

(*ii*) *Transitions in ammonium nitrate*[4]

The transitions which occur in ammonium nitrate crystals arise from orientational disorder. Both ammonium and nitrate ions exhibit disorder in various crystal phases, and in combination they produce a particularly complex system with many transitions. Investigations have been complicated by the fact that some of the transitions are sluggish, and their occurrence and characteristics depend on how dry the sample is and on its thermal history. The most common transitions are shown in Fig. 2.8.2.

(*iii*) *Eutectic composition*

Let us consider a mixture such as triphenylmethane and stilbene (trans-sym-diphenylethene). The melting of pure triphenylmethane produces a sharp peak near 367 K. If a little stilbene is added and the mixture heated again, a new peak appears at about 350 K. This is due to the melting of the eutectic mixture of the stilbene and a little of the triphenylmethane. The subse-

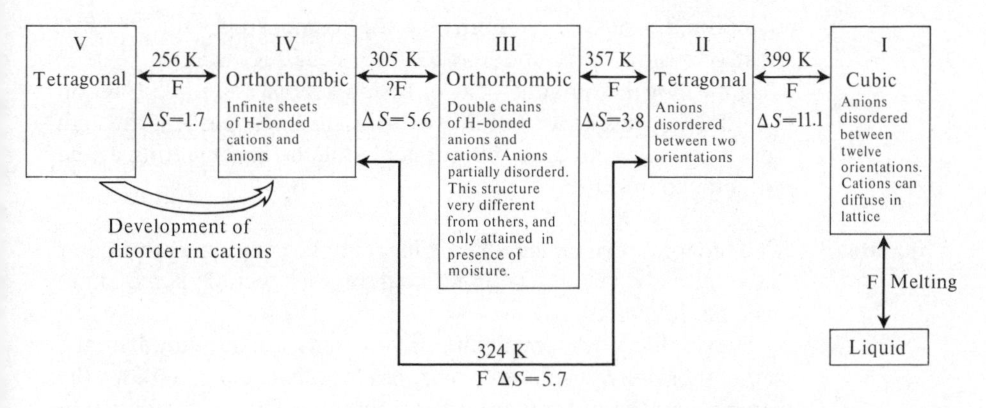

Fig. 2.8.2. Transitions in ammonium nitrate

quent fusion above 350 K is due to the melting of the rest of the triphenylmethane, which gives a broader peak than in the pure compound and a maximum at a slightly depressed temperature, Fig. 2.8.3.

When further stilbene is added to the mixture, more of the eutectic mixture is formed. Therefore the peak at 350 K increases in size, although it does not shift on the temperature axis. Further increases in the percentage of stilbene accentuate this effect, until the fusion is of the stilbene/triphenylmethane eutectic mixture, and the excess stilbene. With 100 per cent stilbene, there is a single peak at the melting point of the pure

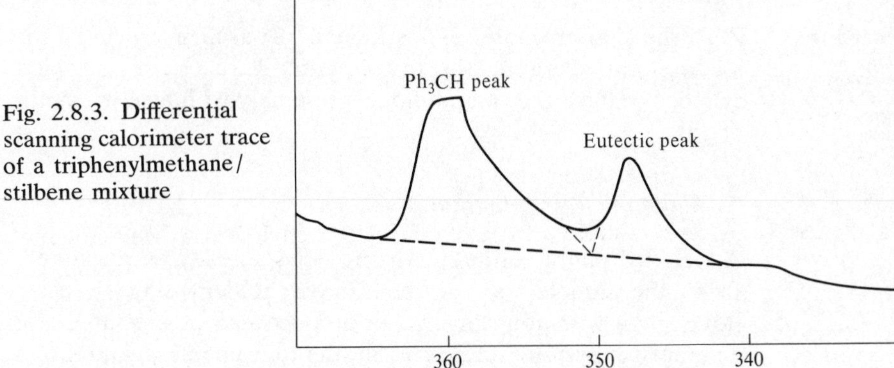

Fig. 2.8.3. Differential scanning calorimeter trace of a triphenylmethane/stilbene mixture

compound, and consequently at this composition the eutectic heat of fusion (area under the curve at 350 K) is zero.

If we plot the eutectic heat of fusion against the mole fraction of stilbene present we obtain two straight lines, passing through zero for 0 per cent and 100 per cent stilbene and meeting at the eutectic composition.

Apparatus

The general principles of a differential scanning calorimeter have already been described, and in this section some more specific details are given.

Two holders for small aluminium pans are mounted in the same thermally insulated environment. One pan contains the sample and the other is used for reference. The heat capacity of the reference must be of comparable magnitude to that of the sample, and in these experiments an empty pan suffices. The pans are sealed by crimping on a lid with a special press.

Embedded in each pan holder are a heater and a platinum resistance thermometer or thermocouple for temperature measurement. The heaters are operated by two control systems; one keeps the average temperature of the sample and reference rising at a predetermined rate, and the other equalizes the temperatures of the pans by altering the relative heating of each. One some commercial instruments, not only does the chart recorder plot the difference in heating of the pans, but also marks temperature intervals with a second pen (typically in kelvins with a gap every ten). The temperature of the sample is shown on a dial, although in practice this has to be calibrated by recording the transitions of several known pure substances over the range of the experiment.

To prevent decomposition products fouling the delicate components, the sample holder assembly is continuously flushed with a slow stream of nitrogen.

Procedure

Note that all changes recorded in this experiment are endothermic.

Zero the chart recorder as instructed.* Consult instructions for loading and operating the instrument.* Turn on the nitrogen cylinder to flush the sample holder assembly with a slow stream of gas.

(*i*) *Calibration with indium*

A weighed sample of indium is provided, already encapsulated. Place the sample pan in the sample holder using tweezers (touching it with your fingers can produce spurious results). Use an empty pan as the reference. Set the calorimeter to 400 K and wait 10 min for the sample to reach thermal equilibrium. Then

Remember that supercooling is a major problem in thermochemistry experiments; take great care to ensure that the samples have reached thermal equilibrium before each run.

scan through the melting point (429 K). Check that this agrees with the temperature indicated on the instrument to within the expected accuracy (see *Results*). Also obtain a trace of the fusion peak, and make a careful note of the sensitivity (or range) setting and chart speed at which it was measured.

(*ii*) *Ammonium nitrate transitions*

Encapsulate about 5 mg of ammonium nitrate using the press provided. Measure the three or four transition temperatures between 300 and 450 K. Obtain a complete peak of a good size for each transition, and note each range setting and chart speed.

(*ii*) *Eutectic composition of a binary mixture*

Make up mixtures of several grammes of triphenylmethane and stilbene containing accurately known mole fractions x of stilbene of about 0.1, 0.2, 0.3, 0.4, 0.6, and 0.8. Gently melt each mixture in a boiling tube so that the components are thoroughly mixed together. Allow each mixture to solidify. Gently grind it into a powder with a glass rod while resting the boiling tube on the bench.

Weigh a sample pan and lid very accurately. Add about 5 mg of one of the powdered mixtures, crimp the lid onto the can, and reweigh accurately. Scan the mixture over the temperature range 300–380 K.

Repeat the procedure for the other mixtures and for the pure components. Make careful note of the compositions and weights, and the range setting and chart speed for each peak.

(*iv*) *Shut down*

Switch off and unload the instrument, switch off the chart recorder, and turn off the nitrogen supply.

Calculation

Measurement of peak areas

To find the areas of the peaks on the calorimeter trace, first interpolate the 'baseline' as shown in Fig. 2.8.3. Then use one of these three methods, which are listed in order of increasing accuracy:

(a) Trace the area to be measured onto a piece of graph paper and count the squares.

(b) Cut out the area and weigh the paper.

(c) Use a *planimeter*. Trace around the area with the small wheel, and read off the area from the vernier scale. The measurement may need a little practice, using the instructions supplied.*

(*i*) *Calibration with indium*

Measure the area of the indium fusion peak. Given that the heat of fusion of indium is 3.26 kJ mol^{-1} = 28.4 J g^{-1}, use eqn (2.8.1) to find the calibration constant k in units appropriate to the method you have used to find the peak area.

(*ii*) *Ammonium nitrate transitions*

Measure the temperature and peak area of each transition. Multiply each peak area by a factor:

$$\frac{\text{(indium peak chart speed)}}{\text{(mixture peak chart speed)}} \times \frac{\text{(indium peak amplifier sensitivity)}}{\text{(mixture peak amplifier sensitivity)}}.$$

If the amplifier sensitivity is controlled by a 'range' switch the second term will be

(mixture peak range setting)/(indium peak range setting).

Use the resulting areas, together with the measured mass of the mixture and the calculated calibration constant k, to calculate the transition enthalpies by means of eqn (2.8.1).

Calculate the entropies of the transitions. Discuss the transition temperatures and entropies in the light of Fig. 2.8.2.

(*iii*) *Eutectic composition*

Measure the areas of the fusion peaks of the triphenylmethane/stilbene eutectic mixtures at around 350 K, and those of the pure components. Calculate the enthalpies of fusion by the method just described.

Plot the enthalpies of fusion against mole fraction. Two straight lines should be obtained, passing through $\Delta H = 0$ at $x = 0$ and 1, and cutting at the eutectic composition. Hence find the composition and enthalpy of fusion of a eutectic mixture of triphenylmethane and stilbene. Inspect your results, and hence estimate their accuracy.

Results

(i) The fusion temperature of indium as indicated by the microcalorimeter may be several kelvins from the true melting point of

429 K. In this experiment it is not practicable to calibrate the temperature scale with several unknowns, so we can simply note the likely accuracy of the temperatures obtained, and place emphasis on the enthalpies of the transitions. The uncertainty in temperature does not serously affect the accuracy of the entropy measurements.

(ii), (iii) A typical trace is shown in Fig. 2.8.3. Note the construction lines to give the peak area.

(iii) The points will be scattered but should yield a eutectic composition correct to within 10 mole per cent or, if great care is taken, to within 3 mole per cent.

Comment

Differential scanning calorimeters are versatile research instruments, typically operating at temperatures anywhere between 100 and 1000 K. They can measure the temperature of a particular transition, the thermal capacity and hence heat capacity of the sample, and the change in heat content at a transition or phase change. In this experiment we have measured the enthalpies of fusion of pure compounds and a eutectic mixture, and the temperatures, enthalpies, and entropies of some solid state transitions. There are many other properties which can be studied; for example heats of fusion, sublimation, vaporization, and crystallization, reaction (including polymerization, oxidation, and combustion) and reaction activation energies, heats of decomposition or dehydration, and of solution, adsorption, and desorption. The use of small samples increases the accuracy of the technique and allows the study of substances which are rare or expensive.

Technical notes

Apparatus

Differential scanning calorimeter, either commercial[§§] or laboratory built;[††] aluminium sample pans, lids and press; nitrogen supply; indium, stilbene, triphenylmethane, ammonium nitrate; [planimeter with instructions].

References

[1] *Differential scanning calorimetry*. J. L. McNaughton and C. T. Mortimer, MTP Int. Rev. Sci. *Physical chemistry series* 2, **10,** 1 (1975).

[2] *Experimental physical chemistry*. F. A. Bettelheim. Saunders, Philadelphia (1971), Expt 23.

[3] *Experiments in physical chemistry*. D. P. Shoemaker, C. W. Garland, J. I. Steinfeld and J. W. Nibler. McGraw-Hill, New York, 4th edn (1981), Expt 50.

[4] *Disorder in crystals*. N. G. Parsonage and L. A. K. Staveley. Clarendon Press, Oxford (1978), pp. 338–350.

2.9 Surface tension of a liquefied gas by the capillary rise method

The most apparent feature of the equilibrium between a liquid and its vapour is the liquid surface itself, the properties of which may be characterized by the quantity known as the *surface tension*.

Surface tension arises because the resultant attractive forces between molecules in a liquid are greater than those they experience in the vapour phase. The forces acting on a molecule at the surface are therefore asymmetric, being much greater across the surface and into the liquid than out into the vapour. This causes the surface of the liquid to act as if it were a membrane under tension. Its shape can also be affected by the surface tension of the walls of a vessel if the liquid is in a container, and by gravity.

When a liquid is placed in a tube, surface tension gives rise to two familiar effects. The liquid forms a curved meniscus, and also, if the tube is narrow, the liquid may climb up (or be pushed down) a certain distance, a phenomenon known as *capillary rise* (or *depression*.) If two narrow tubes of different radii are connected in parallel to form a surface tension cell, the difference in heights of the menisci caused by capillary rise can be used to measure the surface tension of the liquid.

Liquids formed by the condensation of simple gases are easier to understand than more complicated liquids with higher boiling points.

The capillary rise method may be used for any liquid which wets the walls of the tubes. Here we set out to find the surface tension of liquid oxygen, which has to be condensed by liquid nitrogen (b.p. 77.4 K). To measure the density of the liquid, oxygen gas of known pressure, volume, and temperature is condensed into a specially calibrated tube.

Theory

Consider a spherical drop of liquid surrounded by its vapour at equilibrium, Fig. 2.9.1. The boundary of the liquid forms the *surface of tension* between the two phases. The fact that the surface is curved shows that the liquid pressure p_1 and vapour pressure p_2 are different, and that there exists an excess pressure $(p_1 - p_2)$. We now derive an expression for surface pressure in terms of surface tension and the radius of the drop.

The surface area of the spherical drop is $4\pi r^2$. Suppose we increase the volume of the drop very slightly so that its radius becomes $r + \delta r$, as shown in the figure. Then to a good approximation the increase in volume will be $4\pi r^2\,\delta r$, and the work necessary to bring about this change in volume will be $(p_2 - p_1)4\pi r^2\,\delta r$.

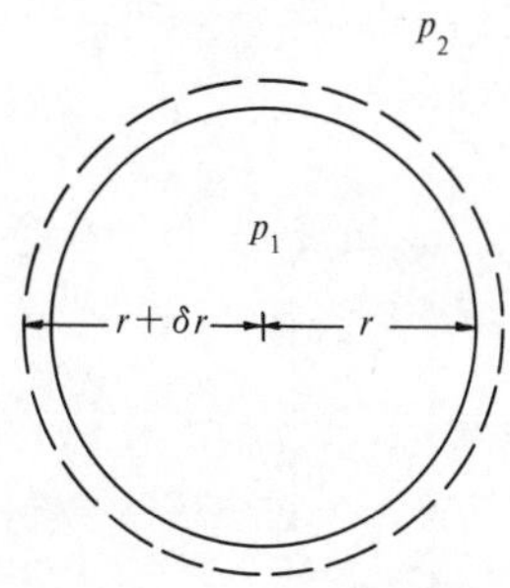

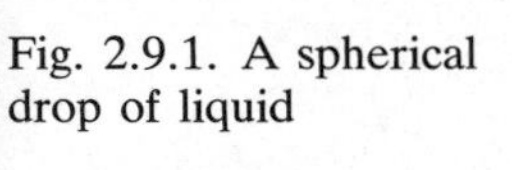
Fig. 2.9.1. A spherical drop of liquid

To increase the area of the surface by an amount δA, a quantity of work $\gamma\delta A$ is required. γ is the *surface tension* of the liquid, defined as the work required to expand an infinite surface by unit area, which is equivalent to the force exerted in the plane of the surface per unit length. The increase in the area is

$$\delta A = 4\pi(r+\delta r)^2 - 4\pi r^2 \simeq 8\pi r\,\delta r, \tag{2.9.1}$$

and the work required to overcome the surface tension is therefore $\gamma 8\pi r\,\delta r$.

If equilibrium is maintained, the total work done is zero and

$$(p_2 - p_1)4\pi r^2\,\delta r + \gamma 8\pi r\,\delta r = 0. \tag{2.9.2}$$

It follows that the excess pressure is given by

$$p_1 - p_2 = 2\gamma/r. \tag{2.9.3}$$

This is *Laplace's equation.* It includes an implicit assumption that the surface of tension is a structureless membrane sited at the radius at which the surface tension acts.

In the surface tension cell, Fig. 2.9.2, we consider a curved meniscus instead of a drop of liquid. Between the meniscus and the wall of the tube is an angle θ, known as the *contact angle*, determined by the mutual surface tensions of the liquid, wall and vapour.[1] This angle is between 10° and 30° for water in slightly dirty glass tubes, and for mercury is greater than 90°, leading to the familiar convex meniscus. In this experiment, however, we can assume it is zero, and that each meniscus is a hemisphere with the same radius as the capillary tube. The surface pressures of the left- and right-hand capillary tubes are therefore $2\gamma/r_1$ and $2\gamma/r_2$ respectively.

The pressure arising from the difference in heights of the liquid columns is $\Delta h(\rho_l - \rho_g)g$, where Δh is the height difference, ρ_l and ρ_g the density of the liquid and vapour respectively, and g the local value of the acceleration due to gravity, which can be

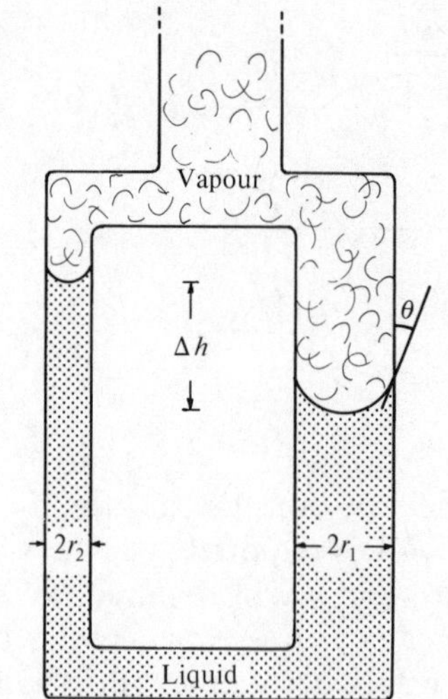

Fig. 2.9.2. The surface tension cell

assumed to be the standard value. The small correction to allow for the curvature of the menisci may be ignored. Since ρ_l is very much greater than ρ_g, to a good approximation the pressure term is $\Delta h\, \rho_l g$. At equilibrium this balances the difference in surface pressures, so that

$$2\gamma\left(\frac{1}{r_1}-\frac{1}{r_2}\right)=\Delta h\, \rho_l g. \tag{2.9.4}$$

Apparatus[2] The oxygen is manipulated in a vacuum line as shown in Fig. 2.9.3. The storage globe is already filled with oxygen to a pressure of about one atmosphere. Other features of the line are the surface tension cell, the calibrated capillary tube, and thermostatted globe of known volume for the density measurement, and a blow-off, vacuum pump with air bleed, and gas inlet. There is also a manometer with a thin layer of silicone oil on the mercury to prevent oxidation.

The heights of the liquid oxygen menisci in the various capillary tubes are measured with a cathetometer—an accurately vertical column with a vernier scale indicating the height of a horizontal telescope with crosswires in the eyepiece. The Dewar flasks are specially constructed with unsilvered strips so that the menisci can be viewed through them and illuminated from behind.

Procedure (*i*) *Setting up*

Check that the cathetometer telescope remains level when swivelled in a horizontal plane at various heights. Switch on the lights behind the Dewar vessels.

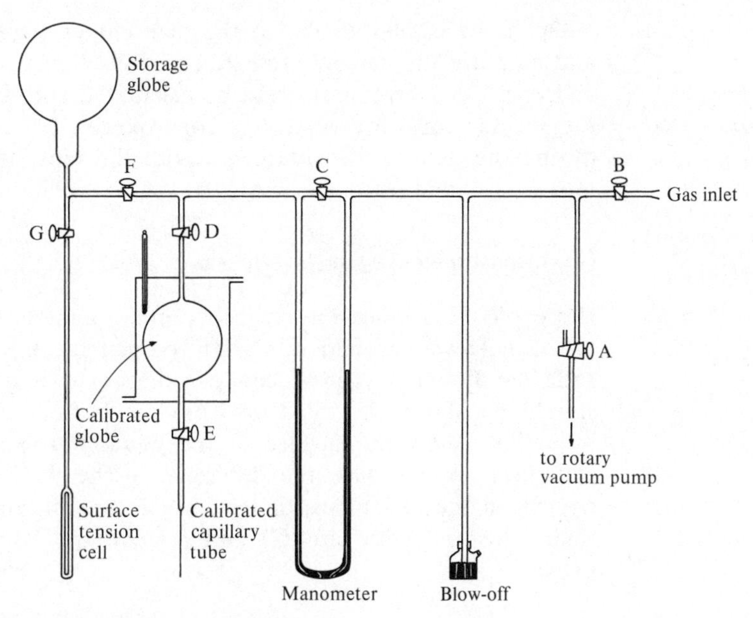

Fig. 2.9.3. The vacuum line

Circulate water slowly around the calibrated volume.

Evacuate the arm of the manometer vessel nearest the vacuum pump (tap A open to line, taps B and C closed). Continue pumping it throughout the experiment.

(*ii*) *Measurement of density*

First measure the temperature and pressure of the oxygen in the calibrated globe. Read the temperature with the thermometer in the water jacket before and after the pressure measurement. Measure the pressure p_1 on the manometer with the globe open to the calibrated capillary tube but closed off from the surface tension cell (taps D and E open, taps C and F closed). Close the tap above the calibrated globe (tap D) and then raise the special Dewar vessel filled with liquid nitrogen round the capillary tube.

Allow 5 or 10 minutes for attainment of equilibrium, then close off the capillary (tap E).This is to reduce the volume of vapour above the liquid, and so minimize the effect of small temperature fluctuations on the liquid level. The liquid oxygen meniscus must be below the liquid nitrogen level. By training the cathetometer through the liquid nitrogen, measure the position of the bottom of the meniscus and the etch mark on the capillary tube. Repeat these measurements several times.

Note that increasing the dead space above the liquid oxygen will not reduce its vapour pressure; by the 'cold finger principle' the vapour pressure is that relating to the liquefied oxygen at its own temperature regardless of the temperature or volume of the manometer.

Open the capillary tube to the manometer (taps D and E) and measure the vapour pressure p_2 of the liquid oxygen. This will enable the temperature to be calculated and will also give the residual pressure of the gaseous oxygen left in the globe during the density measurement. With the taps still open, remove the liquid nitrogen Dewar from the capillary tube.

(iii) Measurement of surface tension

Close off the surface tension cell (tap G) and raise the liquid nitrogen Dewar around it. Carefully open the tap again until sufficient liquid oxygen is condensed; the level of the liquid should be about one third up from the bottom of the wider tube, but below the surface of the liquid nitrogen. With the cathetometer, measure the difference in height of the liquid oxygen in the two tubes, and repeat the measurement several times. Re-open the tap (G) and remove the liquid nitrogen Dewar.

(iv) Shut down

Switch off the vacuum pump and immediately open its air bleed (tap A).

Calculation

Measurements have been made for you of the volume of the calibrated bulb at 20 °C, the calibrated capillary tube at 20 °C, the capillary to the etch mark at 78 K, and the radii of the various capillary tubes at 78 K.* The second virial coefficient of oxygen at 20 °C is -1.8×10^{-5} m^3 mol^{-1}.

(i) Density

First calculate the density of the liquid oxygen by the following steps:

(a) Calculate the mass of oxygen originally present in the calibrated bulb and capillary tube. Use your measurements of the pressure and temperature of the gas, and the volumes and second virial coefficient given (eqn (1.2.2), p. 9).

(b) Calculate the mass of the gas which remained in the bulb after the rest of the gas was condensed with liquid nitrogen.

(c) Calculate the mass of gas still in the capillary tube after condensation, assuming the temperature of the gas below the liquid nitrogen level to be 78 K, and above this level to be room temperature. Since the density of gas still in the capillary tube is small, the gas may be assumed to be perfect.

(d) Find the mass of the liquid from (a), (b), and (c).
(e) Calculate the volume of liquid below the bottom of the meniscus.
(f) Calculate the volume of liquid above the bottom of the meniscus, assuming the meniscus is hemispherical.
(g) Calculate the density of liquid oxygen.

(ii) Temperature

Calculate the temperature of the liquid oxygen from its vapour pressure p_2 by means of a linear interpolation of the experimental data listed in Table 2.9.1.

Table 2.9.1. Vapour pressure of liquid oxygen as a function of temperature[3]

$\log_{10}(p/\text{mmHg})$	1.9724	2.0106	2.0488	2.0869	2.1250	2.1631	2.2012
T^{-1}/K^{-1}	0.01350	0.01340	0.01330	0.01320	0.01310	0.01300	0.01290
$\log_{10}(p/\text{mmHg})$	2.2392	2.2772	2.3150	2.3527	2.3904	2.4280	
T^{-1}/K^{-1}	0.01280	0.01270	0.01260	0.01250	0.01240	0.01230	

(iii) Surface tension

Find the surface tension of the liquid oxygen at the calculated temperature. Compare your result with the surface tension of water at 20°C of $7.3\times10^{-2}\ \text{N m}^{-1}$. What causes the difference in the values?

Results

Specimen results for N_2O (liquid density similar to O_2) using CF_2Cl_2 as refrigerant.
$B_{N_2O}(20\,°\text{C}) = -1.35\times10^{-4}\ \text{m}^3\ \text{mol}^{-1}$. $M = 44.01\ \text{g mol}^{-1}$. $V_{\text{bulb}}(20\,°\text{C}) = 7.23\pm0.03\times10^{-4}\ \text{m}^3$; $V_{\text{capillary}}(20\,°\text{C}) = 1.818\pm0.008\times10^{-6}\ \text{m}^3$; $V_{\text{capillary etch mark}} = 6.980\pm0.015\times10^{-7}\ \text{m}^3$.
Radii of capillary tubes: $r_1 = 1.0038\pm0.0005\times10^{-3}\ \text{m}$, $r_2 = 2.003\pm0.001\times10^{-4}\ \text{m}$. $\Delta h = -1.30\times10^{-2}\ \text{m}$.

Density

(a) 2.128 g; (b) 1.470 g; (c) 2.15×10^{-3} g; (d) 0.656 g; (e) $4.857\times10^{-7}\ \text{m}^3$; (f) $r\times\pi r^2-\frac{2}{3}\pi r^3 = 3.482\times10^{-9}\ \text{m}^3$; (g) Total volume of liquid $= 4.892\times10^{-7}\ \text{m}^3$, so density $= 0.656\times10^{-3}/4.892\times10^{-7} = 1.342\times10^3\ \text{kg m}^{-3} = 1.342\ \text{g cm}^{-3}$.

Temperature

$T = 182.4\ \text{K}$ for $p_{N_2O} = 669\ \text{mmHg}$

Surface tension

From eqn (2.9.4), $\gamma = 2.14 \times 10^{-2}\,\mathrm{N\,m^{-1}} = 21.4\,\mathrm{dyn\,cm^{-1}}$.

Accuracy

Likely uncertainties: (i) impure gas sample = $0.3\,\mathrm{g\,mol^{-1}}$ in M, (ii) pressure = 2 mmHg (iii) volume = $3 \times 10^{-6}\,\mathrm{m^3}$ (iv) temperature = 0.1 K (v) $\Delta h = 5$ per cent. Uncertainties of 1.3 per cent on density calculations (a) and (b), lead to 8 per cent maximum in (d). Therefore density is accurate to ±10 per cent and surface tension to ±15%.

Comment

Experiments concerned with subtle surface tension effects in the absence of gravity are currently being developed in the NASA space shuttle.

There are several alternative ways of measuring surface tension, for example the determination of the pressure in a bubble, the size of a liquid drop, and the force required to lift a ring through the surface of a liquid.[4] However, unlike the capillary rise technique, these methods require prior calibration with a liquid of known surface tension. Surface tension measurements can be used to elucidate the structures of liquids and liquid–vapour boundaries.[1] Measurements of density, which were also necessary in this experiment, have themselves been used to study intermolecular forces in liquids.[2] Surface tension is the only force which we shall encounter weak enough to be affected by gravity, and makes no significant contribution to the phase equilibria investigated in the rest of this section.

Technical notes

Apparatus

Vacuum line as in Fig. 2.9.3 with surface tension cell and calibrated capillary tube (for suitable dimensions, see *Results*, but note that initial pressure need only be ~1 atm for oxygen)†; Dewar vessels with unsilvered strips made by taping before silvering††, cathetometer$, rotary vacuum pump$, liquid nitrogen$.

Calibration

Fill bulb with water, then weigh. For capillaries, weigh measured threads of mercury and correct volume to 80 K.

References

[1] *Molecular theory of capillarity.* J. S. Rowlinson and B. Widom, Clarendon Press, Oxford (1982), p. 7.

[2] *Density and surface tension of liquid xenon and theory of corresponding states for the inert gases.* A. J. Leadbetter and H. E. Thomas, *Trans. Far. Soc.* **61.1,** 10 (1965), but see also Expt. 1.3, Ref. 5.

[3] *Vapour pressure and fixed points of oxygen and heat capacity in the critical region.* H. J. Hoge. *J. Res. Nat. Bur. Standards* **44,** 321 (1950).

[4] *Findlay's practical physical chemistry.* Revised by B. P. Levitt. Longman, London, 9th edn (1973), sect. 6D.

2.10 Adsorption of gas on charcoal by pressure measurements

Adsorption is the process whereby atoms or molecules of *adsorbate* become attached to a surface. Adsorption may in principle occur at all surfaces, but its magnitude is particularly noticeable when a porous solid such as charcoal, which has a high surface area, is in contact with gases or liquids. The forces which bind the foreign particles to the surface may be either physical or chemical in nature, and the adsorption processes are sometimes referred to as *physisorption* and *chemisorption.*

In this experiment we study the physisorption of 1,2-dichlorotetrafluoroethane ($CClF_2-CClF_2$) on activated charcoal. From measurements of the change in volume of the gas above the charcoal with pressure we estimate the effective surface area of the solid, and attempt to explain the way in which the gas molecules occupy the sites on the surface.

Theory[1,2]

An *adsorption isotherm* describes the dependence of adsorption on the pressure or concentration of gas at constant temperature. The simplest such equation, known as the *Langmuir isotherm*, employs the following five assumptions:

(a) the adsorbate in the bulk gas phase behaves as a perfect gas,

(b) the particles adsorbed are confined to a monomolecular layer of single sites, and never adsorb one on top of the other,

(c) every part of the surface has the same energy of adsorption,

(d) no adsorbate–adsorbate interaction is taken into account, and therefore every site is assumed to be equivalent,

(e) the adsorbed molecules are localized, i.e. they have definite points of attachment to the surface.

Assumption (e) is implicit in the derivation which follows, but is only used explicitly in statistical derivations of the Langmuir isotherm.[3]

According to Langmuir, the rate of adsorption is proportional first to the number of molecules that strike the surface per second, calculated from the kinetic theory for a perfect gas, secondly to the *sticking probability*, *f*, i.e. the probability that the molecules will adsorb on the surface after striking it, and thirdly to the fraction $(1-\theta)$ of the surface not already occupied by adsorbed molecules, where θ is the fraction of occupied sites. Thus

$$\text{rate of adsorption} \propto \frac{p}{(2\pi m k_B T)^{1/2}} f(1-\theta), \qquad (2.10.1)$$

where p is the pressure of the adsorbate, and m the mass of one of its molecules.

The rate of desorption is proportional to the fraction of the surface occupied by the adsorbate and to the fraction of the total number of adsorbed molecules that have enough energy to leave the attractive forces of the surface expressed as a uniform molar adsorptive energy U. Thus

$$\text{rate of desorption} \propto \theta e^{-U/RT}. \tag{2.10.2}$$

At equilibrium, the rate of adsorption and desorption are equal, and therefore

$$\frac{p}{(2\pi m k_B T)^{1/2}} f(1-\theta) = k_0 \theta e^{-U/RT} \tag{2.10.3}$$

where k_0 is a constant. We may express this in the form

$$bp = \theta/(1-\theta) \tag{2.10.4}$$

where the *Langmuir constant* b is given by

$$b = f/[k_0(2\pi m k_B T)^{1/2} e^{-U/RT}]. \tag{2.10.5}$$

On rearranging eqn (2.10.4), we obtain the *Langmuir isotherm*

$$\theta = bp/(1+bp). \tag{2.10.6}$$

At saturation ($\theta = 1$), the whole surface of the solid is covered with a monomolecular film of adsorbate, of equivalent volume in the gas phase V_{sat}. If V is the volume of gas adsorbed at a lower equilibrium pressure p, then $\theta = V/V_{sat}$. Substituting this relation into eqn (2.10.6) and rearranging, we find that

$$\frac{p}{V} = \frac{p}{V_{sat}} + \frac{1}{bV_{sat}}. \tag{2.10.7}$$

According to this equation, a graph of p/V against p will yield a straight line of slope of $1/V_{sat}$, and an intercept at $p = 0$ of $1/bV_{sat}$.

Apparatus

The vacuum line is shown in Fig. 2.10.1. The connections in the line are made with capillary tubing to reduce the 'dead volume' in the apparatus. 1,2-dichlorotetrafluoroethane, referred to here by its British Oxygen trade name of Arcton 114, is supplied from a cylinder. A blow-off prevents the pressure in the line from rising above atmospheric. The volume of gas in the line is measured in a gas burette by carefully balancing the mercury levels in the burette and levelling-tube. The burette is surrounded by water at room temperature. The pressure of gas

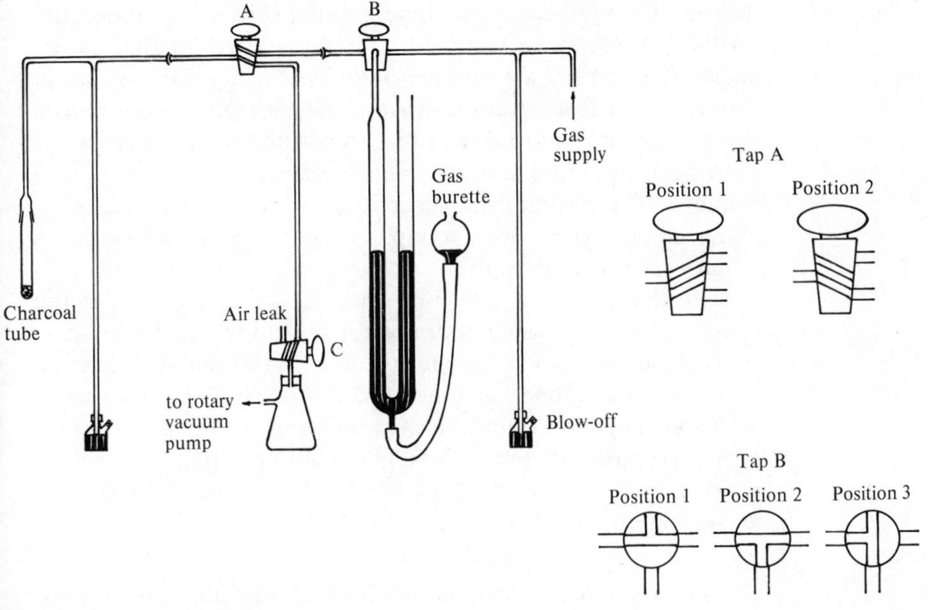

Fig. 2.10.1. The vacuum line

is measured on a capillary manometer. Both the readings of volume and pressure depend on the ambient atmospheric pressure, which is checked on a barometer several times during the experiment.

A sample tube containing the activated charcoal is connected to the apparatus by means of a ground glass joint. A small electric heater can be placed around the sample tube.

The line is connected to the rotary vacuum pump via a trap and air leak.

Procedure (*i*) *Set-up*

With the sample tube removed from the line, turn the three-way taps A and B so that gas from the cylinder will blow out of the sample tube joint (positions 1 in Fig. 2.10.1). Turn on the gas supply and blow through a little Arcton to flush out air in the line to the right of tap A.

Then flush out the gas burette with Arcton as follows. Shut tap A, and turn tap B so that all three tubes are connected (position 2 in Fig. 2.10.1). Gas should be bubbling gently through the blow-off. Lower slightly the mercury reservoir so as to fill the gas burette. Then turn tap B to position 3 so that the gas burette is only connected to tap A, and turn tap A to open the burette to the sample tube joint (position 1). Raise the mercury reservoir carefully to vent the Arcton to the atmosphere. *Do not raise the mercury in the burette too high—it should remain well below tap B and well below the top of the open arm.* Repeat this procedure three times to ensure that all traces of air have been displaced from the gas burette.

Shut tap A and refill the gas burette from the cylinder. After filling, connect the burette to tap A only (tap B at position 3).

Turn off the gas supply.

Take the empty sample tube and clean the grease from the ground glass cone using cotton wool soaked in trichloroethene. Also clean the socket on the gas line. Wash out the charcoal tube with propanone (acetone) and dry it. Apply a thin ring of Apiezon 'T' grease round the *wide end* only of the ground glass cone, and push the sample tube gently into position. There is no need to twist it since it will suck into place when the apparatus is evacuated.

(ii) Determination of volume of charcoal tube and manometer

The capillary manometer will read several millimetres below the ambient atmospheric pressure because of capillary depression (p. 66).

Switch on the pump and turn taps A and C so that the manometer and charcoal tube are evacuated (tap A at position 2). Continue until the manometer nearly reaches barometric height. Then close tap A. If the apparatus is vacuum-tight the manometer should now remain steady. When you have checked this adjust the mercury reservoir until the levels in burette and levelling-tube are exactly equal. Read the volume V_1 of Arcton in the burette. Now open tap A very carefully so as to admit Arcton from the gas burette to the adsorption tube and manometer. When the pressure as indicated by the manometer has increased by about 100 mmHg, shut tap A. Adjust the mercury levels in the burette and read the volume V_2 of Arcton remaining in the burette. Read the manometer to obtain the pressure p of the Arcton in the apparatus, i.e. the difference between the manometer reading with the apparatus evacuated and the reading obtained after each addition of Arcton. Read the atmospheric pressure p_{atm} on a barometer.

Make further determinations by admitting increasing volumes of Arcton from the burette to the apparatus, and repeating the measurements as above.

(iii) Measurement of adsorption of Arcton 114 on charcoal

Fill the apparatus with Arcton at atmospheric pressure, remove the charcoal tube, and clean the grease off both parts of the ground joint. Put into the tube an accurately weighed quantity of active charcoal granules (about 0.5 g), grease the joint carefully as described above, and put the tube into position. Pump out the apparatus. Place the furnace round the charcoal tube, plug the *open* end loosely with ceramic fibre, and heat for *1 hour* with the pump still running, so as to drive off adsorbed gas from the charcoal. Allow the charcoal to cool with the pump still running; then close tap A, reverse tap C so as to admit air to the pump, and switch off the pump.

Place a large beaker (not less than 600 cm^3) filled with water round the charcoal tube, so as to maintain it at room temperature. Now admit successive measured quantities of Arcton (about 10 cm^3) to the charcoal tube using the same procedure, and taking the same measurements, as described in the previous section. The charcoal adsorbs Arcton very strongly at first, and therefore in order to admit a suitably small volume, open tap A very slowly until gas just begins to pass. It will probably be necessary to wait 15 to 20 minutes after each addition before equilibrium is reached and the manometer remains steady. Continue adding Arcton in approximately 10 cm^3 steps until a pressure of nearly 1 atm has been reached. This will require at least one refill of the burette.

Use a barometer to check the ambient atmospheric pressure p_{atm}.

(iv) Shut down

Switch off the rotary vacuum pump and immediately open the air leak (tap C). Turn off the Arcton at the cylinder.

Calculation

(i) Volume of sample tube and manometer

Let the volume of the sample tube and manometer be V'. The volume of gas admitted to this part of the apparatus is $(V_1 - V_2)$, these volumes being measured at the ambient atmospheric pressure p_{atm}. When the gas enters the volume V' it creates a pressure p, and therefore

$$pV' = p_{atm}(V_1 - V_2). \qquad (2.10.8)$$

Plot a graph of V' against the height of the mercury meniscus in the manometer. This should be linear and will show the value of V' corresponding to any manometer reading.

The volume of unadsorbed gas V_u is equal to V' less the volume of the sample of charcoal, the latter being calculated on the assumption that the density of the charcoal is 1.8 g cm^{-3}.

(ii) Adsorption of Arcton on charcoal

Tabulate the volumes of gas admitted to the sample tube at each stage. Calculate the volume of gas unadsorbed by the charcoal from the equilibrium pressure and the appropriate value of V_u. Hence calculate the volume of Arcton adsorbed by the charcoal at each stage, expressed in cm^3 of gas at s.t.p.

Plot an adsorption isotherm of volume V cm^3 of Arcton adsorbed per gramme of charcoal, against the equilibrium pressure p in mmHg. The upper part of the curve will be seen to approach asymptotically the saturation value V_{sat}.

Test the validity of the Langmuir isotherm by plotting p/V against p. Calculate V_{sat} from the slope.

Use the value of V_{sat} obtained from the Langmuir plot to make a rough estimate of the effective surface area of 1 g charcoal, as follows. First employ the perfect gas equation (p. 4) and Avogadro's constant L to calculate the number of gas molecules n_{sat} in the volume V_{sat}. If we assume that the molecules are packed cubes of side length σ, then the area occupied by a surface layer will be $n_{sat}\,\sigma^2$. A value of σ may be estimated from the molar volume of the liquid gas $\bar{V}_{liq}$ if it is assumed that the molecules pack in the same way as on the surface, so that $\bar{V}_{liq} = L\sigma^3$. $\bar{V}_{liq}$ is calculated from the liquid density, which is 1.44 cm^{-3} for the Arcton at 30 °C.

Results

The surface area of activated charcoal is very large and it will be found that the quantity of gas adsorbed corresponds to a liquid volume almost equal to the volume of charcoal

Accuracy

Assumption (a) on p. 73 is often reasonable, but (b) is unreasonable for physisorption, and (c) and (d) are never correct. No surface is ever uniform, and even in a noble gas there is enough adsorbate–adsorbate interaction to account for about 25 per cent of the measured heat of adsorption at half saturation of the surface. However, since the effect of surface non-uniformity is to cause the energy of adsorption to decrease with θ, whereas the adsorbate–adsorbate interactions cause it to increase, the two effects tend to cancel each other.

Comment

This experiment provides an introduction to surface chemistry, but it is a very basic introduction for two reasons. The first is that we have not tested isotherms which are founded on more complex models of the surface interactions. The BET (Brunauer, Emmett, and Teller) isotherm, for example, allows for multilayer adsorption. The Temkin isotherm assumes that the enthalpy of adsorption changes linearly with pressure, and the

Freundlich isotherm, first used as a purely empirical relation, incorporates a logarithmic change. The second difficulty with experiments of this type is that we have no description of the surface of the solid onto which the adsorption is taking place. The charcoal is clearly very porous, and owes this characteristic to its natural origin. However, current studies of surfaces tend to use graphite, which has a better characterized structure. Alternatively metals such as tungsten, platinum, and iridium are used because they can be cleaned by heating in a vacuum, and because specific crystal planes can be identified by diffraction techniques and exposed to the adsorbate species.

The properties of surfaces have also been widely studied by spectroscopic techniques, electric and magnetic measurements, and by experiments using photons, ion beams, and low energy electrons.[1]

Technical notes

Apparatus

Gas line etc. as in Fig. 2.10.1;$ activated charcoal; $CClF_2$–$CClF_2$ (Arcton 114); barometer; electric heater for sample tube; Apiezon 'T' grease; trichloroethene.

References

[1] *Chemisorption: an experimental approach.* G. Wedler. Butterworths, London (1976), ch. 3.
[2] *Physical chemistry of surfaces.* A. W. Adamson. John Wiley, New York, 4th edn (1982), ch. 16.
[3] Ref 2, p. 523.

Section 3
Thermodynamics of chemical reactions

The five experiments in this section involve the measurement of thermodynamic changes accompanying various chemical reactions. The reactions are the combustion of solid organic acids (Expt 3.1), the dimerization of ethanoic acid (Expt 3.2), the dissociation of dinitrogen tetroxide (Expt 3.3), the reduction of carbon dioxide (Expt 3.4) and the formation of amine complexes (Expt 3.5). Measurements are made of the change in either temperature, vapour density, pressure, or the concentrations of the reactants and products, and the results used to calculate the enthalpies of the reactions.

In the first experiment, the calculation of the enthalpy of combustion requires a knowledge of the First Law of thermodynamics only. In experiments 3.2 and 3.3, the enthalpy change is calculated from the free energy change, which in turn is calculated from the equilibrium constant using the van't Hoff Isotherm. So these investigations require a knowledge of the Second Law of thermodynamics. Experiments 3.4 and 3.5 also employ the Second Law. In Expt 3.4, standard entropies are used as an additional source of information, and in Expt 3.5 the entropies of complex formation are determined.

The science of thermodynamics does not include time as a variable, nor does it explicitly tell us anything about the structure of chemical compounds. Thus none of the experiments in this section involves a direct investigation of the mechanism of the reaction. Only in the last experiment do we start to associate the entropy changes accompanying the formation of various chelate compounds with the nature of the reaction taking place.

3.1 Enthalpies of combustion by bomb calorimeter

The enthalpy of combustion of a compound may be determined by burning a known amount of the substance in a *bomb calorimeter* and measuring the rise in temperature.

A bomb calorimeter is a sealed vessel which can be pressurized with oxygen. No products can escape, and the excess oxygen ensures that rapid combustion takes place to form products which are completely oxidized. The compound is ignited by passing a current through a wire sealed through the wall of the vessel. The heat capacity of the calorimeter assembly is determined by igniting a known mass of a substance of accurately known enthalpy of combustion. Heat loss to the surroundings can be calculated by use of a correction curve (similar to that on p. 110) or it may be prevented altogether by arranging a jacket around the calorimeter which is maintained at the same temperature as the calorimeter itself.

In this experiment we find the heat capacity of the calorimeter assembly by burning a sample of benzoic acid as a standard, and then find the enthalpies of combustion of maleic and fumaric acid. From the results, we deduce which compound is *cis* and which is *trans*.

Theory

(*i*) *The thermodynamics of combustion*

The quantity we require from our experiment is the standard enthalpy of combustion $\Delta H^{\ominus}_{cb}$ defined as the enthalpy change which accompanies the reaction of the compound to form specified combustion products such as $CO_2(g)$, $H_2O(l)$, and $N_2(g)$, all reactants and products being in their respective standard states under specified conditions, usually a constant temperature of 298 K and constant pressure of 1 atm (101 kPa).

Processes during which no heat enters or leaves the system are termed *adiabatic*. The reaction which takes place in the bomb calorimeter may therefore be regarded as an adiabatic combustion at constant volume with a change in internal energy ΔU_{cb}. To relate this term to $\Delta H^{\ominus}_{cb}$, we use the *First Law of thermodynamics*:

$$\Delta U = q + w, \tag{3.1.1}$$

where q is the heat entering the system, and w is the work done on the system. For the adiabatic reaction, $q = 0$. Considering only pV work,

$$w = -p\,\Delta V. \tag{3.1.2}$$

It may seem strange that there is no internal energy change during a combustion in an adiabatic bomb calorimeter, but remember that ΔU_{cb} is the total internal energy change of both reactants and products.

At constant volume $\Delta V = 0$ and therefore w is also zero. Thus for the overall reaction in the bomb calorimeter,

$$\Delta U_{cb} = 0. \tag{3.1.3}$$

Considering the combustion of both the sample and the cotton fuse,

$$\Delta U_{\text{cb, sample}} + \Delta U_{\text{cb, cotton}} = 0. \tag{3.1.4}$$

After the adiabatic combustion, the temperature of the products T_2 is higher than the initial temperature T_1. To relate the adiabatic combustion with this temperature change to the isothermal combustion at temperature T_1, the energy change on cooling the products and bomb calorimeter from T_2 to T_1 at constant volume must be calculated. This is demonstrated for benzoic acid in Fig. 3.1.1. The energy change is given by

$$\Delta U_{\text{cooling}} = -C_V(T_2 - T_1) = -C_V\,\Delta T, \tag{3.1.5}$$

where C_V is the heat capacity at constant volume of the products and bomb. *Does C_V also include the heat capacity of the outer jacket?*

The process of adiabatic combustion followed by cooling of the products corresponds to an isothermal combustion at constant volume and constant temperature T_1 with internal energy change $\Delta U_{\text{cb}}(T_1)$. *Hess's law* tells us that we can equate the energy changes associated with these two routes from reactants to products:

$$\Delta U_{\text{cb}} + \Delta U_{\text{cooling}} = -C_V\,\Delta T$$
$$= \Delta U_{\text{cb}}(T_1) = \Delta U_{\text{cb, sample}}(T_1) + \Delta U_{\text{cb, cotton}}(T_1). \tag{3.1.6}$$

Assuming that ΔU_{cb} is independent of pressure and tempera-

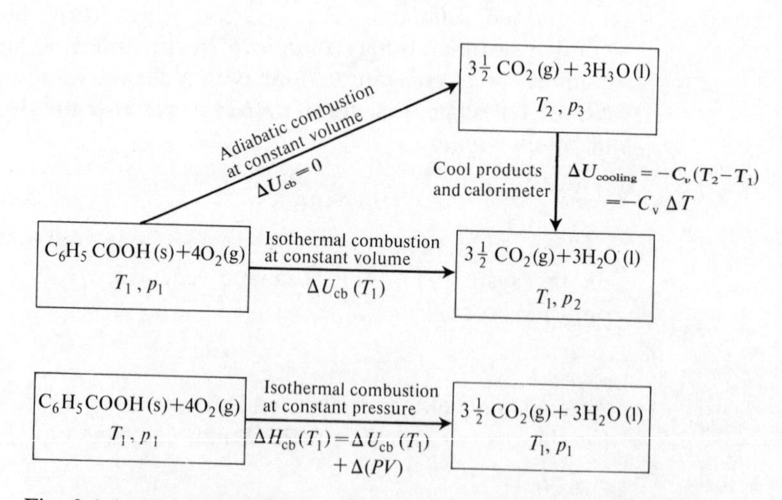

Fig. 3.1.1. Thermodynamic relation between adiabatic combustion at constant volume and isothermal combustion at constant pressure

ture, $\Delta U_{\text{cb}}(T_1) = \Delta U^{\ominus}_{\text{cb}}$. It follows that

$$\Delta U^{\ominus}_{\text{cb, sample}} + \Delta U^{\ominus}_{\text{cb, cotton}} = -C_V\,\Delta T. \tag{3.1.7}$$

By definition, a change in enthalpy ΔH is related to the corresponding change in internal energy ΔU by the equation:

$$\Delta H = \Delta U + \Delta(pV). \tag{3.1.8}$$

Assuming that the gaseous products obey the perfect gas law and that the $\Delta(pV)$ terms of solids and liquids are negligible:

$$\Delta(pV) = \Delta n_{\text{gas}} RT, \tag{3.1.9}$$

where Δn_{gas} is the change in the number of moles of gas during the combustion. On incorporating these relations into eqn (3.1.7), we obtain the working equation for the experiment:

$$\begin{aligned}\Delta H^{\ominus}_{\text{cb, sample}} &= \Delta U^{\ominus}_{\text{cb, sample}} + \Delta n_{\text{gas}} RT_1 \\ &= -\Delta U^{\ominus}_{\text{cb, cotton}} - C_V\,\Delta T + \Delta n_{\text{gas}} RT_1.\end{aligned} \tag{3.1.10}$$

The heat capacity term C_V may be found by using the combustion of benzoic acid as a standard, since $\Delta U^{\ominus}_{\text{cb}}$ for benzoic acid has been very accurately measured by international standards laboratories. If $\Delta U^{\ominus}_{\text{cb, cotton}}$ for the fuse is known, C_V may then be calculated from eqn (3.1.7). C_V is the heat capacity at constant volume of both the combustion products and the calorimeter assembly. Since the calorimeter has a very much greater heat capacity than the reaction products, it is reasonable to assume that C_V is constant for different samples, and to use the value obtained from the combustion of the benzoic acid standard in all the calculations.

(*ii*) *Enthalpies of formation*

The chemical formula of both maleic and fumaric acid is $C_4H_4O_4$. Figure 3.1.2 shows a thermodynamic cycle linking the standard enthalpy of combustion $\Delta H^{\ominus}_{\text{cb}}$ of maleic and fumaric acid with their standard enthalpies of formation $\Delta H^{\ominus}_{\text{f}}$. By Hess's Law, the sum of the change in enthalpy over the complete cycle must be zero, and therefore

$$\Delta H^{\ominus}_{\text{f}} = \Delta H^{\ominus}_{1} - \Delta H^{\ominus}_{\text{cb}}, \tag{3.1.11}$$

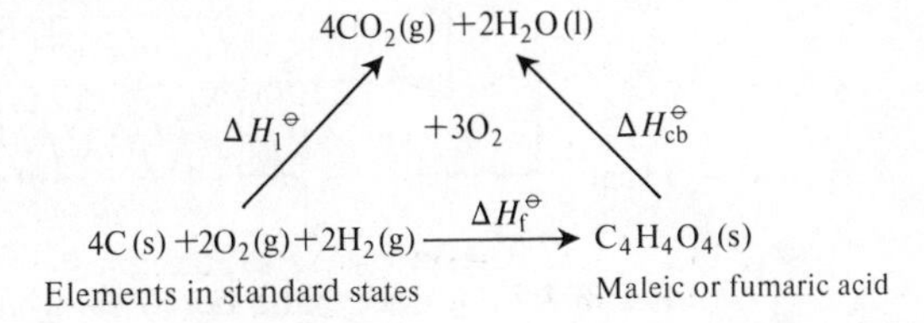

Fig. 3.1.2. Hess's Law cycle linking the standard enthalpy of combustion of maleic or fumaric acid $\Delta H^{\ominus}_{\text{cb}}$, with the standard enthalpy of formation $\Delta H^{\ominus}_{\text{f}}$

where $\Delta H_1^\ominus$ is the standard enthalpy of combustion of the appropriate number of moles of carbon and hydrogen to form carbon dioxide and water.

Apparatus Adiabatic bomb calorimeters have various shapes and arrangements, but all have the features labelled in Fig. 3.1.3. The bomb is placed in a water can which contains a precision thermometer calibrated to 0.01 °C. The can is itself placed inside an outer water jacket from which it is separated by an air gap. The outer jacket is heated electrically or has a device for entry of hot water from an external source. The apparatus also has stirrers, and thermometers or thermistors, as shown in the diagram.

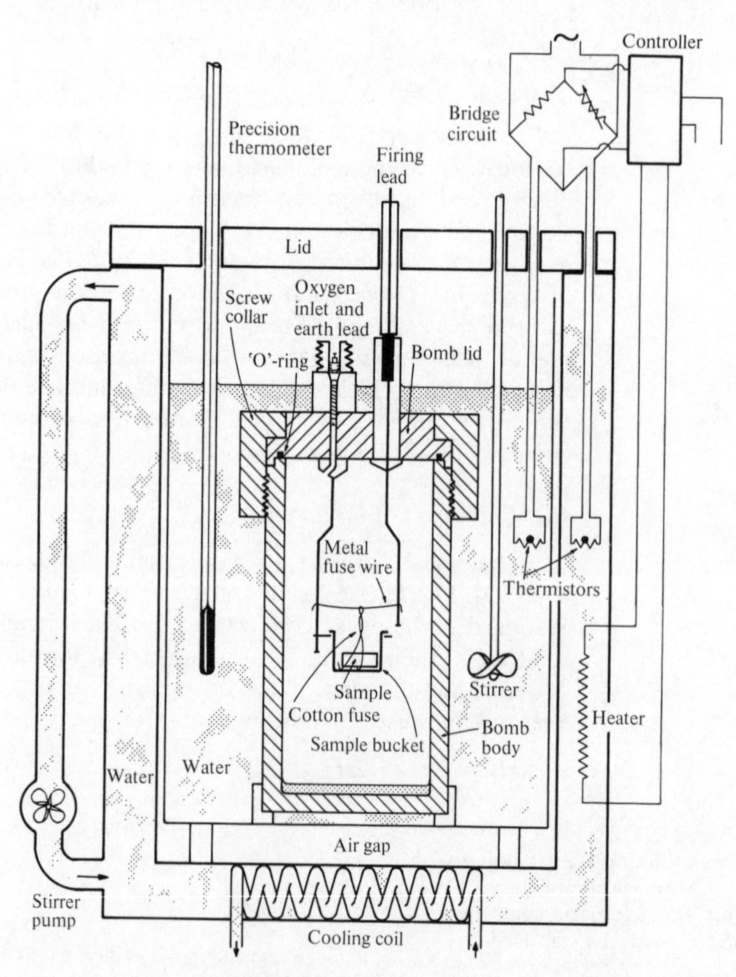

Fig. 3.1.3. An adiabatic bomb calorimeter

A control unit supplies electricity to the stirrer motors and the ignition circuit. Some units include circuitry to maintain automatically the outer jacket to within a small fraction of a degree of the temperature of the can and bomb, employing a pair of thermistors and a powerful electrical heater. There may also be a device to test the electrical continuity of the bomb before firing, and an illuminated reader for the precision thermometer with an electrically operated vibrator to prevent sticking of the very fine mercury column.

Subsidiary items of apparatus include tools and accessories for the bomb, fuse wire, cotton, a pellet press, and a stand or cloth to keep the parts of the bomb free from dirt.

Procedure

There is a danger in the operation of bomb calorimeters both from explosion and electric shock. It is essential that the following precautions are taken:

(a) *The bomb must be kept clean and when dismantled should be placed on a clean folded towel or stand if provided. Be careful not to scratch, dent, or drop the bomb.*
(b) *Switch off all electricity supplies before assembling or dismantling the bomb.*
(c) *The amount of sample must not exceed* 1.5 g.
(d) *The pressure of oxygen must not exceed* 25 *atm.*
(e) *If the stirrers are belt driven, check that their direction of rotation will impel water downwards.*
(f) *Do not fire the submerged bomb if bubbles of gas show that it is leaking (less than one bubble every* 10 s *is insignificant).*
(g) *Stand well clear of the bomb as it is fired and for at least* 20 s *afterwards.*
(h) *In some earlier models, the heaters which maintain the temperature of the outer jacket work by passing a heavy current directly through the water. To avoid overheating check that the jacket is nearly full of water, but do not top it up with anything other than de-ionized water. Beware of the electrical terminals if they are exposed.*
(i) *Refer to additional instructions at points marked * in the text and for any further warnings necessary for the particular apparatus used. Bomb calorimeters require skill to operate successfully, so do not hesitate to ask for the assistance of an instructor.*

(*i*) *Preparation of samples*

Roughly weigh out two 1 g portions of benzoic acid, two of maleic acid, and two of fumaric acid. Form each into a pellet with the press,* place it in a labelled weighing bottle, and weigh to 0.1 mg (4 decimal places).

(ii) Assembly of the bomb and peripherals

Cut a piece of fuse wire 3 cm longer than the distance between the rods which go down inside the calorimeter. Cut a piece of cotton exactly 1 m long and weigh it to 0.1 mg.

A secondary cotton fuse is a good ignition method for use by beginners.

Stretch the wire between the rods and form a good contact at each end.* Do not leave the wire ends protruding. Check that the crucible is in position. Cut off enough cotton from the weighed length to act as a fuse, as described below. Measure this length, and calculate its mass.

Knot the middle of the cotton around the centre of the wire, dangle the two ends down into the crucible, and form them into coils there. Place the first pellet on the cotton. Weigh its empty weighing bottle to find the mass of the pellet.

Introduce about 1 cm^3 of purified water into the bomb. *How does this ensure that we know the final state of any water given off in the combustion?*

With the sealing ring in position, assemble the bomb.* If a rubber O-ring is used, tighten the bomb with hand pressure only. If the bomb incorporates a lead sealing gasket, tighten it with the wrench provided.

If a testing circuit is incorporated in the control box, or if a test meter is available, check for continuity between the two terminals (or terminal and case) to ensure that the fuse wire is connected properly.

Connect the oxygen filling pipe to the bomb.* Open the main cylinder valve and fill the bomb to 25 atm* controlling the flow of gas with the *small* control valve on the cylinder. Close the main cylinder valve. Disconnect the sealed bomb.*

At the end of the next step, the bomb should be in its can immersed in a known quantity of hot and cold water at a temperature at the bottom of the scale of the precision thermometer.* (If a stream of bubbles rises from the bomb, remove and depressurize it,* re-adjust or replace the sealing ring, and reassemble. Seek advice if necessary.)

Place the can in the outer jacket assembly, and position the peripheral apparatus, namely the lid, stirrer, precision thermometer, ignition leads, and thermistor if fitted.*

(iii) Combustion of sample

Adjust the temperature of the water in the outer jacket to within 0.1 K of the water in the can. In the case of manually controlled outer jackets, do this by circulating hot or cold water as necessary. If the jacket is automatically controlled and its temperature is too low, switch on the electric heater.* If too high, circulate cold water through the cooling coil.

Record the temperature of the inner can with as great a precision as possible every 30 s for 5 min. Tap the thermometer *very gently* before each reading, or use the electrical shaker.

There is no noise when the bomb is fired—'bomb' refers to the steel vessel, not the reaction taking place in it!

Note the time and depress the firing switch. Release the switch if there is an indicator which shows that current has ceased to pass and that ignition has therefore taken place. Otherwise release the switch after 10 sec. Continue to record the temperature every 30 s for a further 15 min. It should rise by a degree or more after the combustion, and then begin to fall again.

If combustion does not take place, switch off all electrical circuits and disassemble and reassemble the peripheral apparatus.* Seek advice if necessary.

(*iv*) *Dismantling the bomb*

Switch off all electrical circuits. Disassemble the peripheral apparatus.* Depressurize the bomb.* Open the bomb and check that all the sample has been burnt; there should be no specks of carbon on the walls. Clean and dry all parts of the bomb.

Repeat the procedure described in paragraphs (ii), (iii), and (iv), using the other pellets.

Calculation

For each experiment, plot a graph of temperature against time, making a break in the axis so that the scale for the initial and final temperatures can be expanded as much as possible (Fig. 3.1.4). Draw best-fit straight lines through the initial temperatures, and through the temperatures after the heating due to combustion. The intersections of these lines with the vertical at the time of ignition give T_1 and T_2.

Given that $\Delta U^{\ominus}_{\text{cb, benzoic acid}} = -26.434\ \text{kJ g}^{-1}$,[1] and $\Delta U^{\ominus}_{\text{cb, cotton}} = -18\ \text{kJ g}^{-1}$, calculate the heat capacity term C_V (eqn (3.1.7)). Hence calculate the standard enthalpies of combustion of maleic acid and fumaric acid (eqn (3.1.10)).

Given that the standard enthalpies of combustion of carbon and hydrogen to gaseous carbon dioxide and liquid water are $-393\ \text{kJ mol}^{-1}$ and $-285\ \text{kJ mol}^{-1}$ respectively, calculate the standard heats of formation of maleic and fumaric acid by Hess's Law.

The structural formula of both acids is HOOC.CH=CH.COOH. The isomer which has the greatest internal steric

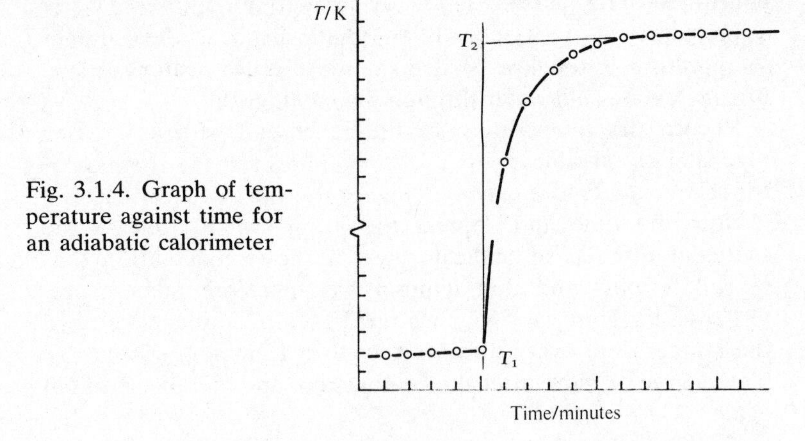

Fig. 3.1.4. Graph of temperature against time for an adiabatic calorimeter

hindrance will be the least stable; decide whether this is the *cis* or *trans* form. Hence deduce from the enthalpies of formation, of maleic and fumaric acids, which is *cis* and which is *trans*.

Results

See Table 3.1.1.

Table 3.1.1. Specimen results for benzoic acid

Sample/g	1.1731	cotton fuse/g	0.0068
T_1/°C	21.363	ΔU_{sample}/kJ	−31.01
T_2/°C	24.340	ΔU_{cotton}/kJ	−0.12
ΔT/K	2.977	C_V/kJ K^{-1}	10.46

Accuracy

The greatest inherent source of inaccuracy is in the inexactness of the working equation, which does not properly describe all the processes occurring. For example, we have not considered the small fraction of the iron wire which is oxidized, and the fact that some of carbon dioxide produced in the reaction dissolves in the water in the bomb. *What other factors have not been considered?*[2] Nevertheless you should be able to attain an accuracy of ±1 per cent on a fairly simple apparatus with manually controlled jacket heating, and ±0.5 per cent with an automated apparatus.

Comment

The calorific value of food is calculated by subtracting the heat

This experiment provides a good example of the use of the First Law of thermodynamics. If all the correction terms are included, results of high accuracy can be obtained with bomb of combustion of the appropriate human waste, measured in a bomb calorimeter, from the heat of combustion of the food itself.[3]

calorimeters.[2] Heats of combustion are of great interest to technologists, industrialists and dietitians. Because of their relationship to heats of formation, they also provide valuable information for experimental and theoretical chemists.

Technical notes

Apparatus

Bomb calorimeter with can, lid, outer jacket, heater, stirrers, control unit;† $ or $$ thermometer calibrated to 0.01 °C; thermometer for outer jacket; tools and parts for bomb; fuse wire; cotton; pellet press; large oxygen cylinder with regulator, pressure gauge and connector to bomb;$ benzoic acid and sample compounds stored in desiccator; [scales for weighing can].

Design note

Ignition circuits must supply about 5 A at less than 30 V. Higher voltages are dangerous.

References

[1] e.g. *The enthalpies of combustion and formation of ortho- and parafluorobenzoic acid.* W. H. Johnson and E. J. Prosen. *J. Res. Nat. Bur. Standards* **79A,** 481 (1975).
[2] *Standard states for bomb calorimetry.* E. W. Washburn. *Bur. Standards J. Res.* **10,** 525 (1933).
[3] *McCance and Widdowson's 'The Composition of Foods'*, A. A. Paul and D. A. T. Southgate. HMSO, London and Elsevier/North Holland, Amsterdam, 4th edn (1978).

3.2 Enthalpy of dimerization by Dumas' vapour density method

There are two well-known methods of measuring vapour density, one developed by Victor Meyer, and the other by Dumas.[1] Dumas' method is more suited for use at several different temperatures. Its principle is straightforward. Liquid is placed in a glass vessel known as a *Dumas bulb.* The bulb is heated in a furnace at a known temperature so that the vapour evaporates and displaces all the air. The bulb is then sealed, cooled, its volume measured and the amount of vapour determined by chemical analysis.

At temperatures just above its boiling point, the vapour of ethanoic (acetic) acid consists largely of *hydrogen-bonded dimers,* with the degree of dimerization decreasing at higher temperatures. In this experiment we employ Dumas' method over a range of temperatures to find values of the equilibrium constant for the association reaction, and hence find the enthalpy of formation of the dimers.

Theory

(*i*) *Equilibrium constant for ethanoic acid dimerization*

The equilibrium for n_0 moles of ethanoic acid associating to an extent α may be written

$$CH_3COOH \leftrightharpoons \tfrac{1}{2}(CH_3COOH)_2.$$

Moles of vapour $\quad n_0(1-\alpha) \quad n_0\alpha/2 \quad$ Total: $n_0(1-\alpha/2)$.

The experiment is carried out at the ambient atmospheric pressure, p_{atm}. Assuming that the vapour is a perfect gas,

$$p_{atm}V = n_0(1-\alpha/2)RT. \tag{3.2.1}$$

n_0 is found by titration, and the volume, temperature, and atmospheric pressure are also measured. Thus α may be calculated.

The partial pressures of unassociated ethanoic acid, p_1, and dimers, p_2, are

$$p_1 = \frac{n_0(1-\alpha)}{n_0(1-\alpha/2)} p_{atm}, \tag{3.2.2}$$

and

$$p_2 = \frac{n_0\alpha/2}{n_0(1-\alpha/2)} p_{atm}. \tag{3.2.3}$$

The dimensionless equilibrium constant for the reaction is given by

$$K_p^{\ominus} = \frac{(p_2/p^{\ominus})^{1/2}}{(p_1/p^{\ominus})} \tag{3.2.4}$$

where $p^{\ominus}$ is the standard pressure (1 atm).

(*ii*) *Thermodynamic equations*[2]

A careful consideration of a gas phase reaction in which all the components obey the perfect gas law leads to the *van't Hoff isotherm*

$$\Delta G^{\ominus} = -RT \ln K_p^{\ominus}. \tag{3.2.5}$$

$\Delta G^{\ominus}$ is the standard free energy change when 1 mole of reaction takes place with both reactant and product remaining in their standard state at a pressure of 1 atm.

The *Gibbs–Helmholtz equation* is derived from the definition of free energy, $G = H - TS$:

$$\left(\frac{\partial(\Delta G/T)}{\partial T}\right)_p = -\Delta H/T^2. \tag{3.2.6}$$

On differentiating eqn (3.2.5) at constant pressure and substituting the expression above, we obtain the *van't Hoff isochore*:

$$\left(\frac{\partial \ln K_p^\ominus}{\partial T}\right)_p = \frac{\Delta H^\ominus}{RT^2}. \tag{3.2.7}$$

K_p is independent of pressure, and the equation can therefore be written more simply as

$$\frac{d \ln K_p^\ominus}{dT} = \frac{\Delta H^\ominus}{RT^2}. \tag{3.2.8}$$

Assuming the standard enthalpy change $\Delta H^\ominus$ is constant over the temperature range studied, the equation may be integrated:

$$\ln K_p^\ominus = -\frac{\Delta H^\ominus}{RT} + \text{const.} \tag{3.2.9}$$

Apparatus

It is salutary to take note of the problems faced by Dumas in his original experiments; constant temperature could only be maintained by immersing the bulb in boiling alloy in a smelting furnace, and the temperature measured using air thermometers requiring up to 65 kg of mercury.[3]

The apparatus comprises a set of Dumas bulbs and, to heat them, an electric tube-furnace with heating controller, and temperature monitoring device. Subsidiary items are required for making the bulbs if they are not supplied, for sealing them, and for the titration of the ethanoic acid.

Procedure

Glacial ethanoic (acetic) acid is caustic. Avoid contact with skin and eyes. Wear safety glasses. Avoid breathing vapour. Do not use a glassblowing torch for the first time without the advice of an instructor.

The Dumas bulbs have dimensions to suit the furnace. Their ends are drawn out into fine capillary tubes and these are bent through two right-angles (Fig. 3.2.1) by holding each section horizontally and heating it *very gently* with a bunsen burner. This causes the free end of the tube to fall by gravity. If no Dumas bulbs are supplied, or if they are provided but you would like to try making some yourself, ask an instructor for advice.* About six bulbs are required.

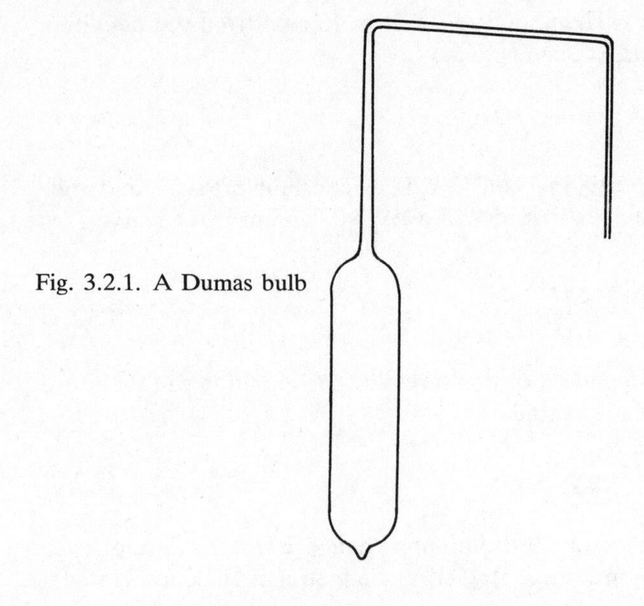

Fig. 3.2.1. A Dumas bulb

Heat the furnace to 150 °C.*

Hold one of the Dumas bulbs in a clamp stand. Carefully pour some glacial ethanoic acid into a small beaker. Hold the beaker so that the drawn end of the Dumas bulb dips into it. By gentle heating and cooling of the bulb, introduce about 2 cm^3 of the acid into it.

Prepare a hand blowtorch or *small* bunsen burner ready for sealing off the Dumas bulb.

Check that the temperature of the furnace is steady. Place the bulb in the furnace* and insulate it at the top of the furnace by packing some ceramic fibre around it. The capillary end of the Dumas bulb should be pointing downwards and be open to the air. Collect the ethanoic acid in a test tube as it distils out. As soon as vapour ceases to issue from the drawn end, showing that there is no more liquid in the bulb (as may be confirmed visually if there is a window in the furnace), very carefully and thoroughly remove any drops of acid from the capillary by careful flaming with a bunsen burner. Then with the hand blowpipe or small bunsen burner, seal off the capillary end of the bulb at the point at which it enters the fibre plug.

Check the temperature of the furnace. Carefully remove the tube and allow it to cool.

Meanwhile measure the ambient pressure of the atmosphere with a barometer. Also remove dissolved air and carbon dioxide from about 1 dm^3 of purified water. Do this by warming the water in a round-bottomed flask while evacuating the flask with an aspirator (water suction pump). Allow the water to cool while still under vacuum.

When the sealed Dumas bulb is cold, first make sure that there is no ethanoic acid in its sealed tip. Warm the point slightly to displace the acid if necessary. Pour about 200 cm^3 of the degassed water into a wide beaker. Then immerse in this the sealed end of the bulb and break off the tip using the pair of side-cutters provided. The bulb should fill with water almost completely; if it does not, air must have been present in the bulb, so the determination is invalid and should be repeated.

Find the point on the capillary tube which is a few centimetres from the body of the bulb and where the diameter is about 5 mm. Make a short scratch with a glass knife. Place the tube with its body tilting downwards, and the scratch uppermost and positioned in a steel V. Strike the end of the tube a sharp tap with the back of a glass knife. The end of the tube should break off cleanly. Shake out the contents into a flask. Wash the tube out with a little distilled water and add the washings to the flask.

Titrate the contents of the flask against barium hydroxide solution of an accurately known concentration around 0.01 M, with phenolphthalein as indicator. If you overshoot the end-point, add 1 cm^3 of 0.025 M ethanoic acid with a pipette and take a new end-point.

Finally measure the volume of the bulb by running in water from a burette.

Repeat the whole procedure at four or five more temperatures to cover the range 150–250 °C.

Calculation

From your results calculate the values of α, the fraction of the ethanoic acid molecules which have dimerized. Find the partial pressures of unassociated molecules, p_1, and dimers, p_2, at each temperature. Hence calculate the values of the dissociation constant $K_p^\ominus$. Plot these in such a way that you obtain the standard enthalpy of association $\Delta H^\ominus$.

Results

See Table 3.2.1.

Accuracy

Suppose that you estimate the minimum overall systematic error in the experimental measurements to be 1 per cent. At higher temperatures α is small, and thus very sensitive to error when calculated by means of eqn (3.2.1). The resulting uncertainty in $\Delta H^\ominus$ is correspondingly magnified to ~20 per cent. Lack of skill in handling the Dumas bulbs can increase the total error in $\Delta H^\ominus$ up to ±40 per cent.

Table 3.2.1. Specimen results

Observations		Calculations	
T/K	557	T^{-1}/K^{-1}	1.795×10^{-3}
Titre/(cm^3 0.0119 M $Ba(OH)_2$)	52.6	n_0/mol	1.252×10^{-3}
Bulb volume/cm^3	53.8	α	0.109
p/mmHg	764.3	p_1/atm	0.948
		p_2/atm	0.058
		$K_p^{\ominus}$	0.2541
		$\ln K_p^{\ominus}$	−1.370

Comment

For reasons explained above, this experiment does not produce results of high accuracy. However, it is a historically interesting and elegant method of studying association in a vapour, which employs some important theory and encourages good experimental technique. Results of higher accuracy can be obtained by measuring directly the pressure of a known mass of vapour in a constant volume, as in the next experiment.

Technical notes

Apparatus

Tube furnace (preferably with window onto base of bulb, device for lowering and raising the bulb, digital temperature read-out and automatic temperature control),$^{\$\dagger}$ Dumas bulbs,† ceramic fibre (glass wool is hazardous), small and standard bunsen burners, glass-blowing apparatus, steel V, side-cutters, 1 dm^3 flask, other flasks, wide 500 cm^3 beaker, burette, glacial ethanoic acid, phenolphthalein, ~0.01 M $Ba(OH)_2$ checked by back titration, 0.025 M ethanoic acid.

Barium hydroxide solution

Instructions to make a 4 dm^3 batch of standardized solution:— Boil 4 dm^3 distilled H_2O in 5 dm^3 flat bottomed flask for 10 min to remove CO_2. Slowly add ~70 g of $Ba(OH)_2.8H_2O$ to give ~0.05 M solution. Connect to apparatus constructed to allow the remaining operations to be carried out in the absence of CO_2. Transfer appropriate quantity to measuring cylinder. Dilute to required volume with boiled, distilled water in absence of CO_2. Transfer solution to stock bottle fitted with soda lime tube. Fit two-way burette with soda lime tube to stock bottle. Allow solution to stand for 12 hr. Titrate with potassium tetroxalate using phenolphthalein as indicator. If necessary, add more CO_2-free water or barium hydroxide solution and re-titrate.

References

[1] *An advanced treatise on physical chemistry*. J. R. Partington. Longmans, London (1967), vol. 1, pp. 758 and 762.

[2] *Basic chemical thermodynamics*. E. B. Smith, Clarendon Press, Oxford, 3rd edn (1982), p. 52.
[3] *Sur la densité de la vapeur du phosphore*. M. J. Dumas. *Ann. de Chimie et de Physique* XLIX, 210 (1832).

3.3 Enthalpy of dissociation by pressure measurement

In a simple gas phase equilibrium reaction in which the number of moles of gas alters from reactants to products, the position of the equilibrium may be determined from a direct measurement of pressure, provided that we also know the volume of the vessel and the amount of substance inside it. A series of pressure measurements at different temperatures will give the temperature coefficient of the equilibrium constant, which in turn yields the enthalpy of the reaction. This is the basis of the present experiment, in which we set out to determine the enthalpy of dissociation of dinitrogen tetroxide to nitrogen dioxide (nitrogen IV oxide).[1]

Theory

The equilibrium for n_0 moles of dinitrogen tetroxide dissociating to an extent α may be written

$$\begin{array}{ll} N_2O_4 & \leftrightharpoons 2NO_2 \\ n_0(1-\alpha) & 2n_0\alpha. \end{array}$$

The theory required for the study of this system is very similar to that for the dimerization of ethanoic acid, pp. 90–91. Here, however, we choose to consider the equilibrium in terms of dissociation rather than association. With this in mind, use the earlier discussion to help you (i) find the equation for determining α from the measurements of pressure, volume, temperature, and n_0, (ii) find expressions for the partial pressures of the gases and (iii) derive the expression

$$K_p^{\ominus} = 4\alpha^2 p/p^{\ominus}(1-\alpha^2). \qquad (3.3.1)$$

Also remind yourself of the source of the integrated form of the *van't Hoff isochore*

$$\ln K_p^{\ominus} = -\Delta H^{\ominus}/RT + \text{const.} \qquad (3.3.2)$$

Apparatus

The central item of apparatus is a glass vessel into which is sealed a *glass helical Bourdon pressure gauge*[2] (Fig. 3.3.1). This is constructed from a thin-walled glass tube which is partially flattened and then wound into a helix, with one end sealed. If

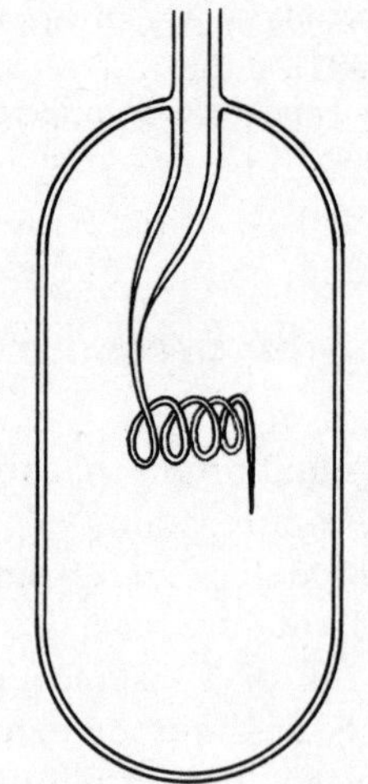

Fig. 3.3.1. A glass helical pressure gauge

the pressure inside the helix is p_i, and the pressure outside is p_o, it can be readily shown that the net outward radial force is $(p_o - p_i)2\pi td$, where d is the axial depth and t the thickness as shown in Fig. 3.3.2. If the pressure inside the helix is greater than the pressure outside, the force tends to unwind it, whereas if p_i is less than p_o, there is a net inward force tending to wind up the helix. The helix may be either horizontal or vertical, and to its end may be attached either a mirror or a pointer. If a pointer is used, reference pointers are arranged either side of it.

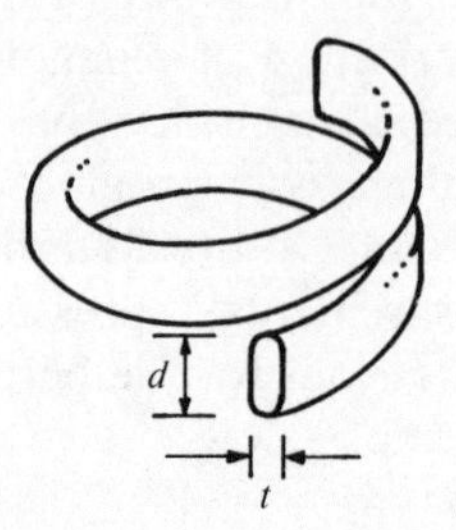

Fig. 3.3.2. Detail of the pressure gauge

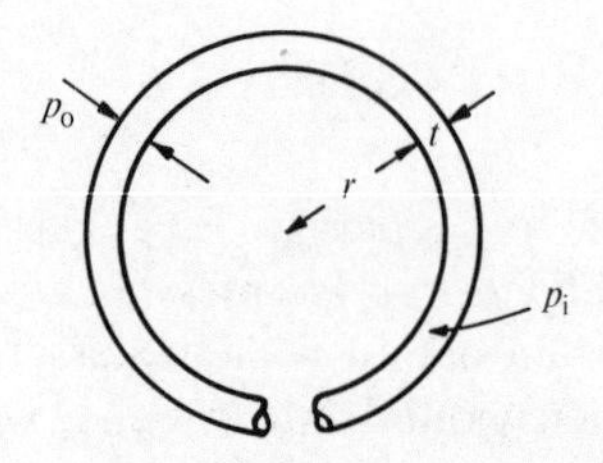

If a mirror is employed, a beam of light is shone on to it, and the extent of movement of the mirror measured by the deflection of the reflected beam along a scale. If the gauge has been previously calibrated it may then be used to measure the pressure directly. In this experiment, however, the device is used simply as an *isolating gauge.* It isolates the reaction vessel from the vacuum line and mercury manometer connected to it, thus keeping the volume of the reaction vessel effectively constant and preventing the gas from contaminating the mercury and line. The pressure gauge is used solely to indicate when the pressure on the mercury manometer is the same as the pressure in the reaction vessel. This occurs when the light beam illuminates the null point marked on the scale, or when the pointer at the end of the helix lies exactly between the two fixed pointers.

The complete apparatus is shown in Fig. 3.3.3. The helix tube of the pressure gauge is connected via a tap or needle valve (C) to a globe which has been previously pumped out by an aspirator connected via a trap and air leak. This allows the pressure in the gauge to be reduced. To increase the pressure the gauge is also connected to a tap or needle valve open to the atmosphere, and to measure the pressure it is joined, via a ballast volume, to a mercury manometer.

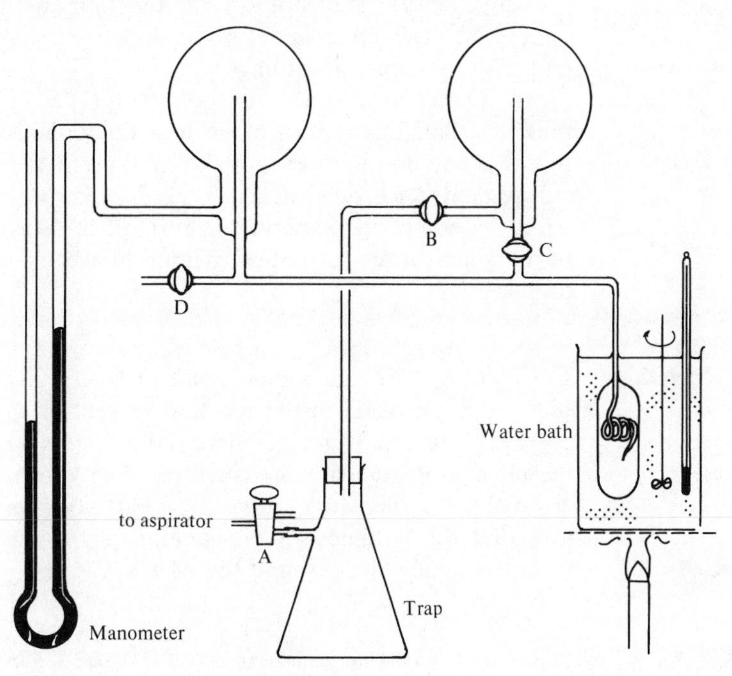

Fig. 3.3.3. The complete apparatus

Other configurations are possible; for example there may be a separate reaction chamber joined to the helical tube of the gauge, with the gauge chamber connected to a pump and manometer.

Procedure

With its air leak open (tap A), turn on the aspirator. Then carefully evacuate the globe (tap or needle valve C closed, tap B opened slowly). Close off the globe from the aspirator (tap B), open the air leak (tap A), and turn off the aspirator.

Check that the tap to the air (tap D) is closed. Slowly open the tap or needle valve (C) to let air from the gauge into the evacuated globe until the Bourdon gauge is balanced. If the balance point cannot be reached, close the tap and follow the procedure in the previous paragraph to re-evacuate the globe. Then continue as before. If you overshoot the balance point, close the tap to the evacuated globe and slowly open the tap to the air (tap D). By careful manipulation of both the taps to the evacuated globe and to the air, the gauge can be balanced.

Measure the pressure on the mercury manometer and the ambient atmospheric pressure with a barometer. Note the temperature of the water bath.

The N_2O_4/NO_2 mixture becomes so dark above 80 °C that it hides the gauge.

Switch on the stirrer and light the bunsen burner. At five degree intervals up to about 65 °C, balance the Bourdon gauge and take manometer readings.

Allow the water bath to cool, and take a further set of readings. Small quantities of ice may be added to a water bath to assist cooling, but never make measurements while there is still unmelted ice present.

At the end of the experiment, turn off the stirrer and bunsen burner, and bring the entire vacuum line back to atmospheric pressure.

Calculation

Calculate n_o, the number of moles of N_2O_4, given the mass of dinitrogen tetroxide in the reaction vessel.*

At each temperature, calculate the pressure p in the reaction vessel. Hence calculate the corresponding values of α and K_p, given the volume V of the vessel.* Plot your results in such a way that the enthalpy of dissociation $\Delta H^\ominus$ may be obtained from a straight line through the points.

Results

Suppose $V = 328\ cm^3$ and mass of $N_2O_4 = 0.535$ g. Then $n_0 = 0.535/92.0 = 5.82 \times 10^{-3}$ moles. $p_{atm} = 757.3$ mm Hg.
See Table 3.3.1.

Table 3.3.1. Specimen results

T/°C		75.0	p/mmHg	677
T/K		348.2	p/atm	0.891
$10^3 \times T^{-1}$/K^{-1}		2.872×10^{-3}	α	0.756
man. levels/cm	line	50.9	$K_p^{\ominus}$	4.754
	atm	42.9	$\ln K_p^{\ominus}$	1.559

Accuracy

Likely uncertainties: (i) impure gas sample = 1 per cent in mass, (ii) pressure = 2 mm Hg on manometer + 0.7 per cent change in Hg density over temperature range, (iii) volume = 1 per cent, (iv) temperature = 1 K. Express your own estimates as error bars on your graph and estimate the error of the slope.

Comment

This is an interesting experiment because it demonstrates one of the most direct methods of finding the position of a chemical equilibrium, and provides another example of how to use the van't Hoff isochore to obtain the enthalpy of a reaction. In this case, the enthalpy can also be obtained photometrically, because N_2O_4 is brown and NO_2 is colourless.[3]

The present experiment gives results of good accuracy, and is simple to carry out once the difficulties of constructing and loading the reaction vessel and its gauge have been overcome. The same apparatus can be used for measuring vapour pressures, and a similar arrangement with a bellows isolating gauge has been used in research.[4]

Technical notes

Apparatus

Reaction vessel gauge,††[2] vacuum line and assorted apparatus (Fig. 3.3.3),† research grade nitrogen dioxide.

Setting up

Construct helical gauge and reaction cell, but with an inlet tube connected to body of cell. Weigh. Carefully fill with water to proposed seal-off point. Reweigh to find volume. Empty. Connect to vacuum line and pump out. Cool NO_2 to give straw-coloured liquid N_2O_4 and pour into out-gassed bulb. Freeze to solid with liquid nitrogen. Pump out. Seal off reaction vessel. Reweigh vessel and discarded tube to find mass of N_2O_4.

References

[1] *Advanced inorganic chemistry.* F. A. Cotton and G. Wilkinson. Wiley Interscience, New York, 4th edn (1980), p. 426.

[2] *Glass spirals for use in sensitive pressure gauges*. S. J. Yorke, *J. Sci. Instr.* **22,** 196 (1945).
[3] *Findlay's practical physical chemistry*. rev. B. P. Levitt. Longman, London, 9th edn (1973), p. 189.
[4] *Vapour pressure of cyclohexane*, 25° to 75 °C. A. J. B. Cruickshank and A. J. B. Cutler. *J. Chem. Eng. Data* **12,** 326 (1967).

3.4 Enthalpy of CO_2 reduction by a flow method

Equilibria in reactions involving a single solid phase and a gas phase of varying composition can be conveniently studied by a flow method. The technique involves passing a slow stream of gas through a bed of granules of the solid which is maintained at a constant temperature. Provided the flow rate is slow enough for reaction to be complete, simple analysis of the effluent gas mixture gives the necessary data for calculation of the equilibrium constant. No steps need to be taken to 'freeze' the equilibrium while the gas stream is cooling, since no further reaction can take place when the effluent is no longer in contact with the solid.

In this experiment we study the reaction

$$CO_2 + C \rightleftharpoons 2CO.$$

Carbon dioxide is passed over carbon granules in a reaction tube inside a furnace, and the resulting mixture of carbon monoxide and unchanged carbon dioxide is analysed to give the equilibrium constant $K_p^{\ominus}$. The enthalpy of reaction is then calculated from the temperature variation of the equilibrium constant. The result is compared with enthalpies calculated at each temperature from $K_p^{\ominus}$ and the known standard entropies of the reactants and products.

Theory The dimensionless equilibrium constant of the reaction

$$CO_2 + C \rightleftharpoons 2CO$$

is given by

$$K_p^{\ominus} = \frac{(p_{CO}/p^{\ominus})^2}{(p_{CO_2}/p^{\ominus})} \tag{3.4.1}$$

where $p^{\ominus}$ is the standard pressure of 1 atm.

The *van't Hoff isochore* (p. 91) relates the temperature coefficient of $K_p^{\ominus}$ to $\Delta H^{\ominus}$, the standard enthalpy of reaction relating to the substances in a pure state at 1 atm pressure:

$$\frac{\mathrm{d}\ln K_{\mathrm{p}}^{\ominus}}{\mathrm{d}T}=\frac{\Delta H^{\ominus}}{RT^{2}}. \tag{3.4.2}$$

Alternatively we may choose not to use the information contained in the temperature coefficient of $K_{\mathrm{p}}^{\ominus}$. Instead we can calculate $\Delta G^{\ominus}$ at each temperature from the *van't Hoff isotherm*:

$$\Delta G^{\ominus}=-RT\ln K_{\mathrm{p}}^{\ominus}. \tag{3.4.3}$$

We then calculate $\Delta S^{\ominus}$ from known values of the standard entropies of the reactants and product. These standard entropies are related to $\Delta S^{\ominus}$ by the stoichiometry of the reaction, so that

$$\Delta S^{\ominus}=2S_{\mathrm{CO}}^{\ominus}-(S_{\mathrm{CO_2}}^{\ominus}+S_{\mathrm{C}}^{\ominus}). \tag{3.4.4}$$

$\Delta H^{\ominus}$ may then be calculated from the relation

$$\Delta G^{\ominus}=\Delta H^{\ominus}-T\,\Delta S^{\ominus}. \tag{3.4.5}$$

Apparatus

The reaction takes place in a silica tube containing a bed of charcoal granules packed between silica wool plugs, as shown in Fig. 3.4.1. The tube is arranged so that the charcoal bed lies in the central, constant temperature zone of an electric furnace. The temperature of this zone is monitored by a thermocouple linked to an electronic controller which switches the furnace on and off to maintain a constant temperature. The temperature of the charcoal itself is monitored by another thermocouple as shown. This may either be linked to a digital temperature

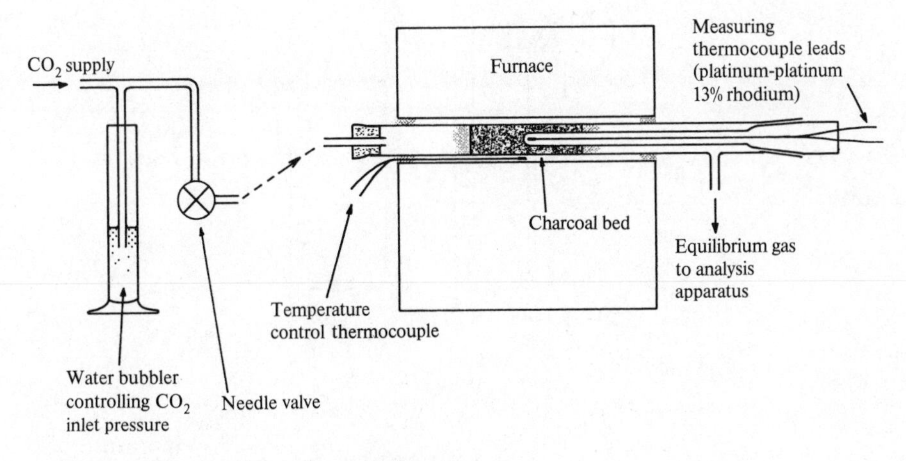

Fig. 3.4.1. The reaction tube and furnace

display device, or to a cold junction in melting ice and a potentiometer. If the latter arrangement is employed, standard tables must be used to relate e.m.f. to temperature.

The carbon dioxide is supplied from a cylinder. The pressure of the supply is controlled firstly by the cylinder regulator, and secondly by a blow-off which reduces the pressure to a maximum equivalent to about 1 cm water (~1 mm Hg). The flow rate is adjusted to about 4 cm^3 per minute by means of a precision needle valve.

The mixture of gases emerging from the reaction zone is analysed by a simple device incorporating two detergent-bubble flow meters, as shown in Fig. 3.4.2. The sample gas passes through a two-way tap and into the bottom of the first flowmeter. If the rubber teat filled with detergent solution is squeezed momentarily, a detergent film will form across the graduated tube. By timing the motion of this film, the flow rate of CO and CO_2 can be measured accurately. At the top of the flowmeter, the film is burst by a needle. The mixture of gases then passes through sodalime, which absorbs the CO_2. It then passes through a second flowmeter, in which the rate of flow of the CO is measured.

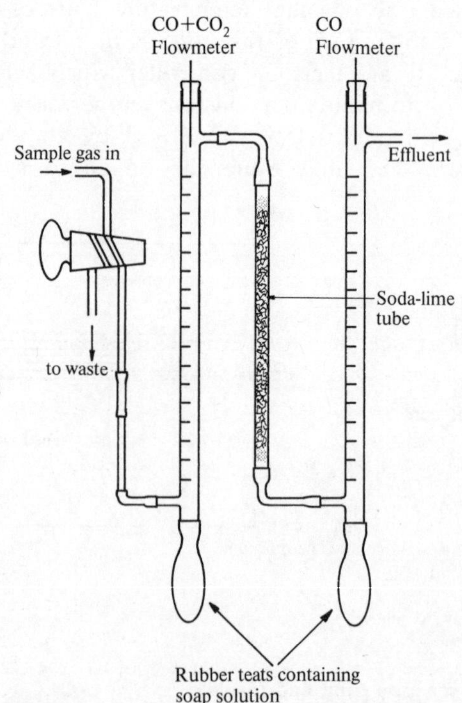

Fig. 3.4.2. CO/CO_2 gas analyser

Procedure

(i) Set up

Switch on the furnace and its temperature controller, and set the controller to about 800 °C.*

Always pour off the excess water from ice/water slushes used for 0 °C cold junctions, as pools of water at up to 4 °C can exist below ice.

If you are measuring the temperature of the furnace by means of a thermocouple which requires a cold junction at 0 °C, make up a slush of water and crushed ice, and surround the cold junction with this.

If using a potentiometer to measure the thermocouple e.m.f., set up the potentiometer as directed.*

The sodalime between the flowmeters is of the self-indicating variety, which turns purple when exhausted. When the purple colour extends about one third of the way down the tube, it should be emptied and refilled.

Check that the needle-valve is closed. Turn on the CO_2 supply and adjust the cylinder regulators to give a bubble rate through the water blow-off of several bubbles per second. Open the needle valve. Send detergent films up both flowmeters. Until the walls of the flowmeter are wetted all the way up, the detergent films will burst prematurely. Continue making detergent films until both flowmeters are working correctly. Adjust the flow rate, as registered on the $CO+CO_2$ flowmeter, to about $4\ cm^3\ min^{-1}$. Then turn the two-way tap to vent the gases. In general do not pass gas through the flowmeters until a measurement is required, for otherwise the sodalime will be used up very quickly.

(ii) Measurements

Measure the relative rates of flow of $(CO+CO_2)$ and CO at temperatures in the range 1050–1200 K. At each temperature, wait until the reading of the thermocouple in the charcoal column is reasonably steady. Also remember that there is a *dead-space* between the reaction zone and the flowmeters. If this dead-space has a volume of $40\ cm^3$, then at a flow rate of $4\ cm^3\ min^{-1}$ it will take 10 min for the gases to reach the flowmeters. So allow enough time before making measurements. Make a number of determinations of the two flow rates at each temperature.

Calculation

Convert the relative flow rates of the gases to relative partial pressures p, assuming that the total pressure of the gases is constant at 1.0 atm. Hence calculate $K_p^{\ominus}$ at each temperature, and find $\Delta H^{\ominus}$ using the van't Hoff isochore (eqn (3.4.2)).

Values of the absolute entropies of the reactants and products

Table 3.4.1. Standard absolute molar entropies $S^{\ominus}/\mathrm{J\,K^{-1}\,mol^{-1}}$

T/K	C (graphite)	CO (gas)	CO_2 (gas)
1000	24.4	234.8	269.2
1100	26.5	238.0	274.5
1200	28.5	240.9	279.3
1300	30.3	243.7	283.8

are listed in Table 3.4.1. Plot these values, and read off the entropies at the temperatures used in the experiment. Hence find $\Delta H^{\ominus}$ at each temperature by means of the van't Hoff isotherm (eqn (3.4.3)) and the definition of $\Delta G^{\ominus}$ (eqn (3.4.5)).

Results

Accuracy

Compare the values of $\Delta H^{\ominus}$ at each temperature with the value obtained from the van't Hoff isochore. Discuss the accuracy of the two types of determination.

Comment

This experiment gives practice in the use of standard thermodynamic relations, and illustrates an elegant way of studying a solid–gas equilibrium. *Could the same apparatus be used to determine the rate of the reaction*?

Technical notes

Apparatus

As shown in Figs. 3.4.1 and 3.4.2;[†$] self-indicating sodalime.

Soap solution

Concentration should be such that detergent bubbles are easily formed, and always punctured by the needle at the top of each flow tube. Try 1:1 washing-up liquid + water.

Temperature range

Below 1050 K the rate of reaction may be too small for equilibrium to be established. Above 1200 K the equilibrium lies so far towards products that there is very little CO_2 in the products and measurement of relative partial pressures becomes difficult.

3.5 Enthalpies and entropies of complex formation in a solution calorimeter

The enthalpy of a reaction in solution is measured by allowing a known amount of substance to react inside a *solution calorimeter* and measuring the consequent temperature rise.

There are two reactant solutions in the solution calorimeter which initially are kept separate. This is achieved by pouring one solution into a mixing device which is immersed in the other solution. The two solutions are then mixed by operating the device from outside the calorimeter. *Why is the second solution not simply poured into the calorimeter to start the reaction*? A stirrer ensures that the reaction goes quickly to completion. The best way of measuring the small temperature rise (less than 1 K) is with a thermistor. A precision thermometer is also satisfactory, but has the disadvantage of a much larger heat capacity because of its sizeable mercury bulb. To minimize the heat loss to the surroundings, the calorimeter is in the form of a Dewar flask (vacuum flask) with lid. The heat capacity of the container and its contents is found by passing a known quantity of electricity through a heater inside the calorimeter, and noting the temperature rise.

In the present experiment we measure the enthalpies of formation of two complexes. If the relevant equilibrium constants are known, the free energy changes may be calculated, and using the measured enthalpies we can then calculate the entropy changes which accompany the formation of the complexes.

Theory

(i) Stability constant for complex formation

Copper(II) ions react with ammonia in solution to form the cuprammonium complex:

$$Cu^{2+}(aq) + 4NH_3(aq) \leftrightharpoons [Cu(NH_3)_4]^{2+}(aq). \tag{3.5.1}$$

With 1,2-diaminoethane (ethylenediamine), they react to form a chelate compound, i.e. a co-ordination compound in which the metal ion combines with more than one donor group in the same molecule:

Ethylenediaminetetracetic acid, also known as EDTA, or edetate when anionic, is a well-known chelating agent for heavy metal ions; sodium calcium–edetate is prescribed as an antidote for lead poisoning.[1]

$$Cu^{2+}(aq) + 2NH_2CH_2CH_2NH_2(aq)$$

$$\rightleftharpoons \left[(CH_2)_2 \begin{matrix} NH_2 \\ NH_2 \end{matrix} Cu \begin{matrix} NH_2 \\ NH_2 \end{matrix} (CH_2)_2 \right]^{2+} (aq) \tag{3.5.2}$$

The resulting complex is usually written $[Cu\ en_2]^{2+}$. The ammonia and 1,2-diaminoethane molecules are acting as *ligands*, because they are donor partners in coordinate bonds. If we denote metal ions as M and ligands as L, both reactions may be written in the general form

$$M + \nu L \leftrightharpoons ML_\nu. \tag{3.5.3}$$

The corresponding equilibrium constant or *stability constant* K in terms of concentrations is

$$K = \frac{[\mathrm{ML}_\nu]}{[\mathrm{M}][\mathrm{L}]^\nu}. \tag{3.5.4}$$

The standard stability constant $K^\ominus$ is a dimensionless constant obtained by dividing each concentration term by the standard concentration, which in this and most cases is 1 mol dm^{-3}.

(*ii*) *Measurement of enthalpy of reaction*

Let the temperature rise on the formation of n moles of complex be $\Delta T_{\text{reaction}}$. Since the reaction occurs at constant pressure, it is the enthalpy of formation of a mole of complex, $-\Delta H$, which is measured, and

$$-\Delta H = \frac{\Delta T_{\text{reaction}} C_{\text{p}}}{n}, \tag{3.5.5}$$

where C_{p} is the heat capacity of the calorimeter and its contents.

If, instead of the reaction, a current I at voltage V is switched through the heater for time t, then the temperature rises by an amount ΔT_{heater}. The energy E put into the system by the heater is

$$E = VIt = \Delta T_{\text{heater}} C_{\text{p}}. \tag{3.5.6}$$

In practice, V and I will vary slightly with time, whereupon

$$E = \langle VI \rangle t \approx \langle V \rangle \langle I \rangle t, \tag{3.5.7}$$

where $\langle\ \rangle$ represents the average value over time t.

From eqns (3.5.5), (3.5.6) and (3.5.7)

$$-\Delta H \approx \frac{\Delta T_{\text{reaction}} \langle V \rangle \langle I \rangle t}{\Delta T_{\text{heater}} n}. \tag{3.5.8}$$

If an uncalibrated thermistor of resistance R is used to measure the temperature changes, then since ΔT is approximately proportional to $-\Delta \log_{10} R (\neq -\log_{10} \Delta R)$, the working equation becomes

$$\Delta H \approx \frac{(\Delta \log_{10} R)_{\text{reaction}} \langle V \rangle \langle I \rangle t}{n (\Delta \log_{10} R)_{\text{heater}}}. \tag{3.5.9}$$

(*iii*) *Thermodynamic equations*

A consideration of the chemical potentials of reactants and products in the equilibrium reaction leads to the *van't Hoff isotherm*

$$\Delta G^{\ominus} = -RT \ln K^{\ominus}, \tag{3.5.10}$$

where $\Delta G^{\ominus}$ is the standard free energy change of a mole of reaction. Thus if the value of $K^{\ominus}$ is known, $\Delta G^{\ominus}$ may be calculated. If ΔH has been measured, and it is assumed that $\Delta H = \Delta H^{\ominus}$, $\Delta S^{\ominus}$ may be calculated from the defining equation for free energy,

$$\Delta G^{\ominus} = \Delta H^{\ominus} - T\,\Delta S^{\ominus}. \tag{3.5.11}$$

Apparatus[2,3] The body of the calorimeter is in the form of a Dewar flask with lid as shown in Fig. 3.5.1. Through the lid is mounted a thermistor in a thin protective stainless steel tube. Also passing through the lid is a stirrer, a heating coil, and a mixing device. The latter can be in the form of a siphon from which the solution is blown over with a rubber bulb, or it can comprise a vessel with a small hole at the bottom which is plugged with tap grease, the plug being blown out by compressed air. Alternatively it can simply be a test-tube with plunger, or a glass phial which is broken open to mix the reactants.

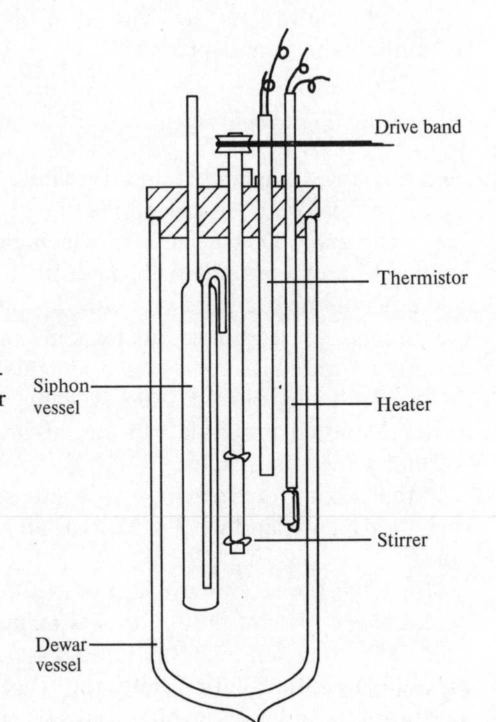

Fig. 3.5.1. A solution calorimeter with a thermistor and a siphon mixing device

Current is supplied from an adjustable, stabilized power supply. A change-over switch connects this either to the calorimeter heater or to a dummy heater of the same resistance. The use of a dummy heater allows the current to be adjusted and to settle down before the calorimeter itself is heated. The calorimeter heater should never be switched on in air, as it will burn out.

If a thermistor is used for temperature measurement, its resistance is determined with a Wheatstone bridge. This can either be manually operated, or can incorporate a linearized amplifier.

Procedure

Take care when using solutions of 1,2-*diaminoethane* (*ethylenediamine*) *and ammonia, since they have harmful vapours and are irritating to the skin, eyes and respiratory system.*

(*i*) *Reactant solutions*

Stock solutions of ammonium hydroxide of concentration 1 M or greater evolve significant amounts of ammonia gas while standing at room temperature, and should always be checked by titration before use.

Make up accurately 500 cm^3 of 0.2 M copper sulphate solution using pure $CuSO_4.5H_2O$ and de-ionized water. Check that the concentration of the ammonium hydroxide solution provided is between 3.5 and 5 M by titration against a standardized strong acid with methyl red as indicator. A 2 M solution of 1,2-diaminoethane is also provided.

(*ii*) *Setting up the calorimeter*

Find out how to assemble and dismantle the calorimeter.* Note that some calorimeter assemblies have a lid which is clamped and a moveable Dewar flask, in which case do not move the lid or the apparatus mounted through it.

Wash out the Dewar flask with de-ionized water. Also wash the mixing device, stirrer, heater coil, and thermistor case with de-ionized water. If appropriate, do this by raising a beaker of the water up to the lid. Remove any water from the mixing device, although traces left in the mixing device or calorimeter will not affect the results.

If the mixing device employs a grease plug, dry the device very thoroughly, and work a plug of tap grease into the capillary hole.

Using the funnel or syringe provided, carefully fill the mixing device three quarters full with 4 M ammonium hydroxide solution. This reactant will be in excess. Pipette 50 cm^3 of the copper sulphate solution into the Dewar flask. Pour into a measuring cylinder enough de-ionized water to fill the calorime-

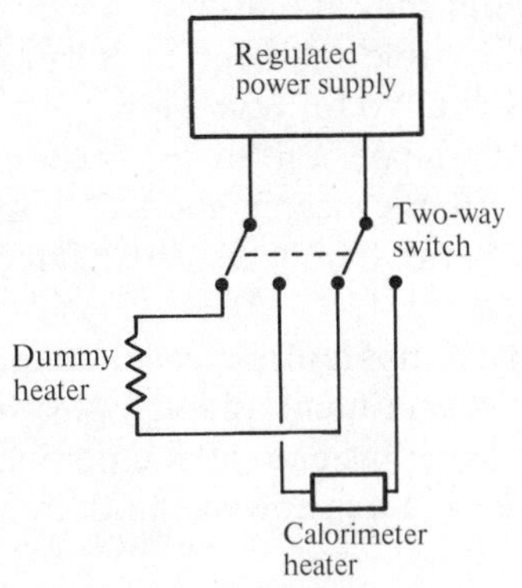

Fig. 3.5.2. The dummy heater arrangement

ter.* Adjust the temperature of this to between 24 and 25 °C by running hot or cold water over the outside of the measuring cylinder. Add the de-ionized water to the Dewar vessel.

Check that the solutions will not mix while the calorimeter is being assembled, and that the stirrer blades are rotating freely.

Assemble the calorimeter.

Connect the stabilized power supply and dummy heater as shown in Fig. 3.5.2. If a thermistor is used, connect it to its resistance bridge circuit and galvanometer.

Switch on the stirrer, and the thermistor circuit if used.* Leave both running throughout the experiment. Check that the change-over switch is connected to the dummy heater. Switch on the power supply to the dummy heater and adjust the current if necessary.*

(*iii*) *Temperature measurements*

First observe the temperature change in the calorimeter, due to heat transfer from or to its environment, by recording and plotting its temperature at 2-min intervals for 10 min. If a thermistor is used, measure its resistance to four significant figures. If using a thermometer, read it with a magnifying glass or reader, and tap it gently before each measurement. If after 10 min the temperature drift is greater than 0.005 K min^{-1} (or the corresponding change in thermistor resistance*), continue the observations until the drift settles within this range.

Operate the mixing device so that the solutions form a deep blue chelate compound.

Continue the observations at 2 min intervals until the temperature drift settles to within the stated range, showing that thermal equilibrium has again been reached. This will take at least 10–15 min.

Then carry out the calibration run with the electrical heater as follows. At the same moment as starting a timer, operate the change-over switch to turn on the heater. Continue the observations. Leave the heater on for the recommended time (typically 10 min) or until the recommended temperature rise occurs.* Note accurately the time when the heater is turned off with the change-over switch.

Continue the temperature measurements for another 15 min.

Repeat the experiment (paragraphs (ii) and (iii)) with 2 M 1,2-diaminoethane instead of ammonium hydroxide solution. Then carry out at least one of the determinations again.

(*iv*) *Shut down*

Switch off and disconnect all the electrical circuits. Wash out the calorimeter and associated apparatus with de-ionized water.

Calculation Draw graphs of increasing temperature or decreasing $\log_{10} R$ against time for each of the experiments. Find the initial and final temperatures from the intersections of straight lines drawn through the points as in Fig. 3.5.3. From these calculate the values of $\Delta T_{\text{reaction}}$ and ΔT_{heater}, or $(\Delta \log_{10} R)_{\text{reaction}}$ and $(\Delta \log_{10} R)_{\text{heater}}$. So find the enthalpies of formation of the cuprammonium and $[\text{Cu en}_2]^{2+}$ complex at the temperature of the experiment. Assume that this is equal to the standard enthalpy of formation at 25 °C.

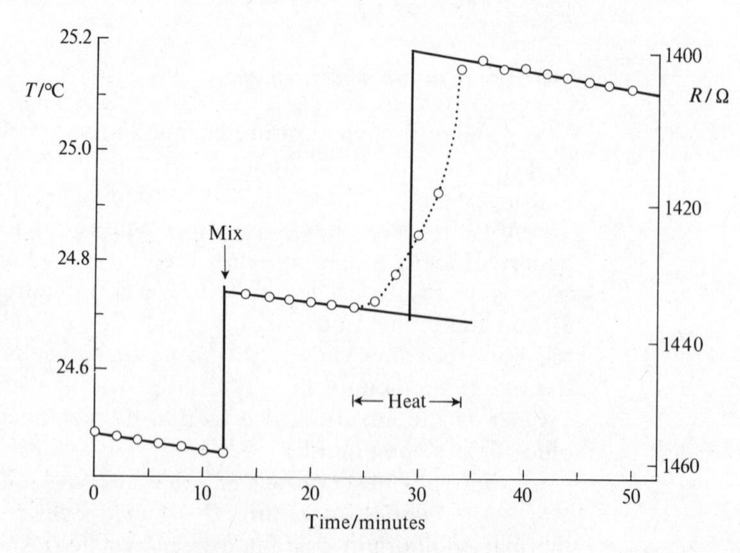

Fig. 3.5.3. Solution calorimeter heating curve

The stability constants of coloured complexes can also be measured spectroscopically (Expt 5.9).

The stability constants of the cuprammonium and $[Cu\,en_2]^{2+}$ complexes at 25 °C, in solutions of similar ionic strength to those used in the experiment, are given by $\log_{10} K^{\ominus} = 13.0$ and 20.0 respectively.[4] Use these data to calculate the free energies of the reactions and so find the entropy of formation of the complexes.

As an example of a different system, the value of $\Delta S^{\ominus}$ for the formation of $[Cu\,tn_2]^{2+}$ is given below. Explain why the entropies of formation become less negative and more positive through the series $[Cu(NH_3)_4]^{2+}$, $[Cu\,tn_2]^{2+}$, $[Cu\,en_2]^{2+}$.[5]

Results

(Sample = $[Cu\,tn]^{2+}$ where tn = $NH_2(CH_2)_3NH_2$).
Given: $\log_{10} K^{\ominus} = 17.2$.[4] Mix at 12 min. Heat from 24 min for 600 s exactly. $\langle V \rangle = 6.15$ V, $\langle I \rangle = 0.363$ A. Results shown in Fig. 3.5.3. Intercepts: $\Delta T_{reaction} = 24.842 - 24.547 = 0.295$ °C, $\Delta T_{heater} = 25.212 - 24.802 = 0.410$ °C; *or* $(\Delta \log_{10} R)_{reaction} = \log_{10} 1432 - \log_{10} 1458 = -0.0078$, $(\Delta \log_{10} R)_{heater} = \log_{10} 1401 - \log_{10} 1435.5 = -0.0106$. Equation (3.5.8): $\Delta H = -(0.295 \times 6.15 \times 0.363 \times 600)/(0.410 \times 0.01) = -96.4$ kJ mol^{-1}, *or* eqn (3.5.9): $\Delta H = (-0.0078 \times 6.15 \times 0.363 \times 600)/(0.0106 \times 0.01) = -98.6$ kJ mol^{-1}.

Precision

ΔT is measured to precision of about 2 per cent with the thermometer or thermistor.

Accuracy

If the experiment is carried out with sufficient care, and the temperature drift minimized by allowing a long enough time for the temperature to settle, the measurement of enthalpy using a precision thermometer or calibrated thermistor is capable of an accuracy approaching the precision of ±2 per cent. Use of the relation $\Delta T = -\Delta \log_{10} R$ can introduce an error of another 2 per cent. $\Delta S^{\ominus}$ is calculated from the small difference between the large terms $\Delta G^{\ominus}$ and $\Delta H^{\ominus}$ (eqn (3.5.11)). As a consequence, even the minimum 2 per cent error in $\Delta H^{\ominus}$ gives rise to an error in $\Delta S^{\ominus}$ very nearly as large as $\Delta S^{\ominus}$ itself. A feature of any quantity which is close to zero relative to its experimental error is that its value may not only be of entirely the wrong magnitude, but also of the wrong sign. Errors of this size also occur in published entropy data.[4]

Literature values for $[Cu\,tn]^{2+}$: $\Delta H^{\ominus} = -95.4$ kJ mol^{-1}, $\Delta S^{\ominus} = 8$ J K^{-1} mol^{-1}.[4]

Comment

This experiment illustrates the use of a solution calorimeter, which can be used for the determination of the enthalpies of dilution, mixing, precipitation, and a host of other reactions in solution. In the particular case of the formation of complexes, enthalpies can be measured with reasonable accuracy using the present apparatus. However, the entropies of formation are small by comparison and their uncertainties very large. Nevertheless it is the change in entropy, rather than enthalpy, which is the more interesting when complexes are studied. Thus even if the measured values are not very accurate, the correct trend mentioned previously is worthy of explanation and discussion.

Technical notes

Apparatus

Dewar vessel of about 500 cm^3 capacity; lid with mixing device, [funnel or syringe for filling mixing device], stirrer, heater (around 15 Ω), and thermistor in stainless steel can or precision thermometer,† stand allowing easy dismantling of Dewar from lid; 6 V 1/2 A stabilized power supply with accurate voltage and current meters;† change-over switch with dummy heater of same resistance (high wattage) as calorimeter heater;† resistance bridge for thermistor if required; copper(II) sulphate pentahydrate; 2 M solution of 1,2-diaminoethane; recently checked 4 M NH_4OH solution.

Note that a moderately priced commercial solution calorimeter is now available (Parr Model 1451).[6]

References

[1] *Textbook of pharmacology*. W. C. Bowman and M. J. Rand. Blackwell, Oxford, 2nd edn (1980).

[2] *Linearized thermistor temperature bridges for calorimetry*. S. R. Gunn. *J. Chem. Educ.* **50,** 515 (1973).

[3] *A solution calorimeter and thermistor bridge for undergraduate laboratories*. R. A. Bailey and J. W. Zubrick. *J. Chem. Educ.* **58,** 732 (1981).

[4] *Stability constants of metal-ion complexes*. L. G. Sillén and A. E. Martell. Chemical Society Special Publication No. 17 and No. 25, The Chemical Society, London (1964 and 1971).

[5] *Advanced inorganic chemistry*. F. A. Cotton and G. Wilkinson. Wiley Interscience, New York, 4th edn (1980), p. 71.

[6] *Solution calorimetry in the advanced laboratory*. R. W. Ramette, *J. Chem. Educ.* **61,** 76 (1984).

Section 4
Ions at equilibrium

This section comprises six experiments which involve the study of ionic solutions at equilibrium. Just as in previous sections, there are no measurements of times or rates; the properties under investigation depend on the steady ion concentrations at equilibrium, and may therefore be described by means of the law of mass action or by equating the electrochemical potentials of ions in different phases.

Experiment 4.1 introduces a direct method of studying the properties of ionic solutions, which is to measure the solubility of an ionic salt by a straightforward titration procedure. The properties are described either in terms of simple concentration equilibria, or by means of the more complicated Debye–Hückel theory of ionic solutions.

The most common and widely applicable method of investigating ionic solutions is to immerse reversible electrodes in them, and equilibrium electrochemistry experiments of this type form the remainder of the section. Experiment 4.2 introduces the theory of equilibrium electrochemistry, and also the fundamental experimental techniques such as the preparation of electrodes, the setting up of cells, and the measurement of e.m.f. The experiment then demonstrates how a number of thermophysical properties may be obtained from various electrochemical cells, and Expt 4.3 gives further examples based on a particularly useful cell first employed by Harned.

In Expt 4.4 the thermodynamic quantities associated with a cell reaction are found by measuring the extent to which the e.m.f. of the cell varies with temperature. In Expt 4.5 we turn to a redox system in which both the reactants and products are charged, and for which the e.m.f. is measured as a function of titrant volume. The system is characterized in terms of formal redox potentials, which take into account the non-ideality of the solution at the equivalence point due to the high concentration of ionic species which it contains.

Finally there is an experiment which introduces five different types of pH titration, carried out with a glass electrode. It also involves an additional experimental skill, because the first part of the exercise is to program in BASIC a microcomputer which is connected to a simple mechanical titrator. It is hoped by this means to provide an introduction to the microprocessor-controlled analytical instruments found in many industrial and research laboratories.

Although the ionic systems in this section are in equilibrium, the ions themselves are in motion, and the explicit study of ionic motion is the subject of the latter part of Section 7.

4.1 Solubility of ionic salt by titration

A convenient method of investigating the properties of ionic solutions is to measure the solubility of ionic salts under various conditions. As we might expect, the solubility of an ionic salt increases in the presence of a substance which combines with it, and this effect can be explained by the simple equilibrium theory of ionic solutions employed by Arrhenius in 1887. However, it also changes when a salt with a *common ion* is added to the solution. Furthermore, if a so-called *inert salt* containing entirely different ions is added, the solubility increases, a phenomenon known as *salting-in.* The Arrhenius theory gives a qualitative interpretation of the common ion effect, but not of salting-in. It is unable to account for either effect quantitatively, and so for this purpose we employ the Debye–Hückel theory formulated in 1923.

Often compounds of low solubility also have a low rate of solution; the great advantage of KIO_4 is that it is not only sparingly soluble, but also yields a saturated solution at 20 °C after only 2 min of shaking.

In this experiment we measure the change in solubility of potassium periodate (potassium iodate (VII), KIO_4) when common-ion or inert salts are added to the solution. We choose potassium periodate because it is *sparingly soluble* and therefore gives solubility changes which can be more readily measured than those of either an almost insoluble or very soluble salt. The concentrations of periodate in each saturated solution are obtained by using acidified iodide solution to reduce the IO_4^- to I_2,

$$\underset{\text{periodiate}}{IO_4^-} + 8H^+ + 7I^- \rightarrow 4I_2 + 4H_2O,$$

and then by titrating the iodine against sodium thiosulphate:

$$\underset{\text{thiosulphate}}{2S_2O_3^{2-}} \rightarrow \underset{\text{tetrathionate}}{S_4O_6^{2-}} + 2e^-,$$

$$I_2 + 2e^- \rightarrow 2I^-.$$

Theory

(*i*) *Simple ionic equilibria*

A simple example of an equilibrium in aqueous solution is a saturated solution of a slightly soluble 1:1 salt, MX, in contact with excess solid. The equilibrium may be written

$$MX(solid) \leftrightharpoons M^+(aq) + X^-(aq).$$

The equilibrium constant for this system is

$$K' = [M^+][X^-]/[MX(solid)], \qquad (4.1.1)$$

where the ionic concentration terms refer to aqueous species. However the concentration of all pure solids and liquids is constant, and therefore

$$K'[MX(solid)] = K_s' = [M^+][X^-]. \qquad (4.1.2)$$

K_s' is called the *concentration solubility product*, or simply the *solubility product*, of MX. Equation (4.1.2) indicates that *at any one temperature* the product of the concentrations of cations and anions in a saturated solution of a slightly soluble electrolyte is a constant.

The *solubility* s of a salt is defined as the amount of solid which can be dissolved in a stated amount of liquid at a stated temperature. For the salt MX, $[M^+] = [X^-] = s$, and therefore

$$K_s' = s^2. \qquad (4.1.3)$$

It is worth noting the corresponding equations for other types of electrolyte, even though we do not use them in this experiment. For example for a 2:1 electrolyte MX_2, such as $PbCl_2$, the dissociation equilibrium is

$$MX_2 \leftrightharpoons M^{2+} + 2X^-.$$

Since $[M^{2+}] = s$ and $[X^-] = 2s$, it follows that for a 2:1 electrolyte,

$$K_s' = [M^{2+}][X^-]^2 = s(2s)^2 = 4s^3. \qquad (4.1.4)$$

Another important phenomenon is the *common ion effect.* Suppose that our 1:1 electrolyte MX is dissolved in a solution which already contains a concentration x of X^- ions. The solubility of MX is now s', and the concentration of X^- ions is $(s' + x)$. Therefore

$$[M^+][X^-] = K_s' = s'(s' + x) = s^2. \qquad (4.1.5)$$

As x is non-zero, $s' < s$ and the solubility is reduced by addition of a common ion.

(*ii*) *Definitions*

Before proceeding to the Debye–Hückel theory, we must consider the definitions of various parameters associated with ionic solutions. All the equations we shall be using refer to an ionic solution at equilibrium, which implies that if solid is present the solution is saturated.

First we express the chemical potential μ_i of the ion i in an *ideal solution*, as

$$\mu_i = \mu_i^{\ominus} + RT \ln c_i, \tag{4.1.6}$$

where c_i is the molarity of the ion in the solution, and hence $\mu_i^{\ominus}$ is the standard chemical potential of the ion in a 1 M solution. However, this equation only holds near infinite dilution. When the ion i is surrounded by other ions, we need a more general equation

$$\mu_i = \mu_i^{\ominus} + RT \ln a_i, \tag{4.1.7}$$

where a_i is the *activity* of i. The concentration and activity are related by the *activity coefficient* γ_i:

$$a_i = \gamma_i c_i. \tag{4.1.8}$$

It follows that γ_i is unity in an ideal solution.

In an ionic solution, the charges on positive and negative ions balance. It is not possible to make separate measurements of the activity coefficient of each species in solution, so we define a *mean ionic activity coefficient* $\gamma_\pm$, which for a binary monovalent salt has the form:

$$\gamma_\pm = (\gamma_+\gamma_-)^{1/2} = (a_+a_-)^{1/2}/(c_+c_-)^{1/2} = a_\pm/c_\pm, \tag{4.1.9}$$

where γ_+ and γ_- are the ionic activity coefficients of the cations and anions, $a_\pm$ is the mean ionic activity and $c_\pm$ the mean ionic molarity.

Finally we define *ionic strength* I by the relation:

$$I = \tfrac{1}{2} \sum_i c_i z_i^2, \tag{4.1.10}$$

where z_i is the number of charges on the species i, and the summation is over *all* the ions in solution.

(*iii*) *Debye–Hückel theory*[1]

Debye and Hückel considered an ionic solution of point charges attracting each other electrostatically in a homogeneous, structureless solvent. A particular ion will tend to attract *counter-*

ions of opposite charge to form a charged, diffuse sphere known as its *ionic atmosphere*. These favourable ionic interactions occur for all the ions in the solution and lower their chemical potentials; their activities become less than their molarities, corresponding to a fractional value of $\gamma_{\pm}$ (or negative $\ln \gamma_{\pm}$). The effect increases with increasing charge on the ions and with increasing ionic strength. In the Debye–Hückel approach, the number of ions present in the ionic atmosphere at particular radius is calculated from the Boltzmann equation, and the electrostatic potential related to the charge distribution by Poisson's electrostatic equation. The energy of the system may then be calculated, finally leading to the *Debye–Hückel limiting law*:

$$\log_{10}\gamma_{\pm} = -A\,|z_+z_-|I^{1/2}, \qquad (4.1.11)$$

where z_+ and z_- are the charges on the ions of the dissolved salt, and $|\ |$ represents the modulus, i.e. the positive magnitude. The calculated value of the *Debye–Hückel constant* A is 0.509 for aqueous solutions at 25 °C.

For a 1:1 electrolyte eqn (4.1.11) simplifies to

$$\log_{10}\gamma_{\pm} = -As^{1/2}. \qquad (4.1.12)$$

Greater precision may be obtained by using a refinement to the Debye–Hückel limiting law sometimes referred to as the *complete Debye–Hückel equation* or the *Guntelberg equation*:

$$\log_{10}\gamma_{\pm} = -As^{1/2}/(1+s^{1/2}). \qquad (4.1.13)$$

(*iv*) *Solubility*[2]

We are now in a position to derive a more rigorous expression for solubility in ionic solutions. Consider once again the salt MX dissolving to form an aqueous solution of ions:

$$\text{MX(solid)} \rightleftharpoons \text{M}^+\text{(aq)} + \text{X}^-\text{(aq)}.$$

At equilibrium, when the solution has become saturated, the chemical potentials of the solid and dissociated ions must be equal so that

$$\mu_{\text{solid}} = \mu_+^{\ominus} + RT\ln a_+ + \mu_-^{\ominus} + RT\ln a_-. \qquad (4.1.14)$$

The chemical potential of the solid is taken as constant, and the standard chemical potentials $\mu_+^{\ominus}$ and $\mu_-^{\ominus}$ are also constants. This fixes the term $[RT\ (\ln a_+ + \ln a_-)]$. We define the *true thermodynamic solubility product*

$$K_s = a_+a_-, \qquad (4.1.15)$$

which thus is also constant at any particular temperature. This is a more exact expression than that for the concentration solubility product K_s', although the latter is often the more convenient quantity. For salts of infinitesimally small solubility, the interactions between the ions are minimal, so that $\gamma_\pm \approx 1$, and $K_s' \approx K_s$.

We have seen that for a binary monovalent salt dissolved in a solution containing no other ions,

$$s = c_+ = c_-. \tag{4.1.16}$$

Therefore in this case, from eqns (4.1.9) and (4.1.15),

$$s\gamma_\pm = K_s^{1/2}. \tag{4.1.17}$$

Apparatus

The experiment requires standard glassware, but no special equipment.

Procedure

Prepare 1.5 dm^3 of sodium thiosulphate solution of an accurately known concentration of about 0.1 M. The most reliable method is to dissolve fresh solid sodium thiosulphate, because stock solutions tend to change concentration on standing. Add a trace of sodium carbonate to raise the pH.

Dilute 300 cm^3 of the thiosulphate solution exactly twofold.

In each of eleven labelled and stoppered bottles, place about 1 g of potassium periodate in about 100 cm^3 of the following solvents: (a) distilled water, (b) 0.200 M $NaNO_3$, (c) 0.100 M $NaNO_3$, (d) 0.050 M $NaNO_3$, (e) 0.020 M $NaNO_3$, (f) 0.010 M $NaNO_3$, (g) 0.200 M KNO_3, (h) 0.100 M KNO_3, (i) 0.050 M KNO_3, (j) 0.020 M KNO_3, and (k) 0.010 M KNO_3.

Shake each bottle vigorously for at least 3 min and allow the solutions to settle. Check the temperature of each solution. Filter each solution through a dry filter paper and funnel into a dry, labelled flask. Do these operations on the same day, even if you cannot complete all the subsequent titrations at the same time.

As the end-point is approached, the periodate solution goes from red-brown to pale yellow to clear; the end-point is made clearer by adding starch solution to the periodate when it is pale yellow.

Titrate each of the solutions by the following procedure. To 20 cm^3 of the solution add about 2 g of solid KI and enough bench sulphuric acid to give a slight excess of H^+ ions (i.e. ~4 cm^3 of 1 M acid). Then titrate this solution against the sodium thiosulphate solution, using the iodine itself as indicator. At the end point, add more acid and check that there is no change in the end-point. Solutions containing I^- are slowly oxidized by oxygen to I_2 so the yellow colour may slowly return if the solution is left standing. Titrate solutions (a)–(f) against the more concentrated thiosulphate solution, and solutions (g)–(k) against the weaker solution. Obtain at least two concordant results for each titration.

Calculation Draw up a table showing the titres for each solution, the resulting estimates of the IO_4^- concentrations and total concentrations of K^+, and the concentration solubility products K_s' (eqns (4.1.2) and (4.1.5)).

Plot K_s' against $\sqrt{I}$ for each solution, remembering that the ionic strength includes *all* the ions in a particular solution.

The Debye–Hückel limiting law (4.1.11) predicts that $\gamma_\pm \to 1$ as $\sqrt{I} \to 0$, and consequently $K_s' \to K_s$ as $\sqrt{I} \to 0$. Draw a line through the points for solutions (a)–(f), and extrapolate it back to $\sqrt{I} = 0$ to obtain K_s. Also extrapolate the line for solutions (a), (g), (h), (i), (j), and (k) in the same way, to give what should be an identical value of K_s.

Calculate the value of $\gamma_\pm$ in each solution by means of eqn (4.1.17). Comment on the magnitude of the values you obtain.

Plot lines on the graph corresponding to the Debye–Hückel limiting law (eqn (4.1.11)) and the extended Debye–Hückel equation (4.1.13). How closely do your results agree with these theories? Comment on any discrepancies.

Discuss the effect on the solubility, concentration solubility product, and true thermodynamic solubility product of KIO_4 of the different concentrations of $NaNO_3$ and KNO_3. Explain how your experiments illustrate the common ion effect and salting-in.

Explain the phenomenon of salting-in in terms of the Debye–Hückel theory.

Results Concentration solubility product of KIO_4 at 25 °C = 4.9×10^{-4} mol^2 dm^{-6}, molar solubility = 0.022 mol dm^{-3}.[3] At 18 °C, molar solubility = 0.017 mol dm^{-3}. Typical values of $\gamma_\pm$ are ≈0.77 for 0.1 M solutions and ≈0.72 for 0.2 M solutions.[4]

Accuracy

The most usual cause of inaccuracy is careless experimental technique.

Comment This experiment provides a simple and instructive means of testing the Debye–Hückel theory of ionic solutions.

Technical notes

Apparatus.

Pipettes, burettes, twenty 100 cm^3 conical flasks, stoppers for flasks, filter funnel and filter papers. Potassium periodate, sodium carbonate, potassium iodate, potassium iodide, *either* fresh solid sodium thiosulphate *or* fresh stock solution of sodium thiosulphate ≥0.2 M labelled with exact concentration, 0.200 M sodium nitrate soln, 0.200 M potassium nitrate soln, starch indicator, bench sulphuric acid.

Further experiment

This experiment can also be used for demonstrating the variation of K_s with temperature. The variation of the Debye–Hückel constant A with temperature is given by $A = \text{const.}\ T^{-3/2}$.

References

[1] *Modern electrochemistry*, J. O'M. Bockris and A. K. N. Reddy. Macdonald, London (1970), vol. 1, ch. 3.
[2] *Ions in solution* 2. J. Robbins. Clarendon Press, Oxford (1972), ch. 1.
[3] *Solubilities of inorganic and metal organic compounds*, W. F. Linke, rev. A. Seidell, American Chemical Society, Washington D.C., 4th edn (1965), vol. II, p. 239.
[4] *The physical chemistry of electrolytic solutions*. H. S. Harned and B. B. Owen. Reinhold, New York, 3rd edn. (1958), p. 731.

4.2 Standard electrode potentials, solubility product, and water dissociation constant from e.m.f. measurements

An *electrochemical cell* is a device in which the free energy change accompanying a spontaneous chemical reaction is made available as a source of electrical work. The *maximum useful work* is only obtained from a cell if it is discharged infinitely slowly, i.e. with almost no current flowing. The electric potential difference under these conditions, known as the *electromotive force* or *e.m.f.*, can be measured with a *potentiometer*, or a *digital voltmeter* (*DVM*) with a very high internal resistance. In this experiment we set up a number of different electrochemical cells and measure their e.m.f.'s. From the e.m.f.'s we obtain a range of useful information, namely the standard electrode potentials of zinc, copper and silver, the solubility product of silver chloride and the dissociation constant of water.

The Theory and Apparatus sections which follow introduce the fundamental concepts and techniques of simple electrochemistry, which are required for the remaining experiments in this section and for Expts 7.6 and 7.7.

Theory[1,2]

(*i*) *Electrode potentials and the Nernst equation*

Suppose we dip a silver wire into a solution of silver nitrate. The silver wire is called an *electrode*—i.e. it is a body which conducts electricity dipping into an electrolyte solution which also conducts. An equilibrium will be set up between the silver ions in the solution, electrons from the metal electrode, and the silver electrode itself:

$$Ag^+ + e^- \leftrightharpoons Ag.$$

The equilibrium will be partly determined by the respective *chemical potentials* μ_{Ag^+}, μ_{e^-} and μ_{Ag}. The free energies of the silver ions and the electrons, although not of the uncharged metal atoms, will also be affected by the *electric potentials* of the solution ϕ_S and metal ϕ_M. At equilibrium the combination of the chemical and electric potential, known as the *electrochemical potential,* will be the same for the reactants and products. Using square brackets to indicate the electrochemical potential of each species, we may express this equality in the form:

$$[\mu_{Ag^+} + zF\phi_S] + [\mu_{e^-} - zF\phi_M] = [\mu_{Ag}]. \tag{4.2.1}$$

The factor zF has been introduced to convert the electric potentials to the free energies of the ions and electrons involved in the reduction of one mole of ions at a potential ϕ. The electrons' free energy at potential ϕ_M is negative because of their negative charge. If we call Avogadro's number of electrons a 'mole' of electrons, z is the number of moles of electrons which would be transferred in the reaction being considered. Therefore $z = 1$ for Ag^+. F is the *Faraday constant,* a positive quantity equal to the magnitude of the charge on a mole of electrons.

The *electrode potential* E_{Ag} is the potential difference between the metal and solution:

$$E_{Ag} = \phi_M - \phi_S. \tag{4.2.2}$$

The chemical potential μ of a species is related to its activity a by the equation

$$\mu = \mu^{\ominus} + RT\ln(a/a^{\ominus}), \tag{4.2.3}$$

In experimental work it is convenient to take $a^{\ominus} = 1$ mol dm^{-3}, although theoretical derivations tend to use $a^{\ominus} = 1$ mol kg^{-1}.

where $\mu^{\ominus}$ is the standard chemical potential at unit activity. $a^{\ominus}$ is the standard activity $= 1$ mol dm^{-3} or 1 mol kg^{-1}. In the equations which follow, all activities are assumed to be in the appropriate units, whereupon a is written as a dimensionless quantity and $a^{\ominus}$ omitted.

It follows from eqns (4.2.1), (4.2.2), and (4.2.3) that

$$E_{Ag} = \frac{1}{F}(\mu_{e^-} - \mu_{Ag} + \mu^{\ominus}_{Ag^+} + RT\ln a_{Ag^+}). \tag{4.2.4}$$

By convention, the chemical potential of an element in its standard state is zero, so $\mu_{Ag} = 0$. The *standard electrode potential* $E^{\ominus}_{Ag}$ between the electrode and solution is the potential difference when the reactants (and products) have unit activity.

Thus it follows that:

$$E^{\ominus}_{Ag} = \frac{1}{F}(\mu_{e^-} + \mu^{\ominus}_{Ag^+}), \qquad (4.2.5)$$

and

$$E_{Ag} = E^{\ominus}_{Ag} + \frac{RT}{F}\ln a_{Ag^+}. \qquad (4.2.6)$$

Equation (4.2.6) is the *Nernst equation* for the silver electrode, and relates the activity of the silver ions and the standard electrode potential to the electric potential difference. Unfortunately, however, the equation is unusable in its present form because the properties of a single ion cannot be measured directly. Electroneutrality dictates the presence of a counter ion, and the combined effects of ions and counter ions can only be differentiated by the use of a *model* of the system.

An example of such a model is the approximation applied to eqns (4.2.14) and (4.2.21), namely that the ionic activity coefficient of a cation or anion is equal to the geometric mean of the cationic and anionic activity coefficients.

A further problem is that a potentiometer or DVM connected between the silver electrode and solution will record an electric potential difference which is also dependent on the type of probe dipped into the solution to complete the circuit, so that there is no way of measuring the *half-cell potential* E_{Ag} independently. A practical solution to this problem is to measure the half-cell potential relative to the *standard hydrogen electrode.* We assign this a standard electrode potential $E^{\ominus}_{H_2}$ of zero, and standard conditions are employed, as explained overleaf, so that E_{H_2} is also zero. The electric potential difference E is then the difference between that of the silver and standard hydrogen electrode:

$$E = E_{Ag} - E_{H_2} = E_{Ag} = E^{\ominus}_{Ag} + \frac{RT}{F}\ln a_{Ag^+}. \qquad (4.2.7)$$

E is measured at negligible current flow, and is thus the e.m.f. of the cell.

The equation shows that a standard electrode potential may be determined by measuring the e.m.f. of a cell in which an electrode with its ions at unit activity is connected to a standard hydrogen electrode. Now we need to know what electrodes are available for constructing cells, how to connect them, and the sign conventions that are used for standard electrode potentials.

(*ii*) *Types of electrode*[3]

It cannot be too highly emphasized that electrodes used to determine standard electrode potentials must be *reversible.* The reversibility of an electrode relies on fast electrode kinetics

The requirement of reversibility determines, to a large extent, which electrodes can be employed in simple electrochemistry experiments; for example the Ag | Ag^+ electrode is usually reversible, whereas the Pb | Pb^{2+} electrode is usually not reversible.

compared to diffusion in solution, and the equilibrium theory employed in this experiment is only valid when this condition is fully satisfied.

There are five common types of electrode, all of which are reversible with certain substances under the right conditions. We shall use three of the types in this experiment.

Our silver wire dipping into a solution of silver nitrate is an example of a *metal | metal ion electrode* or *metal electrode.* The vertical line | indicates a phase boundary. In this form of electrode an equilibrium is set up between the metal and its ions in solution. It may be constructed by dipping a metal into a solution of one of its soluble salts. Sometimes instead of a pure metal an amalgam is used, as in the Weston standard cell described later and the Clark cell studied in Expt 4.4. In this case the metal is no longer at unit activity unless the amalgam is saturated.

Suppose that instead of dipping the silver wire into unsaturated silver nitrate solution, we dip it into a saturated solution of a very sparingly soluble ('insoluble') silver salt such as silver chloride, thus creating a *metal, insoluble salt* electrode. There will be the same equilibrium at the silver electrode:

$$Ag^+ + e^- \leftrightharpoons Ag,$$

together with the additional solubility equilibrium

$$AgCl \leftrightharpoons Ag^+ + Cl^-.$$

So the overall equilibrium is

$$AgCl + e^- \leftrightharpoons Ag + Cl^-.$$

Because the solution is saturated, the solubility product $a_{Ag^+} \cdot a_{Cl^-}$ (p. 117) will be constant, i.e. there will now be a fixed relationship between the activities of the silver ions and chloride ions. Considering the electrochemical equilibrium as before, it follows that the e.m.f. of this type of electrode can be expressed in terms of the activity of the *anion* according to the equation

$$E_{Ag,AgCl} = \frac{1}{zF}(\mu^{\ominus}_{AgCl} + \mu_{e^-} - \mu^{\ominus}_{Ag} - \mu^{\ominus}_{Cl^-} - RT \ln a_{Cl^-}) \qquad (4.2.8)$$

$$= E^{\ominus}_{Ag,AgCl} - \frac{RT}{F} \ln a_{Cl^-}. \qquad (4.2.9)$$

Because of this, the arrangement is also called an *anion reversible electrode.* It can be made by coating the electrode with a porous layer of the salt, thus ensuring that the solution next to the electrode is saturated. Alternatively a metal can be dipped

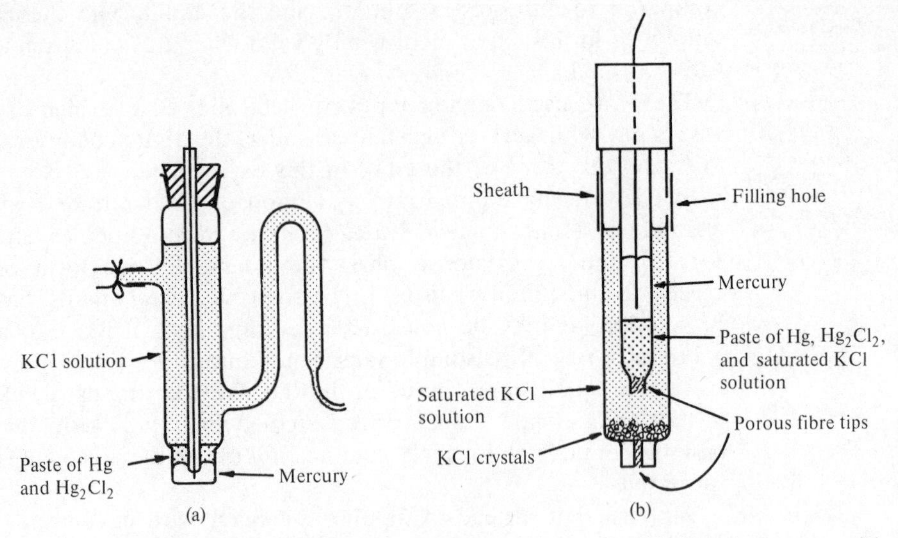

Fig. 4.2.1. Two forms of calomel electrode: (a) laboratory type, (b) commercial type shown saturated

into a saturated solution of one of its sparingly soluble salts, although kinetic considerations show that this is less likely to produce a reversible electrode.

Another important example of this type of electrode is the *calomel electrode.* Calomel is mercurous chloride, Hg_2Cl_2. The electrode is of the form Hg, $Hg_2Cl_2 \mid KCl$ with standard e.m.f. $E^{\ominus}_{cal}$. Two common designs are shown in Fig. 4.2.1. The electrode contains potassium chloride solution, and is really a half-cell rather than just an electrode. It can be dipped directly into solutions other than those such as silver-salt solutions which would block it with insoluble chloride. It is more convenient to use than the hydrogen electrode, and is therefore often employed as a *secondary standard electrode.* Standard conditions are achieved by using potassium chloride solution of known concentration; a saturated solution is easiest to prepare, but the resulting electrode has a higher temperature coefficient than electrodes containing more dilute solutions, Table 4.2.1.

Table 4.2.1. E.m.f. of a calomel electrode at temperature T °C[4]

KCl concentration	E.m.f.
saturated	$0.244-0.0007(T-25)$
1 M	$0.283-0.0002(T-25)$
0.1 M	$0.336-0.00006(T-25)$

Early values of $E^{\ominus}_{cal}$ differ by up to 4 mV from those in current usage.

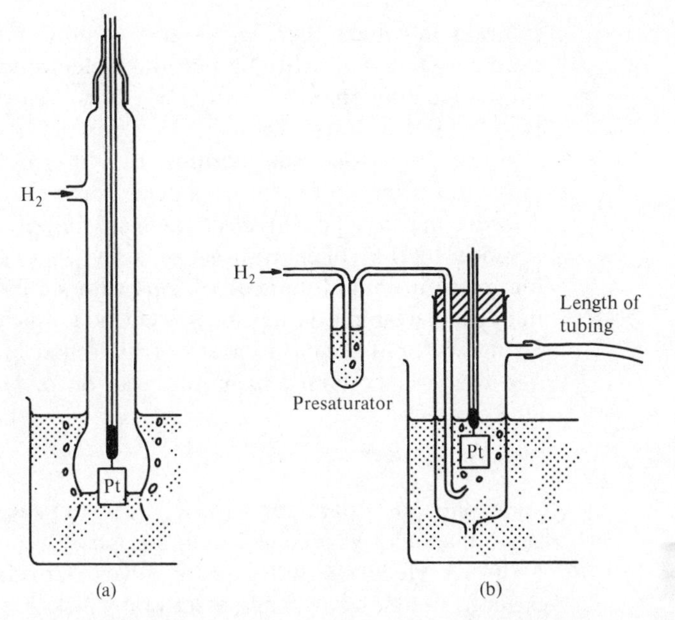

Fig. 4.2.2. Two simple forms of dipping hydrogen electrode: (a) the classical Hildebrand electrode, (b) electrode modified to reduce access of oxygen to electrode surface, and incorporating simple hydrogen presaturator

The hydrogen electrode is the chief example of an *inert metal | gas electrode*, or *gas electrode*. These electrodes are constructed so that a gas bubbles around an inert metal electrode while it is also surrounded by a solution containing ions related to the gas. In the case of the hydrogen electrode (Fig. 4.2.2), hydrogen gas is bubbled round a platinum electrode dipped into an acid, resulting in the equilibrium

$$H^+ + e^- \leftrightharpoons \tfrac{1}{2}H_2,$$

for which

$$E = E^{\ominus}_{H_2} + \frac{RT}{F} \ln \frac{a_{H^+}}{a_{\frac{1}{2}H_2}}. \tag{4.2.10}$$

Standard conditions are achieved by using an acid solution in which the activity of the protons is unity, and bubbling the gas at 1 atm pressure (strictly unit fugacity). Then $E = E^{\ominus}_{H_2} = 0$. Once again, the kinetics of the electrode process must be borne in mind—the corresponding $Pt \mid O_2$ electrode, for example, is not reversible.

A *redox electrode* is an inert metal electrode in contact with a solution in which there is a redox equilibrium (p. 146)—for example it may be bright platinum electrode dipping into a solution containing Fe^{2+} and Fe^{3+} ions. An electrode of this type is used in Expt. 4.5.

A *membrane electrode* monitors the differing activities of ions in solutions on either side of a semi-permeable membrane. *Ion selective* or *specific ion electrodes* are all of this form. They respond to the concentrations or activities of a single type of ion, or to a limited number of ions with similar properties. An important example is the glass electrode which is employed in Expt 4.6 for the measurement of hydrogen ion concentration, and which is explained in more detail on p. 152.

(*iii*) *Salt bridges*

As explained in paragraph (i), we need to use two electrodes to set up an electrochemical cell. Suppose we choose to use a hydrogen electrode and a silver, silver chloride electrode. The latter is an anion reversible electrode which must be dipped into a solution containing chloride ions, whereas the hydrogen electrode requires a solution containing hydrogen ions. Therefore both electrodes can be dipped into a beaker of hydrochloric acid. We write the cell:

$$\mathrm{Pt}\,|\,\mathrm{H_2}\,|\,\mathrm{HCl}\,|\,\mathrm{AgCl, Ag}.$$

This is known as a *Harned cell*, and many useful measurements can be made with it (Expt 4.3).

However, we might wish to use a standard hydrogen electrode with H^+ ions of unit activity, while still being able to alter the chloride ion concentration of the anion reversible electrode:

$$\mathrm{Pt}\,|\,\mathrm{H_2}\,|\,\mathrm{HCl}(a_{\mathrm{H^+}} = 1)\,|\,\mathrm{HCl}\,|\,\mathrm{AgCl, Ag}.$$

This is a *cell with liquid junction*, whereas the previous example was a *cell without liquid junction*. If we are unable to avoid the use of a cell with a liquid junction, then we must solve the problem of how to link the two solutions. We cannot use a wire, because it would affect the e.m.f. A possible but rather inconvenient method would be to use a *flowing junction*, in which the two solutions are poured together steadily and the mixture discarded. Alternatively we could insert a porous disc between the two solutions to allow transport of ions while hindering ordinary diffusion. Both arrangements, however, give rise to *liquid junction potentials* (Expt 7.7) which affect the e.m.f. of the cell.[5] These potentials may be minimized by using a *salt bridge*. This is a bridge of electrolyte solution, held in place by some device as for example in Fig. 4.2.3. Saturated potassium

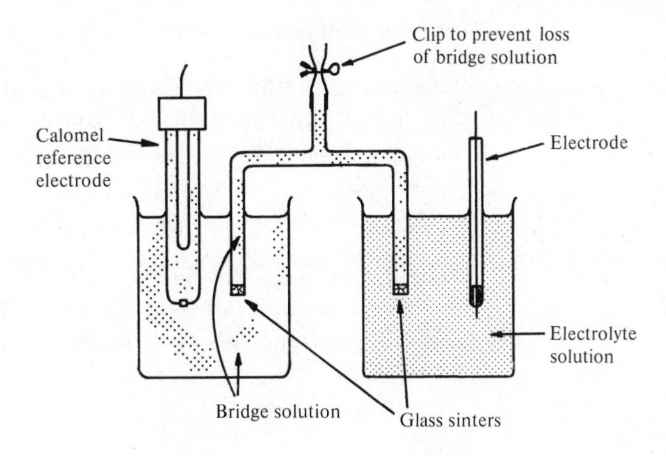

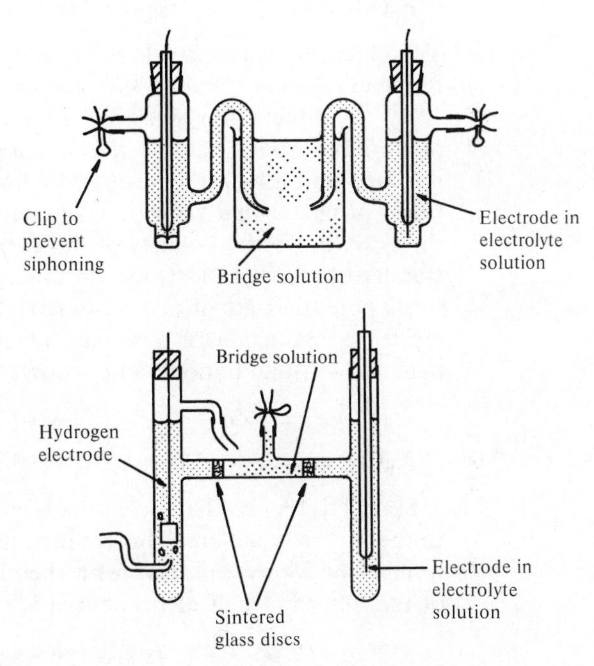

Fig. 4.2.3. Three typical cells with liquid junctions

chloride solution gives rise to the lowest liquid junction potentials, and ammonium nitrate or potassium nitrate can also be used. *For which electrodes would a potassium chloride salt bridge be unsuitable*? When such an arrangement is used, it is obviously essential that the liquid levels on each side are the same so that no siphoning occurs. If siphoning cannot be prevented in any other way, a gel link may be used which contains potassium chloride dissolved in agar jelly.[6] When liquid junction potentials are avoided by the use of salt bridges, the junction in the cell is written with a double line $\|$.

(iv) Sign convention

The polarity of the electrodes in a particular electrochemical cell is found experimentally, or from the electrochemical series. The cell is then written with the positive electrode on the right, for example:

$$-\mathrm{Zn}\,|\,\mathrm{ZnSO_4}\,\|\,\mathrm{HCl}\,|\,\mathrm{AgCl, Ag}+.$$

The directions of the chemical changes may be deduced from the observed polarity of the electrodes. Thus in the present case the equilibria (reactants ⇋ products), written for the transfer of one electron, are

$$\text{Left:}\quad \tfrac{1}{2}\mathrm{Zn} \leftrightharpoons \tfrac{1}{2}\mathrm{Zn^{2+}} + \mathrm{e^-},$$

$$\text{Right:}\quad \mathrm{AgCl} + \mathrm{e^-} \leftrightharpoons \mathrm{Ag} + \mathrm{Cl^-}.$$

In an electrochemical (galvanic) cell the potential of the cathode is more positive than that of the anode, but in an electrolytic cell, the anode is more positive.[7]

The electrode where oxidation occurs is the *anode*: that where reduction occurs is called the *cathode.*

By the Stockholm sign convention, the standard electrode potential is the e.m.f. of a cell comprising the electrode in question and a hydrogen electrode, both under standard conditions, provided that there is no junction potential. The sign of the potential is the polarity of the electrode with respect to the standard hydrogen electrode. In this convention, standard electrode potentials are always written as the potentials corresponding to the *reduction* of a species by one electron. Thus for the negative left-hand anode and positive right-hand cathode:

$$\tfrac{1}{2}\mathrm{Zn^{2+}} + \mathrm{e^-} \leftrightharpoons \tfrac{1}{2}\mathrm{Zn} \qquad E_L^{\ominus} = -0.763\ \mathrm{V},$$

$$\mathrm{AgCl} + \mathrm{e^-} \leftrightharpoons \mathrm{Ag} + \mathrm{Cl^-} \qquad E_R^{\ominus} = +0.222\ \mathrm{V}.$$

So $E_R^{\ominus}$ refers to the correct half-cell reaction, but $E_L^{\ominus}$ refers to the reverse reaction. This is taken into account in the *general form of the Nernst equation* for a chemical cell, in which the sign of the left-hand half-cell potential is changed:

$$E = E_R^{\ominus} - E_L^{\ominus} + \frac{RT}{zF}\ln\left(\frac{a_{\text{react}\,1}\cdot a_{\text{react}\,2}\cdot\ldots}{a_{\text{prod}\,1}\cdot a_{\text{prod}\,2}\cdot\ldots}\right). \qquad (4.2.11)$$

Therefore the Nernst equation for the present example is:

$$E = 0.222 + 0.763 + \frac{RT}{F}\ln\frac{(a_{\mathrm{Zn}})^{1/2}\cdot a_{\mathrm{AgCl}}\cdot a_{\mathrm{e^-}}}{(a_{\mathrm{Zn^{2+}}})^{1/2}\cdot a_{\mathrm{e^-}}\cdot a_{\mathrm{Ag}}\cdot a_{\mathrm{Cl^-}}}$$

$$= 0.985 + \frac{RT}{F}\ln\left(\frac{1}{(a_{\mathrm{Zn^{2+}}})^{1/2}\cdot a_{\mathrm{Cl^-}}}\right)$$

$$= 0.985 - \frac{RT}{2F}\ln(a_{\mathrm{Zn^{2+}}}\cdot a_{\mathrm{Cl^-}}^2). \qquad (4.2.12)$$

(*v*) *Calculations from cell e.m.f.s*

We are now in a position to derive some electrochemical and physical constants from the e.m.f.s of various electrochemical cells.

First consider the cell:

$$\underbrace{-\mathrm{Hg}, \mathrm{Hg_2Cl_2} \,|\, \mathrm{KCl}}_{\text{standard calomel electrode}} \,\|\, 0.1\ \mathrm{M}\ \mathrm{AgNO_3} \,|\, \mathrm{Ag}+.$$

The e.m.f. of the cell is:

$$E = E_{\mathrm{Ag}} - E^{\ominus}_{\mathrm{cal}} = E^{\ominus}_{\mathrm{Ag}} - E^{\ominus}_{\mathrm{cal}} + \frac{RT}{F} \ln a_{\mathrm{Ag}^+}. \tag{4.2.13}$$

By definition

$$a_{\mathrm{Ag}^+} = c_{\mathrm{Ag}^+}\, \gamma_{\mathrm{Ag}^+} \approx c_{\mathrm{Ag}^+}\, \gamma_{\pm \mathrm{AgNO_3}}, \tag{4.2.14}$$

where γ_{Ag^+} is the activity coefficient (p. 116) of the silver ions in 0.1 M $AgNO_3$ solution, and is assumed to be equal to $\gamma_{\mathrm{NO_3}^-}$ so that $\gamma_{\mathrm{Ag}^+} = \gamma_{\pm \mathrm{AgNO_3}}$. Then

$$E = E^{\ominus}_{\mathrm{Ag}} - E^{\ominus}_{\mathrm{cal}} + \frac{RT}{F} \ln(c_{\mathrm{Ag}^+}\, \gamma_{\pm \mathrm{AgNO_3}}). \tag{4.2.15}$$

Thus $E^{\ominus}_{\mathrm{Ag}}$ may be obtained from a measurement of E, provided that $E^{\ominus}_{\mathrm{cal}}$ and γ_{Ag^+} in 0.1 M $AgNO_3$ solution are also known.

Now consider the cell

$$-\mathrm{Ag}, \text{satd}\ \mathrm{AgCl} \,|\, 0.1\ \mathrm{M}\ \mathrm{KCl} \,|\, \underset{\text{salt bridge}}{\text{satd}\ \mathrm{KNO_3}} \,|\, \underbrace{\mathrm{KCl} \,|\, \mathrm{Hg_2Cl_2}, \mathrm{Hg}}_{\text{standard calomel electrode}}+$$

The e.m.f. of this cell is

$$\begin{aligned} E &= E^{\ominus}_{\mathrm{cal}} - E_{\mathrm{Ag,AgCl}} \\ &= E^{\ominus}_{\mathrm{cal}} - E^{\ominus}_{\mathrm{Ag,AgCl}} + \frac{RT}{F} \ln a_{\mathrm{Cl}^-} \\ &= E^{\ominus}_{\mathrm{cal}} - E^{\ominus}_{\mathrm{Ag,AgCl}} + \frac{RT}{F} \ln c_{\mathrm{Cl}^-} \gamma_{\mathrm{Cl}^-}. \end{aligned} \tag{4.2.16}$$

So $E^{\ominus}_{\mathrm{Ag,AgCl}}$ can be found if $E^{\ominus}_{\mathrm{cal}}$ and γ_{Cl^-} in 0.1 M KCl are known.

The magnitude of the term $E^{\ominus}_{\mathrm{Ag,AgCl}} - E^{\ominus}_{\mathrm{Ag}}$ may be obtained from this and the previous measurement, or by setting up a cell to measure it directly. We may then obtain $K_{\mathrm{S}} = a_{\mathrm{Ag}^+} \cdot a_{\mathrm{Cl}^-}$ as follows. It has been shown in eqn (4.2.5) that

$$E^{\ominus}_{\mathrm{Ag}} = \frac{1}{F} (\mu^{\ominus}_{\mathrm{Ag}^+} + \mu_{\mathrm{e}^-}),$$

and it follows from eqn (4.2.8) that

$$E^{\ominus}_{\mathrm{Ag,AgCl}} = \frac{1}{F}(\mu^{\ominus}_{\mathrm{AgCl}} + \mu_{\mathrm{e^-}} - \mu^{\ominus}_{\mathrm{Cl^-}}).$$

In the silver, silver chloride electrode, $\mu^{\ominus}_{\mathrm{AgCl}}$ is the same as μ_{AgCl}, because the silver chloride is a pure solid and thus at its standard state. The solid is also in equilibrium with the dissolved ions, so that

$$\mu^{\ominus}_{\mathrm{AgCl}} = \mu_{\mathrm{AgCl}} = \mu_{\mathrm{Ag^+}} + \mu_{\mathrm{Cl^-}}. \tag{4.2.17}$$

Then

$$\begin{aligned} E^{\ominus}_{\mathrm{Ag,AgCl}} - E^{\ominus}_{\mathrm{Ag}} &= \frac{1}{F}(\mu_{\mathrm{Ag^+}} + \mu_{\mathrm{Cl^-}} + \mu_{\mathrm{e^-}} - \mu^{\ominus}_{\mathrm{Cl^-}} - \mu^{\ominus}_{\mathrm{Ag^+}} - \mu_{\mathrm{e^-}}) \\ &= \frac{1}{F}(\mu_{\mathrm{Ag^+}} + \mu_{\mathrm{Cl^-}} - \mu^{\ominus}_{\mathrm{Cl^-}} - \mu^{\ominus}_{\mathrm{Ag^+}}) \\ &= \frac{RT}{F}(\ln a_{\mathrm{Ag^+}} + \ln a_{\mathrm{Cl^-}}) \\ &= \frac{RT}{F}\ln K_{\mathrm{S}}. \end{aligned} \tag{4.2.18}$$

The solubility S of silver chloride is not simply the square root of the solubility product; how is it related?

The third cell we consider is

$$\mathrm{Pt}\,|\,\mathrm{H_2(1\ atm)}\,|\,0.1\ \mathrm{M\ HCl}\,|\,\underset{\text{salt bridge}}{\text{satd KCl}}\,|\,0.1\ \mathrm{M\ NaOH}\,|\,\mathrm{H_2(1\ atm)}\,|\,\mathrm{Pt}.$$

The half-cell reactions at the two hydrogen electrodes are

Left: $\tfrac{1}{2}\mathrm{H_2(1\ atm)} \leftrightharpoons \mathrm{H^+(0.1\ M\ HCl)} + \mathrm{e^-}$,

Right: $\mathrm{H^+(0.1\ M\ NaOH)} + \mathrm{e^-} \leftrightharpoons \tfrac{1}{2}\mathrm{H_2(1\ atm)}$.

The Nernst equation is therefore:

$$E = E^{\ominus}_{\mathrm{R}} - E_{\mathrm{L}} + \frac{RT}{zF}\ln\frac{a_{\mathrm{H^+}}(0.1\ \mathrm{M\ NaOH}) \cdot a_{\frac{1}{2}\mathrm{H_2}}(1\ \mathrm{atm})}{a_{\frac{1}{2}\mathrm{H_2}}(1\ \mathrm{atm}) \cdot a_{\mathrm{H^+}}(0.1\ \mathrm{M\ HCl})}. \tag{4.2.19}$$

Since the two electrodes are the same, $E^{\ominus}_{\mathrm{R}} = E^{\ominus}_{\mathrm{L}}$, and

$$E = \frac{RT}{F}\ln\frac{a_{\mathrm{H^+}}(0.1\ \mathrm{M\ NaOH})}{a_{\mathrm{H^+}}(0.1\ \mathrm{M\ HCl})}. \tag{4.2.20}$$

The dissociation constant of water K_{W} is $a_{\mathrm{H^+}} \cdot a_{\mathrm{OH^-}}$. So in the present case,

$$\begin{aligned} E &= \frac{RT}{F}\ln\frac{K_{\mathrm{W}}}{a_{\mathrm{H^+}}(0.1\ \mathrm{M\ HCl}) \times a_{\mathrm{OH^+}}(0.1\ \mathrm{M\ NaOH})} \\ &= \frac{RT}{F}\ln\frac{K_{\mathrm{W}}}{0.1 \times \gamma_{\mathrm{H^+}}(0.1\ \mathrm{M\ HCl}) \times 0.1 \times \gamma_{\mathrm{OH^-}}(0.1\ \mathrm{M\ NaOH})}. \end{aligned} \tag{4.2.21}$$

Thus if we assume that $\gamma_{H^+} = \gamma_{\pm}(0.1\ \text{M HCl})$ and that $\gamma_{OH^-} = \gamma_{\pm}(0.1\ \text{M NaOH})$, then the dissociation constant of water may be calculated from the e.m.f. of the cell and the mean ionic activity coefficients of 0.1 M hydrochloric acid and sodium hydroxide.

Apparatus

(*i*) *Electrodes*

Metal electrodes are usually made from a wire, gauze or spade of the metal fixed through the end of a glass tube. If the metal is expensive, the electrode is joined to a different wire inside the tube, by soldering where possible or with a drop of mercury. The metal can also be plated onto a different substrate, for example silver can be electroplated onto platinum. Silver, copper and zinc electrodes are used in this experiment.

Why is it not practical to make a silver electrode by plating silver onto copper?

A standard calomel electrode (Fig. 4.2.1) is supplied. The strength of the potassium chloride solution makes a substantial difference to the standard electrode potential, and it is essential that it is known accurately. If a saturated solution is used, excess crystals of potassium chloride must be visible. Calomel electrodes with porous plugs are stored in a test-tube containing electrolyte of the correct strength.

The setting up of the silver, silver chloride and hydrogen electrodes is described in the *Procedure* section.

(*ii*) *Measurement of e.m.f.*

The traditional way of measuring e.m.f. is with a potentiometer and standard cell. The simplified circuit diagram of a commercial potentiometer is shown in Fig. 4.2.4. As can be seen, it uses four electrical components in which voltage is applied across a

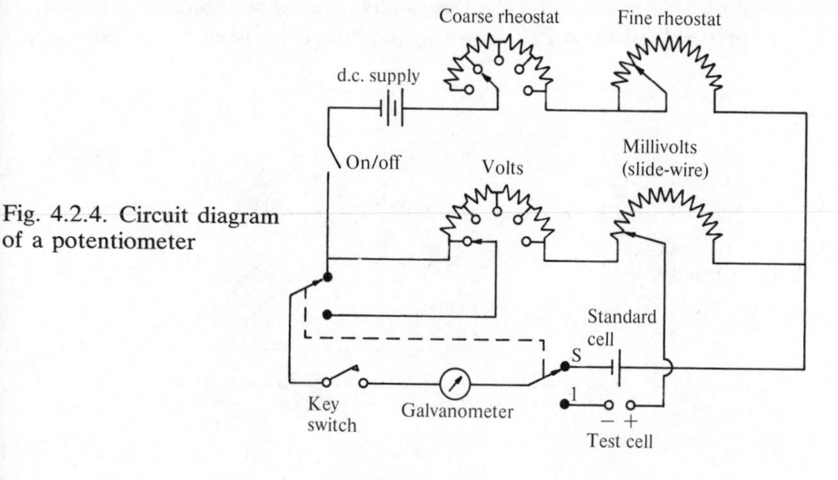

Fig. 4.2.4. Circuit diagram of a potentiometer

resistance wire or a series of resistors, shown as zig-zags. The voltage can be tapped either by changing a switch position, or by sliding a contact over the resistance wire. These components are themselves known as potentiometers, but to avoid confusion we shall refer to them by their older name of rheostat.

If the galvanometer, d.c. supply or standard cell are external, they are first connected to the potentiometer, with the correct polarity. The selector switch is then switched to the 'S' position which connects the standard cell across the central resistors. The d.c. supply is switched on, and made to balance accurately the standard voltage by adjustment of the coarse and fine rheostat controls. The balance point occurs when there is no galvanometer deflection on momentary depression of the key switch. The selector is then turned to position '1'. The test cell is then balanced against the adjusted supply voltage by operation of the volts and millivolts rheostat controls and the key switch, *while the coarse and fine rheostat controls remain untouched.* The volts and millivolts rheostats are calibrated so that the voltage can be read directly once the balance point is obtained. The standard cell is usually a 'Weston standard cell':

Cd(Hg),	$CdSO_4$	| satd $CdSO_4$ soln |	Hg_2SO_4,	Hg
12.5 per cent cadmium amalgam	hydrated cadmium sulphate crystals		mercurous sulphate paste	

This cell has the advantage of giving a stable e.m.f. with only a small temperature coefficient. Various modifications are possible, and the cell should therefore be labelled with its accurate e.m.f.

Another method of measuring voltages is with a digital voltmeter. These generally work by the 'dual slope integration' principle, Fig. 4.2.5. The input voltage E_i is made to charge a

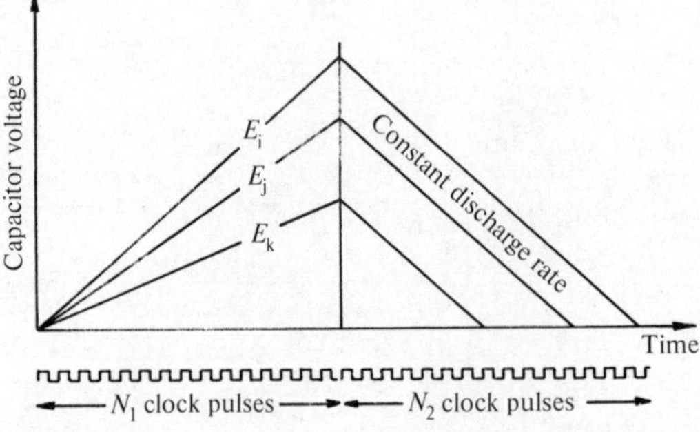

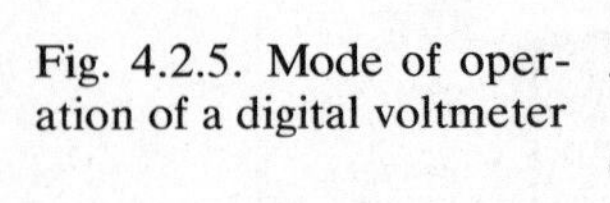
Fig. 4.2.5. Mode of operation of a digital voltmeter

Ordinary 'multimeter' digital voltmeters draw sufficient current (typically 10^{-7} A) to cause significant errors in the measurement of cell e.m.f.s; they should only be used if an op-amp voltage follower is connected between the cell and the DVM to reduce the current to acceptable levels.

capacitor through an operational amplifier in an integrating configuration. The charging occurs for a fixed time of N_1 internal-clock pulses so that the final voltage across the capacitor is dependent on the initial voltage, as shown. The input of the amplifier is then connected to a reference voltage E_{ref} of opposite polarity, which causes the capacitor to discharge at a rate proportional to the reference voltage. The time for complete discharge, N_2 clock pulses, is then measured. Provided the operational amplifier is linear, it can be shown quite simply that $E_i/E_{ref} = N_2/N_1$. With the correct choice of E_{ref} and N_1, N_2 can be displayed directly as a voltage even though it is in fact a time measurement. The reference voltage is supplied by a solid–state band-gap device mounted in a circuit which gives it a negligible temperature coefficient.

Potentiometers are still used in research for the most accurate e.m.f. measurements. DVMs have the advantage that they provide a rapid, continuous read-out and can be used for computer-assisted data acquisition.

(*iii*) *The cell vessel*

The simplest cell vessel is a small beaker into which both electrodes dip. The beaker may be supplied with a lid to support them. If a salt bridge is necessary, it can be placed between two beakers with an electrode in each, or two half-cells can be dipped into a beaker of bridge solution, Fig. 4.2.3. The third type of cell vessel shown in the figure is in the form of a glass H-tube, with one electrode corked into each arm. A liquid junction may be formed either side of a sintered glass disc fitted into the crosspiece, or bridge solution may be inserted between two discs as shown. H-tubes are particularly suitable for use with hydrogen electrodes, which work best if enclosed. A thermometer should be mounted in or near the cell.

Procedure

All electroplating operations and handling of concentrated acids must be carried out in a fume cupboard. Take care with hydrogen gas, and do not expose it to sparks or naked flames.

(*i*) *Preparation of electrodes*

The electrical characteristics of all electrodes depend crucially on the exact state of their surfaces; great care must be taken in preparing and setting them up.

The activity of a metal depends on whether or not it is mechanically stressed, and standard e.m.f.s are usually quoted for polished surfaces. Simple wire electrodes should therefore be rubbed with fine emery paper before use. (Other types of electrode are too difficult to polish, or may be damaged by the process.) It is also necessary, of course, to ensure that the electrodes are entirely clean. Freshly polished and cleaned surfaces are reactive, and so must not be exposed to air. Unplated electrodes should be used immediately, while electroplated electrodes benefit from standing in purified water for a while to leech out any remaining plating solution which might alter the potential.

Clean one copper and two silver wire electrodes by rubbing them gently with emery paper until the entire surface is shiny, dipping them into dilute nitric acid for a moment, washing thoroughly with water, and then rinsing with the solution to be used in the cell. Clean a zinc electrode in the same way but without dipping into nitric acid.

Prepare a coated silver,silver chloride electrode by first soaking one of the cleaned silver electrodes in concentrated ammonium hydroxide solution in a fume cupboard for at least 5 min. Then wash it very thoroughly with purified water. Place a small beaker of 0.1 M hydrochloric acid on an electric stirrer in a fume cupboard. Immerse in it the silver electrode as anode and a platinum or other inert electrode as cathode. (Remember the difference between the cathode and anode in an electrolytic and galvanic cell.) Estimate the surface area of the cathode in cm^2 and pass a current of the order of $0.4\,mA\,cm^{-2}$, or 2 mA if greater, for 30 min. Stir the solution *slowly*, making sure that the stirrer flea is well away from the electrodes. Remove the silver, silver chloride electrode, wash it very thoroughly in purified water, and store in purified water away from strong sunlight.

Before the two platinum | hydrogen electrodes are prepared, they must first be cleaned in aqua regia made from three volumes of concentrated hydrochloric acid and one volume of concentrated nitric acid. There is no need to dismantle platinum electrodes if they are already mounted in their glass sheaths. Wash the electrodes thoroughly in purified water. Estimate the *total* surface area of the electrodes and electrolyse them in platinizing solution at a current density of the order of $15\,mA\,cm^{-2}$ or 2 mA, whichever is the greater. Every ten minutes or so, reverse their polarity. The deposition of platinum gives rise to a hardly visible grey film, which may go darker with time. The film should appear after about 30 min electrolysis of each electrode. Provided that the platinizing solution has not been contaminated, return it to its bottle.

Set up one hydrogen electrode in a small beaker of 0.1 M hydrochloric acid and the other in an adjacent small beaker of 0.1 M sodium hydroxide. Bubble presaturated hydrogen through

Hydrogen electrodes are particularly sensitive to poisoning and to the presence of oxidizable species such as dissolved oxygen; the e.m.f.s of contaminated hydrogen electrodes are very dependent on the rate of hydrogen bubbling and may not reach equilibrium.

each electrode for 10 min to remove oxygen from the solutions. (Ideally the hydrogen should be passed through several wash bottles to remove oxygen and wash it—in practice it is sufficient to presaturate the hydrogen with some of the solution from the cell, to prevent it from changing the concentration of the solution in the cell itself.) Turn down the bubbling rate before making measurements.

(ii) Measurement of emfs

Set up the following cells, using the prepared hydrogen electrodes and salt bridges of the correct type where appropriate. Connect a potentiometer as instructed,* or a DVM.

Find the true cell e.m.f.s by taking readings every 2 min for at least 10 min and until the drift is less than 2 mV min^{-1}. If using a potentiometer, restandardize it frequently against the standard cell. If the e.m.f. of a cell has not settled in 30 min, the cell should be dismantled and the electrodes freshly prepared. Note the temperature of each cell.

The cells to be studied are:

$Zn \mid 0.1\,M\,ZnSO_4 \mid KCl \mid Hg_2Cl_2, Hg$

$Hg, Hg_2Cl_2 \mid KCl \mid 0.1\,M\,CuSO_4 \mid Cu$

$Hg, Hg_2Cl_2 \mid KCl \parallel 0.1\,M\,AgNO_3 \mid Ag$

$Ag, AgCl \mid 0.1\,M\,KCl \parallel KCl \mid Hg_2Cl_2, Hg$

$Pt \mid H_2 \mid 0.1\,M\,HCl \parallel 0.1\,M\,NaOH \mid H_2 \mid Pt.$

Calculation

Given the e.m.f.s of calomel electrodes listed in Table 4.2.1, and the mean ionic activity coefficients shown in Table 4.2.2, use your results to calculate the standard electrode potentials of zinc, copper and silver, the solubility product and solubility of silver chloride, and the dissociation constant of water.

Table 4.2.2. Mean ionic activity coefficients $\gamma_{\pm}$ of 0.1 molal (assumed 0.1 molar) salt solutions at 25 °C[8]

$ZnSO_4$	0.16	KCl	0.77
$CuSO_4$	0.15	HCl	0.80
$AgNO_3$	0.73	NaOH	0.77

Results

Accuracy

The accuracies of the results of this experiment are dictated by the care with which the cells are prepared and their e.m.f.s measured. Results to within a few mV of the correct e.m.f.'s can be achieved, but careless experimentation will cause errors up to 0.3 V, with corresponding discrepancies from the well-known literature values of the derived electrochemical and physical constants.

Comment

We have seen how standard electrochemical potentials and dissociation constants may be measured, while using literature values of the relevant mean ionic activity coefficients. The next experiment demonstrates the determination of similar quantities without the use of additional data.

In this and the remaining experiments in this section, electrochemical cells, and the basic methods of measuring their e.m.f.s, are adequately described in terms of equilibrium electrochemistry. Over the last 20 years, however, attention has focussed on the problems of electrode kinetics and interfacial structure. The investigation of these effects requires much more powerful theories which involve quantum mechanical and kinetic arguments, and which describe the ionic interactions at the surface of an electrode in terms similar to those of the Debye–Hückel theory.[9,10] Some of these theories are employed in the study of liquid junction potentials in Expt 7.7.

Technical notes

Apparatus

Zinc electrode;† copper electrode;† 2 silver wire electrodes;† inert (platinum) electrode;† labelled calomel reference electrode; 2 hydrogen electrodes Fig. 4.2.2;[11]† salt bridge (Fig. 4.2.3) or gel link;[6] potentiometer; fume cupboard containing concentrated hydrochloric and nitric acid, with external adjustable 6 V d.c. power supply with meter f.s.d. 50 or 100 mA; stirrer; hydrogen cylinder + lines to bench; hydrogen pre-saturator.

Solutions

0.1 M $ZnSO_4$, $CuSO_4$, $AgNO_3$, KCl, NaOH, HCl; satd KNO_3; satd KCl; platinizing solution, i.e. 0.072 M (3.5 per cent) chloroplatinic acid + 1.3×10^{-4} M (0.005 per cent) lead ethanoate (acetate)[12]–chloroplatinic acid (Pt (IV) chloride) can be made from scrap platinum.[11]

References

[1] *Physical chemistry*. P. W. Atkins, Oxford University Press, Oxford, 3rd edn. (1986), sec 11.3, 11.4 and 12.1.

[2] *Elementary electrochemistry*. A. R. Denaro, Butterworths, London, 2nd edn (1978), ch. 4 and 6.

[3] *Vogel's textbook of quantitative inorganic analysis*. rev. J. Bassett *et al.* ELBS and Longman, Harlow, 4th edn. (1982) ch. 14.

[4] *Findlay's practical physical chemistry*. rev. B. P. Levitt, Longman, London, 9th edn (1973), p. 284.
[5] *Electrochemistry*. C. W. Davies. George Newnes, London (1967), p. 154.
[6] *Experimental electrochemistry for chemists*. D. T. Sawyer and J. L. Roberts Jr. John Wiley, New York (1974), p. 31.
[7] Ref. 1, sect 11.3(a).
[8] *The physical chemistry of electrolytic solutions*. H. S. Harned and B. B. Owen. Reinhold, New York, 3rd edn. (1958), p. 564 and Appendix A.
[9] *Modern electrochemistry*, J. O'M. Bockris and A. K. N. Reddy. Macdonald, London (1970), vol. 1, ch. 1.
[10] *The teaching of electrochemistry at the tertiary level*. Education Division, Royal Society of Chemistry, London. Articles by W. J. Albery, p. 3, and G. J. Hills, p. 21.
[11] *Reference electrodes*, D. J. G. Ives and G. J. Janz. Academic Press, New York and London (1961), p. 107.
[12] *Platinized platinum electrodes*. A. M. Feltham and M. Spiro. *Chem. Revs*. **71,** 177 (1971).

4.3 Standard electrode potential, activity coefficients and acid dissociation constant from Harned cell e.m.f.s

In this experiment we use an electrochemical cell of a type developed by Harned to find the standard electrode potential of a silver,silver chloride electrode, $E^{\ominus}_{Ag,AgCl}$, the mean ionic activity coefficient of hydrochloric acid $\gamma_{\pm HCl}$ and the dissociation constant K_{HA} of a weak monobasic acid such as ethanoic (acetic) acid. A knowledge of the Debye–Hückel theory is required (pp. 116–17), as well as an understanding of cells and of the particular electrodes we are about to use (pp. 120–6).

Theory

(*i*) *The Harned cell*[1]

The *Harned cell* is a *cell without liquid junction*, and is constructed from a hydrogen electrode and a silver,silver chloride electrode dipping into hydrochloric acid:

$$\mathrm{Pt}\,|\,\mathrm{H_2}\,|\,\mathrm{HCl}\,|\,\mathrm{AgCl, Ag.}$$

The reduction at the hydrogen electrode is

$$\mathrm{H^+ + e^- \leftrightharpoons \tfrac{1}{2}H_2} \qquad E_{H_2},$$

and at the silver,silver chloride electrode

$$\mathrm{AgCl + e^- \leftrightharpoons Ag + Cl^-} \qquad E_{Ag,AgCl}.$$

The overall cell reaction is

$$\mathrm{\tfrac{1}{2}H_2 + AgCl \leftrightharpoons Ag + H^+ + Cl^-},$$

and so the *Nernst equation* for the cell takes the form (p. 128),

$$E = E_{\mathrm{Ag,AgCl}} - E_{\mathrm{H_2}}$$

$$= E^{\ominus}_{\mathrm{Ag,AgCl}} - E^{\ominus}_{\mathrm{H_2}} + \frac{RT}{F}\ln\frac{a_{\frac{1}{2}\mathrm{H_2}}}{a_{\mathrm{H^+}}\cdot a_{\mathrm{Cl^-}}}$$

$$= E^{\ominus}_{\mathrm{Ag,AgCl}} - \frac{RT}{F}\ln a_{\mathrm{H^+}}\cdot a_{\mathrm{Cl^-}}, \qquad (4.3.1)$$

assuming that the hydrogen is bubbled at 1 atm pressure so that to a good approximation $a_{\frac{1}{2}\mathrm{H_2}} = 1$.

(*ii*) *Determination of* $E^{\ominus}_{\mathrm{Ag,AgCl}}$ *and* $\gamma_{\pm\mathrm{HCl}}$

By definition (p. 116),

$$a_{\mathrm{H^+}}\cdot a_{\mathrm{Cl^-}} = \gamma_{\mathrm{H^+}}\cdot c_{\mathrm{H^+}}\cdot \gamma_{\mathrm{Cl^-}}\cdot c_{\mathrm{Cl^-}}$$

$$= \gamma^2_{\pm\mathrm{HCl}}\cdot c_{\mathrm{H^+}}\cdot c_{\mathrm{Cl^-}}. \qquad (4.3.2)$$

Since hydrochloric acid is a strong electrolyte, the concentrations of the ions $c_{\mathrm{H^+}}$ and $c_{\mathrm{Cl^-}}$ are equal to the concentration of the acid c_{HCl} used to form the solution. Therefore

$$a_{\mathrm{H^+}}\cdot a_{\mathrm{Cl^-}} = \gamma^2_{\pm\mathrm{HCl}}\cdot c^2_{\mathrm{HCl}}. \qquad (4.3.3)$$

Substituting this expression into the Nernst equation for the cell:

$$E = E^{\ominus}_{\mathrm{Ag,AgCl}} - \frac{2RT}{F}\ln(\gamma_{\pm\mathrm{HCl}}\cdot c_{\mathrm{HCl}}). \qquad (4.3.4)$$

Grouping together the cell e.m.f. and concentration terms:

$$E + \frac{2RT}{F}\ln c_{\mathrm{HCl}} = E^{\ominus}_{\mathrm{Ag,AgCl}} - \frac{2RT}{F}\ln \gamma_{\pm\mathrm{HCl}}. \qquad (4.3.5)$$

The *Debye–Hückel limiting law* (p. 117) states that

$$\log_{10}\gamma_{\pm} = -A\,|\,z_+z_-\,|\,I^{1/2}. \qquad (4.3.6)$$

For a univalent electrolyte this expression simplifies to

$$\log_{10}\gamma_{\pm} = -Ac^{1/2}, \qquad (4.3.7)$$

where A is the Debye–Hückel constant.

Therefore from eqn (4.3.5),

$$E + \frac{2RT}{F}\ln c_{\mathrm{HCl}} = E^{\ominus}_{\mathrm{Ag,AgCl}} + 2\times 2.303\times\frac{RT}{F}Ac^{1/2}_{\mathrm{HCl}}. \qquad (4.3.8)$$

Thus if the e.m.f. of the Harned cell is measured for various concentrations of hydrochloric acid, a graph of $(E+$

$2RT \ln c_{\mathrm{HCl}}/F$) against $c_{\mathrm{HCl}}^{1/2}$ should be a straight line of slope $4.606RTA/F$, and intercept $E^{\ominus}_{\mathrm{Ag,AgCl}}$. Once $E^{\ominus}_{\mathrm{Ag,AgCl}}$ is known, $\gamma_{\pm\mathrm{HCl}}$ at any particular concentration may be found from eqn (4.3.5).

(*iii*) *Determination of* K_{HA}[2]

The e.m.f.s of the cells used in this experiment vary with pH; graphs depicting this variation are known as *Pourbaix diagrams*,[3] and are used for the study of the pH regions of stability, corrosion, etc.

To find the dissociation constant of a weak monobasic acid HA, we use a modified Harned cell:

$$\mathrm{Pt}\,|\,\mathrm{H_2}\,|\,\mathrm{HA, NaA, NaCl}\,|\,\mathrm{AgCl, Ag}.$$

The concentration of hydrogen ions in the cell is determined by the buffer system of the acid HA and its sodium salt NaA. The acid dissociates according to the equilibrium:

$$\mathrm{HA} \rightleftharpoons \mathrm{H^+ + A^-},$$

for which the acid dissociation constant K_a is given by:

$$K_a = \frac{a_{\mathrm{H^+}} \cdot a_{\mathrm{A^-}}}{a_{\mathrm{HA}}}. \tag{4.3.9}$$

Substituting this expression into the Nernst equation for the cell, eqn (4.3.1), we obtain

$$\begin{aligned} E &= E^{\ominus}_{\mathrm{Ag,AgCl}} - \frac{RT}{F} \ln \frac{K_a \cdot a_{\mathrm{HA}} \cdot a_{\mathrm{Cl^-}}}{a_{\mathrm{A^-}}} \\ &= E^{\ominus}_{\mathrm{Ag,AgCl}} - \frac{RT}{F} \ln \frac{K_a \cdot \gamma_{\mathrm{HA}} \cdot c_{\mathrm{HA}} \cdot \gamma_{\mathrm{Cl^-}} \cdot c_{\mathrm{Cl^-}}}{\gamma_{\mathrm{A^-}} \cdot c_{\mathrm{A^-}}}. \end{aligned} \tag{4.3.10}$$

In the buffered system the concentration of dissociated acid $c_{\mathrm{A^-}}$ is largely determined by the amount of salt added, c_{NaA}. The concentration of acid, c_{HA}, is to a good approximation the same as the concentration of acid added to the cell, and the chloride ion concentration $c_{\mathrm{Cl^-}}$ is determined by the concentration of sodium chloride in the cell, c_{NaCl}. Thus rearranging the previous equation:

$$E - E^{\ominus}_{\mathrm{Ag,AgCl}} + \frac{RT}{F} \ln \frac{c_{\mathrm{HA}} \cdot c_{\mathrm{NaCl}}}{c_{\mathrm{NaA}}} = -\frac{RT}{F} \ln K_a + \frac{RT}{F} \ln \frac{\gamma_{\mathrm{A^-}}}{\gamma_{\mathrm{HA}} \cdot \gamma_{\mathrm{Cl^-}}}. \tag{4.3.11}$$

As the ionic strength I (p. 116) tends to zero, all the activity coefficients tend to unity. Therefore K_a may be obtained from an extrapolation of the term on the left-hand side of eqn (4.3.11) to zero ionic strength.

Apparatus

The electrodes used in this experiment have been described on pp. 123, 125, and 134. They are inserted into the cell vessel,

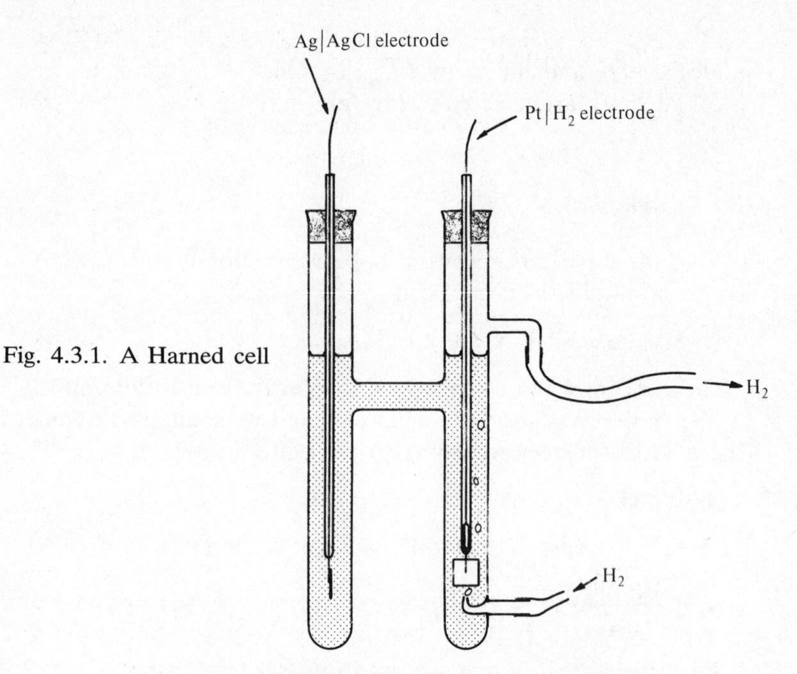

Fig. 4.3.1. A Harned cell

which can be in the form of an H-tube, Fig. 4.3.1, or one of the arrangements shown on p. 127. The e.m.f. of the cell is measured with a digital voltmeter (p. 132) or a potentiometer with standard cell (p. 131).

Procedure

(*i*) *Preparation of electrodes*

All electroplating operations and handling of concentrated acids must be carried out in a fume cupboard. Take care with hydrogen gas, and do not expose it to sparks or naked flames.

Prepare the Ag, AgCl, and Pt | H_2 electrodes as described on p. 134, but connect the Pt | H_2 electrode as cathode with an inert (Pt) anode and do not reverse the polarity.

Measure the volume of the cell vessel.

(*ii*) *Measurements with hydrochloric acid*

Make up suitable volumes of four solutions of hydrochloric acid at concentrations covering the range 0.01 to 0.1 M.

Fill the cell and the hydrogen saturator with the first solution. Bubble hydrogen through the cell for at least ten minutes to remove oxygen from the cell. Reduce the bubbling rate. Connect the potentiometer or digital voltmeter as instructed.*

Measure the e.m.f. of the cell every 2 min for at least another 10 min and until the e.m.f. drifts by less than 2 mV min^{-1}.

Repeat the determinations with the other solutions. Rinse the cell and electrodes thoroughly with each new solution before filling. Do not allow the electrodes to dry.

(iii) Measurements with ethanoic acid

Make up suitable volumes of four solutions containing accurately known quantities of ethanoic acid, sodium ethanoate, and sodium chloride in approximately equimolar proportions. The four solutions should range from 0.01 M in each component to 0.1 M in each.

Measure the e.m.f. of the cell with each solution as before.

Calculation

First check the slope of your graph based on eqn (4.3.8), given that the calculated value of the Debye–Hückel constant A is 0.50 for aqueous solutions at room temperature. If the slope of your graph differs from the calculated value, to what value of A does it correspond?

From your measurements, find the standard electrode potential $E^{\ominus}_{Ag,AgCl}$, $\gamma_{\pm HCl}$ in 0.01 and 0.1 M solutions, and the dissociation constant K_a of ethanoic acid.

Results

Accuracy

The accuracy of the results is determined by the care with which the experiment is performed; you should be able to obtain results to within a few mV of the true value of $E^{\ominus}_{Ag,AgCl}$, and to within 20 per cent of K_a.

Comment

This experiment demonstrates how absolute determinations of electrode potentials, activity coefficients and dissociation constants can be made from cell e.m.f. measurements. Harned cells in particular have been widely used to study the dissociation constants of weak acids and their complex salts.[4]

If the dissociation constant of an acid is determined from cell e.m.f.s over a range of temperatures, the enthalpy of ionization can be calculated. Other effects of temperature on cell e.m.f. are studied in the next experiment.

It is interesting to note that the enthalpy of ionization of ethanoic acid is zero at room temperature.[5]

Technical notes

Apparatus

Hydrogen electrode,† hydrogen cylinder with line to bench, hydrogen pre-saturator;† inert (platinum) electrode;† silver (wire) electrode;†

digital voltmeter or potentiometer with standard cell; fume cupboard with 6 V d.c. power supply with meter 50 or 100 mA f.s.d.; stirrer; [cell vessel].

Solutions

0.1 M HCl, 0.1 M solution of ethanoic (acetic) acid, 0.1 M NaOH, plantinizing solution as described in Technical notes on p. 136.

References

[1] *Electrolyte solutions.* R. A. Robinson and R. H. Stokes. Butterworths, London, 2nd edn (rev) (1970), p. 198.
[2] *The physical chemistry of electrolytic solutions.* H. S. Harned and B. B. Owen. Reinhold, New York, 3rd edn. (1958), Ch 11.
[3] *Modern electrochemistry.* J. O'M. Bockris and A. K. N. Reddy, Macdonald, London (1970), vol. 2, p. 1121.
[4] *The determination of stability constants.* F. J. C. Rossotti and H. Rossotti. McGraw-Hill, New York (1961), p. 140.
[5] *The dissociation constant of acetic acid* from 0 to 60° Centigrade, H. S. Harned and R. W. Ehlers, *J. Am. Chem. Soc.* **55,** 652 (1933).

4.4 Thermodynamics of a chemical reaction from temperature coefficient of a cell e.m.f.

The measurement of the e.m.f.s of electrical cells under reversible conditions, and the variation of cell e.m.f. with temperature, are valuable and commonly used methods of obtaining thermodynamic quantities for chemical reactions.

In this experiment we obtain the free energy change ΔG, the enthalpy change ΔH and the entropy change ΔS for the reaction

$$Zn(s) + Hg_2SO_4(s) + 7H_2O(l) \rightleftharpoons ZnSO_4.7H_2O(s) + 2Hg(l).$$

We do this by measuring the e.m.f. E, and the temperature coefficient at constant pressure $(\partial E/\partial T)_p$, of the Clark cell:

$$Zn \mid ZnSO_4.7H_2O(s) \mid \begin{matrix}\text{saturated aqueous} \\ \text{solution of } ZnSO_4\end{matrix} \mid Hg_2SO_4(s) \mid Hg(l).$$

Theory[1]

The free energy change ΔG for one mole of the cell reaction is related to the cell e.m.f. E by the relation

$$\Delta G = -zFE, \qquad (4.4.1)$$

where F is the Faraday constant, and z refers to the number of electrons transferred in the reaction (p. 121).

Since $(\partial \Delta G/\partial T)_p = -\Delta S$, it follows that

$$\Delta S = zF(\partial E/\partial T)_p. \tag{4.4.2}$$

By definition, $\Delta H = \Delta G + T\,\Delta S$, and therefore

$$\Delta H = -zFE + zFT(\partial E/\partial T)_p. \tag{4.4.3}$$

Apparatus[2]

The Zn amalgam will be thermodynamically equivalent to solid zinc when (and only when) it is saturated with zinc.

The Clark cell, shown in Fig. 4.4.1, is supplied ready-assembled. The zinc is in the form of a ~40 per cent w/v amalgam of zinc in mercury, which makes the electrode potential more reproducible. When the cell is being cooled down again after heating, hydrated zinc sulphate crystallizes out of solution. As it does so, there is a tendency for air bubbles to form, which, if they accumulate sufficiently, will prevent the cell developing an e.m.f. The insertion of glass wool into the side-arm minimizes this effect, but does not entirely prevent the development of air locks over a period of time.

A convenient way of maintaining the cell at a stable, known temperature is to mount it in a Dewar flask, and pour in water which has been heated or cooled to the required temperature.

The e.m.f. of the cell is measured with a potentiometer or digital voltmeter (DVM) (p. 131).

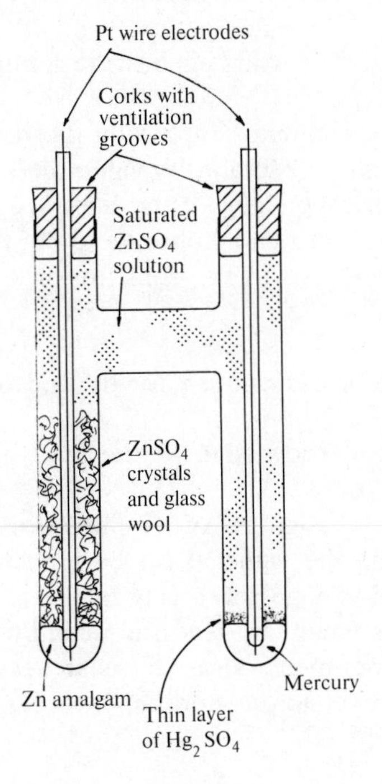

Fig. 4.4.1. A Clark cell

Procedure

If using a potentiometer, standardize it and connect it as instructed.*

Connect the potentiometer or DVM to the cell. It is important that the cell be kept *upright* at all times.

Measure the e.m.f. of the cell at room temperature. If using a potentiometer, confirm that the cell is functioning reversibly. Do this by checking that, in two consecutive measurements, the e.m.f. is independent of the side from which the balance point is approached.

Mount the Clark cell in the Dewar flask. Heat some water in a beaker to 41 °C, and pour it around the cell. Cover the cell, but keep the water well away from the corks and electrodes. Allow time for thermal equilibrium to be achieved.

The criterion for judging whether thermodynamic equilibrium has occurred within the cell is the constancy of the cell e.m.f. Attainment of thermal equilibrium is not the only requirement for this; time must also be allowed for mass transport, especially when the temperature is increased.

Measure the e.m.f. of the cell at 40 °C. It should be within 15 mV of 1.400 V. If it is not, *gently* wiggle the electrode which dips into the amalgam. If this does not give the correct e.m.f., seek advice.

Repeat the procedure for nine temperatures spaced over the range 0 °C to 40 °C. Do not study the temperature variation sequentially, but instead change the order of temperature so as to randomize any systematic errors associated with increasing or decreasing the temperature.

At the end of the experiment, ensure that the wires are disconnected from the electrodes of the cell.

Calculation

Plot the e.m.f. E in volts against the temperature T of the cell in kelvins.

Calculate or estimate graphically the best-fitting straight line through the points. Obtain the value of E at 298 K by interpolation. Measure the slope of the line $(\partial E/\partial T)_p$.

From your experimental observation of the polarity of the cell and your knowledge of the constitution of its electrodes, state the direction of the cell reaction given previously. Show that $z = 2$.

Calculate ΔG, ΔH and ΔS for the cell reaction (eqns (4.4.1), (4.4.2) and (4.4.3)).

Employ your knowledge of thermodynamics to answer the following questions:[1]

(a) Why is the sign of ΔG negative for the cell reaction?

(b) How can the value of ΔS be interpreted in terms of the physical changes occurring in the reaction?

(c) What is meant by reversible conditions in the context of this experiment, and why is it necessary to measure E under reversible conditions in order to evaluate ΔG and ΔS?

(d) When the cell is operating reversibly, what is the magnitude of the heat exchanged with the surroundings and the direction of its flow?

(e) Under what conditions would all the energy of the reaction appear as heat lost to the surroundings? What is the magnitude of this heat loss?

Results

Accuracy

The cell e.m.f. may alter if non-negligible currents have been drawn from it, or if air locks have formed after it has been in service for some time.

Comment

This experiment demonstrates a convenient way of measuring the thermodynamic changes accompanying a chemical reaction. The limitation on the method is that the reaction must occur in a reversible cell at the temperature of interest, and many electrodes are not reversible (p. 122). The questions provide an interesting exercise in the application of the First and Second Laws of thermodynamics.

Technical notes

Apparatus

Clark cell (Fig. 4.4.1), including Pt wire electrodes (p. 131), and Zn amalgam;[$†] potentiometer with galvanometer and standard cell, or DVM; [Dewar flask]; electrical leads; beaker, bunsen burner, gauze.

Preparation of Zn amalgam

Allow an excess quantity of fresh zinc shavings to dissolve in mercury in a stoppered flask for two or three days.

References

[1] *Basic chemical thermodynamics.* E. B. Smith. Clarendon Press, Oxford, 3rd edn (1982), p. 105.

[2] *A laboratory manual of experiments in physical chemistry.* D. Brennan and C. F. H. Tipper. McGraw-Hill, London (1967).

4.5 Redox potentials and equivalence points from potentiometric titrations

An *electrode potential* is the potential associated with the reduction of a species by electrons, relative to that of the reduction of protons by electrons. *Redox potentials* refer to reductions in which both the reactants and products are charged species in solution:

$$M^{y+} + ze^- \leftrightharpoons M^{(y-z)+}.$$

The potential of an inert electrode placed in a redox solution will be determined by the relative concentrations, or more strictly activities, of the charged reactants and products. The greater the relative concentration of the oxidized species, M^{y+}, the more positive will be the redox potential.

The relative concentrations of the ions may be altered by adding known quantities of a solution of an oxidizing or reducing agent from a burette. The end point of the *potentiometric titration* may be found from the resulting changes in e.m.f.

In this experiment we investigate the well-known system

$$\mathrm{Fe(III)} + e^- \leftrightharpoons \mathrm{Fe(II)}.$$

The potential is measured with a bright platinum electrode, with a standard calomel electrode to complete the cell. Fe(II) ions are first titrated with Ce(IV) using 'ferroin' as indicator, and a comparison made of the redox and colorimetric end-points or *equivalence points.* The experiment is then repeated in the presence of phosphoric acid.

The ferric/ferrous equilibrium is reproducible and fairly quickly achieved, unlike that of many redox systems.

Theory[1–3]

This experiment requires an understanding of electrodes and electrochemical cells (pp. 122–33).

We shall study the redox equilibria:

$$\mathrm{Fe(III)} + e^- \leftrightharpoons \mathrm{Fe(II)} \qquad \text{sample,}$$

$$\mathrm{Ce(IV)} + e^- \leftrightharpoons \mathrm{Ce(III)} \qquad \text{titrant,}$$

$$\mathrm{Fe(II)} + \mathrm{Ce(IV)} \leftrightharpoons \mathrm{Fe(III)} + \mathrm{Ce(III)} \qquad \text{overall.}$$

After each addition of titrant, the reaction proceeds until at equilibrium the redox potentials of the sample and titrant become equal:

$$E^{\ominus}_{\mathrm{Fe}} + \frac{RT}{F}\ln\frac{a_{\mathrm{Fe(III)}}}{a_{\mathrm{Fe(II)}}} = E^{\ominus}_{\mathrm{Ce}} + \frac{RT}{F}\ln\frac{a_{\mathrm{Ce(IV)}}}{a_{\mathrm{Ce(III)}}}, \qquad (4.5.1)$$

where $E^{\ominus}_{\mathrm{Fe}}$ is the standard redox potential of the Fe(III)/Fe(II) couple, and $E^{\ominus}_{\mathrm{Ce}}$ is that of the Ce(IV)/Ce(III) couple. $E^{\ominus}_{\mathrm{Ce}}$ is more positive than $E^{\ominus}_{\mathrm{Fe}}$ by an amount sufficient for there to be effectively complete oxidation of Fe(II) by Ce(IV), i.e. the overall equilibrium shown above is very far to the right.

Since the redox potentials of the two couples are equal, we may study whichever is convenient. Before the equivalence point the concentration of Ce(IV) is immeasurably small, and so we study the system:

$$\underbrace{\text{Hg, Hg}_2\text{Cl}_2\,|\,\text{KCl}\,|}_{\substack{\text{standard calomel}\\ \text{electrode}}}\,\text{Fe(III), Fe(II)}\,|\,\text{Pt}$$

The corresponding Nernst equation (p. 128) is

$$E_{obs} = E^{\ominus}_{Fe} - E^{\ominus}_{cal} + \frac{RT}{F}\ln\frac{a_{Fe(III)}}{a_{Fe(II)}}$$

$$= E^{\ominus}_{Fe} - E^{\ominus}_{cal} + \frac{RT}{F}\ln\frac{\gamma_{Fe(III)}\cdot c_{Fe(III)}}{\gamma_{Fe(II)}\cdot c_{Fe(II)}},$$

where E_{obs} is the measured e.m.f. of the cell, and $E^{\ominus}_{cal}$ is the standard electrode potential of the calomel electrode.

Since $E^{\ominus}_{cal}$ is known, $E^{\ominus}_{Fe}$ could be found by extrapolating e.m.f. measurements with equimolar quantities of Fe(III) and Fe(II), ($c_{Fe(III)} = c_{Fe(II)}$), to zero ionic strength ($\gamma_{Fe(III)} = \gamma_{Fe(II)} = 1$). This procedure was carried out in Expt 4.3 to find $E^{\ominus}_{Ag,AgCl}$ (p. 138). However it is not usually possible to calculate the e.m.f. of a practical redox system from the standard redox potential. This is because a system such as the one studied here, containing the redox system, oxidizing agent, and acid, has an ionic strength too high for calculations based on simple Debye–Hückel theory. It is therefore common practice not to measure the standard redox potential, but to use instead the *formal redox potential* $E^{\ominus'}$. This is defined as the potential observed experimentally in a solution containing equal numbers of moles of the oxidized and reduced species together with other specified substances at specified concentrations. In dilute solutions in which there is a significant concentration of Fe(II) ions, the appropriate Nernst equation then becomes:

$$E_{obs} = E^{\ominus'}_{Fe} - E^{\ominus}_{cal} + \frac{RT}{F}\ln\frac{c_{Fe(III)}}{c_{Fe(II)}}. \qquad (4.5.3)$$

Formal redox potentials may have quite different magnitudes to standard redox potentials, but are experimentally much more useful.

The quantity $\gamma_{Fe(III)}/\gamma_{Fe(II)}$ varies with concentration, and thus a comparison with eqn (4.5.2) underlines the point that eqn (4.5.3) is *only* valid when the composition of the solution is the same as that quoted for the formal redox potential $E^{\ominus'}_{Fe}$.

The corresponding equation for the Ce(IV)/Ce(III) couple is

$$E_{obs} = E^{\ominus'}_{Ce} - E^{\ominus}_{cal} + \frac{RT}{F}\ln\frac{c_{Ce(IV)}}{c_{Ce(III)}}. \qquad (4.5.4)$$

By definition the equivalence point occurs when

$$\frac{c_{Fe(II)}}{c_{Fe(III)}} = \frac{c_{Ce(IV)}}{c_{Ce(III)}}. \qquad (4.5.5)$$

Adding eqns (4.5.3) and (4.5.4) and substituting eqn (4.5.5), we find that the potential at the equivalence point is approximately

Equation (4.5.6) implies that the formal redox potential at an equivalence point is independent of the concentration of the reactants, but this is only true because by definition formal redox potentials apply to a specific composition of solution.

$$E_{\text{equiv}} = \tfrac{1}{2}(E^{\ominus\prime}_{\text{Fe}} + E^{\ominus\prime}_{\text{Ce}}) - E^{\ominus}_{\text{cal}}. \tag{4.5.6}$$

A different expression applies to systems with more complicated stoichiometry.[3]

The end-point may also be found colorimetrically. A suitable indicator is 'ferroin', a 3 : 1 complex of 1 : 10-phenanthroline with Fe(II). This undergoes a striking colour change when oxidized:

$$\underset{\text{deep red}}{[Fe(C_{12}H_8N_2)_3]^{2+}} \rightleftharpoons \underset{\text{pale blue}}{[Fe(C_{12}H_8N_2)_3]^{3+}} + e^-.$$

By what simple method could the formal redox potential of this system, $E^{\ominus\prime}_{\text{ind}}$, be found? If a sufficiently small quantity of indicator is added to the redox system the colour change will occur when $E^{\ominus\prime}_{\text{ind}} = E_{\text{equiv}}$.[4]

Redox potentials may be altered by the addition of a complexing agent. If a ligand stabilizes the lower oxidation state with respect to the higher, the redox potential increases. If a ligand is used which stabilizes the higher oxidation state with respect to the lower, it is possible for a positive redox potential to change sign; for example the redox potential of the Fe(III)/Fe(II) system in the presence of EDTA is −0.12 V.

Apparatus

The apparatus comprises a saturated calomel electrode and a bright platinum electrode dipping into a beaker. Ideally the cell is enclosed with some method of flushing out the air with an inert gas to prevent oxidation of the Fe(II) ions. Ce(IV) sulphate solution is added from a burette above the beaker, and the solution is stirred continuously.

The electrodes have been described on p. 124. The potential between them is measured with a digital voltmeter (DVM), p. 132.

Procedure

Concentrated acids can cause severe burns. They must be used with great care in a fume cupboard.

Clean the bright platinum electrode by dipping it into a small beaker of aqua regia made from three volumes of concentrated hydrochloric acid and one volume of concentrated nitric acid. Keep it in purified water until required.

Obtain or prepare fresh solutions of 0.1 M Ce(IV) sulphate in 1 M sulphuric acid, and 0.1 M Fe(II) ammonium sulphate in 1 M sulphuric acid. Fill a burette with the Ce(IV) sulphate solution.

Set up the apparatus as described in the section above, with lid and gas supply if provided. Use a 100 cm^3 beaker. Ensure that the stirrer flea is well away from the electrodes when it is in motion.

Set up the DVM and calibrate if necessary.*

Dilute a few cm^3 of ferroin solution to about 50 cm^3 with water. Find $E^{\ominus'}_{\text{ind}}$ for this solution by the method you were asked to devise previously.

Rinse the cell, and then fill it, with 25 cm^3 of 0.1 M Fe(II) ammonium sulphate solution in 1 M sulphuric acid. Add a drop of ferroin as indicator.

Titrate with the Ce(IV) solution, stirring gently and measuring the e.m.f. at each stage. Add the oxidizing agent a few cm^3 at a time to begin with, and then dropwise as the equivalence point is approached. Plot the e.m.f. against added volume of oxidizing agent as the experiment proceeds. Also record the visual equivalence point. Discontinue the titration when the e.m.f. becomes approximately constant for the second time.

Repeat the experiment in the presence of 1 cm^3 of concentrated phosphoric acid, which should be added carefully *while the solution is being stirred.*

Dissolve a small quantity of an Fe(III) salt in water, add carefully a few drops of concentrated phosphoric acid from a dropping pipette, and note the colour change.

Calculation

Comment on the shape of your curve of e.m.f. against titrant volume, guided by the comment about a different system in the Results section.

Check that the visual equivalence point indicates the correct relative strengths of the ceric (Ce(IV)) sulphate and ferrous (Fe(II)) ammonium sulphate solutions.

Plot graphs of e.m.f. against $\ln(c_{\text{Fe(III)}}/c_{\text{Fe(II)}})$, and e.m.f. against $\ln(c_{\text{Ce(IV)}}/c_{\text{Ce(III)}})$. Hence find the formal electrode potentials $E^{\ominus'}_{\text{Fe}}$ and $E^{\ominus'}_{\text{Ce}}$ (eqns (4.5.3) and (4.5.4)), using the appropriate calomel electrode e.m.f. from Table 4.2.1 (p. 124). Use the formal electrode potentials to calculate the equivalence point in the presence of sulphuric acid.

Also find the equivalence point directly from the graph of e.m.f. against titrant volume. Do this first by drawing straight lines through the points which form the regions of slowly changing e.m.f. well before and after the region of the equivalence point. Join one line to the other with a vertical line, but position the latter so that it is bisected by the titration curve. Take the equivalence point as the intersection of the vertical and the titration curve. Compare these results with the visual equivalence point.

Determine the equivalence point in the presence of both sulphuric and phosphoric acid by the same three methods.

Discuss the effect of the phosphoric acid on both the colorimetric end-point and the Fe(III)/Fe(II) redox potential, and thus decide whether you think it is advantageous to carry out the redox titration in the presence of phosphoric acid.

Results

Some results for a different system, Fe(II) oxidized by Mn(VII) in the presence of sulphuric acid, are given here to guide your calculations.

25 cm^3 ~0.1 M ferrous ammonium sulphate titrated with 0.020 M $KMnO_4$. Colorimetric end-point at 24.6 cm^3 $KMnO_4$. Therefore ferrous ammonium sulphate is $24.6 \times 0.020 \times 5/25 = 0.0984$ M. Results up to the end-point are shown in Table 4.5.1. (You will need an equivalent set of results for Ce(III)/Ce(IV) after the end-point as well.)

If you plot the e.m.f./vol. curve of Table 4.5.1, you will notice that it is symmetrical about the end-point. This suggests that the potential is determined by the Mn(III)/Mn(II) couple immediately beyond the end-point,[3] rather than the Mn(VII)/Mn(II) couple as expected.[2]

Graph of e.m.f./$\ln(c_{Fe(III)}/c_{Fe(II)})$ gives $E_{obs} = 0.396$ V at $\ln(c_{Fe(III)}/c_{Fe(II)}) = 0$ (i.e. when $c_{Fe(III)} = c_{Fe(II)}$). Therefore $E_{Fe}^{\ominus'} - E_{cal}^{\ominus} = 0.396$ V (eqn (4.5.3)). Hence $E_{Fe}^{\ominus'}$ may be found.

Table 4.5.1. Results for oxidation of Mn(VII) by Fe(II)

Vol $KMnO_4/cm^3$	E.m.f./mV	$c_{Fe(III)}$ (see text for units)	$c_{Fe(II)}$	$c_{Fe(III)}/c_{Fe(II)}$	$\ln(c_{Fe(III)}/c_{Fe(II)})$
8.0	0.3762	8.0	16.6	0.48	−0.730
10.0	0.3850	10.0	14.6	0.69	−0.378
12.0	0.3935	12.0	12.6	0.95	−0.049
14.0	0.4020	14.0	10.6	1.32	0.278
16.0	0.4110	16.0	8.6	1.86	0.621
18.0	0.4210	18.0	6.6	2.73	1.003
20.0	0.4338	20.0	4.6	4.35	1.470
22.05	0.4520	22.05	2.55	8.65	2.157
24.0	0.4960	24.0	0.6	40.0	3.689
24.5	0.7580	24.5	0.1	245.0	5.501
24.7	0.8750				
24.9	0.9270				
25.1	0.9455				
25.7	0.9750				
26.9	1.0150				

Accuracy

Your consideration of the accuracy of your own results should include a discussion of the factors which may affect the ionic equilibria cited at the beginning of the Theory section.[3]

Comment This experiment employs the fundamental theory and experimental techniques of equilibrium electrochemistry, and introduces standard and formal redox potentials. Potentiometric titrations give an important quantitative insight into the nature of equivalence points, and to the pH titrations carried out in the next experiment.

Technical notes

Apparatus

Potentiometer and standard cell; labelled calomel electrode; bright platinum electrode. *Either* fresh 0.1 M Ce(IV) sulphate solution in 1 M sulphuric acid and 0.1 M Fe(II) ammonium sulphate solution in 1 M sulphuric acid *or* their separate constituents; ferroin = 1,10-phenanthroline-iron (II) complex solution; fume cupboard containing conc. phosphoric acid, conc. nitric acid, conc. hydrochloric acid and dropping pipette.

References

[1] *Physical chemistry*. P. W. Atkins. Oxford University Press, Oxford, 3rd edn (1986), sect 12.4(b).
[2] *Vogel's textbook of quantitative inorganic analysis*. rev. J. Bassett *et al.* ELBS and Longman, Harlow, 4th edn (1978), ch. 14.
[3] *Chemical analysis*. H. A. Laitinen and W. E. Harris. McGraw-Hill, New York, 2nd edn (1975), ch. 15.
[4] Ref. 3, p. 290.
[5] Ref. 2, p. 597.

4.6 Acid dissociation constants and analysis by microcomputer-controlled pH titrations

The purpose of this experiment is twofold; firstly, to demonstrate the use of pH titrations in the analysis of solutions, and secondly to demonstrate how a microcomputer can be made to control and improve the reliability of chemical measurements.

The strict definition of the pH of a solution is that

$$\mathrm{pH} = -\log_{10} a_{\mathrm{H}^+}, \tag{4.6.1}$$

where a_{H^+} is the activity of the hydrogen ion in solution. Later on, however, we shall ignore differences between activities and concentrations, and assume that

$$\mathrm{pH} = -\log_{10} [\mathrm{H}^+], \tag{4.6.2}$$

where $[\mathrm{H}^+]$ is the hydrogen ion concentration.

The pH of a solution may be measured by an electrode which has an e.m.f. dependent on a_{H^+}. The most obvious electrode to use would be the Pt | H_2 electrode (p. 125). However, oxidizing and reducing solutions, sulphur, and proteins all interfere with the use of platinized platinum, and to avoid these difficulties it is usual to employ the more convenient *glass electrode*, which is really a half-cell. The electrochemical cell is completed by the test solution and a *reference electrode* such as Ag,AgCl. The two electrodes are often incorporated into a single *combination electrode*. The e.m.f. between the electrodes is measured by a millivoltmeter known as a *pH meter*, which is calibrated to display the pH of the solution directly.

pH titrations may be used for several different types of analysis, and this is demonstrated by the use of sodium hydroxide as titrant for five different solutions. The first three are a strong acid (hydrochloric) and two weak acids (ethanoic and malonic). These titrations emphasize the advantages of potentiometric end-points over colorimetric end-points deduced from indicators, and illustrate the calculation of the dissociation constants of weak monoprotic acids and polyacids. Finally two precipitation titrations are carried out. The first, of acidified aluminium sulphate solution, demonstrates the formation of sodium aluminate, and the second the estimation of magnesium ions in the presence of calcium ions.

The experiment is carried out by means of a simple titration apparatus controlled by a microcomputer. The microcomputer is connected to the output of the pH meter, and on the basis of the pH, can be instructed to add either a small volume or a larger volume of titrant. The initial part of the experiment involves programming the microcomputer in BASIC so that the titrations may be carried out automatically, and therefore requires a rudimentary understanding of this language.

Theory

(*i*) *The glass electrode*[1]

A typical commercial combination electrode is shown in Fig. 4.6.1. Measurements are made by immersing the combination electrode, or a glass electrode and standard calomel electrode, into the test solution, whereupon the following cell is set up:

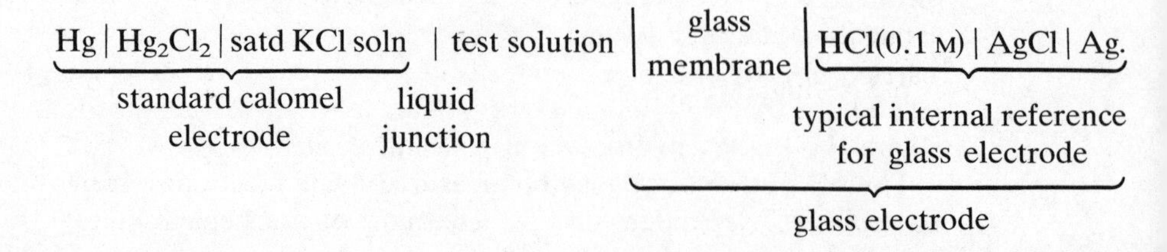

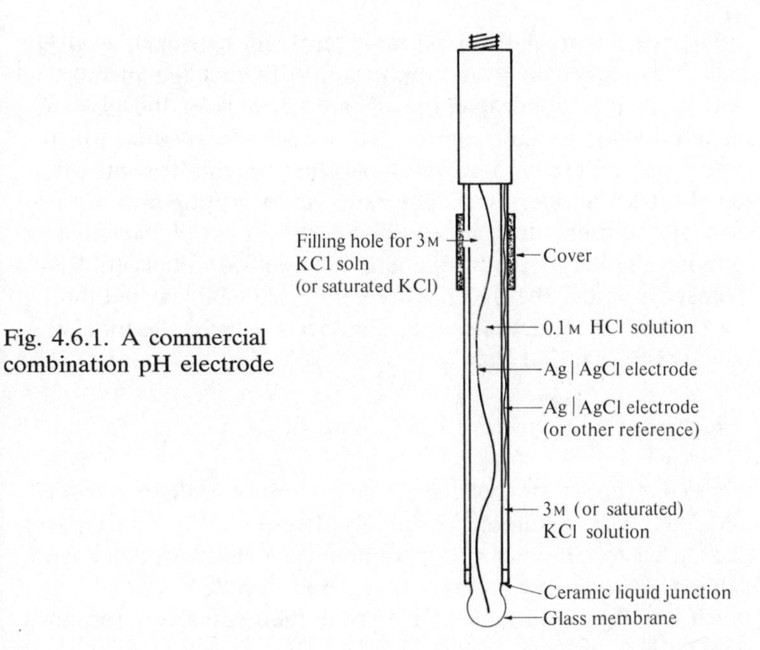

Fig. 4.6.1. A commercial combination pH electrode

The properties of the test solution affect the overall e.m.f. of this cell at two points. One is the liquid junction between the saturated KCl and the test solution. The use of saturated KCl minimizes the liquid junction potential (p. 126), and we assume that this potential is small and constant. The remaining contribution arises from the effect of the test solution on the potential across the glass membrane. If the membrane is selective towards H^+ ions, the cell potential is

$$E = \text{const.} + \frac{RT}{F} \ln a_{H^+}, \tag{4.6.3}$$

where a_{H^+} is the activity of the H^+ ions in the test solution. The constant term incorporates the other interfaces in the cell, all of which feature phases of constant composition.

The *standardizing* of the electrode involves immersing it in a buffer solution of known a_{H^+} or pH. The constant term in eqn (4.6.3) may then be evaluated, and the pH of other solutions deduced from the e.m.f. generated when the combination electrode is immersed in them. In practice the standardization is carried out by adjusting a control on the pH meter so that the pH of the buffer solution is registered correctly. For reasons which we shall now discuss, eqn (4.6.3) is approximate, and therefore the pHs of the standard and test solutions need to be as close as possible.

The idea that glass membranes are semi-permeable to H^+ ions is a dangerous over-simplification. In fact the membrane potentials appear because the silicate network of the glass has an affinity for certain cations. These are adsorbed within the structure, and create a charge separation which alters the interfacial potential difference. The rates of adsorption and desorption are altered, and an equilibrium is finally achieved in a manner similar to that for liquid junction potentials (p. 377). Transport across the dry glass region is probably carried out by Na^+ ions. If equations are employed which describe this mechanism, we find that

This equation accounts for the *alkali error* observed in glass electrodes; at high pH values (low a_{H^+}) the potential of a glass electrode becomes increasingly dependent on sodium ion concentration, and may, for example, register a pH of 12.8 in a solution 0.01 M in Na^+ at pH 12.0.

$$E = \text{const.} + \frac{RT}{F} \ln(a_{H^+} + k_{H^+,Na^+} \cdot a_{Na^+}), \qquad (4.6.4)$$

where k_{H^+,Na^+} is the *potentiometric selectivity coefficient* for H^+ and Na^+. This parameter is largely determined by the type of glass used for the electrode; commercial pH electrodes have a value of k_{H^+,Na^+} low enough to make the product $k_{H^+,Na^+} \cdot a_{Na^+}$ much less than a_{H^+}, and the electrode then effectively responds only to H^+.

(*ii*) *Dissociation constants of weak monoprotic acids*

Consider the dissociation of a weak monoprotic acid HA to yield an anion and a hydrated proton:

$$\underset{x}{HA} \rightleftharpoons A^- + H^+ \qquad \text{dissociation}$$

where x is the initial concentration of acid (i.e. the concentration in the solution before dissociation). The acid dissociation constant K_a is

$$K_a = a_{H^+} \cdot a_{A^-}/a_{HA}, \qquad (4.6.5)$$

although for the purposes of this experiment we assume that

$$K_a = [H^+][A^-]/[HA]. \qquad (4.6.6)$$

Let us consider the titration curve shown in Fig. 4.6.2 which relates to the neutralization of the acid by a strong alkali;

$$HA + \underset{y}{OH^-} \rightarrow A^- + H_2O \qquad \text{neutralization}$$

where y is the concentration of base added. Then the concentration of anions A^- arises from both the dissociation of the acid, as shown earlier, and the neutralization of the acid by OH^-. Thus it follows that:

$$[A^-] = [H^+] + y. \qquad (4.6.7)$$

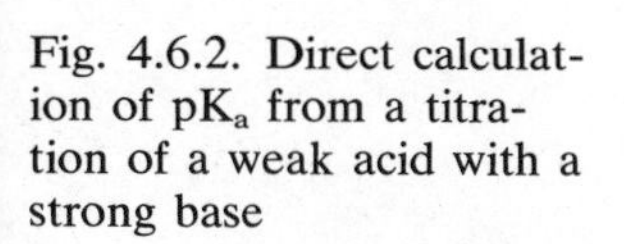
Fig. 4.6.2. Direct calculation of pK_a from a titration of a weak acid with a strong base

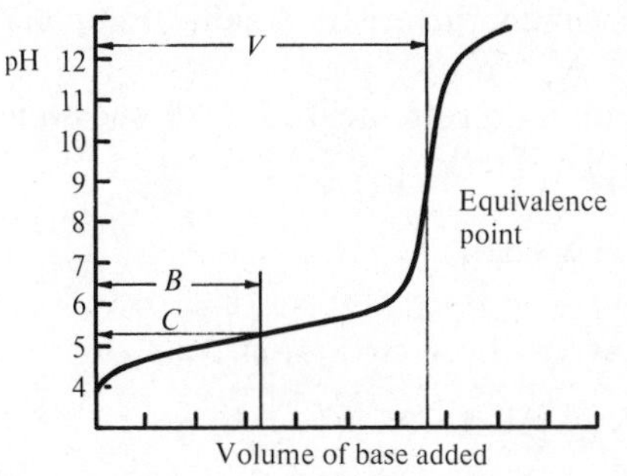

The concentration of undissociated acid, [HA], will be equal to the initial concentration x less the concentration dissociated and the concentration neutralized:

$$[HA] = x - [H^+] - y. \tag{4.6.8}$$

Therefore from eqn (4.6.6):

$$K_a = [H^+]([H^+] + y)/(x - [H^+] - y). \tag{4.6.9}$$

This expression may be used at several positions along the titration curve to calculate K_a.

There is, however, a much simpler method of calculating K_a. If we examine the chemical equation for neutralization shown above, we see that when half the volume of alkali necessary for complete neutralization has been added,

$$[A^-] = [HA]. \tag{4.6.10}$$

It follows from eqn (4.6.6) that at this point

$$K_a = [H^+],$$

or

$$pK_a = pH, \tag{4.6.11}$$

where pK_a is $-\log_{10}K_a$. The pK_a value of an acid may therefore be read directly from the titration graph. If V is the volume of base required for neutralization (Fig. 4.6.2), then an intercept is drawn at $B = V/2$, and the pK_a value corresponds to the pH at the point C.

(*iii*) *Dissociation constants of weak polyacids*

Polyacids, i.e. those which contain several replaceable hydrogen atoms, dissociate step by step. Each step corresponds to the tendency of one of the hydrogen atoms to ionize, and possesses its own characteristic dissociation constant $K_1, K_2, K_3 \ldots$ where $K_1 > K_2 > K_3 \ldots$.

For a diprotic acid H_2A dissociating in two stages

$$H_2A \leftrightharpoons H^+ + HA^-,$$
$$HA^- \leftrightharpoons H^+ + A^{2-},$$

we write the dissociation constants in terms of concentrations:

$$K_1 = [H^+][HA^-]/[H_2A],$$
$$K_2 = [H^+][A^{2-}]/[HA^-]. \qquad (4.6.12)$$

For many acids, including malonic acid, K_1 is more than two orders of magnitude greater than K_2, and the first step of the dissociation is practically complete before the second begins. The titration curve is therefore in the form of two effectively independent steps. If a volume V of NaOH is required for the first dissociation, it follows that a volume $2V$ will be required for the second step. Thus pK_1 can be read off from the intercept at $V/2$, and pK_2 from the intercept corresponding to $([HA^-] = [A^{2-}])$, i.e. half-way from the first to the second neutralization, at $3V/2$.

If there is less difference between K_1 and K_2, these parameters may be found by measuring $[H^+]$ in two different solutions and solving a series of four simultaneous equations.[2]

(iv) Precipitation titrations

The progress of a precipitation which is dependent on $[H^+]$ may also be studied by means of a pH titration. The precipitations may be expressed by means of equations similar to those used for the neutralization reactions.

Figure 4.6.3 shows the shape of the titration curve obtained when an acidified solution of aluminium sulphate is titrated with sodium hydroxide solution. The pH remains relatively steady during the precipitation of a particular species, and then rises again after precipitation is complete. The first step in the curve

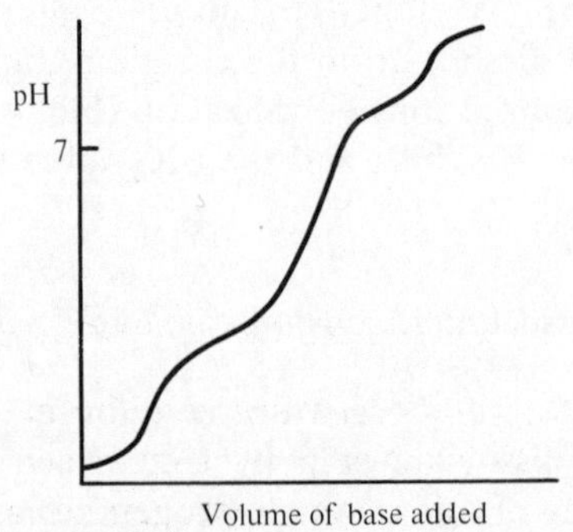

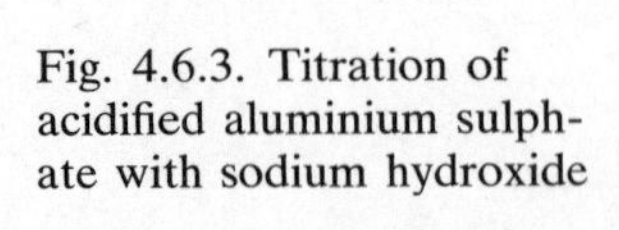
Fig. 4.6.3. Titration of acidified aluminium sulphate with sodium hydroxide

occurs after the neutralization of the excess acid and the third step indicates the formation of sodium aluminate.

The last titration carried out in this experiment illustrates a method of estimating Mg^{2+} in the presence of Ca^{2+} ions. With NaOH as titrant, this corresponds to the precipitation of an insoluble hydroxide ($Mg(OH)_2$) in the presence of a less insoluble hydroxide ($Ca(OH)_2$). It may be shown that precipitation of the magnesium hydroxide continues while

The preferential precipitation of $Mg(OH)_2$ from a solution containing both Mg^{2+} and Ca^{2+} ions is a basis of the commercial manufacture of $Mg(OH)_2$ from sea water. The $Mg(OH)_2$ so produced is dehydrated to form MgO which is used as a refractory.

$$\frac{[Mg^{2+}]}{[Ca^{2+}]} > \frac{K_{sp,Mg(OH)_2}}{K_{sp,Ca(OH)_2}}, \tag{4.6.13}$$

where K_{sp} is the solubility product of the species. Since $K_{sp,Mg(OH)_2}$ is a factor of 10^7 less than $K_{sp,Ca(OH)_2}$, the precipitations are effectively independent, and the titration can be used for the quantitative estimation of Mg^{2+} in the presence of Ca^{2+}.

Apparatus

A suitable apparatus for this experiment is shown in Fig. 4.6.4.[3] It has four parts.

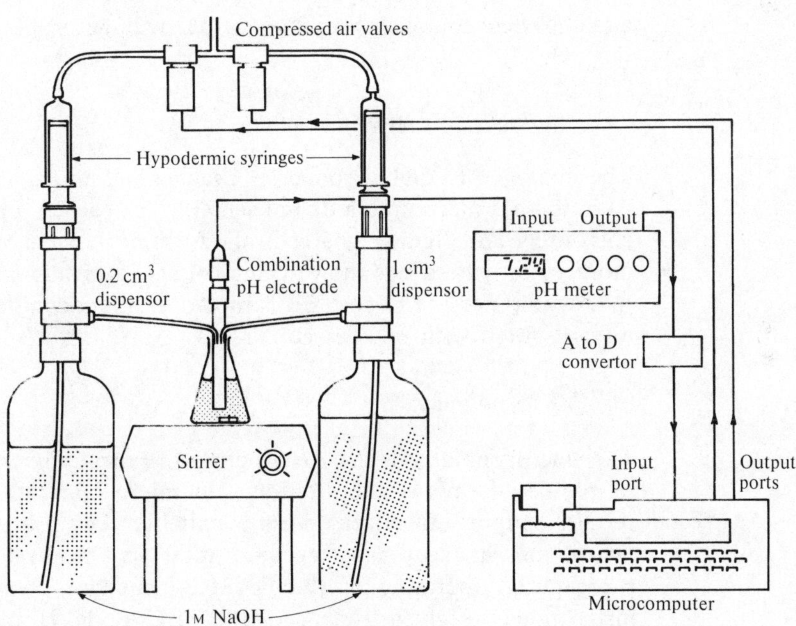

Fig. 4.6.4. The titration apparatus

(i) *The titration system*

The sodium hydroxide solution used in the titrations is delivered from two bottles. Each is fitted with a *dispenser* in the form of a calibrated ground-glass syringe which delivers a preset volume of liquid from a stock bottle on full depression of a plunger. One dispenser is set to deliver 1.00 cm^3, the other 0.20 cm^3. Thus alkali can be added in 1 cm^3 portions when it causes the pH to change slowly, and in 0.2 cm^3 portions when the pH changes rapidly (e.g. in the region of an end point).

The plunger on each dispensor is operated by compressed air acting on a piston made from the body of a hypodermic syringe. The compressed air supply is controlled by two solenoid valves, one for each dispensor, which are operated by the microcomputer.

The bottles are arranged on either side of a magnetic stirrer, on which is placed a 250 cm^3 conical flask containing the solution to be titrated.

(*ii*) *pH meter*

During the titration the pH of the solution is monitored by a combination (glass/reference) electrode connected to a pH meter. As well as displaying the pH, the meter also provides an analogue (continuously variable) output voltage proportional to the pH.

(*iii*) *Analogue to digital converter*

The analogue to digital converter changes the analogue voltage from the pH meter into a digital signal which can be read by the computer. The digital signal is in the form of 0 or $+v$ V on 8 or more wires, where v is the working voltage, typically 5 V. Some microcomputers (e.g. the BBC model B microcomputer) are already fitted with this device.

(*iv*) *Microcomputer*

The microcomputer is used to operate the titration system and to monitor the pH of the solution. Any microcomputer with the required input and output ports is suitable. The programming instructions in the present account apply specifically to a microcomputer operating in the BASIC language. Two sets of instructions are given—set A for the BBC model B microcomputer, and set B for the Rockwell Aim 65 microcomputer.

Procedure

(*i*) *Setting up of apparatus*

Place a conical flask under the outlet tubes of the titrator in case the dispensers operate while you are setting up the apparatus.

Switch on the pH meter and allow it to warm up.

Calibrate the pH meter using buffer solutions of pH 7 and pH 4.*

(ii) Setting up of computer

A. BBC B microcomputer

Switch on the computer, which is ready to be programmed in BASIC.

The pH meter is connected to channel 1 of the computer's analogue to digital convertor. Check that the computer is reading in from this channel by running the short program:

```
10 *FX16,1
20 *FX17,1
30 X=ADVAL(1)
40 PRINT X
```

The computer should print a five-figure number.

Enter the command

```
?&FE62=&FF
```

to connect the output port.

B. Rockwell Aim 65

Note that all commands which are underlined refer to a single key.

Switch on the computer. In order to use BASIC, type 5, then type RETURN in response to the questions 'MEMORY SIZE?' and 'WIDTH?'. The computer is then ready to be programmed in BASIC.

Check that the computer is reading in from the pH meter by typing PRINT PEEK (40691) RETURN. The computer should respond with an integer between 1 and 255. If the number is zero, switch off the computer for a few moments and then try again.

Enter the command POKE 40962,255 to connect the output ports of the computer to the valves.

(iii) Titrations

Switch on the solenoid valves and turn on the compressed air supply.

Devise a BASIC program to carry out titrations automatically, i.e. operate the appropriate dispenser, and print out the pH as a function of the volume of alkali added. Information to help you do this is given in sections (iv), (v) and (vi).

Carry out titrations with 1 M sodium hydroxide solution of the solutions in the list below. Just before each titration, remove any air bubbles from the dispenser tubes by operating the plungers manually a few times.

(a) 1 M hydrochloric acid,
(b) 1 M ethanoic acid,
(c) 0.5 M malonic acid,
(d) 0.5 g aluminium sulphate $Al_2(SO_4)_3$ dissolved in distilled water containing a little dilute sulphuric acid,
(e) about 0.5 g each of magnesium and calcium carbonate, dissolved in dilute hydrochloric acid, and boiled carefully to remove all carbon dioxide.

In each case use 25 cm^3 of solution, and add sufficient distilled water to cover the end of the electrode.

At the end of the experiment, *unscrew the dispensers from the bottles and pump purified water through them.*

(*iv*) *Designing the program*

You will find it useful, if not essential, to construct a *flow diagram* of the proposed computer program, of the type shown on p. 342. Bear in mind that the following factors need to be considered:

(a) the computer must read in and print out the pH, and print out the volume of alkali that has been added,
(b) the rate of change of the pH with titrant volume must be monitored so that the computer has warning of when an equivalence point is approaching, and it is useful for the program to print out the monitoring parameter (see below),
(c) the dispensors must be operated to deliver the appropriate amount of liquid (1 cm^3 or 0.2 cm^3),
(d) the program must end at a specified point, e.g. when 30 cm^3 of alkali has been added.

(*v*) *Reading pH and determining equivalence points*

The digital signal read by the computer from the A to D convertor is a five-figure integer in the case of the BBC B microcomputer, and an integer between 1 and 255 in the case of the Rockwell Aim 65. The integer must be converted back into a pH value. In order to detect the approach of an equivalence point, pH readings need to be stored so that each reading can be compared with the previous one. This is carried out by placing them in a matrix, the dimensions of which need to be given early in the program. The gradient, G, of the line joining two consecutive points of the titration curve must be calculated, and when $|G|$ reaches a certain value (e.g. if $|G|>5$) the volume increment should be decreased from 1 cm^3 to 0.2 cm^3. Each dispenser is operated by a solenoid valve which controls the compressed air supply. Sufficient time needs to be allowed (a) for the plunger to be depressed after the valve has opened, and (b) for the plunger to rise again after the valve has been closed. Various program lines to carry out these operations are given below.

DIM P(50)	Assigns 50 places in a matrix, P.
G = (P(J) − P(J − 1))/D	Calculates the gradient G of the line joining P(J) and P(J − 1), where D is the volume increment.
P = P(J)*C	Converts the number to pH. C is a conversion factor which you will have to calculate by comparing the pH displayed on the pH meter with the number read in on the analogue to digital convertor channel.

FOR K = 1 TO N	This subroutine generates a pause. N will probably need to
NEXT K	be in the range 1000–10 000.

A. BBC B Microcomputer

?&FE60 = &0001	Opens valve to add 1 cm^3.
?&FE60 = &0002	Opens valve to add 0.2 cm^3.
?&FE60 = &0000	Closes both valves.

B. Rockwell Aim 65

POKE 40960, 1	Opens valve to add 1 cm^3.
POKE 40960, 2	Opens valve to add 0.2 cm^3.
POKE 40960, 0	Closes both valves.

(vi) Further hints on programming in BASIC

Having drawn the flow diagram, write out the program in BASIC, numbering the lines in increments of 10. Then type it into the computer, ending every line by pressing the RETURN key. If you make a mistake, simply retype the appropriate line number and the corrected line. Lines may be inserted by typing in a line which has a number between numbers of the two neighbouring lines. It is good practice to put in more PRINT statements than will ultimately be necessary, to check on the calculation of all the important variables. Where the program branches, or if you employ subroutines, mark the line numbers on the flow diagram.

Ambitious programmers may like to include in the program a routine for compensating for linearity error in the pH meter (see Results), and for calculating pK_a values—but remember that a purpose of this experiment is to show how microprocessors can save time, not waste it....

When you have finished typing the program into the computer, obtain a copy of it by giving the command LIST, and then RUN the program.

Some further operating commands will be needed for the AIM 65, and these are given in the Technical Notes.

Calculation

Plot the titration curves for each of the five systems, displaying repetitions on the same graph.

Calculate K_a and pK_a for ethanoic acid, and K_1, K_2, pK_1 and pK_2 for malonic acid.

Explain the shapes of the curves for the two precipitation titrations.

Results

See Figs. 4.6.2 and 4.6.3.

Accuracy

You may find that at high and low pH values there is a difference between the pH meter reading and the pH reading from the computer. The difference arises from errors in the analogue voltage from the pH meter, or in the linearity of the amplifier. No serious error should result, but it can nevertheless be corrected by making a manual adjustment, or by writing a simple function into the program. Having considered this, you should then estimate carefully whether the accuracy of the results in this experiment is limited by the precision of the titration. Calculate the precision of the analogue-to-digital convertor from its number of bits (p. xviii): the convertor in the BBC micro acts effectively as a 12-bit convertor in a laboratory environment, whereas a laboratory-assembled convertor will probably be 8-bit. Remember also that the precision of the titrant volume cannot be greater than the volume added by the smallest dispensor.

Comment

pH titrations are a reliable method of finding end-points because they rely on a whole series of measurements, rather than one ill-defined point as in indicator titrations. They are very suitable for use with small samples, and can be adapted for use with many other systems, for example the concentration of sulphides can be estimated by employing a lead electrode and titrating with lead salts.

The importance of carrying out the titration with a microcomputer lies in the fact that most analytical instruments used in industry and research now incorporate microprocessors, which are the essential components of computers. The next stage in using the present system would be to convert the whole program to *machine language* and store it permanently in a plug-in memory known as an *EPROM*. The titrator could then be switched on, and would simply 'get on with the job', finally displaying the end-point, pK_a, and any other information required.

Technical notes

Apparatus

See Fig. 4.6.4. Microcomputer[$] (see Apparatus (iv)); combination electrode, or glass electrode and reference electrode; stirrer; pH meter with digital output, or with analogue output connected to analogue-to-digital convertor in computer or separate (e.g. RS (RadioSpares) 303–454); titrator[†] (see below). 1 M hydrochloric acid, 1 M ethanoic acid, 0.5 M malonic acid (5.203 g/100 cm^3). *Either* aluminium sulphate, magnesium carbonate, calcium carbonate, dil. sulphuric acid, dil. hydrochloric acid, conical flask, bunsen, tripod, gauze *or* solution (d) and fresh solution (e), (p. 160). 1 M sodium hydroxide.

Design note

The present titrator makes use of a low-pressure compressed air supply. This is fed to glass syringes (plastic syringes stick) via electrically operated valves (e.g. RS 348-380 + relays RS 348-582 driving RS 346-974), which depress dispensers set at 1.0 cm^3 and 0.2 cm^3 (e.g. Oxford Adjustable Dispensor, by Lancer Lab. Prods. Int. Inc., available in U.K. from Boehringer Corp., Lewes BN7 1LG). In the absence of compressed air, an electric motor and cam arrangement could be used, with springing to prevent damage to the dispensors. A syringe and stepping motor arrangement could also be used.[4] Peristaltic pumps are not suitable, since they have an accuracy of 3 per cent at best.

Choice of computer

The BBC B microcomputer is recommended as a computer which is easy to set up, which students may already be familiar with, and for which much software and general expertise is available. The Rockwell Aim 65 is recommended as a cheaper, dedicated machine which has a display and printer already incorporated. Ref. 4 refers to the Apple II plus microcomputer.

Length of experiment

The experiment can be shortened by giving students an unfinished program.[3]

Further AIM 65 commands

5	Enter BASIC. (Type RETURN in answer to the two questions). The program may then be typed in.
6	Re-enter BASIC after resetting without losing previous program.
F1	Stop the program while it is running.
CTRL PRINT	Press these two keys together to turn the printer OFF or ON. This saves paper when printing is unnecessary.

	An exclamation mark used in a PRINT statement will cause printing to occur even when the printer is off, e.g. PRINT! 'VOLUME'.
DEL	Delete the previous character.
?	Shorthand for PRINT.
LF	Line Feed. Feeds paper without printing.
RESET	(White button to the left of the display.) Use this if the computer seems lost in a program and doesn't respond to anything else. Type 6 to re-enter BASIC without losing program.

References

[1] *Electrochemical methods.* A. J. Bard and L. R. Faulkner. John Wiley, New York (1980), p. 73.

[2] *Ionic equilibrium. A mathematical approach.* J. N. Butler. Addison-Wesley, Reading, Massachusetts (1964), p. 206.

[3] *Computing advances in the teaching of chemistry.* W. P. Baskett and G. P. Matthews. *School Sci. Rev.* **66,** 19 (1984).

[4] *A simple and inexpensive pH-stat and autotitrator based on the Apple II plus computer.* R. D. Cornelius and P. R. Norman. *J. Chem. Educ.* **60,** 98 (1983).

Part II Structure

Section 5
Structure of atoms and simple molecules

This section contains thirteen experiments in which the structures of various chemical species are studied by spectroscopic methods, and, in the last instance, by quantum mechanics. The first task in spectroscopic investigations is the preparation of the sample. All the gaseous samples in the present experiments are assumed to have been made up in advance, although instructors may wish their students to be involved with this. The liquid samples require only the careful quantitative manipulation of solids and solutions for their preparation. Once a sample has been prepared, its spectrum may be measured by use of the appropriate spectrometer. The method of operating such an instrument is a knack which can be acquired by the study of the detailed manuals supplied, but which is generally learned much more quickly and more profitably from an instructor.

Although the practical aspects of these experiments should present no great difficulties, the theory of the effects on which they are based can be complicated. This is true for all of the present experiments, apart from two, 5.5 and 5.9, in which the intensity of a particular absorption is used simply as the measure of the amount of a species in solution.

A loose definition of spectroscopy is that it is the elucidation of the structure of a species by means of a spectrum of intensity against energy. In the case of mass spectroscopy, Expt 5.1, the sample molecules are bombarded by electrons and made to fragment. The spectrum is formed by scanning through the masses of the fragments so formed, and the identity or composition of the sample can be deduced by putting the fragments together like a jig-saw puzzle to recreate the structure of the molecular ion.

In the remaining experiments it is the electromagnetic spectrum which is scanned. Figure 5.0.1 shows the various regions of the spectrum, and the particular structural aspects which may be studied by spectroscopy in each region.

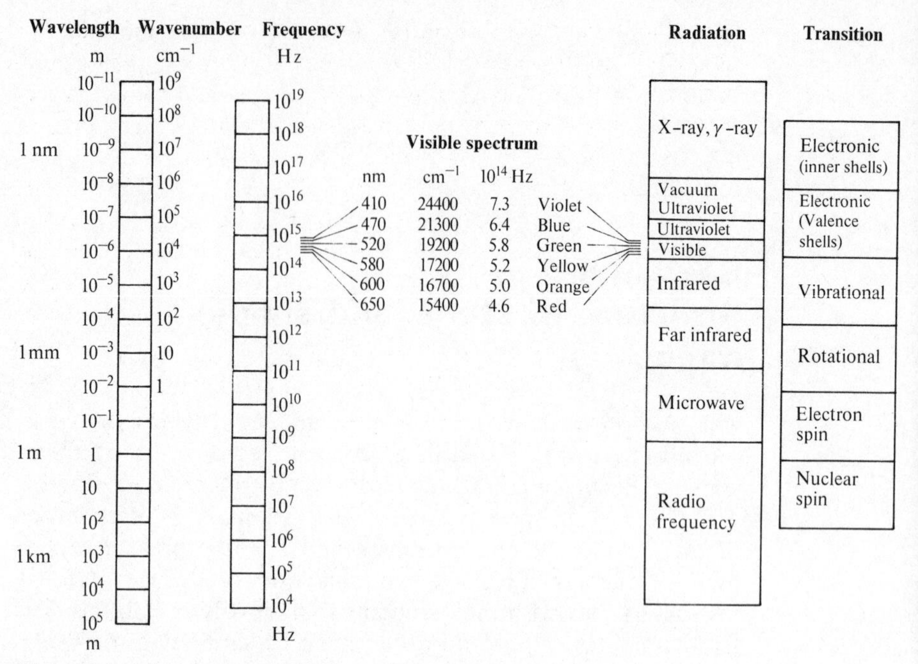

Fig. 5.0.1. The electromagnetic spectrum

Experiment 5.2 concerns the measurement of dipole moments by measuring the capacitance of a sample at radio frequencies, and its refractive index in visible light. The experiment introduces dipole moments, which are useful in the understanding of bonding and solvation. However, the method has now been superseded by the more precise spectroscopic techniques studied in the experiments which follow.

In the case of gaseous atoms, Expt 5.3, the excitation of samples in an electric discharge gives rise to visible and ultraviolet emissions which, when dispersed, form line spectra. The energies of the spectral lines yield the energies of the orbitals in the atoms. Experiment 5.3 also illustrates how the intensity of lines emitted from a flame may be used to measure the flame temperature.

The nine experiments which follow concern the elucidation of the structure of molecules by means of radiation of progressively higher frequency. In Expt 5.4 we use radiation in the far infrared region to examine the pure rotation spectra of HCN and NO. The experiment introduces interferometers, and the technique of Fourier transformation.

Experiments 5.5–5.8 involve radiation in the infrared region, from which information about the vibrations in molecules can be gained. Perhaps the commonest use of both infrared and ultraviolet spectroscopy is in the identification of organic compounds, by using their absorption spectra as a 'fingerprint'. However, such experiments tend to be the province of organic chemistry, and we do not include them here. Apart from simple identification, the most straightforward method of using infrared spectroscopy is to employ the intensity of a particular absorption band as a measure of the amount of substance in solution, linking the two quantities by means of the Beer–Lambert law. This type of investigation is exemplified by Expt 5.5, in which a sample is heated, and the change in intensity of the —O—H absorption with temperature yields the enthalpy of hydrogen bond formation between the sample and its solvent.

In Expt 5.6 we examine the fine structure of a vibrational spectrum caused by rotational transitions. The spectra are analysed to give structural constants for HCl, DCl and ethyne (acetylene). In Expt 5.7 we use quantum mechanics to analyse the infrared spectrum of chloromethane, and obtain from it the heat capacity of the gas. This is compared with heat capacity measurements from an acoustic interferometer. Experiment 5.8 involves vibrational Raman spectroscopy, where the transitions still occur in the infrared region, but now are due to periodic changes in the polarizability of the molecules, rather than in their dipole moments.

Experiments 5.9–5.11 involve the emission and absorption of radiation in the visible and ultraviolet regions of the spectrum. As shown in Fig. 5.0.1, radiation at these frequencies causes, or arises from, electronic transitions in the molecules. Once again, the simplest type of experiment is to use the intensity of a particular absorption as a measure of the amount of a species in solution, and Expt 5.9 illustrates how the stoichiometry and stability constant of a complex may be determined by this method.

The analysis of the band spectrum of iodine, arising from fine structure in the electronic spectrum, is the subject of Expt 5.10. The positions of the bands are analysed to give the dissociation energy of the molecule. Experiment 5.11 is a study of fluorescence and phosphorescence, which arise from complications in the transitions between vibrational levels in the different electronic states. Part B of the experiment uses a fluorescent species as a 'clock' for the measurement of the rates of electron transfer reactions, and therefore anticipates a sound knowledge of reaction kinetics (Section 8).

Nuclear magnetic resonance occurs at radio frequencies when the sample is placed in a magnetic field. Experiment 5.12 gives a general introduction to n.m.r., and to the characteristics of a simple n.m.r. spectrometer.

Finally, in Expt 5.13, we calculate the conformational energies of various simple molecules by extended Hückel theory. The experiment is based on a computer program which can be run on any reasonably powerful microcomputer.

5.1 Quantitative and qualitative analysis by mass spectrometry

This experiment gives practice in the use of a mass spectrometer, one of the most important and well-known analytical instruments in chemistry. In the apparatus, a sample in the form of a neutral gas or vapour is converted to positive ions by bombardment with electrons. The ions are then analysed according to their charge-to-mass ratio, and a *mass spectrum* is produced from which the composition or identity of the original sample may be deduced.

The experiment is in several parts, each of which illustrates a different use of mass spectrometry. Part A concerns the determination of the relative abundances of isotopes in a pure gaseous sample, and Part B the analysis of a simple gaseous mixture. Part C provides some familiarization with spectra of known simple organic compounds, and with metastable peaks. In Part D the principles acquired are applied to the identification of unknown organic samples from their mass spectra. Finally in Part E, the height of a particular peak in the mass spectrum of methyl cyanide (acetonitrile, ethanonitrile), CH_3CN, is plotted as a function of bombarding electron energy to give the *ionization efficiency curve*, from which thermochemical information may be derived.

Theory[1,2]

In this section we consider the theory of mass analysis using a magnetic mass spectrometer. The layout of a simple mass spectrometer is shown in Fig. 5.1.1; the component parts of the instrument are described in more detail in the Apparatus section.

(*i*) *Trajectories*

The atoms or molecules of the sample are converted into positive ions in the ion source, described below. The ions, of charge e and mass m, are usually produced at thermal energies,

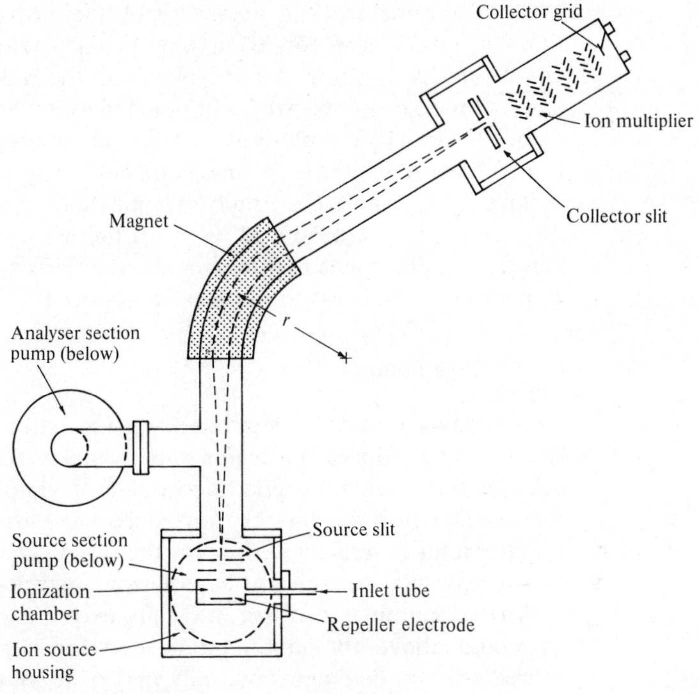

Fig. 5.1.1. A simple mass spectrometer

so that their initial spread of velocities is relatively small. If they are then accelerated through a voltage V, to a good approximation their kinetic energies will be given by

$$\tfrac{1}{2}mv^2 = eV, \tag{5.1.1}$$

where v is their final velocity. Subsequently the ions are then deflected by a homogeneous magnetic field of strength B, which exerts a force Bev on the ions at right angles to both the field and the direction of motion. This force makes the ions follow a circular trajectory of radius r, given by:

$$Bev = mv^2/r, \tag{5.1.2}$$

or

$$mv/e = Br. \tag{5.1.3}$$

Combining eqns (5.1.1) and (5.1.3),

$$m/e = B^2r^2/2V. \tag{5.1.4}$$

The parameters on the left of this equation are characteristics of the ions, and those on the right-hand side are determined by the

A typical mass spectrometer has a field strength B of around 2000 gauss (0.2 Tesla), an accelerating voltage V of a few thousand volts, and a trajectory radius r of a few centimetres.

mass spectrometer. The geometry of the instrument dictates that only those ions following a particular radius of curvature will reach the detector. All others will hit the walls of the vessel and will then be neutralized and pumped away. So for particular values of B and V, only ions of a specific charge to mass ratio m/e will be detected. If the magnetic field B or the accelerating voltage V is scanned, a graph of abundance against charge to mass ratio may be produced, which is the mass spectrum of the compound. Note that if B is scanned, the mass scale is crowded at high mass, because m is proportional to B^2.

(*ii*) *Fragmentation*

The removal of single electrons from the parent molecules of the sample produces *molecular ions.* Some of these are stable, but others are highly excited and dissociate into charged *fragments.* The pressure in the ion source and the body of the spectrometer is sufficiently low for the fragmentation process to be unimolecular, unaffected by collisions between the ions.

When a parent ion dissociates, any excess excitation energy over and above the minimum needed for bond breaking is shared among the fragments, only part of it going into relative kinetic energy. Some primary ion fragments then have enough energy to dissociate further, producing secondary fragments. In propanone, for example, the following process may take place:

$$CH_3COCH_3 + e^- \rightarrow \underset{\text{parent ion}}{CH_3COCH_3^+} + 2e^-,$$

$$CH_3COCH_3^+ \rightarrow \underset{\text{primary fragment}}{CH_3CO^+} + CH_3,$$

$$CH_3CO^+ \rightarrow \underset{\text{secondary fragments}}{CH_3^+ + CO}.$$

The overall fragmentation pattern is the result of such competing consecutive unimolecular dissociations. The mass spectrum can be interpreted by recognizing characteristic ions or neutral fragments, and putting them together like a jig-saw puzzle to reconstitute the molecular ion.

(*iii*) *Metastable ions*

If an ion, whether a parent, primary or secondary fragment, has only just enough energy to overcome the activation barrier to dissociation, then its unimolecular dissociation rate may be sufficiently slow (10^5 to $10^6\,s^{-1}$) that some ions dissociate after acceleration, but before entering the analyser magnetic field. After such a dissociation, which we may write

$$m_1^+ \rightarrow m_2^+,$$

the daughter ion will have the velocity characteristic of its parent ion. We may express this phenomenon, using eqn (5.1.1), by the relation

$$v_2 = v_1 = (2eV/m_1)^{1/2}. \tag{5.1.5}$$

In the analysing magnet the daughter ion follows a path of radius r:

$$m_2 v_1/e = Br. \tag{5.1.6}$$

Hence from eqns (5.1.5) and (5.1.6),

$$m_2^2/m_1 e = B^2 r^2/2V. \tag{5.1.7}$$

By comparing this expression with eqn (5.1.4), it can be seen that such ions will appear at an apparent mass $m^* = m_2^2/m_1$. Furthermore, because the point at which these are formed is not precisely defined, and also because their kinetic energy will deviate from the nominal value by any energy released in dissociation, the ions give rise to broad, diffuse peaks in the mass spectrum. These *metastable peaks* are extremely useful in interpreting spectra, because they directly demonstrate the relationships between parent and daughter ions. They can be recognized by their characteristics of being: (a) extremely weak (maximum sensitivity is needed); (b) broader (sometimes much broader) than normal peaks; and (c) at non-integral mass.

(*iv*) *Ionization efficiency curves*

The maximum energy that can be transferred to a molecule by an electron is the kinetic energy of the electron itself. Any lesser energy can also be transferred, however, in which case the electron continues in its course with diminished velocity. Processes such as ionization and fragmentation, which require definite minimum energies, can only occur if the energy of the ionizing electron is greater than the relevant threshold. At the threshold the probability of reaction is zero, but at higher energies it rises continuously as a function of electron energy. In practice there are no very sharp changes in plots of ion yield versus energy, known as *ionization efficiency curves*, because the thermionic electrons used for ionization have a broad spread of energies. Nevertheless, such plots can be used to make qualitative and approximate quantitative deductions about the thermochemistry of ionic reactions.

Apparatus[1] There are a great number of modifications and arrangements of mass spectrometers, and the description below is limited to the ion source and detector used in one common type of instrument.

(*i*) *Ion source*

The most common ion source is an *electron impact source.* A jet of gas or vapour is directed into an ionization chamber at a pressure of about 10^{-5} Torr. There it is intercepted by a well-collimated beam of electrons produced by a hot filament, commonly of tungsten. Some of the electrons collide with the sample molecules to produce positive ions, and those which do not are collected in a trap at the other side of the chamber. The positively charged molecular ions and fragments are driven out of the ionization chamber by a repeller electrode, and then accelerated through the source slits towards the mass analysing section of the instrument. In sources of this type the energy of the bombarding electrons is not very sharply defined, but can be altered easily. The sources produce ions which have a low spread of energies, particularly useful in simple mass spectrometers.

(*ii*) *Mass selector*

The instrument described in the present experiment uses a magnetic sector, with an electromagnet of which the field can be scanned. Resolution can be varied by adjusting the width of a slit. The shape of the field has been calculated to provide what is known as *first order focusing.* This means that all ions of the same mass and kinetic energy come to a focus at the collector slit, irrespective of the angle at which they leave the source slit.

(*iii*) *Detector*

The standard method of detecting ions in a modern mass spectrometer is by means of an *ion-electron multiplier.* This type of detector comprises a series of plates or *dynodes,* often shaped like Venetian blinds, which are maintained at different voltages. The positive ions, with energies in the range 5000–10 000 eV, strike the first dynode and typically release two or three electrons. The multiplier is shaped so that these electrons strike the next dynode. Each electron then produces another two or three, and so the effect multiplies down the chain of dynodes to produce a signal current at the collector grid proportional to the number of ions which originally hit the detector.

Procedure *General instructions*

(i) Start up and sample handling

Turn on the mass spectrometer following the instructions supplied.*

Samples are usually stored in cylindrical or spherical glass vessels, or in a length of glass tubing between two taps, and attached to the sample handling section of the spectrometer by means of a socket and cone joint. Read any instructions about the particular samples you will be using.* Be sure to inform a demonstrator if you know that air has been admitted to a sample (which will be revealed as an m/e 32 peak in the spectrum), if a sample vessel is empty or if the grease on a tap is streaky. Liquid samples can also be injected through a septum, and all samples are expanded into a reservoir before being admitted to the mass spectrometer.

(ii) General method for recording mass spectra

Search the spectrum manually at a series of diminishing sensitivities until you find the largest peak. Adjust the amplifier sensitivity so this peak is as large as it can be without going off-scale on the chart recorder.*

Write the amplifier sensitivity setting on the chart recorder paper, and switch on the paper feed. Set the spectrometer to scan from about $m/e = 12$ over about half of the scale, or over a convenient smaller range for the problem in hand. Adjust the scan speed (i.e. time for a scan) to the fastest at which the pen is able to follow the peaks to their true heights. (If you scan too fast, the pen recorder cannot follow the peaks, and the apparent heights are meaningless.) When a useful spectrum has been obtained, write the date, name of sample and any vital conditions on the chart paper *at the time.*

Turn the amplifier to higher sensitivity and repeat the procedure. The large peaks will now go off scale, but the smaller ones will be recorded much more accurately. To record all doubly-charged ions and metastable peaks, make one run at maximum sensitivity.

The full procedure for reducing a mass spectrum to standard form is given below. An example of the calculations is given in the Results section.

(a) Number the mass scale by counting from known peaks such as those given in the specimen blank spectrum. Interpolate the scale in the blank regions of the spectrum.

(b) Measure the heights of all the reasonably large peaks in millimetres from the baseline. Carry out the measurement on the spectrum on which the particular peak is largest but not off-scale.

(c) Now multiply or divide each peak height by a factor related to the amplifier sensitivity at which it was measured, thus obtaining a series of peak heights relating to the sensitivity setting of the blank spectrum.

(d) Tabulate your results as a list of peak heights against mass number. Subtract the blank spectrum from the sample spectrum. Multiply the resulting peak heights by a normalizing factor so that the maximum height is 100 units, and draw the spectrum as a line graph.

Go through this full procedure for *one* of your mass spectra, and produce a neat, well labelled line graph, with the main peaks identified. Analyses of the other spectra can be done using the annotated raw spectra themselves.

(*iii*) *Blank spectrum*

Record the *blank spectrum*, following the procedure described above. This is the spectrum with no sample in the instrument, and is caused by residual material such as air, water vapour and hydrocarbons from the pump oil. The peaks should become smaller as pumping out continues and the ionization chamber warms up; if this has not occurred after 30 min, consult an instructor.

A typical blank spectrum is shown in the Results section. Learn to recognize the masses of the main peaks. The current fed to the electromagnet can be calibrated to give a rough indication, but *masses must be checked by counting from a known peak*. In the blank spectrum there will always be a substantial peak at m/e 18 due to residual water (H_2O^+) together with a peak at m/e 17 due to the fragment ion OH^+ derived from it. From residual traces of air there will be a peak of m/e 28 due to nitrogen $(^{14}N^{14}N)^+$ together with some CO^+ (also m/e 28) formed by reaction of water vapour with traces of carbon on the filament. A peak at m/e 32 is given by molecular oxygen. As not many other substances give an ion of mass 32 a peak at this mass number usually indicates contamination of the sample by air, especially if associated with an m/e 28 peak about five times the size of the m/e 32 peak. In addition to these peaks at m/e 17, 18, 28 and 32, the blank will show a pattern of smaller peaks up to the top of the scan at about m/e 200. These are due to minute residual traces of hydrocarbons and are always present in systems, evacuated by conventional pumps, which have not been baked out at high temperature. Allocate mass numbers to the larger peaks by reference to the annotated specimen spectra.

Note the groups of peaks corresponding to fragments with 1, 2, 3, 4, 6 . . . carbon atoms. The alkyl ions $(C_nH_{2n+1})^+$ tend to be the largest in each group. There is a general tendency for hydrocarbon fragment ions to show an alternation in intensity. *Why?*

When the filament is turned on after being off overnight the blank will be large until the hydrocarbons, which are desorbed from the ionization chamber region as it warms up, have been pumped away. Record the blank mass spectrum at intervals and note the intensities at m/e 41 and 43 (the largest peaks given by residual traces of hydrocarbon oils).

A. *Measurement of isotopic abundance*

Record the mass spectra of the sample or samples supplied.

Suitable samples are neon, argon and xenon. The spectrum of neon should be recorded from m/e 12 to 45 on several sensitivities. Identify the isotopes of neon and look for impurities in the sample.

In the case of argon, there are some isotopes of low abundance at mass numbers at which there are contributions from the blank spectrum. So record the blank spectrum both before and after the argon spectrum, and subtract it carefully by the procedure described previously.

Xenon has many isotopes at masses between 126 and 145: identify them all and find their relative abundances.

B. *Qualitative analysis of air*

Measure the mass spectrum of a sample of air and compare it with a sample of respired air (your breath) blown through a sample tube.

C. *Spectra of small organic molecules*

Measure the mass spectra of some of the known samples supplied. Find and interpret examples of metastable peaks, rearrangement ions and doubly-charged ions.

D. Identification of unknown organic compounds

Measure the mass spectra of two or three of the marked samples provided, and suggest possible identifications.

E. Dissociation energetics of the CH_3CN^+ ion

Methyl cyanide has a low vapour pressure, so admit two or three doses of its vapour into the spectrometer. Tune the spectrometer to the $CHCN^+$ peak (m/e 39). Consult the operating instructions* and measure the ionization efficiency curve, i.e. the height of this peak as a function of bombarding-electron energy. Examine your curve *at once*, and decide whether an absolute calibration of the energy scale is necessary for the interpretation. If so, record the ionization efficiency curve of the parent ion, CH_3CN^+, in the same way.

Calculation

A. Isotopic abundances

Measure the isotopic abundances in the one or more samples which you have studied. Since the sensitivity of the detector is the same for different isotopes of the same element, the relative abundance of two isotopes is equal to the ratio of the heights of their peaks.

B. Gas mixture

Identify the components in the samples of air. Start with the largest peaks, and determine whether these can be assigned as the parent peaks of possible components. The blank spectrum shown in Fig. 5.1.2 will help in the interpretation of the air

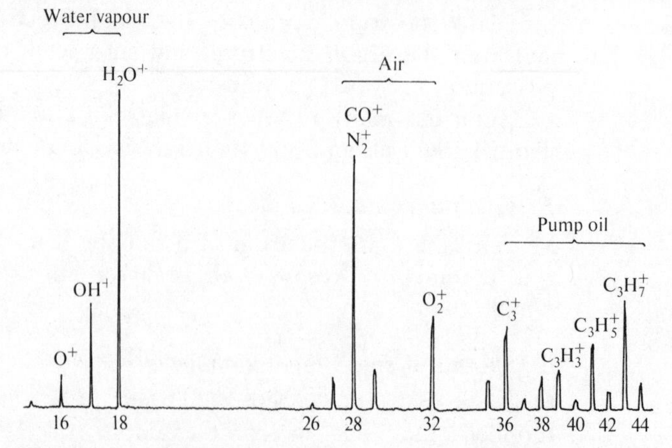

Fig. 5.1.2. A blank spectrum

spectrum. Tabulate the spectra of fresh and respired air using $N_2^+ = 100$ as a basis. Is any loss of O_2 in respired air compensated by an increase in CO_2? Is all the O_2 used up?

C and D. Identification of polyatomic samples

The identification of organic molecules from their mass spectra is a refined art which can be learned from tutorial texts on the subject.[1,3] The following rules, together with the worked example in the Results section, are sufficient to identify straightforward samples.

(a) Molecular ion

The molecular ion, formed by removal of one electron from a molecule, must have an odd number of electrons. If the ion at the highest m/e value is an even-electron ion it cannot be the molecular ion. If a fragment appears at a chemically impossible interval below the ion thought to be the parent (e.g. an interval of 7–11, 20–23 mass numbers), this is evidence that the suspected parent ion is itself a fragment.

(b) Deduction of molecular formula

Having identified the parent or molecular ion, calculate by difference the masses of the neutral fragments lost to give the more abundant ions. For example, in the case of n-butane, the ion at m/e 43 is formed from the parent by loss of $(58-43) =$ 15 units, i.e. a methyl group, and that at m/e 29 by loss of 29 units, i.e. C_2H_5. It is often useful, having arrived at the molecular formula, to write down all the possible structures and then see which agrees best with the other features of the spectrum. The following 'Rules' are often of help in assigning a formula on the basis of a mass spectrum:

The Nitrogen Rule: If a molecular ion has an odd m/e value it must contain an odd number of nitrogen atoms. If the m/e value is even, the compound contains an even number of nitrogens (zero is even).

The nitrogen rule is broken only by some organo-metallic compounds of metals of variable valency.

The Hydrogen Rules: (i) If the m/e value of a C, H, N, O compound is even, so is the number of hydrogen atoms it contains. (ii) If its m/e value is divisible by 4 so is the number of H atoms it contains.

The index of hydrogen deficiency for the molecule $C_6H_5 . CH_2 . CH_2 = CH_2$ is 5, i.e. 4 for the benzene ring and 1 for the double bond.

The Hydrogen Deficiency Rule: For an empirical formula $C_xH_yN_zO_n$, the *index of hydrogen deficiency*, i.e. the number of *pairs* of hydrogen atoms that must be added to the molecule to make it saturated, is given by: $x - \frac{1}{2}y + \frac{1}{2}z + 1$ (counting halogens as H).

(c) Recognition of elements other than C, H, N and O
Fluorine gives an ion at *m/e* 19, and the presence of a unit with this unusual mass is generally easily recognized. Elements such as sulphur, chlorine and bromine may be recognized by their characteristic isotopic abundance, as described in the next paragraph.

(d) Isotopic abundance
The elements which make up chemical samples are not isotopically pure. Naturally occurring elements not only give the expected major 'A' peaks, but may also give characteristics $A+1$ or $A+2$ peaks (i.e. one or two mass numbers higher) of lower abundance. The natural isotopic abundances of some common elements are shown in Table 5.1.1. Chlorine and bromine, for example, are recognizable from their 3:1 and 1:1 peaks two mass numbers apart. If a molecular ion contains more than one carbon atom, the probability of one of these atoms being a ^{13}C isotope will be correspondingly increased. For a molecular ion containing j carbon atoms, the relative abundance $(A+1):A$ will be $(1.1\times j)$ per cent.

Table 5.1.1. Relative natural isotopic abundances of some common elements

Element	A		$A+1$		$A+2$	
	Mass	Nominal abundance	Mass	Abundance relative to A	Mass	Abundance relative to A
H	1	100	2	0.016		
C	12	100	13	1.1		
N	14	100	15	0.36		
O	16	100	17	0.04	18	0.20
F	19	100				
S	32	100	33	0.78	34	4.4
Cl	35	100			37	32.5
Br	79	100			81	98.0

(e) Fragment ions
The most abundant fragment ions in a spectrum usually correspond to simple bond breaking, but exceptions also occur. Rearrangements involving hydrogen atom transfer are quite common. In the spectrum of isocyanic acid, HNCO, for example, there is a very intense peak for HCO^+ as well as for the NH^+ and CO^+ ions that can be formed by direct cleavage. In ring compounds single bond breakage does not lead to fragmentation, and rearrangements of ions are common.

(f) Doubly-charged ions
Ions with a double charge are formed quite easily by electron impact, but are not abundant in spectra of most molecules because the dissociation process

$$m_1^{2+} \rightarrow m_2^{+} + m_3^{+}$$

is usually favoured. Since the mass spectrometer actually measures mass/charge, doubly charged ions of odd-integer mass number appear in the spectrum at half-unit mass numbers, where they are easily recognized. When such ions are seen, they indicate that the parent structure is unusually stable, for example a structure which contains an aromatic or other ring species.

E. Ionization efficiency curves

As a first approximation we may assume that the yield $[X^+]$ of an ionic product X^+ is linearly proportional to the excess electron energy above its production threshold E_0:

$$[X^+] = \text{const.}(E - E_0). \quad (5.1.8)$$

If this is true, thresholds can be derived from ionization efficiency curves by linear extrapolation of straight sections either to the baseline or to lower linear segments. Application of this procedure to the CH_3CN data will give an apparent ionization potential for CH_3CN and one or more appearance potentials for $CHCN^+$. The absolute values may be inaccurate because of extraneous fields in the ion source causing scale shifts: adjust the scale to agree with the known ionization potential of CH_3CN of 12.22 eV.

The heat of formation of a positive ion is the enthalpy of formation of the ion and an electron in its standard state, i.e. in vacuum an infinite distance away.

To calculate thermochemical thresholds for $CHCN^+$ production use the following enthalpies of formation and dissociation, which relate to a spectroscopic temperature of 0 K: $\Delta H_f(CH_3CN) = 95\ \text{kJ mol}^{-1}$, $\Delta H_f(CHCN^+) = 1550\ \text{kJ mol}^{-1}$, and $D(H-H) = 432\ \text{kJ mol}^{-1}$.[4]

Results

A typical blank spectrum is shown in Fig. 5.1.2.

As an example of the interpretation of the mass spectrum of an organic compound, suppose that we obtain the spectrum shown in the first three columns of Table 5.1.2. Note that we measure the peaks at the highest sensitivity possible without them going off-scale. We then normalize the peak heights to a common sensitivity (column 4), and subtract the blank spectrum corrected to the same sensitivity to give the final spectrum for interpretation. It is useful to draw a line graph of the final spectrum, as shown in Fig. 5.1.3.

At first glance, the peak at $m/e = 96$ seems most likely to be that of the molecular ion. We then notice the nearly equal abundance at m/e 94, and the lower equal abundances at m/e 79 and 81. We

Table 5.1.2. Example of a mass spectrum

Mass number	Peak height/mm	Sensitivity setting	Peak height at sensitivity = 1	Blank at sensitivity = 1	Peak − blank	Normalized relative abundance
12	34	0.2	7		7	1.3
13	39	0.2	8		8	1.5
14	51	1	51	32	19	3.6
15	70	5	350	36	314	60.0
16	171	0.2	34	31	3	0.6
39.5	10	0.2	2		2	0.4
40.5	8	0.2	2		2	0.4
46	46	0.2	9	2	7	1.3
46.5	8	0.2	2		2	0.4
47	73	0.2	15	3	12	2.3
47.5	5	0.2	1		1	0.2
48	42	0.2	8	2	6	1.1
79	65	1	65	7	58	11.1
81	67	1	67	10	57	11.0
91	144	0.2	29	7	22	4.2
92	101	0.2	20	6	14	2.7
93	185	0.2	37	3	34	6.5
94	106	5	530	8	522	100.0
95	129	0.2	26	8	18	3.4
96	100	5	500	4	496	95.0
97	84	0.2	17	10	7	1.3

therefore deduce from Table 5.1.1 that Br is present. Thus the 94 and 96 peaks are the molecular ion with ^{79}Br and ^{81}Br respectively, and the rest of the molecule has a mass of 15, confirmed by a large peak. Comparing the abundance ratio of the 96 and 97 peaks suggests one carbon atom, and that the molecule is CH_3Br. This is confirmed by the

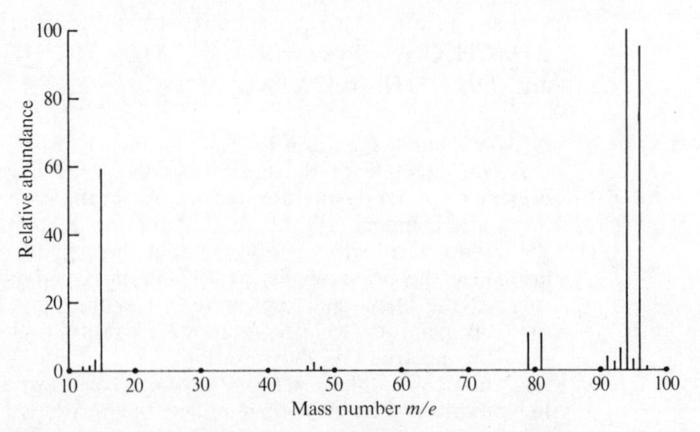

Fig. 5.1.3. Line graph of the CH_3 Br mass spectrum

other fragments in the spectrum. The peaks at 39.5–48 are due to doubly charged ions.

Comment

Mass spectrometers are vital analytical tools in many branches of chemistry. Modern instruments tend to be *double-focusing*. The molecular ions and fragments pass through a magnetic sector, and then through an electric sector. The high resolution obtained by this method, typically 1 part in 10^4, allows individual isotopes and allotropes to be identified. Modified mass spectrometers of this type can also be used for the study of the fragmentation patterns of individual mass peaks. Further developments include the use of chromatographic columns with mass spectrometers, computerized data acquisition and interpretation, the application of mass spectrometry to organometallic and complex natural products, and its use for carbon dating.

A highly sophisticated mass spectrometer has recently been constructed for the measurement of ^{12}C/^{13}C isotope ratios; this will allow the dating of tiny samples of archaeological material such as the much-publicised Shroud of Turin.[5]

Technical notes

Apparatus

Single-focusing sector mass spectrometer.$$ A: sample for isotopic abundance measurement, e.g. Ar, Ne or Xe. B: simple gas mixture. C: simple named organic molecules, e.g. propenenitrile (acrylonitrile) which gives a metastable peak and, in commercial samples, an identifiable impurity. D: organic sample with spectrum interpretable by rules given, e.g. 4-octanone (butyl-propyl ketone), or thiophene. E: methyl cyanide. Appropriate annotated specimen spectra.

References

[1] *An introduction to mass spectrometry*. J. H. Beynon and A. G. Brenton. University of Wales Press, Cardiff (1982).
[2] *Introduction to mass spectrometry*. H. C. Hill, Heyden, London 2nd edn (1972).
[3] *Interpretation of mass spectra*. F. W. McLafferty, University Science Books, C.A., 3rd edn (1980).
[4] *Energetics of gaseous ions*. H. M. Rosenstock, K. Draxl, B. W. Steiner and J. T. Herron. *J. Phys. and Chem. Ref. Data* **6,** Suppl. 1, (1977).
[5] *New directions in carbon-14 dating*. R. Hedges. *New Scientist* **77,** 599 (1978).

5.2 Dipole moment from measurements of relative permittivity and refractive index

In this experiment we measure the relative permittivity (dielectric constant) and refractive index of a solution of a chloronitrobenzene in cyclohexane. From the results we calculate the dipole moment of the sample, and hence deduce whether it is ortho, meta or para substituted.

Theory[1,2]

(i) Dipole moments and polarization

In some molecules, the centre of action of negative charge arising from the electrons is not coincident with the centre of action of the positive charge associated with the nuclei. Such molecules are called *polar* and are said to possess a *permanent electric dipole moment* p. If the two resultant charges have magnitudes $-q$ and $+q$, and are separated by a distance r, the dipole moment has magnitude

$$p = qr. \tag{5.2.1}$$

Physicists regard the dipole as directed from the negative towards the positive charge. Organic chemists prefer the opposite convention, from positive to negative, since this indicates the direction of electron drift. The symbol $\nrightarrow$ is then used to denote the dipole.

Let us consider a parallel-plate capacitor, as shown in Fig. 5.2.1. Let the charge per unit area on one plate be $+\sigma$, and let it be $-\sigma$ on the other. If the plates are of area A, then the charges on the plates are σA and $-\sigma A$ respectively. Electrostatic theory then gives the field between the plates, when the capacitor is in a vacuum, as σ/ε_0, where ε_0 is the permittivity of a vacuum.

Now consider the effect of introducing a *dielectric* between the plates, which does not conduct electricity but in which the field causes a separation of positive and negative charges. If the dielectric contains molecules that are permanent dipoles, they tend to align as shown in Fig. 5.2.2. However, as shown, the dipoles are not perfectly aligned, because the random thermal motions of the molecules oppose the orienting action.

The field is said to *polarize* the dielectric, and the *polarization* P is defined as the average dipole moment per unit volume. The polarization causes charge to appear on the surface of the dielectric, which opposes the charge on the plates themselves,

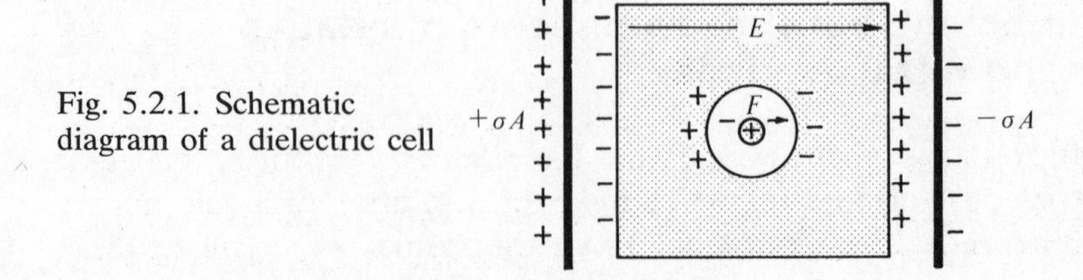

Fig. 5.2.1. Schematic diagram of a dielectric cell

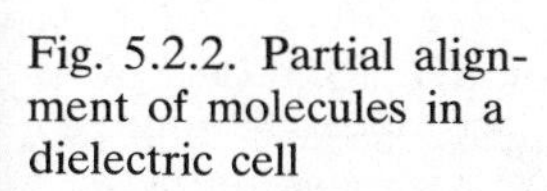

Fig. 5.2.2. Partial alignment of molecules in a dielectric cell

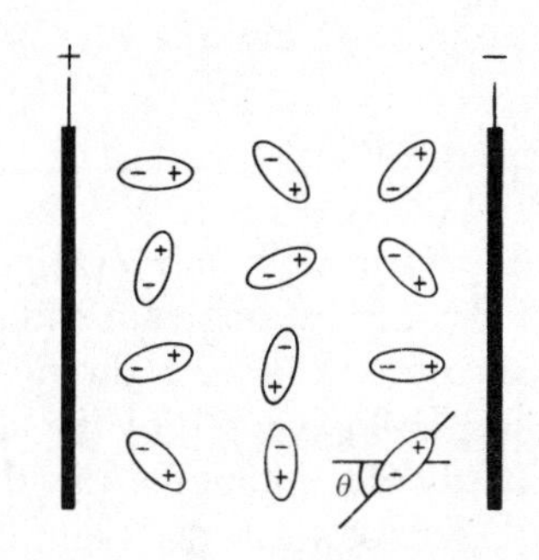

Fig. 5.2.1. The field E inside the dielectric is lowered from σ/ε_0 to σ/ε, where ε is the permittivity of the sample. We also define *relative permittivity* (or *dielectric constant*) ε_r as $\varepsilon/\varepsilon_0$. Thus

$$E = \sigma/\varepsilon = \sigma/\varepsilon_0\varepsilon_r. \tag{5.2.2}$$

Furthermore it can be shown that ε_r is equal to the capacitance C of a cell filled with sample divided by the capacitance C_0 of the cell when evacuated:

Capacitance is increased by the presence of a polarizable medium; typical values of ε_r are 1.0006 (air), 5.94 (chlorobenzene), 15.5 (liquid ammonia), and 80 (water at 20 °C).

$$\varepsilon_r = C/C_0. \tag{5.2.3}$$

The reduction of the field E inside the dielectric caused by the polarization reduces the effective charge on the plates from σ to $\sigma - P$, and thus

$$E = (\sigma - P)/\varepsilon_0. \tag{5.2.4}$$

Therefore from eqns (5.2.2) and (5.2.4)

$$P = E\varepsilon_0(\varepsilon_r - 1). \tag{5.2.5}$$

The relationship of the *polarizability* α to the polarization P is:

$$\alpha = \frac{V}{N} \cdot \frac{P}{E} = \frac{M}{L\rho} \cdot \frac{P}{E} \tag{5.2.6}$$

where $(N/V) = (L\rho/M)$ is the number of molecules per unit volume, and ρ is the density of the solution.

There is a complication which prevents us from making any further exact derivations (except for gases at low pressure). The difficulty is that a single molecule somewhere inside the dielectric experiences not only the field E but also an additional field arising from the surrounding molecules. The calculation for polar solutes in non-polar solvents (as well as dense gases and non-polar liquids) involves the consideration of the local field F inside a small hollow sphere which surrounds the molecule, Fig. 5.2.1. From this model is derived the *Clausius–Mossotti equation*:

$$\frac{\varepsilon_r - 1}{\varepsilon_r + 2} \cdot \frac{M}{\rho} = \frac{4\pi L}{3} \alpha = P^M. \qquad (5.2.7)$$

P^M is the molar polarization or *molar polarizability*. So far it includes only the contribution from induced dipoles, and we must therefore incorporate a term due to permanent dipoles.

The molar polarizability caused by the alignment of molecules with permanent dipole moments is the *molar orientation polarizability* P^M_{or}. As shown in the Fig. 5.2.2, the molecules do not orientate themselves in a completely ordered fashion, but instead take up a range of positions at an angle θ to the electric field. The potential energy U of each molecule is given by

$$U = -pE \cos \theta. \qquad (5.2.8)$$

According to Boltzmann's law, the number of molecules distributed with the axes of their dipoles pointing in the directions within a solid angle $d\Omega$ is $Ae^{-U/kT}\,d\Omega$, where A is a constant depending on the number of molecules considered. From this expression we may obtain the average moment per molecule in the form of a series expression. Taking the first non-zero term only, and considering the relation between the mean moment and the molar polarizability, we obtain the *Debye equation*:

$$P^M_{or} = \frac{Lp^2}{9\varepsilon_0 kT}. \qquad (5.2.9)$$

The molar polarizability arising from the induced dipoles comprises an *electronic* term, P^M_{el}, arising from the displacement of the electron cloud with respect to the nuclei, and a *distortion* term, P^M_d, which arises from the stretching and bending of the nuclear framework of the molecule as a result of the effective charges on the atoms being different (e.g. $\overset{\delta^-}{O}—\overset{2\delta^+}{C}—\overset{\delta^-}{O}$). The dependence of the three components of the molar polarizability on frequency is as shown in Fig. 5.2.3. At low frequencies, all three effects are present. As the frequency is increased into the infrared region, the molecules no longer have time to orient themselves before the field reverses, and the P^M_{or} contribution drops out. In the visible region, the applied frequency is too high for molecular distortion to contribute to the molar polarizability, and the only contribution is electronic, P^M_{el}. At optical frequencies this is the only component of the molar polarizability, and there is a simple relationship between relative permittivity and refractive index n:

$$\varepsilon_r = n^2. \qquad (5.2.10)$$

At X-ray frequencies even the electronic contribution to the molar polarizability, P^{M}_{el}, drops out. As a consequence, the refractive index of materials for X-rays is almost unity, and direct X-ray microscopy is impracticable because no lenses are available.

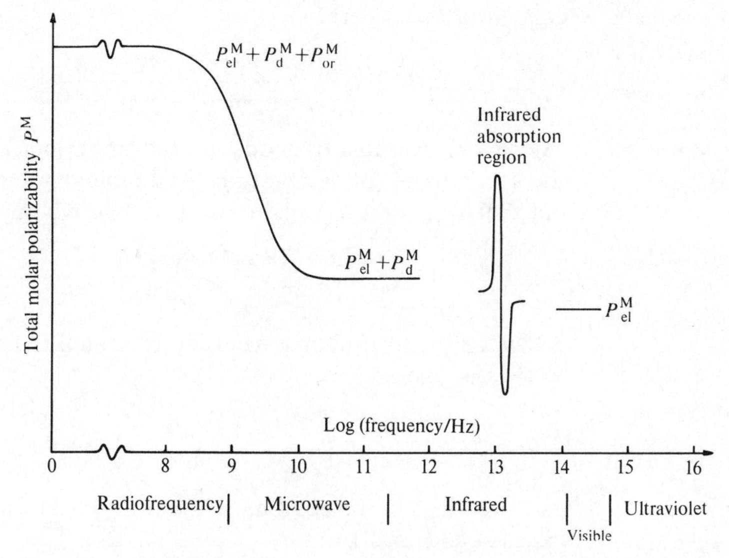

Fig. 5.2.3. Schematic diagram showing change in polarization with frequency

P^{M}_{el} may therefore be obtained by substituting this expression into the Clausius–Mosotti relation (5.2.7), to give the *Lorentz equation*:

$$P^{M}_{el}=\frac{n^2-1}{n^2+2}\cdot\frac{M}{\rho}. \tag{5.2.11}$$

We have seen that the total molar polarizability P^{M} is the sum of the orientation, electronic and distortion polarization:

$$P^{M}=P^{M}_{or}+P^{M}_{el}+P^{M}_{d}. \tag{5.2.12}$$

It follows from eqn (5.2.9) that

$$P^{M}-P^{M}_{el}=P^{M}_{or}+P^{M}_{d}=\frac{Lp^2}{9\varepsilon_0 kT}+P^{M}_{d}. \tag{5.2.13}$$

(ii) *Dipole moment in solution*[3]

The equations derived so far strictly apply only to pure substances in the gaseous state, since any short-range interactions of neighbouring polar molecules have been neglected. If this approximation is applied to a dilute solution of polar molecules in a non-polar solvent, then both ε_r and $1/\rho$, where ρ is the density of the solution, are almost linear functions of solute

A consequence of ignoring short-range interactions in the simplest theories is that electric dipole moments derived from measurements on gases and solutions are in general not equal.

concentration. Therefore

$$P^{\mathrm{M}} = x_1 P_1^{\mathrm{M}} + x_2 P_2^{\mathrm{M}} = \frac{\varepsilon_r - 1}{\varepsilon_r + 2} \cdot \frac{x_1 M_1 + x_2 M_2}{\rho}, \tag{5.2.14}$$

where x is the mole fraction, and the subscripts 1 and 2 refer to the solvent and solute respectively. Employing the same reasoning that was used to obtain eqn (5.2.11), we find that

$$x_1 P_{\mathrm{el},1}^{\mathrm{M}} + x_2 P_{\mathrm{el},2}^{\mathrm{M}} = \frac{n^2 - 1}{n^2 + 2} \cdot \frac{x_1 M_1 + x_2 M_2}{\rho}. \tag{5.2.15}$$

The concentration of solvent, c_1, is related to its mole fraction x_1 by the expression

$$c_1 = \frac{x_1 \cdot \rho}{(x_1 M_1 + x_2 M_2)}, \tag{5.2.16}$$

and there is a corresponding relation for c_2. From eqns (5.2.14), (5.2.15) and (5.2.16):

$$(P_1^{\mathrm{M}} - P_{\mathrm{el},1}^{\mathrm{M}}) c_1 + (P_2^{\mathrm{M}} - P_{\mathrm{el},2}^{\mathrm{M}}) c_2 = \frac{\varepsilon_r - 1}{\varepsilon_r + 2} - \frac{n^2 - 1}{n^2 + 2}$$

$$= \frac{3(\varepsilon_r - n^2)}{(\varepsilon_r + 2)(n^2 + 2)}. \tag{5.2.17}$$

Applying eqn (5.2.13) to each component in eqn (5.2.17), and noting that for the solvent there is no dipole moment,

$$\frac{3(\varepsilon_r - n^2)}{(\varepsilon_r + 2)(n^2 + 2)} = \frac{L p_2^2}{9 \varepsilon_0 k T} c_2 + P_{\mathrm{d},1}^{\mathrm{M}} \cdot c_1 + P_{\mathrm{d},2}^{\mathrm{M}} \cdot c_2. \tag{5.2.18}$$

We are now faced with the problem of estimating the distortion terms. The most satisfactory assumption is that they are small and nearly constant, since distortion polarizability will vary roughly with the size of the molecule. Thus a plot of the term on the left against c_2 will yield a line which will have a slope from which the dipole moment can be estimated.

Apparatus[2]

(i) Capacitance measurement

One way of measuring the capacitance of a dielectric cell is by the *heterodyne beat method* illustrated in Fig. 5.2.4. One oscillator provides a fixed frequency of, say, 100 kHz, which is fed to the Y plates of an oscilloscope. The other is a variable frequency oscillator, and is connected to the X plates of the oscilloscope. (The time base of the oscilloscope is not used.) The

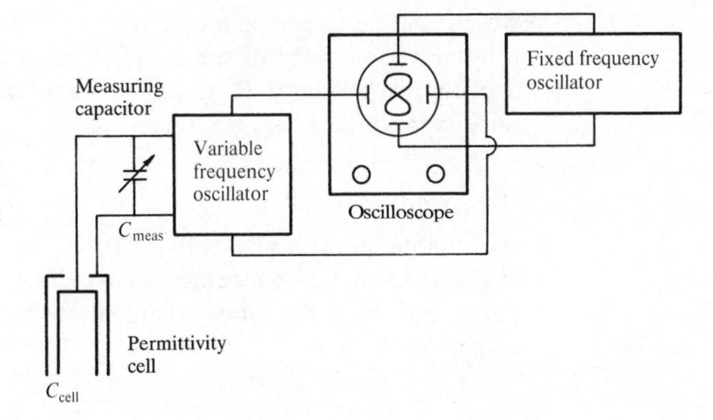

Fig. 5.2.4. Heterodyne beat apparatus for the measurement of relative permittivity

frequency of the variable oscillator is adjusted until a figure 8 (known as a Lissajous figure) is traced on the oscilloscope screen. This marks the point at which the variable frequency oscillator is giving an output exactly double the frequency of the fixed one.

The frequency of the variable oscillator is controlled by the total capacitance C_{tot} of an external circuit, as shown in the figure. C_{tot} is equal to the sum of the capacitance of the cell, C_{cell}, that of the precision measuring capacitor C_{meas}, and the residual capacitance C_{resid} of the leads and of the measuring capacitor at its zero position:

$$C_{tot} = C_{cell} + C_{meas} + C_{resid}. \tag{5.2.19}$$

Each determination involves the adjustment of the measuring capacitor until the Lissajous figure 8 is displayed, whereupon we know that C_{tot} has reached a fixed value. C_{resid} is also constant, so from eqns (5.2.3) and (5.2.19) it follows that

$$C_{cell} + C_{meas} = C_0\varepsilon_r + C_{meas} = \text{const.} \tag{5.2.20}$$

The most accurate type of measuring capacitor is an earthed stepped rod which can be moved axially within a live cylinder by a micrometer screw. It is constructed so that the capacity C_{meas} is directly proportional to the micrometer reading R. If the proportionality constant is k', then

$$C_0\varepsilon_r + k'R = \text{const.} \tag{5.2.21}$$

We measure R for air, the solvent and the solution. It follows from eqn (5.2.21) that:

$$\frac{R_a - R}{R_a - R_1} = \frac{\varepsilon_r - \varepsilon_{r,a}}{\varepsilon_{r,1} - \varepsilon_{r,a}}, \tag{5.2.22}$$

where the subscript a refers to air. This equation allows the relative permittivity of the solution ε_r to be obtained from the capacitance readings R, given the relative permittivities of the pure components, $\varepsilon_{r,a}$ and $\varepsilon_{r,1}$.

(*ii*) *Permittivity cell*

A suitable permittivity cell is shown in Fig. 5.2.5. It is constructed from two concentric glass tubes coated with platinum paint, and the inner tube is thermostated at 25 °C by water from a thermostat bath.

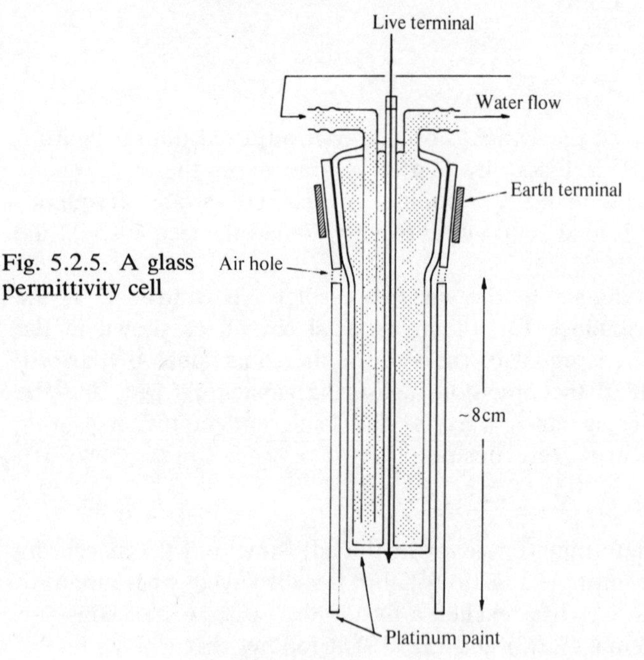

Fig. 5.2.5. A glass permittivity cell

(*iii*) *Refractive index*

The refractive index is measured by a precision Abbé refractometer. The refractometer is thermostatted and uses monochromatic sodium light so that no corrections are needed for either temperature variation or dispersion effects.

Procedure

Water is the most serious threat to the accuracy of this experiment. Both solute and solvent must be dried thoroughly, and exposure of the samples to the atmosphere must be minimized.

Cyclohexane vapour is explosive when mixed with air. Return all cyclohexane solutions to the residues bottle provided—do not pour any down the sink. Use the cyclohexane wash bottle with care.

You are supplied with a sample X, which is a chloronitrobenzene of molar mass 157.5.

Make up as accurately as possible four solutions of X, using the following procedure. Clean, dry and weigh a labelled

100 cm^3 volumetric flask. Accurately weigh about 2 g of X in a weighing bottle. Tip the contents into a clean, dry beaker, and re-weigh the weighing bottle. Dissolve the X in about 50 cm^3 of cyclohexane in the beaker. Note that the chloronitrobenzene dissolves very slowly in cold cyclohexane but quite easily if the temperature is raised a few degrees by dipping the bottom of the beaker into a larger beaker containing some hot water. Transfer the solution *quantitatively* from the beaker to the volumetric flask, and make up to the mark at 25 °C in the thermostat bath.

Repeat the procedure with approximately 3, 4 and 5 g of X in cyclohexane.

Measure the relative permittivities of the solvent and solutions.*

Measure the refractive indices of the solution and of pure cyclohexane using the Abbé refractometer.

Calculation

Given that the relative permittivity of air, $\varepsilon_{r,a}$, is 1.00, and that of cyclohexane, $\varepsilon_{r,1}$, is [2.023–0.0016 $(T-20)$] where T is the temperature in °C, calculate the relative permittivity of the solutions by means of eqn (5.2.22).

Use these values and your measurements of refractive index to calculate the dipole moment of the chloronitrobenzene (eqn (5.2.18)).

The *sense* of a molecular dipole (i.e. whether it is +− or −+) is very difficult to measure, and it is usually assumed to be as predicted by elementary valence theory.

Dipole moments are vectors, i.e. they have both magnitude and direction. The dipole moment of chlorobenzene is 1.60 D in the direction C↔Cl, and that of nitrobenzene is 4.01 D in the direction C↔NO_2. Calculate the dipole moments of ortho, meta and para chloronitrobenzene by vector addition, solving the vector triangle shown in Fig. 5.2.6 either by the cosine rule or by scale drawing.[2] Assume that the chlorobenzene and nitrobenzene dipole moments are the same as the dipole moments of the functional groups. Compare your results with your measured moment and find whether X is the ortho, meta or para isomer.

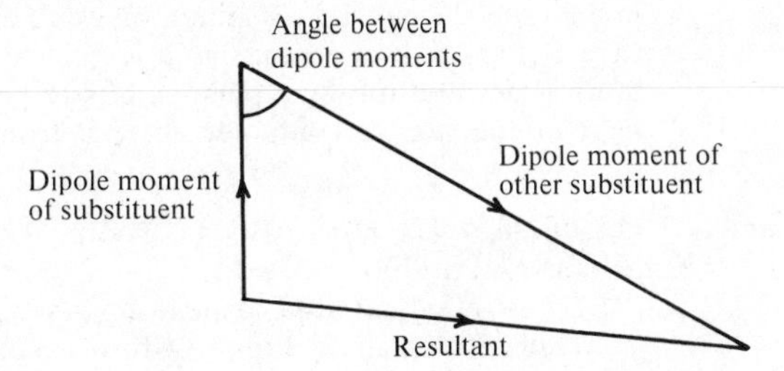

Fig. 5.2.6. Vector triangle for the calculation of resultant dipole moments

Results

Accuracy

The approximation of ignoring short-range interactions, mentioned previously, is a fundamental source of error. The present Guggenheim procedure for analysing the solution dipole moments avoids the chore of precise measurement of solution densities necessary in the so-called Halverstadt–Kumler procedure, but requires correspondingly precise measurement of refractive indices.[3] Experimentally, water is the greatest source of error, since it is polar, associates with the solute, and has a low molar mass so that a small mass can have a considerable effect. The concentration plot to determine the dipole moment is not always a good straight line, but should yield an unambiguous value in terms of the functional position.

Comment

Measurements of dipole moments have been used extensively to yield structural information about organic molecules.[4] Such methods have now been superseded by spectroscopic and diffraction techniques, because, as demonstrated in the following experiments, they give much more specific information. Nevertheless a familiarity with dipole moments is still important in understanding solvent effects and organic reaction mechanisms.

Technical notes

Apparatus

Heterodyne beat capacitance meter,†$ oscilloscope, permittivity cell,†† thermostat bath, Abbé refractometer; a chloronitrobenzene stored in a dessicator, dry cyclohexane.

Design note

The old-style vaned variable capacitor often used as a dielectric cell for student experiments is unsatisfactory. The cell shown in Fig. 5.2.5 works well, and can be made by a skilled glassblower. The glass tubes are joined by a standard ground-glass joint at the top, and must be accurately concentric. The inner tube is coated with Pt paint which is connected to the central wire by metallized glue. The earth terminal is connected to the Pt paint covering all surfaces of the outer tube by a metal band which also supports the cell. Suitable dimensions are o.d. of inner tube 2.2 cm, i.d. of outer tube 2.7 cm, height of paint on inner tube 5 cm. The best arrangement is to have the cell stationary, enclosed at the sides and top by a perspex box to prevent damage, and a device to raise the sample container up to it from below.

References

[1] *Physical chemistry*. W. J. Moore, Longmans, London, 5th edn (1972), p. 700.
[2] *Some electrical and optical aspects of molecular behaviour*. M. Davies, Pergamon Press, Oxford (1965), ch. 1–3.

[3] *The determination of dipole moments in solution.* H. B. Thompson, *J. Chem. Educ.* **43,** 66 (1966).
[4] *Determination of organic structures by physical methods.* Vol. 1, ed. F. C. Nachod and J. J. Zuckerman. Academic Press, New York (1971), ch. 9 by L. E. Sutton.

5.3 Ionization potentials and flame temperature from atomic spectra

A consequence of the wave character of the electrons in atoms is that their energy levels are quantized, i.e. they have only certain discrete values $E_1, E_2, E_3, \ldots$. If the electrons move between these levels, they absorb or radiate quantized amounts of energy ΔE. Since for all electromagnetic radiation

$$\Delta E = h\nu = hc/\lambda, \tag{5.3.1}$$

where h is Planck's constant and c the speed of light, there are corresponding discrete values of the spectral frequencies ν, and wavelengths λ. The most direct manifestation of this is that if we project the quantized radiation through a prism or grating to deflect the different energies by different amounts, we observe a *line spectrum* (Fig. 5.3.1).

The simplest line spectrum is that of atomic hydrogen, because it possesses only one electron. Part A of this experiment involves the measurement of the wavelengths of the visible lines in the spectrum of atomic hydrogen, and the derivation of the hydrogen Rydberg constant and ionization potential. There are then three optional parts: (B) measurement of the visible line

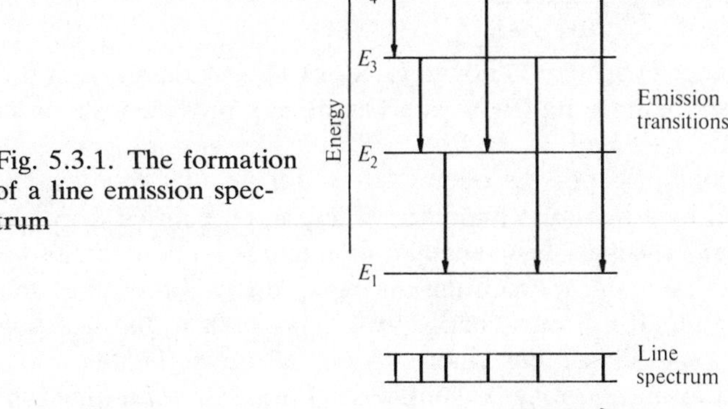

Fig. 5.3.1. The formation of a line emission spectrum

spectrum of sodium, and of potassium if available, and estimation of the ionization potential, (C) measurement of the spectrum of helium and calculation of its ionization potential, and (D) measurement of the temperature of a sodium flame. Part B introduces angular momentum and the splitting of the electronic levels caused by electron spin, and Part C the complications of a two-electron spectrum. Part D illustrates one method in which the intensities of lines in simple absorption and emission spectra may be used, and also introduces the laws of thermal radiation.

Theory

A. Hydrogen

The hydrogen atom is a sufficiently simple system for its energy levels to be almost exactly calculable. The solution of the appropriate *Schrödinger equation* yields

$$E_n = -\left(\frac{me^4}{32\pi^2\varepsilon_0^2\hbar^2}\right)\frac{1}{n^2} = -\frac{\mathcal{R}_H}{n^2}, \tag{5.3.2}$$

where $\mathcal{R}_H$ is the *Rydberg constant* of the hydrogen atom. It is a peculiarity of the hydrogen atom that at this level of approximation its electronic energy levels depend only on a single number n known as the *principal quantum number.* In dropping from an energy level with quantum number n_2 to an energy level with quantum number n_1, the energy change of the electron is

$$\Delta E = -\mathcal{R}_H\left(\frac{1}{n_1^2}-\frac{1}{n_2^2}\right) = -h\nu, \tag{5.3.3}$$

where ν is the frequency of the emitted radiation. The frequency is more often expressed as a *wavenumber* $\tilde{\nu}$ in units of reciprocal centimetres. The *Rydberg formula for hydrogen* then takes the form:

$$\tilde{\nu} = \frac{1}{\lambda} = \frac{\nu}{c} = \mathcal{R}_H\left(\frac{1}{n_1^2}-\frac{1}{n_2^2}\right). \tag{5.3.4}$$

The Rydberg constant is now a factor of hc less than $\mathcal{R}_H$ in the previous equation, but there is no ambiguity provided the units of $\mathcal{R}_H$ are specified. n_1 and n_2 may differ by any integer.

Electrons also possess quantized *orbital angular momentum,* governed by a quantum number l. The photon emitted during a transition possesses spin angular momentum and to conserve the overall angular momentum during an *allowed transition,* the quantum number l must change by 1. This leads to the *selection rule* $\Delta l = \pm 1$. Transitions where $\Delta l = 0, \pm 2$ are *forbidden.*

This information may be summarized on a *Grotrian diagram,* Fig. 5.3.2. The values of l from 0 to 4 are given their traditional

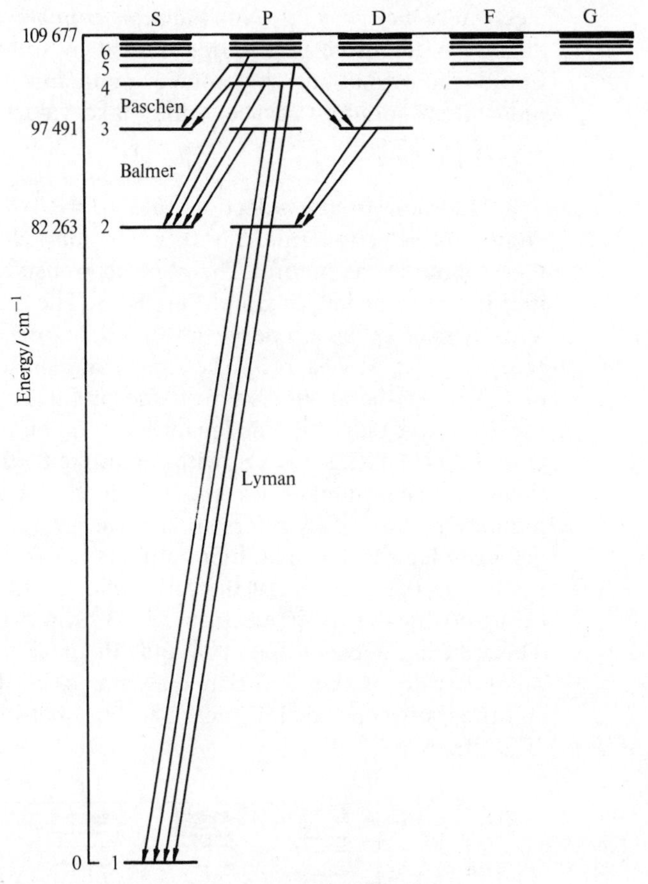

Fig. 5.3.2. Grotrian diagram for H

labels S, P, D, F, and G. The emission transitions form named series as shown. In this experiment we measure the first three lines of the *Balmer series* for which $n_1 = 2$.

B. Sodium[2]

The spectra of alkali metal atoms are more complicated than that of hydrogen. The transitions studied are still those of a single electron, but its permitted energies are now affected to a much greater degree by the orbitals of other, inner electrons. The shielding of the nuclear charge by the inner electrons varies according to the type of orbital occupied by the single valence electron, with the result that its energy increases (becomes less negative) with increasing l. Furthermore, the transitions are found to be split into doublets by the spin of the valence electron which has a *spin quantum number* $s=\frac{1}{2}$. The total angular momentum is now determined by the vector combination of the orbital and spin angular momenta. The *total angular momentum quantum number j* may take values

$$j=l+s,\ l+s-1,\ l+s-2\ldots|l-s|. \tag{5.3.5}$$

In addition to the selection rule $\Delta l=\pm 1$, there is now an additional selection rule $\Delta j=0, \pm 1$. n may change by any integral amount, as before. The allowed transitions are summarised in the Grotrian diagram, Fig. 5.3.3. The diagram shows the *term symbol* of each energy level. As before, S, P, D, and F refer to $l=0$, 1, 2, and 3. The right-hand subscript is the value of j. The left-hand superscript is the *multiplicity* equal to twice the resultant electron spin quantum number plus 1, or in this case $(2\times\frac{1}{2})+1=2$. Levels with multiplicity 2 are known as *doublets*. The number on the left is the principal quantum number n. So $3\,^2S_{1/2}$ refers to the energy level for which the valence electron has quantum numbers $n=3$, $l=0$ and $s=1/2$, i.e. it is in the s orbital of the 3rd shell. All the lower levels are occupied by the inner electrons, so $3\,^2S_{1/2}$ is the *ground state.* The splitting between the $^2P_{3/2}$ and $^2P_{1/2}$ states is imperceptible on the scale of the Grotrian diagram, as is the even smaller splitting between the $^2D_{5/2}$ and $^2D_{3/2}$ and between the $^2F_{7/2}$ and $^2F_{5/2}$ states.

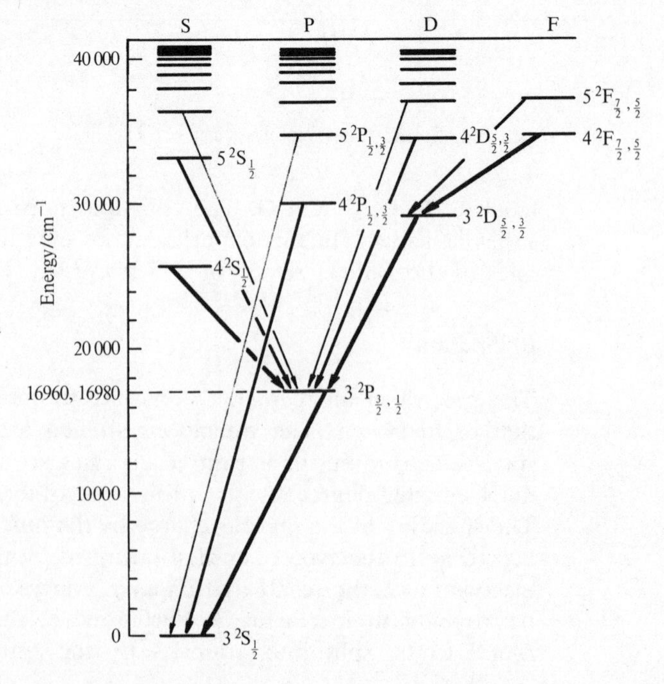

Fig. 5.3.3. Grotrian diagram for Na

The thickness of the lines gives a guide to their intensities in emission. The two intense yellow emissions arising from the $3\,^2P_{3/2,1/2} \rightarrow 3\,^2S_{1/2}$ transitions are known as the *sodium D lines.* The factors which affect intensity are discussed in part D.

The Grotrian diagram also shows transitions from more highly excited electronic states down to the $3\,^2P_{3/2,1/2}$ state. The splitting of the higher levels is either zero or much less than that between $3\,^2P_{1/2}$ and $3\,^2P_{3/2}$, so the spacing of each pair of lines is almost the same as in the yellow doublet. The transitions have frequencies which obey the *general Rydberg formula* for hydrogen-like atoms:

$$\begin{aligned}\tilde{\nu} &= \mathcal{R}\left(\frac{1}{(n_1-\delta_1)^2}-\frac{1}{(n_2-\delta_2)^2}\right)\\ &= \tilde{\nu}_\infty - \frac{\mathcal{R}}{(n_2-\delta_2)^2},\end{aligned} \qquad (5.3.6)$$

where δ_1 and δ_2 are known as the *quantum defects,* $(n_1-\delta_1)$ and $(n_2-\delta_2)$ are the *effective principal quantum numbers, and* $\tilde{\nu}_\infty$ is the wavenumber of a *series limit.*

To obtain a *rough* estimate of the ionization potential, δ_2 can be ignored, and the wavenumbers of the lines in the $n\,^2S \rightarrow 3\,^2P$ series or in the $n\,^2D \rightarrow 3\,^2P$ series plotted against $1/n^2$. The limit at $1/n^2 = 0$ *plus* the excitation energy of the $3\,^2P$ state corresponds to the approximate ionization potential.

C. Helium[3]

The atomic spectrum of helium is formed by the transitions of both its electrons. Their individual angular momenta and spins combine to give overall values referred to as S, L, and J. The ground state for helium has the closed shell configuration $1s^2$, for which the resultant spin S is zero. The term symbol for the ground state is therefore $^1S_0$.

Now consider the electronic configuration 1s 2s. The individual orbital angular momenta are both zero, and so $L = 0$. However the spins of 1/2 can combine to give $S = 0$ or $S = 1$. The multiplicities are 1 and 3 respectively, i.e. there are *singlet* and *triplet* states. The corresponding values of J are 0 and 1, and the term symbols are therefore $^1S_0$ and $^3S_1$.

Other available levels are drawn in the Grotrian diagram, Fig. 5.3.4. To prevent ambiguity, both the term symbols and configurations are shown. In some Grotrian diagrams, the principal quantum number of the excited electron is quoted instead of the full configuration. There are selection rules $\Delta L = \pm 1$ and $\Delta J = 0, \pm 1$. In addition, transitions for which $\Delta S \neq 0$ are forbidden;

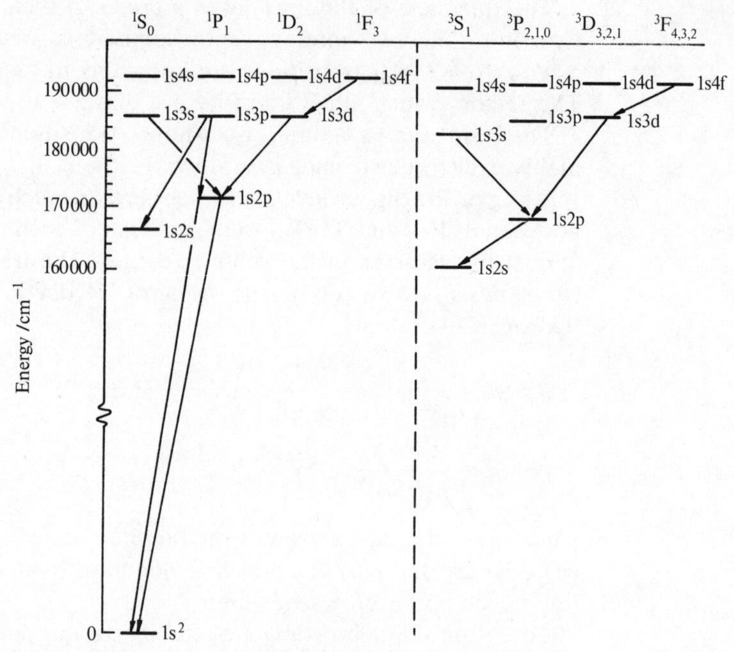

Fig. 5.3.4. Grotrian diagram for He

the singlet and triplet states are therefore drawn separately, because there are no spectroscopic transitions between them. Such transitions may be made, however, via ionization to He^+.

The $1s\,2s\,^3S_1$ state is higher than the ground state, but nevertheless there are no transitions from it to any lower state. This state is therefore said to be *metastable. Which other state of helium is metastable?*[4]

The lines in the helium spectrum follow the general form of the Rydberg expression.

D. Flame temperature[5]

The intensity of a spectral line is determined by the rate at which electrons undergo the relevant transition. This rate is proportional to the number, or *population*, of electrons in the initial state, multiplied by the probability of the transition. For an allowed transition, the population n_i at a temperature T is given by the general form of the *Boltzmann distribution law*

$$\frac{n_i}{n_0} = \frac{g_i}{g_0}\,e^{-\varepsilon_i/k_B T}. \qquad (5.3.7)$$

n_0 is the ground state population at the temperature T. g_0 is the *statistical weight* or *degeneracy* of the ground state. g_i is the degeneracy of level i, and ε_i is the energy required to excite electrons into the energy level from the ground state.

Thermal radiation is the radiation emitted by a system in which the various quantum states participating in the emission process are in thermodynamic equilibrium, and have populations in accord with the Boltzmann distribution, eqn (5.3.7). We define the *brightness* of a thermally radiating body as

$$B = \frac{(\mathrm{d}E/\mathrm{d}t)}{\mathrm{d}A\,\mathrm{d}\omega \cos\theta}, \tag{5.3.8}$$

where $\mathrm{d}E/\mathrm{d}t$ is the rate at which energy is radiated from an element $\mathrm{d}A$ of the surface of a luminous body within a solid angle $\mathrm{d}\omega$ at an angle θ to the normal. In this experiment, the brightness observed is not over the whole spectrum, but only that at the wavelength of the sodium D line. When spectrally resolved in this fashion, B is referred to as *spectral brightness.*

In this part of the experiment, a comparison is made between the brightness of the sodium lines excited in a flame, and the brightness of a tungsten filament lamp viewed through the flame. Both emissions are focussed on to a spectrometer. If the lamp is less bright than the flame, the sodium D lines appear in absorption, whereas if the lamp is brighter than the flame, they appear in emission. When an uninterrupted band spectrum is observed, the brightness of the lamp is equal to the brightness of the flame plus the brightness of the lamp transmitted through the flame:

$$B^0_{\text{lamp}} = B_{\text{flame}} + B^0_{\text{lamp}}(1 - a_{\text{flame}}). \tag{5.3.9}$$

The superscripts indicate that the lamp is assumed to be a black body. a_{flame} is the *spectral absorptivity* of the flame at the sodium D wavelength, defined as the fraction of incident radiation absorbed by it.

The *spectral emissivity* e of any substance is defined as the ratio of its brightness at a particular wavelength to the brightness at the same wavelength of a black body at the same temperature. By *Kirchhoff's law* of emissivity

$$a_{\text{flame}} = e_{\text{flame}} = B_{\text{flame}}/B^0_{\text{flame}}, \tag{5.3.10}$$

and substituting this expression into eqn (5.3.9) we find that

$$B^0_{\text{flame}} = B^0_{\text{lamp}}. \tag{5.3.11}$$

If corrections are made for the fact that neither the flame nor the lamp are black bodies, the known relationship between the

Table 5.3.1. Black body corrections to tungsten lamp temperature[5]

Effective brightness temperature	True temperature
1500	1587
1600	1699
1700	1812
1800	1927
1900	2043
2000	2160
2100	2278
2200	2397
2300	2517
2400	2639

temperature and brightness of a black body may be used to find the temperature of the flame from the temperature of the lamp. Table 5.3.1 incorporates such corrections, with an additional allowance of a 10 per cent loss in brightness at the lens between the lamp and the flame. The temperature of the lamp is measured with an optical pyrometer. This indirect method of measuring flame temperature is known as the *line reversal method.*[5,6] *Why is it not possible to measure the temperature of the flame directly with the pyrometer*?

Note that the property measured is the *electronic temperature* of the flame, i.e. the temperature associated with the distribution of the electrons amongst the electronic energy levels. A thermometer would measure the *translational temperature* of the gas molecules in the flame. Other experiments show that the two temperatures are close, although in general this need not be so if chemiluminescent reactions occur.

Apparatus

There are three types of spectrometer in common use. *Simple prism spectrometers* measure the wavelength of light in terms of the angle at which the eyepiece telescope must be set to see the line in the prism. A calibration chart is used to convert the angles to wavelengths. In a *direct-reading constant deviation spectrometer*, the positions of the collimator and telescope are fixed (Fig. 5.3.5). The prism is moved by a drum marked with wavelengths on a fine vernier scale. In a *spectrograph* (p. 258) the collimator and prism are fixed, and the detector is a photographic plate or film. The positions of the photographed lines are measured by a *comparator*, a travelling microscope with a vernier scale. Alternatively the lines on the plate or film may simply be projected onto paper by means of a photographic enlarger, and the marked positions measured with a ruler.

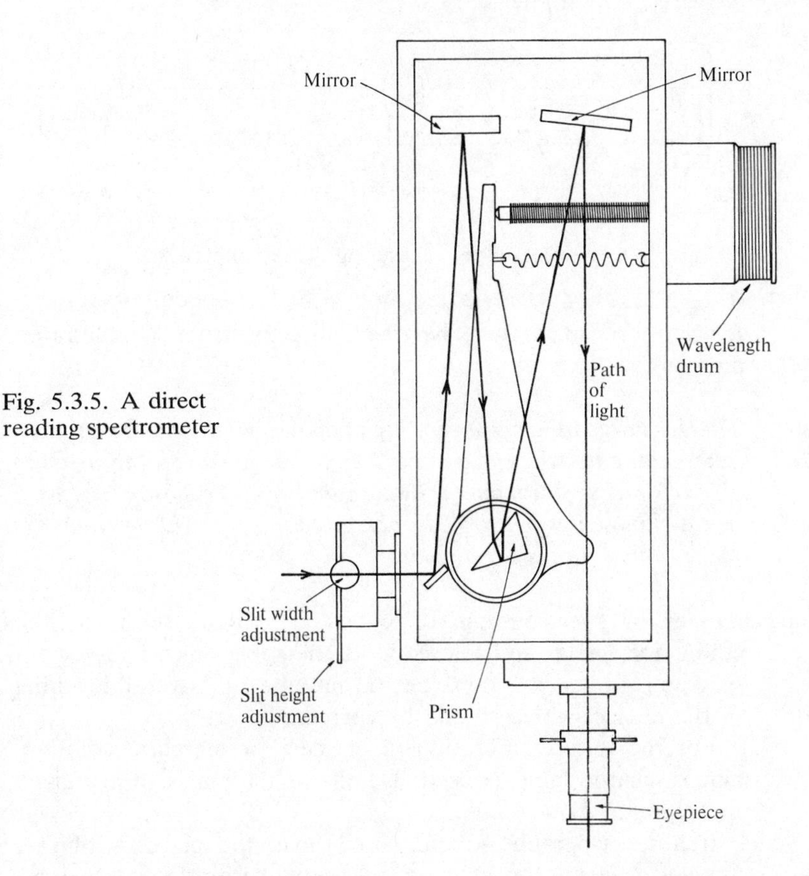

Fig. 5.3.5. A direct reading spectrometer

Direct reading spectrometers must always be calibrated; typically they are only accurate to between 0.5 and 5 nm, depending on the wavelength measured.

Whichever apparatus is used, it will need to be calibrated with lines of known wavelength.

The emission spectra are created by passing a high voltage discharge through the appropriate vapour. Hydrogen is, of course, a molecular gas. However it is dissociated in the discharge, and by careful adjustment of pressure the atomic emissions can be made brighter than the more numerous molecular lines.

The arrangement of the apparatus for part D is shown in Fig. 5.3.6. The sodium flame is a gas flame which is either formed in a special burner designed to suck in dilute sodium chloride solution,[7] or simply in a bunsen burner into which is placed sodium chloride in the form of powder or a hard stick. The flame experiment also employs a tungsten filament lamp, the temperature of which is measured with an *optical pyrometer*. In this device, electric current is passed through a wire at a level which makes it glow at the same intensity as the object. The

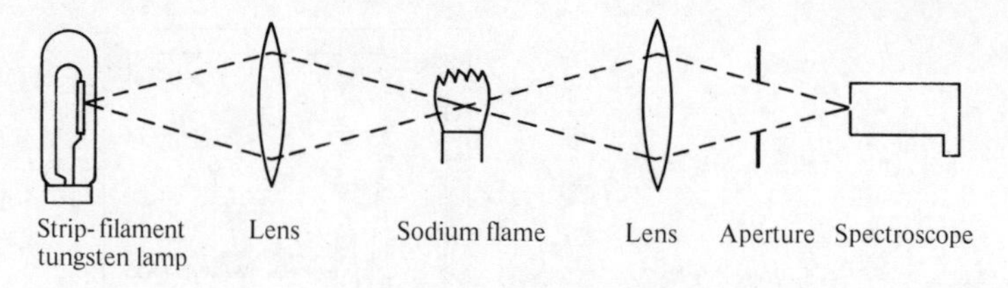

Fig. 5.3.6. Apparatus for measuring flame temperature

temperature may then be read directly from a calibrated ammeter.

Procedure *The discharge lamps use very high voltages which can be lethal—take great care when operating them. Lamps containing mercury give off intense ultraviolet radiation and must be surrounded by a metal chimney or pyrex glass jacket. Also use shielding chimneys for the other lamps if provided.*

Use of spectrometer or spectrograph. If requested,* align and focus the collimator, prism and telescope of the spectrometer and spectrograph. In general, however, do *not* alter the overall levelling of the device, which should be correct already.

For the initial calibration, a mercury or mercury/cadmium/zinc discharge lamp is used. Details of the emission are given in Table 5.3.2 on p. 205.

If a spectrograph is being used, load the plate or film as directed.* Focus the light onto the collimator slit with a lens, and adjust the slit width.* Block off part of the slit or rack the film up or down so that only a strip of the film is exposed.* Expose the film.*

If a spectrometer is used, the emission may either be focussed onto the collimator slit with a lens, or the discharge lamp simply placed close to the slit. With the latter arrangement, the lamp must be properly in line with the collimator and the slit adjusted to give fine but visible lines.

If a simple prism spectrometer is used, identify the lines listed in the table and note the angles at which they appear. Plot a calibration graph of angle against wavelength, or check the calibration graph if one is provided.

When using a direct reading spectrometer, always rotate the drum in the same direction to prevent backlash errors. Identify the lines, and note their wavelengths. Then find the differences $\Delta\lambda$ between the observed wavelengths and the true wavelengths, as shown in Table 5.3.2. Plot a graph of $\Delta\lambda$ against λ, so that all your other readings may be corrected.

A. Hydrogen spectrum. Replace the calibration lamp with a hydrogen lamp, or if a mirror is used to deflect the light from different lamps, move it to the required position.

If a spectrometer is used, find the true wavelengths of the first three lines of the Balmer series, which are red, greenish-blue, and violet in colour. Determine the position of the 4th line by the method described in the Calculation section. *Is the 4th line of the Balmer series visible*?

If a spectrograph is used, photograph the emission on a strip of the film or plate. A calibration spectrum should also be photographed before the film or plate is removed and developed.* Be careful not to fog the plate by exposure to incident room light.

B, C. Alkali metal and helium spectra. Depending on which experiments are to be carried out, replace the calibration lamp with a sodium, potassium or helium discharge lamp, or move the mirror. For sodium and helium, find the accurate wavelengths of the transitions shown in the appropriate Grotrian diagrams. The potassium spectrum is more complicated, and should be examined with the help of your instructor and the corresponding Grotrian diagram.*

D. Temperature of a sodium flame.[5] The arrangement of the apparatus for this experiment was shown in Fig. 5.3.6.

Gross errors can arise in this experiment if the apparatus is not carefully set up and aligned.

Switch on the tungsten lamp. Mount a condensing lens midway between the tungsten lamp and the flame, and adjust it so that the image of the filament falls in the plane of the burner. Turn off the lamp.

Light the flame, and feed it with a *dilute* (~0.01 M) solution of sodium chloride. Place a lens mid-way between the flame and the spectroscope slit, and check that the flame is focussed on the slit. Adjust the spectroscope so that the lines at 589.0 and 589.6 nm are in view and clearly separated. Turn off the flame.

Place a variable aperture mid-way between the spectroscope and the nearest lens. Switch on the tungsten lamp again. Adjust the aperture until the image of the tungsten lamp completely and uniformly fills it. Check that the tungsten band spectrum can be seen in the spectroscope.

Turn on the flame again. Adjust the brightness of the lamp with the variable transformer until the sodium lines of the flame disappear into the band spectrum, appearing neither in emission or absorption.

Switch on the low voltage supply to the optical pyrometer and check that the pyrometer is set up correctly.* Adjust the optical pyrometer so that its filament is the same brightness as that of the tungsten lamp. Note the temperature reading.

Turn off all the apparatus.

Calculation For each spectrum, convert your readings to true wavelengths via your calibration graph, and then convert these wavelengths to wavenumbers in cm^{-1}.

A. Fit the first three lines of the Balmer series to the appropriate Rydberg formula. Hence find the wavelength of the fourth Balmer line, and the value of the hydrogen Rydberg constant $\mathcal{R}_H$ in cm^{-1}. Compare $\mathcal{R}_H$ to the theoretical value, calculated from eqn (5.1.1), of 1.09678×10^5 cm^{-1}. Was the 4th Balmer line visible, or registered on the film? What is the Rydberg formula of the Lyman series, and in what region of the spectrum will the lines appear?

B. For the sodium spectrum, assign your measured wavenumbers to the various transitions shown in the Grotrian diagram, Fig. 5.3.3. Use the scale on the diagram to help you. Estimate the energy difference between the components $^2P_{1/2}$ and $^2P_{3/2}$. How does this arise? The higher wavenumber component, $^2P_{3/2} - {}^2S$ is stronger than $^2P_{1/2} - {}^2S$. Why is this? Calculate the approximate ionization energy of the sodium atom by the method described in the Theory section.

If a potassium spectrum has been studied, carry out the same calculations as for sodium, if necessary with the help of an instructor. How does the splitting of the $^2P_{3/2,1/2}$ components in potassium compare with that in sodium?

C. Assign the helium lines with the aid of Fig. 5.3.4, again using the scale to help you. Fit the three lines of the series $1s\,nd\ {}^1D - 1s\,2p\ {}^1P$ to the Rydberg formula, given that the quantum defect δ_2 for this series is small. The level $1s\,2p\ {}^1P$ lies at 171130 cm^{-1} above the ground state, $1s^2\ {}^1S$. From this, and your value of $\tilde{\nu}_\infty$, calculate the first ionization potential of the helium atom. Use your measurements of other lines in the spectrum to estimate the difference between $1s\,2p\ {}^3P$ and $1s\,2p\ {}^1P$, given that the terms become more hydrogen-like as n and L increase.

D. Find the temperature of the sodium flame, using the corrections listed in Table 5.3.1.

Results See Table 5.3.2. Other Zn, Cd and Hg lines may also be observed.

Comment This experiment introduces the theory and practice of electronic spectroscopy, which is of supreme importance in the determination of the structures of atoms and molecules. The last part of

Table 5.3.2. Calibration using Zn/Cd/Hg lines

Line	Calibration data			Specimen results	
	true λ_{air}	$\lambda_{vac}-\lambda_{air}$	λ_{vac}	obs λ_{air}	$\Delta\lambda$
Hg	404.7	0.1	404.8	405.6	−0.8
Hg	407.8	0.1	407.9		
Hg	435.8	0.1	435.9		
Cd	467.8	0.1	467.9		
Zn	468.0	0.1	468.1		
Zn	472.2	0.1	472.3		
Cd	480.0	0.1	480.1	482.4	−2.3
Zn	481.1	0.1	481.2		
Cd	508.6	0.1	508.7		
Hg	546.0	0.2	546.2		
Hg	577.0	0.2	577.2		
Hg	579.1	0.2	579.3		
Cd	643.8	0.2	644.0		

All wavelengths in nm. $\Delta\lambda = \text{true } \lambda_{vac} - \text{obs } \lambda_{air}$

the experiment illustrates a standard method of measuring the temperatures of flames.[5]

Technical notes

Apparatus

Parts A, B and C: simple prism spectroscope, or direct reading spectroscope or spectrograph,$ with appropriate ancillary apparatus such as drum illumination and plate developing facilities; Hg or Zn/Cd/Hg, H, He, Na, K discharge lamps as required, with power supplies; lens and mountings; optical bench, [Grotrian diagram for K]. Part D (see Fig. 5.3.6): tungsten lamp preferably with vertical strip filament; variable transformer or rheostat; Méker or Bunsen burner, either with aspirator and supply of dilute (~0.01 M) sodium chloride solution (concentrated solution blocks tube), or with sodium chloride stick, or with gauze and powdered sodium chloride; lenses and mountings.

References

[1] *Physical chemistry*. P. W. Atkins. Oxford University Press, Oxford, 3rd edn (1986), sect. 13.1, 13.2, 13.3 and 15.1.
[2] *Spectroscopy*. Ed. B. P. Straughan and S. Walker. Chapman and Hall, London (1976), vol. 1, ch. 1.
[3] *Structure and spectra of atoms*. W. G. Richards and P. R. Scott. John Wiley and Sons, London (1976), p. 52.
[4] *Atomic spectra and atomic structure*. G. Herzberg. Dover, New York, 2nd edn (1945), p. 65.
[5] *Combustion, flames and explosions of gases*. B. Lewis and G. von Elbe. Academic Press, New York (1961), ch. XII.
[6] *Flames. Their structure, radiation and temperature*. A. G. Gaydon

and H. G. Wolfhard. Chapman and Hall, London, 4th edn (1970), ch. 10.
[7] Ref. 6, p. 9.

5.4 Pure rotation spectra using Michelson interferometer

Rotational absorption spectra arise from the absorption of electromagnetic radiation by molecules with permanent electric dipole moments. The molecules are excited from one rotational level to another, and if no vibrational transitions occur, a *pure rotation spectrum* results.

Pure rotational transitions are generally caused by radiation in the microwave region at wavelengths from 1 mm to 30 cm. For many small molecules, however, the spectra occur in the *far infrared region*, at wavelengths from 0.03 to 1 mm. Although far infrared spectra may be studied by a grating spectrometer, the design of a high resolution instrument presents difficulties, not the least because of the weakness of infrared sources in this region. In recent years, therefore, most pure rotation spectrometry of simple gas molecules has been carried out with *interferometers* which have important advantages in this spectral region. The *interferograms* which these instruments produce are converted to absorption spectra by *Fourier transformation*, as explained below.

In this experiment we measure the pure rotation spectra of HCN and NO, and hence find the rotational constants of these molecules. We express the position of spectral lines, the energies of transitions, and the rotational constants B and D, all in wavenumber units (cm^{-1}) (p. 194). The range of far infrared wavelengths, 0.03 to 1 mm, covers wavenumbers of 300 cm^{-1} to 10 cm^{-1}.

Theory

(*i*) *Rotation spectrum of HCN*[1]

The rotational energies ε_J of a rigid diatomic or linear molecule are given by

$$\varepsilon_J = BJ(J+1). \tag{5.4.1}$$

J is the *rotational quantum number* taking values 0, 1, 2, . . . , and the *rotational constant* B is given by

$$B = h/8\pi^2 Ic, \tag{5.4.2}$$

where I is the moment of inertia. For small molecules B_0, the value of B for the lowest vibrational level, is typically between 1 and 10 cm^{-1}. If the molecule is not rigid, centrifugal distortion will occur as the molecule rotates and further terms must be included in the expression for ε_J:

$$\varepsilon_J = BJ(J+1) - DJ^2(J+1)^2 + \ldots, \tag{5.4.3}$$

where D is the *centrifugal distortion constant.* Higher terms are very small and can be ignored in this experiment. Typically D_0, the value of D for the lowest vibrational level, has a value between 10^{-3} and 10^{-6} cm^{-1}, depending on the strength of the bond. The pure rotational absorption spectrum of the molecule contains lines of wavenumber $\tilde{\nu}_J$ arising from transitions from a particular rotational level of energy ε_J to the next highest level ε_{J+1}, which take place while the molecule is at its lowest vibrational level. Thus:

$$\begin{aligned}\tilde{\nu}_J &= \varepsilon_{J+1} - \varepsilon_J \\ &= B_0[(J+1)(J+2) - J(J+1)] \\ &\quad - D_0[(J+1)^2(J+2)^2 - J^2(J+1)^2] \\ &= 2B_0(J+1) - 4D_0(J+1)^3.\end{aligned} \tag{5.4.5}$$

Therefore

$$\tilde{\nu}_J/(J+1) = 2B_0 - 4D_0(J+1)^2, \tag{5.4.6}$$

and a plot of $\tilde{\nu}_J/(J+1)$ against $(J+1)^2$ will yield a straight line of slope $4D_0$ and intercept $2B_0$.

(*ii*) *Rotation spectrum of NO*

The ground state of NO may be described by *molecular orbital theory*,[2] which supposes that the orbitals in molecules extend around two or more nuclei. Essentially the same rules apply to the filling of these orbitals as apply to the filling of atomic orbitals, and in molecular orbital notation the ground state of NO is

$$\mathrm{KK}(\sigma_g 2s)^2(\sigma_u 2s^*)^2(\sigma_g 2p)^2(\pi_u 2p)^4(\pi_g 2p^*)^1.$$

KK represents the two filled K shells of the separate N and O atoms. The first two terms in each set of brackets represent the symmetry of the particular molecular orbital, i.e. σ or π, gerade (even), or ungerade (odd). The next terms, 2s or 2p, are the atomic orbitals which have combined to form the molecular orbitals. The asterisk represents antibonding orbitals, and the superscript the number of electrons in the particular molecular orbital. We see, for example, that the configuration includes one unpaired electron in a $\pi_g 2p$ antibonding molecular orbital.

The configuration may also be described by *term symbols* analogous to those used for atomic configurations (p. 194). Molecular orbitals with angular momentum 0, 1, 2 ... are referred to as Σ, Π, Δ ..., and the lone electron in the ground-state of NO gives rise to a Π state. The lone electron has spin 1/2, and the spin quantum number S of the molecule is therefore 1/2. The electron gives rise to a multiplicity $(2S+1)=2$, written as a superscript, and couples with the angular momentum to give two states, $^2\Pi_{1/2}$ and $^2\Pi_{3/2}$. The $^2\Pi_{3/2}$ state is about 120 cm^{-1} higher in energy than the $^2\Pi_{1/2}$ state, and both are appreciably populated at room temperature. They have slightly different effective rotational constants B_0, and different effective centrifugal distortion constants D_0.

The energies of the rotational states of the molecule depend on the coupling between the rotational angular momentum and the electronic angular momentum, and in the case of NO give rise to complicated expressions. (The coupling is intermediate between Hund's case (a) and (b).[3,4]) The analysis is greatly simplified, however, if we calculate the mean of the absorption wavenumbers $\tilde{\nu}_{1/2}$ and $\tilde{\nu}_{3/2}$ for the two multiplet components with the same lower J value. This mean wavenumber is related to B_0 and D_0 by the expression

$$[\tilde{\nu}_{1/2}(J)+\tilde{\nu}_{3/2}(J)]/2=2(B_0-2D_0)(J+1)-4D_0(J+1)^3, \tag{5.4.7}$$

which is very similar to eqn (5.4.5) for simple non-rigid diatomic molecules. The doublet components appear as pairs of lines either side of the mean wavenumbers given by this equation, with the splitting between the lines increasing with wavenumber, and the $^2\Pi_{3/2}$ line for a given J weaker than the corresponding $^2\Pi_{1/2}$ line. If we take two neighbouring pairs of lines, add the wavenumbers of each pair, and then take the difference of these sums, we may obtain a rough estimate of B_0 by the relation

$$[\tilde{\nu}_{1/2}(J+1)+\tilde{\nu}_{3/2}(J+1)]-[\tilde{\nu}_{1/2}(J)+\tilde{\nu}_{3/2}(J)]\approx 4B_0. \tag{5.4.8}$$

This approximate value of B_0 may be used to assign J values to the observed transitions.

Apparatus[5,6] The optical layout of a Michelson interferometer is shown in Fig. 5.4.1. A water-cooled mercury discharge lamp acts as a black body in this spectral region and provides the source of

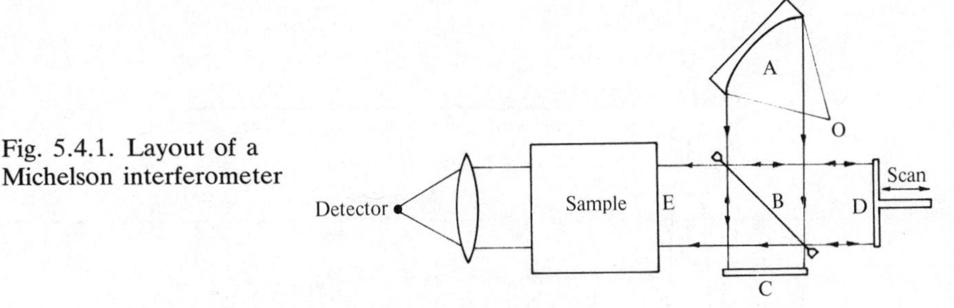

Fig. 5.4.1. Layout of a Michelson interferometer

radiation at 0. The radiation is chopped by a rotating blade, at a frequency to which the detector is locked to reduce *noise* (random interference in the signal). The beam is collimated by the curved mirror A. It then passes through a beam splitter B, made from a thin sheet of plastic film, which divides the beam between a fixed mirror C, and a movable mirror D driven by a micrometer stepping motor. The reflected beams then recombine at the beam splitter, and part of the combined beam reaches the sample, while the rest is lost in the direction of the source. After passing through the sample, the beam falls onto a suitable detector, commonly a solid-state semi-conductor device.

Let us first consider the instrument with no sample in it, and the movable mirror set so that the distances BC and BD in the figure are equal. This is the *zero path difference* position. Under these circumstances all the radiation travelling the path ABCBE will be in phase with the radiation following ABDBE, regardless of frequency.

Now let us consider what happens when the radiation is monochromatic and we move the mirror a distance x. The path lengths ABCBE and ABDBE will now differ by a distance $2x$. So long as the phases of the two reflected beams are the same, constructive interference will result. This condition is met either if the paths are equal, or if they differ by an integral number of wavelengths ($n\lambda = 2x$, where n is an integer and λ the wavelength of the radiation). In between these positions the intensity will drop to a minimum, Fig. 5.4.2. The equivalent graph for radiation covering a range of wavelengths, such as

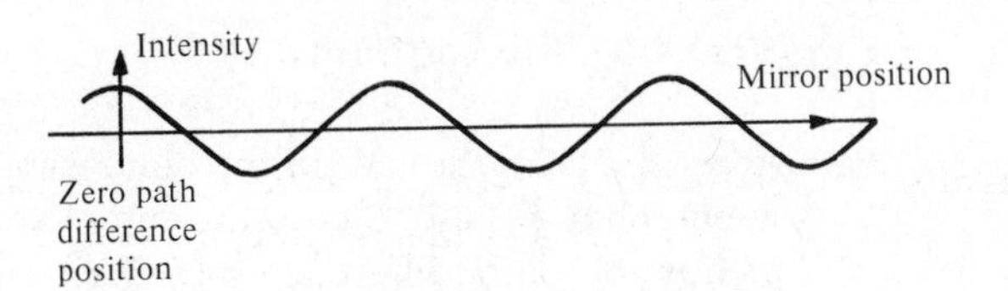

Fig. 5.4.2. Variation of intensity of radiation at E with change in mirror position at D, for monochromatic radiation

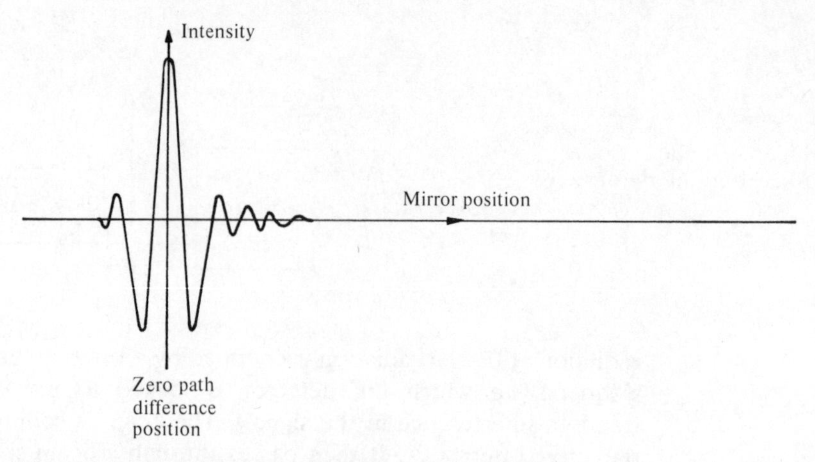

Fig. 5.4.3. Variation of intensity for 'white' radiation

that which is given out by the mercury lamp, is shown in Fig. 5.4.3. This is the *background interferogram.*

Suppose we now place a sample in the position shown in Fig. 5.4.1. If the sample absorbs radiation at the frequencies we are considering, it will have the effect of producing positions of the movable mirror, other than the zero path difference position, for which there is not complete cancellation of the various wave trains. The resulting interferogram thus contains information about the absorption spectrum. In a case where a series of absorption bands is harmonically related, these inverted peaks or *signatures* in the interferogram are particularly clear. The HCN spectrum is like this, since the rotational lines are each spaced by very nearly $2B$ from each other, and the interferogram has signatures which appear at optical path differences equal to $1/2B$. It is wrong to suppose that each signature corresponds to a particular rotational transition; rather each signature is a result of many harmonically related absorptions.

For the results of our experiment to be useful, we need to convert the interferogram into an absorption spectrum. The process by which the interferogram is built from the absorptions in the sample is known as a *Fourier transformation,* and is expressed by the equation[5,7]

$$I(x) = \int_{-\infty}^{\infty} S(\tilde{\nu})\cos(2\pi\tilde{\nu})\,\mathrm{d}\tilde{\nu}, \tag{5.4.9}$$

where $S(\tilde{\nu})$ is the spectrum (intensity S as a function of wavenumber $\tilde{\nu}$) and $I(x)$ is the interferogram (intensity I as a function of path difference x). Curiously, if we carry out a second Fourier transformation, we obtain the original absorption spectrum:

$$S(\tilde{\nu}) = 4\int_{-\infty}^{\infty} I(x)\cos(2\pi\tilde{\nu}x)\,\mathrm{d}x. \tag{5.4.10}$$

In this experiment we employ a computer program to carry out this mathematical process. The resolution of the spectrum is the reciprocal of the maximum path difference, the latter being limited by the distance of travel of the movable mirror. The highest wavenumber, $\tilde{\nu}_{max}$, which can be transformed is $1/2\Delta x$, where Δx is the *sampling interval*, i.e. how far the mirror moves between each reading.

Two major advantages of using interferometers in the far infrared region are now evident. The first is the high throughput of radiation, typically 300 times more than in a grating spectrometer, because of the absence of any slits or gratings on which radiation is dispersed. The second advantage lies in the mode of operation. In a prism or grating spectrometer each spectral element is scanned sequentially. An interferometer, however, effectively allows each spectral element to be observed for the whole duration of the experiment. In simple terms, this results in a spectrum which is much more visible above the detector noise. The *signal-to-noise ratio* is the ratio of the true signal, which forms the spectrum, to the unwanted random interference in the signal. This is proportional to the square root of the observation time, and, for a moderately long experiment, might typically be a factor of 30 better for an interferometer than for a grating spectrometer.

Procedure

There is insufficient gas in the sample cells to be dangerous, but nevertheless exercise reasonable caution when handling them.

Set up the instrument as instructed.*

Evacuate the whole apparatus to below 0.3 Torr, and record a background interferogram without outputting data for Fourier transformation.

Place a quartz disc in the sample compartment and measure the interferogram, again without outputting data for transformation.

Admit air to the sample chamber up to atmospheric pressure and record the water vapour interferogram as instructed,* both on the chart recorder and with an output for Fourier transformation. Note the irregular series of features.

Carefully place the HCN cell in the sample chamber and evacuate the space around it. Record the interferogram* and output data for transformation. Calibrate the trace so that the distances between signatures can be measured in terms of optical path difference. Let the sample chamber up to atmospheric pressure and carefully remove the sample.

Measure the interferogram for NO in a similar manner.*

Calculation Perform Fourier transformations on the interferograms for water vapour, HCN and NO as instructed.*

Calculate an approximate value for B_0 from this spacing of the signatures in the interferogram.

Measure the wavenumbers of the pure rotational absorption lines in HCN. Using the approximate value of B_0, assign the absorptions to lower-state J values. Plot a straight-line graph to find accurately the values of B and D, ideally using a least-squares fit.

Mark the wavenumbers of the absorption lines in NO. Find an approximate value of B_0 and hence assign J values to the transitions. Use a graphical method to find B_0 and D_0. Use your

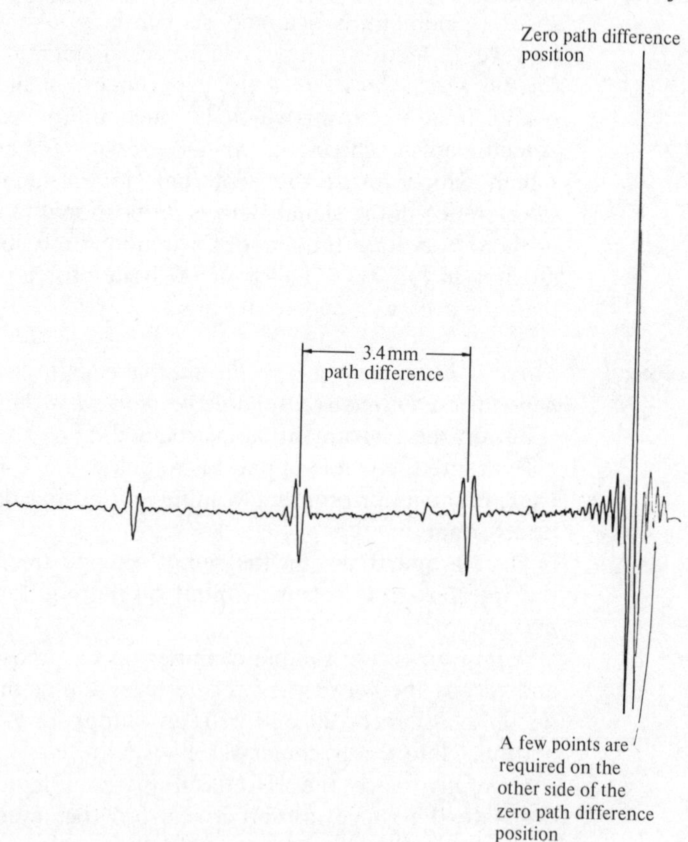

Fig. 5.4.4. Interferogram of HCN at 30 mmHg pressure

value of B_0 to determine the bond length in NO. Draw a scale diagram of the rotational energy levels up to $J = 6\frac{1}{2}$, given that there is an energy separation of 121.1 cm^{-1} between the lowest levels of the components, i.e. between $^2\Pi_{1/2}(J = \frac{1}{2})$ and $^2\Pi_{3/2}$ $(J = \frac{3}{2})$.

Results

Figure 5.4.3 shows a background interferogram, and Fig. 5.4.4 part of the HCN interferogram. Absorptions in the HCN absorption spectrum occur at 29.55 cm^{-1} and other wavenumbers. Pairs of lines in the NO spectrum occur at 21.73 and 22.35 cm^{-1}, 61.82 and 63.38 cm^{-1}, and at other wavenumbers.

Resolution

The resolution could be further improved firstly by reducing the pressure of the samples, thus reducing pressure broadening of the lines. However, such a reduction would also diminish the absorption intensity and so degrade the signal-to-noise ratio. The pressures used here are chosen as a compromise between absorption intensity and pressure broadening. The second improvement would be to use the full optical path available on the instrument. However detector and source noise, and the precision of the data output from some instruments, make longer runs of little value in this experiment.

Comment

Modern interferometers are calibrated by measuring the interferences fringes of laser light sent along a secondary optical path. They are capable of measuring to a resolution of better than 10^{-3} cm^{-1}. The use of fast on-line computing facilities has made interferometers increasingly convenient to use, and they are now employed for precision measurements over the entire infrared region.[6] Typical applications in the far infrared are in the study of hindered rotation in molecules,[8] and of atmospheric and stratospheric species.[9]

Michelson interferometers were mounted in prototypes of the supersonic aeroplane Concorde to measure concentrations of ozone and other minor atmospheric constituents. The study was part of an investigation of the effects of stratospheric aircraft on the Earth's ozone layer.

Technical notes

Apparatus

Michelson interferometer and associated apparatus;$$ data output device; computer with data input device and Fourier transform program (available from author);$ quartz disc in holder for sample spectrum, of stated thickness; 2 absorption cells containing HCN at 30 mm Hg and NO at 150 mm Hg.

Design note

Simple, hand operated interferometers are available which demonstrate very clearly the interference fringes and other features of interferometry.

References

[1] *Fundamentals of molecular spectroscopy.* C. N. Banwell. McGraw-Hill, London, 2nd edn (1972), ch. 2.

[2] Ref. 1, p. 209.

[3] *Pure rotational spectra of CO, NO, and N_2O between 100 and 600 Microns.* E. D. Palik and K. Narahari Rao. *J. Chem. Phys.* **25,** 1174 (1956).

[4] *Molecular spectra and molecular structure 1. Spectra of diatomic molecules.* G. Herzberg. Van Nostrand, Princeton N.J., 2nd edn (1965), p. 218.

[5] *High resolution transmission spectrometry in the far infrared.* J. Fleming. *Chem. in Brit.* **13,** 328 (1977).

[6] *Chemical infrared fourier transform spectroscopy. Chemical Analysis* **43,** P. R. Griffiths, John Wiley, New York (1975), ch. 1 and 2.

[7] Ref. 6, ch. 4.

[8] *Spectroscopy.* Eds. B. P. Straughan and S. Walker. Chapman and Hall, London (1976), Vol. 2, ch. 5.

[9] *Optical systems unravel smog chemistry.* J. N. Pitts Jr., B. J. Finlayson-Pitts, and A. M. Winer. *Environmental Science and Technology* **11,** 568 (1977).

5.5 Thermodynamics of hydrogen bond formation by infrared spectroscopy

Absorption or emission spectra occur at infrared frequencies when the vibrations of a molecule cause a periodic change in its dipole moment. Molecules containing a hydroxyl group show an absorption band in their infrared spectrum at about 3600 cm^{-1} due to the stretching of the —O—H bond. In the presence of acceptor molecules with which hydrogen bonds can be formed a new band appears at a lower frequency (lower wavenumber). This absorption is due to the stretching of hydrogen bonded, and therefore weaker, —O—H bonds. The band is much broader and more intense than that caused by free —O—H groups, as shown in Fig. 5.5.1.

In the system to be studied in this experiment the proton donor is phenol and the acceptor 1,4-dioxan. The equilibrium constant for hydrogen bond formation is calculated from the change in the 3600 cm^{-1} absorption on adding dioxan to a solution of phenol in tetrachloromethane (carbon tetrachloride). The free energy of the hydrogen-bond formation is derived from the equilibrium constant, and measurements of the temperature variation of the equilibrium constant in a heated cell are used to obtain the enthalpy and entropy terms.

Theory

The absorption of infrared radiation obeys the *Beer–Lambert*

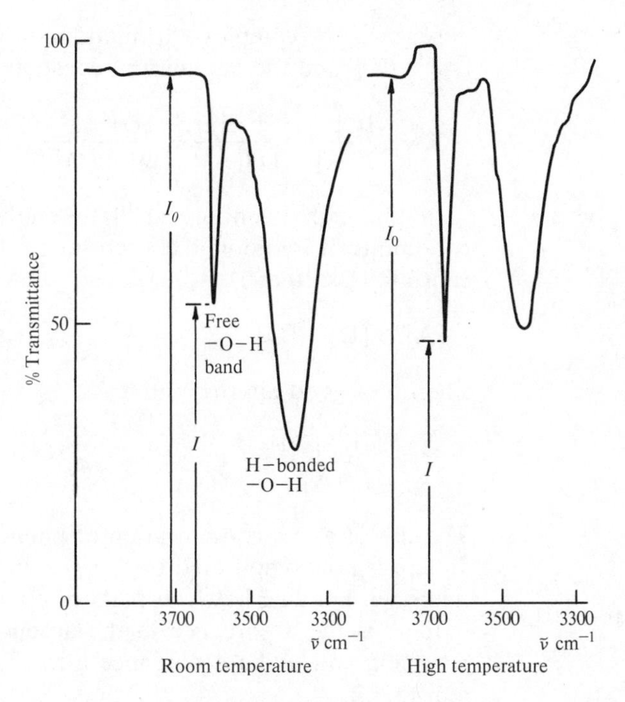

Fig. 5.5.1. Infrared spectra of phenol/dioxan solutions showing the effect of hydrogen bonding on the —O—H absorption

law

$$I = I_0\, 10^{-\varepsilon lc}. \tag{5.5.1}$$

This may also be expressed in the form

$$a = \log_{10}(I_0/I) = \varepsilon lc, \tag{5.5.2}$$

When logarithms to base 10 are used in the Beer–Lambert law, the absorption coefficient is given the symbol ε; when natural (Naperian) logarithms are employed, the symbol is κ.

where a is the optical absorbance, I_0 and I are the incident and transmitted light intensities, ε is the absorption (extinction) coefficient of the hydrogen-bonded complex, l the optical path length in the cell, and c is the concentration of the absorbing species. This relationship holds for each wavenumber (or wavelength), although (I_0/I) and ε are clearly functions of frequency ν (or wavenumber $\tilde{\nu}$). Infrared spectrometers usually record transmittance $\mathcal{T} = I/I_0$, so that the ratio $(I_0/I) = 1/\mathcal{T}$ for the free —O—H bond may be measured directly as shown in Fig. 5.5.1. The optical absorbance a, which is proportional to the concentration of species containing free —O—H, is then calculated from eqn (5.5.2).

The formation of a 1:1 complex C from a donor D and acceptor A can be represented by the equilibrium:

$$D + A \leftrightharpoons C.$$

If $[D]^o$ and $[A]^o$ are the concentrations of D and A in the absence of complex formation, then $[C]=[D]^0-[D]=[A]^0-[A]$, and the equilibrium constant K is given by

$$K=\frac{[C]}{[D][A]}=\frac{[D]^0-[D]}{[D]([A]^0-[D]^0+[D])}. \tag{5.5.3}$$

In this experiment we shall be studying the system under conditions of low donor concentration and high acceptor concentration, so that

$$[A]^0 \gg [D]^0-[D]. \tag{5.5.4}$$

Then to a good approximation

$$K=\frac{[D]^0-[D]}{[D][A]^0}. \tag{5.5.5}$$

The use of a low concentration of phenol also ensures that the system is not complicated by the presence of dimers formed by hydrogen bonding between pairs of phenol molecules.

If a^0 is the absorbance in the absence of hydrogen-bonded complex, and a the absorbance after its formation, then from eqn (5.5.2)

$$K=\frac{a^0-a}{a[A]^0}, \tag{5.5.6}$$

and the dimensionless equilibrium constant $K^\ominus$ is given by

$$K^\ominus=\frac{a^0-a}{a}\frac{c^\ominus}{[A]^0}, \tag{5.5.7}$$

where $c^\ominus$ is the standard concentration of 1 mol dm^{-3}. Thus we do not need to know ε, d, or the concentration of phenol to determine $K^\ominus$. It is sufficient to measure the optical absorbance at the maximum of the sharp free —OH peak for solutions with and without a known concentration $[A]^0$ of dioxan, provided the concentration of phenol is low enough to satisfy eqn (5.5.4).

Use the value of a^0 from a higher temperature spectrum for *all* the calculations; this value of a^0 is a better measure of the free phenol because the phenol is more dissociated at the higher temperature, and the extinction coefficient varies very little with temperature.

If values of $K^\ominus$ are found at two different temperatures, T_1 and T_2, the thermodynamic properties of phenol-dioxan hydrogen bonding may be found by use of various thermodynamic relations. The first is the van't Hoff isochore (p. 91),

$$\frac{d(\ln K^\ominus)}{dT}=\frac{\Delta H^\ominus}{RT^2}. \tag{5.5.8}$$

If we assume that $\Delta H^\ominus$ does not change with temperature, then we may integrate this expression between T_1 and T_2 and then rearrange it to give the equation

$$\Delta H^{\ominus}=R[\ln K^{\ominus}(T_1)-\ln K^{\ominus}(T_2)]\Big/\left(\frac{1}{T_2}-\frac{1}{T_1}\right). \quad (5.5.9)$$

This expression may be used to calculate $\Delta H^{\ominus}$ from the value of $K^{\ominus}$ at the experimental temperatures T_1 and T_2. With the known values of $\Delta H^{\ominus}$ and $\ln K^{\ominus}(T_1)$, it may then be used to find $K^{\ominus}$ at a different T_2, namely the standard temperature of 298 K.

We may then use the van't Hoff isotherm

$$\Delta G^{\ominus}=-RT\ln K^{\ominus} \quad (5.5.10)$$

to calculate $\Delta G^{\ominus}$ at 298 K, and then calculate $\Delta S^{\ominus}$ from the relation

$$\Delta G^{\ominus}=\Delta H^{\ominus}-T\,\Delta S^{\ominus}. \quad (5.5.11)$$

Apparatus

The principles of operation of infrared spectrometers are described on p. 224.* In this experiment, the spectrometer is provided with a variable temperature absorption cell fitted with silica windows, and a Teflon (Fluon) spacer giving a path length of 6 mm. The heating jacket of the cell is supplied from a variable transformer, and its temperature measured by a carefully positioned thermocouple or thermometer. A cell is placed in the reference beam to compensate for absorption by tetrachloromethane or dioxan. This cell is not heated, so the compensation is only partially effective for the high temperature measurements.

Procedure

Infrared cells are fragile; treat them with care and on no account attempt to dismantle them. Do not connect the heating jacket directly to the mains.

Switch on the spectrometer as instructed.*

A solution of phenol in tetrachloromethane of the required strength (about 0.01 M) is provided in a stock bottle containing a drying agent. Filter about 250 cm^3 of the solution into a flask, which must be kept stoppered to minimize changes in concentration due to evaporation losses. This quantity should be sufficient for the whole experiment.

A syringe is provided for transferring solutions to the spectrometer cells. This should be rinsed with tetrachloromethane after each use, and dried with a gentle stream of air if an air line is available.

Fill the variable temperature cell with the filtered solution, and place the cell in its jacket in the sample beam of the spectrometer. The heating current should be switched off. Fill the compensation cell with pure tetrachloromethane, and position it in the reference beam. Adjust the spectrometer to give a transmittance reading of about 98 per cent.* Allow a few minutes for the temperature of the cell to become steady, note the temperature, and then record the spectrum between 3800 and 3300 cm^{-1}.

Switch on the heating current and adjust the variable transformer so that the sample cell attains a steady temperature between 60 and 70 °C. Record the spectrum, using the same paper, but with the wavenumber scale displaced so that the spectra appear side by side (as in Fig. 5.5.1).* Before removing the cells, record the zero transmission point on the chart by *slowly* blanking off the sample beam with a piece of card, releasing the paper drive* and moving the chart manually under the pen. Allow the sample cell to cool.

While the heated cell is cooling, make up two solutions of dioxan, one in the stock phenol solution and one in pure tetrachloromethane, as follows. Half fill a 25 cm^3 standard flask with one of the solvents, and weigh it. Add about 1/200th mole of solute, weigh again, and make up to the mark with solvent to give a solution of an accurately known concentration of around 0.2 M. Make up the second solution with as nearly as possible the same concentration. Keep the standard flask stoppered whenever possible to prevent evaporation of the solvent.

Empty the sample and reference cells, rinse with tetrachloromethane, and dry. Refill the reference cell with dioxan in CCl_4, and the variable temperature cell with the solution of dioxan in CCl_4/phenol. Record the spectrum on a new piece of paper, this time between 3800 and 3200 cm^{-1}. Note the broad —OH band at a lower frequency than the sharper free —OH peak. Record the spectrum again at the same elevated temperature as before.

Repeat the measurements, using two dioxan solutions of equal concentrations of about 0.3 M.

At the end of the experiment you should have six spectra altogether, three at room temperature at dioxan concentrations of 0, 0.2, and 0.3 M, and three at a temperature between 60 and 70 °C.

Calculation

Measure the peak optical density (minimum transmittance I) of the sharp free —OH band on each of the six spectra. Also measure each incident intensity I_0, using the flat part of the spectrum to the high frequency side of the band. In the heated run, there may be a small 'reversed' band at the foot of the —OH band due to over-compensation by the tetrachloromethane in the unheated reference cell. Use the flat part of the curve before this, as shown in Fig. 5.5.1.

From your optical measurements, determine two values of $K^\ominus$ at each temperature. Use the averages of these to calculate $\Delta H^\ominus$, $\Delta G^\ominus$ and $\Delta S^\ominus$ at 25 °C. Comment on the sign and magnitude of the terms.

Results

See Fig. 5.5.1.

Accuracy

Your estimate of the accuracy of your results should be compatible with the difference in the values of $K^\ominus$ for the 0.2 M and 0.3 M solutions. The $K^\ominus$ values at room temperature should differ by no more than 10 per cent.

Comment[1,2]

This simple experiment demonstrates the use of an infrared spectrometer for quantitative intensity measurements, rather than for its more usual analytical purpose. The spectra give direct evidence for hydrogen bonding, and are easily analysed to give a rough estimate of the thermodynamic quantities associated with this phenomenon.

The hydrogen bonding which has been studied in this experiment is of the type which causes intermolecular interactions between pairs of molecules. Hydrogen bonding may also extend over many molecules, as in ice, or cause intramolecular interactions, as in ethanoic acid (Expt 3.2). Hydrogen bonding is a crucial determinant of structure in a great number of biological systems, including proteins, cellulose and nucleic acids.

Technical notes

Apparatus

Infrared spectrometer;[$$] infrared cell with heater,[$ or †] variable power supply and thermometer; unheated reference cell; solution of 1.6 g phenol + ~5 g calcium sulphate drying agent in 2 dm^3 tetrachloromethane; dioxan; tetrachloromethane.

References

[1] *Coulson's valence.* R. McWeeny, Oxford University Press, Oxford, 3rd edn (1979), p. 358.

[2] *The hydrogen bond.* G. C. Pimentel and A. L. McClellan, Freeman, San Francisco (1960).

5.6 Rotation–vibration spectra of simple diatomic and polyatomic molecules

In the last experiment we utilized the fact that, for a sample in solution, an infrared spectrum is produced by vibrations of a molecule which cause a periodic change in its dipole moment. The same is true for a gaseous sample. In this case, however, we may also observe that changes in the rotational energy of a molecule give rise to a *rotational fine structure*. The energies between successive rotational levels are about 1000 times less than between vibrational levels, with the result that a *band spectrum* is observed.

This experiment is concerned with the rotation–vibration spectra of two classes of molecules: simple diatomic molecules, illustrated by hydrogen chloride and deuterium chloride (Part A), and the linear molecule ethyne (acetylene) (Part B). Bands in the spectra are analysed to find the moments of inertia of the molecules, and thus their internuclear separations. The magnitudes of the fundamental vibrational frequencies yield the force constants of the molecules, which are a measure of their bond strengths. The *isotope effect* is demonstrated in Part A.

Theory[1]

A. Diatomic molecules

The simplest way of describing a vibrating diatomic molecule is to assume that it is a harmonic oscillator, with a potential energy which depends quadratically on the change in the distance r between the two nuclei (Fig. 5.10.1, p. 254). The allowed energy levels may be calculated by quantum mechanics,[2] and expressed as term values G:[3]

$$G(v) = \omega_e(v + \tfrac{1}{2}) \tag{5.6.1}$$

where v is the vibrational quantum number having integral values $0, 1, 2, \ldots$ and ω_e is the vibrational wavenumber. In this experiment we shall once again express all the term values, and the constants B and D introduced below, as wavenumbers in units of cm^{-1}.

In practice the vibrations of a diatomic molecule are not those of a harmonic oscillator, and so we need to include an *anharmonicity constant* x_e:

This equation can be used to derive the zero-point energy and dissociation limit of a molecule, as described for the case of iodine in Expt 5.10.

$$G(v) = \omega_e(v + \tfrac{1}{2}) - x_e\omega_e(v + \tfrac{1}{2})^2. \tag{5.6.2}$$

The vibrational wavenumber ω_e is determined by the masses of the atoms and the *force constant* k of the bond, according to the expression

$$\nu_{osc} = \omega_e c = \frac{1}{2\pi}(k/\mu)^{1/2}, \tag{5.6.3}$$

where ν_{osc} is the *classical fundamental vibration frequency* of the molecule and μ is the *reduced mass* $m_1m_2/(m_1+m_2)$. Classically, the spectrum of the molecule would arise from the fundamental frequency ν_{osc}, together with weak *overtone bands*, namely the first overtone (second harmonic) at $2\nu_{osc}$, the second overtone (third harmonic) at $3\nu_{osc}$, and so on. The overtones which are actually observed have frequencies less than $2\nu_{osc}$, $3\nu_{osc}$, ... because of the anharmonicity (Fig. 5.6.1).

The easiest way of describing the rotation of a diatomic molecule is to regard it as a dumbbell or *rigid rotor*. The rotational energies F may then be shown to be[4]

$$F(J) = BJ(J+1), \tag{5.6.4}$$

where the *rotational constant* B is given by

$$B = h/8\pi^2 cI. \tag{5.6.5}$$

J is the rotational quantum number, c is the speed of light, and I is the moment of inertia of the molecule.

In practice *centrifugal distortion* reduces the energies of higher rotational states:

$$F(J) = BJ(J+1) - DJ^2(J+1)^2 \tag{5.6.6}$$

where the *centrifugal stretching constant* D is given by

$$D = 4B^3/\omega^2. \tag{5.6.7}$$

The moment of inertia of a molecule is increased at higher vibrational energies, so the value of B in the ground vibrational state, B_0, is different from that in the first excited state, B_1. In general,

$$B_v = B_e - \alpha(v + \tfrac{1}{2}) \tag{5.6.8}$$

where B_e is the extrapolated value at $r = r_e$, the equilibrium separation, and α is a small constant. *How may α be calculated from B_0 and B_1?*

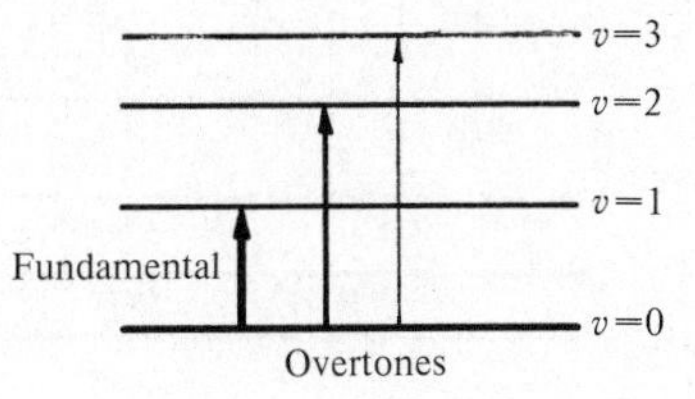

Fig. 5.6.1. Transitions giving rise to the vibrational spectrum of HCl

It follows that the energy levels of a rotating, vibrating diatomic molecule are described by the equation

$$T = G(v) + F(J)$$
$$= \omega_e(v+\tfrac{1}{2}) - x_e\omega_e(v+\tfrac{1}{2})^2 + B_vJ(J+1) - DJ^2(J+1)^2. \qquad (5.6.9)$$

The vibrational spectrum of HCl consists of a single fundamental band, corresponding to $(v=0)\to(v=1)$, though weak overtones can be observed by using long path lengths and high pressures of the absorbing gas. In the fundamental band, the vibrational transition is accompanied by an increase (R branch) or decrease (P branch) in the rotational quantum number, according to the selection rule $\Delta J = \pm 1$ (Fig. 5.6.2). The labelling of the P, Q, R branches is alphabetical in increasing energy. *Verify that if* $\tilde{\nu}_0$ *is the wavenumber of the pure vibrational transition, the wavenumbers of the rotation–vibration transitions (i.e. the difference in wavenumbers of the upper and lower state for each) are given by*

The more general selection rule underlying the rotational transitions is that there must be a change in the parity (+ or −) shown in Fig. 5.6.2.[5]

$$\tilde{\nu} = \tilde{\nu}_0 - (B_0+B_1)J - (B_0-B_1)J^2 + 4DJ^3 \qquad (5.6.10)$$

for the P branch, and

$$\tilde{\nu} = \tilde{\nu}_0 + (B_0+B_1)(J+1) - (B_0-B_1)(J+1)^2 - 4D(J+1)^3 \qquad (5.6.11)$$

for the R branch. In these expressions, J is the rotational quantum number in the *lower* vibrational state, and B_0 and B_1 are the values of B in the ground and first excited vibrational states respectively.

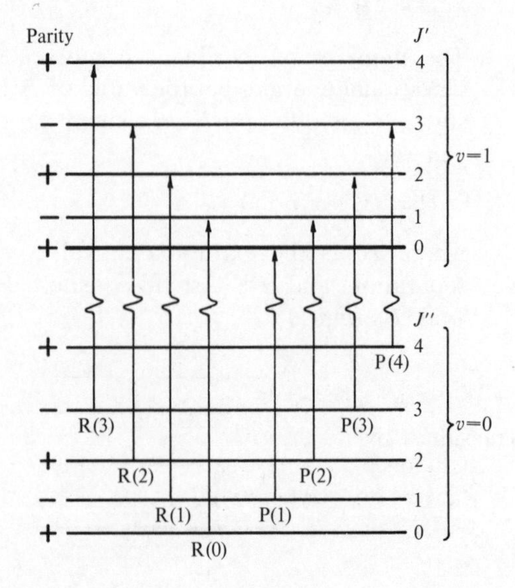

Fig. 5.6.2. Transitions giving rise to rotational fine structure in the vibrational spectrum of HCl

We can represent the whole band by using a running number m, equal to $-J$ in the P branch and $(J+1)$ in the R branch:

$$\tilde{\nu} = \tilde{\nu}_0 + (B_0 + B_1)m - (B_0 - B_1)m^2 - 4Dm^3. \tag{5.6.12}$$

This expression shows that the lines of the band form a continuous series through the P and the R branch.

The lines in the spectrum are labelled with the letter P or R, with the value of J in the *lower* vibrational state in brackets (Fig. 5.6.2). So P(1) represents the rotational transition $J=1$ to $J=0$, and R(3) that from $J=3$ to $J=4$. It can be seen in the figure that the differences $R(J-1)-P(J+1)$, for example R(0)–P(2), have a common upper energy level for any particular value of J, and therefore depend only on the energies of the lower state levels. *Verify from eqn* (5.6.6) *that these differences are given by*:

$$\begin{aligned}\Delta_2 F_0 &= R(J-1) - P(J+1) \\ &= (4B_0 - 6D)(J+\tfrac{1}{2}) - 8D(J+\tfrac{1}{2})^3.\end{aligned} \tag{5.6.13}$$

The two subscripts in the term $\Delta_2 F_0$ remind us that the difference in J is 2, and that the rotational constant is B_0.

The availability of computers for research experiments has now made the use of combination sums and differences less important; combination differences are still very useful where the upper state only suffers from perturbation, but in other cases a simultaneous least-squares fit to all lines is carried out instead.

It is clear that if $\Delta_2 F_0/(J+\frac{1}{2})$ is plotted against $(J+\frac{1}{2})^2$, the slope can be used to determine D, and the intercept to find B_0. This technique of simplifying the calculations, known as the method of *combination differences*, has often been used in the analysis of rotational fine structure. *What combination differences may be used to determine B_1? Combination sums are also useful in the analysis of spectra; show how the band origin, $\tilde{\nu}_0$, and the value of α can be found by plotting the combination sum $R_{J-1}+P_J$ against J^2.*

B. Linear polyatomic molecules

The seemingly random and erratic vibrations of polyatomic molecules can be expressed as a combination of *normal modes.* In each normal mode of vibration all the atoms in a molecule vibrate with the same frequency, and the atoms pass through their equilibrium position simultaneously, and also reach their position of maximum displacement at the same time.

The rotation–vibration spectra of linear polyatomic molecules, containing n atoms, differ in two ways from those of diatomic molecules. Firstly there may be up to $(3n-5)$ normal modes active in the infrared. However, since only vibrations in which there is a change in electric dipole moment are infrared active, and since some modes occur in degenerate pairs, there are in fact always less than this. Nevertheless more than one fundamental band is usually found. The second difference is that in those bands in which the change in dipole is at right angles to the line of the molecule, the so-called *perpendicular bands*, the selection rules allow a change of v without a change in J:

$$\Delta v = \pm 1 \qquad \Delta J = 0, \pm 1.$$

This gives rise to a central Q branch. This does not occur in the *parallel bands*, which are similar in appearance to the fundamental bands of diatomic molecules.

Apparatus[6] The layout of a traditional double–beam infrared spectrometer is shown in Fig. 5.6.3. The source of radiant energy is a length of ceramic tube heated to about 1200 °C by an internal heater. The radiation is divided into two beams, one of which passes through the sample whilst the other serves as a reference. In the sample beam there is a manually controlled attenuator which can reduce its intensity, whilst in the reference beam there is a similar attenuator controlled by a servo motor. The two beams are then combined by a rotating semicircular sector mirror to form a single beam consisting of alternate pulses of radiation from the sample and reference beams. This combined beam passes into the monochromator where it is dispersed by the grating into its spectral components. The grating is slowly rotated so that the dispersed spectrum is scanned across the monochromator exit slit. This slit can be narrowed to improve resolution or widened to improve the *signal-to-noise ratio* (p. 211). The radiation then passes through a filter to remove unwanted (higher order) radiation, and is then focussed onto a thermocouple detector.

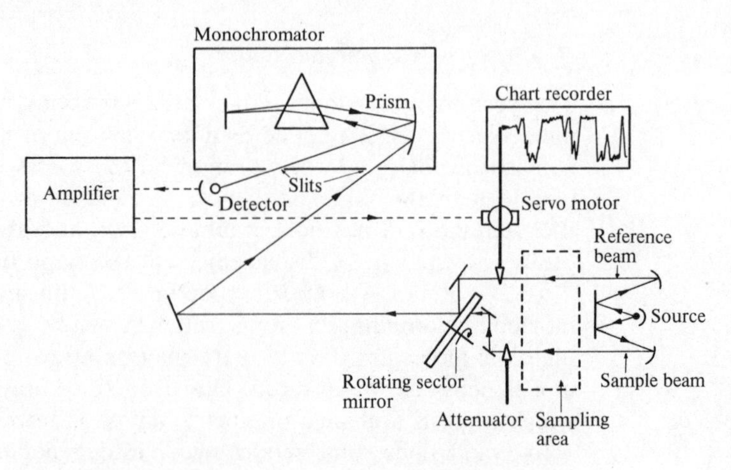

Fig. 5.6.3. Layout of a double-beam infrared spectrometer

If the sample and reference beams are not of the same intensity, the combined beam will flicker at the rotation frequency of the sector mirror. This flicker is detected by the thermocouple, from which a signal is fed to an amplifier. The amplified signal controls the servo motor which moves the beam attenuator in the reference beam until the modulation is zero. The motor is also connected to the recorder pen, which thus records the infrared spectrum.

Modern infrared spectrometers tend to be *ratio-recording*, and use complicated shutters and timing circuits to produce a much more linear response at low transmittance.

Procedure (*A*) *HCl/DCl*

Consult instructions concerning the operation of the spectrometer.*

Fit a sheet of numbered chart paper, and scan the HCl/DCl sample over a wavenumber range of around 625–4000 cm^{-1}.

Fit a roll of graduated but unnumbered paper, and run the scale to just before the onset of the DCl band (band origin ~2090 cm^{-1}). Switch the spectrometer to its high resolution mode as instructed.*

Scan the DCl band. Just before the band starts, stop the scan on a suitable graduation on the wavenumber scale. Lift up the pen carriage, and without moving the paper, make a fine pen or pencil mark at the exact point on the paper at which the pen stopped. Note the wavenumber reading from the scale. Replace the pen on the paper and restart the scan. Do this again at the end of the band. These two marks are the references from which you will measure the wavenumbers of the absorption lines. Label the P and R branches.

(*B*) *Ethyne*

Record the spectrum over a wavenumber range of around 624–4000 cm^{-1} as instructed.*

Using unnumbered paper, record the low wavenumber band (near 700 cm^{-1}) under high resolution.* Mark the P, Q, and R branches.

Calculation[7] *A*(*i*) *HCl/DCl lower resolution spectrum*

Determine the approximate band origins for the HCl and DCl from the low resolution spectrum. Note that they are in the wavenumber ratio of about $2^{1/2}:1$ as expected from eqn (5.6.3).

Table 5.6.1. Overtone frequencies for HCl and DCl, expressed as wavenumbers with units of cm^{-1}

	1st overtone	2nd overtone
HCl	5668	8347
DCl	4125.5	6108.5

Overtone bands for HCl and DCl lie at the wavenumbers shown in Table 5.6.1, which give the mean values for the ^{35}Cl and ^{37}Cl isotopes. Using eqn (5.6.2), obtain a general formula for the vibrational transition $v = 0$ to $v = n$ (i.e. for $G(n) - G(0)$), and call its wavenumber $\tilde{\nu}_n$. Hence use a graphical method to determine ω_e and the product $x_e\omega_e$.

Determine:

(a) the zero point energy ($G(0)$) of HCl,

(b) the force constant in HCl and in DCl (eqn (5.6.3)).

Also make a rough estimate of the heat of dissociation of HCl (dissociation occurs at $d\tilde{\nu}_n/dn = 0$).

A(ii) DCl high resolution spectrum

Assign the lines in the D ^{35}Cl band, as shown in Fig. 5.6.4. Measure the distances of the absorption lines and of the two 'bench marks' from an arbitrary zero. The scale is linear in wavenumber; derive a linear relationship between your measured distances and wavenumber and use this to determine the wavenumbers of the absorption lines.

Tabulate your results in the first four columns of a table which could be set out as shown in Table 5.6.2.

Now use the combination difference method to determine B_0 and D for D ^{35}Cl. Use your B_0 value to obtain the bond length,

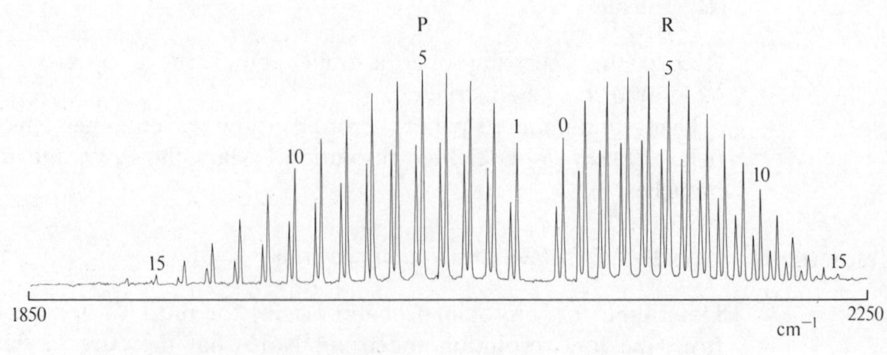

Fig. 5.6.4. Infrared spectrum of DCl

Table 5.6.2. Suggested method of tabulating results

J	R_J position/cm	P_J position/cm	R_J/cm^{-1}	P_J/cm^{-1}	$\Delta_2F_0/(J+1/2)$	$(J+1/2)^2$	$R_{J-1}+P_J$
0						0.25	
1						2.25	
2						6.25	
etc.						etc.	

The value of D, and hence the slope of the plot, is very small, and the scatter in the first few points must be ignored; nevertheless a reasonable estimate of D can be made.

given that $I \simeq \mu r^2$ for a diatomic molecule. Check that your value of D agrees with eqn (5.6.7).

Plot the combination sum to determine ν_0 and α. The value of B_0 could be corrected to B_e by using the α value, thus enabling the calculation of the equilibrium bond length.

B(i) Ethyne low resolution spectrum

Assign the two fundamental bands in your lower resolution ethyne spectrum. The higher frequency band is overlaid by combination and difference bands, which give it an irregular appearance. There is also a marked irregularity in the lower band from the same cause. Lines of the overlying bands can just be seen between the main lines. Note the intensity alternation. The band at 1300 cm^{-1} is the combination ($\tilde{\nu}_4 + \tilde{\nu}_5$).

B(ii) Ethyne high resolution spectrum

The presence of a Q branch in the low frequency ethyne band complicates the assignment of the lines in the P and R branches to specific rotational transitions. There are also some irregularities of position and intensity in the P branch, owing to the presence of Q branches of much weaker difference bands.[8] The assignment can be made by working in to the centre of the band from those parts where the spacing and intensities are regular. We know that:

(a) odd lines are stronger than even, due to the symmetry effects of nuclear spin;[9]
(b) the P branch starts with P(1), the R with R(0);
(c) the P and R branches form a continuous series through the origin with approximately equal spacing . . . P(2) . . . P(1) . . . $\tilde{\nu}_0$ . . . R(0) . . . R(1) although some of the lines are very weak, or obscured by the Q branch or by difference bands;
(d) because the instrument has a finite slit width the origin $\tilde{\nu}_0$ lies under the Q branch, (and is nearer the steeper side).

Table 5.6.3. Ethyne (acetylene)-ν_5—line frequencies and lower state combination differences, expressed as wavenumbers with units of cm^{-1}

J	$(J+1/2)^2$	R_J	P_J	$R_{J-1}-P_{J+1}/(J+1/2)$
0	0.25	—	—	—
1	2.25	733.86	—	—
2	6.25	736.26	724.45	4.7040
3	12.25	738.60	722.10	4.7086
4	20.25	740.93	719.78	4.7156
5	30.25	743.27	717.38	4.7091
6	42.25	745.32	715.03	4.7062
7	56.25	747.91	712.68	4.6933
8	72.25	750.28	710.32	4.7012
9	90.25	752.66	707.95	4.7021
10	110.25	755.05	705.61	4.7067
11	132.25	757.33	703.24	4.7104
12	156.25	759.70	700.88	4.7040
13	182.25	762.02	698.53	4.7059
14	210.25	764.34	696.17	4.7103
15	240.25	766.72	693.72	4.7019
16	272.25	769.06	691.46	4.7042
17	306.25	771.41	689.10	4.7034
18	342.25	773.74	686.75	4.7011
19	380.25	776.07	684.44	4.6995
20	420.25	778.42	682.10	4.6995
21	462.25	780.77	679.73	4.7009
22	506.25	783.07	677.35	4.7004
23	552.25	785.39	675.01	4.6979
24	600.25	787.71	672.67	4.6967
25	650.25	790.07	670.32	4.6945
26	702.25	792.40	668.00	4.6977
27	756.25	794.69	665.58	4.6967
28	812.25	797.01	663.24	4.6933
29	870.25	799.37	660.93	4.6925
30	930.25	801.65	658.58	4.6938
31	992.25	803.99	656.21	4.6924
32	1056.25	806.29	653.84	4.6920
33	1122.25	808.62	651.50	4.6893
34	1190.25	810.91	649.20	4.6896
35	1260.25	813.23	646.83	4.6882
36	1332.25	819.57	644.48	4.6874
37	1406.25	817.86	642.14	4.6877
38	1482.25	820.15	639.78	4.6855
39	1560.25	822.44	637.47	4.6805
40	1640.25	824.75	635.13	4.6857
41	1722.25	827.02	632.76	4.6814
42	1806.25	829.17	630.47	4.6805
43	1892.25	831.65	628.10	4.6765
44	1980.25	833.91	625.74	4.6787
45		836.21	623.45	
46		838.47		

Mark the positions of obscured lines by 'stepping in' from the observed P and R lines, and assign the band origin. Number the lines in the P and R branches; the frequencies are given in Table 5.6.3.

Use a combination difference plot to determine B_0.

The B_0 value for C_2D_2 is 0.8479 cm^{-1}. Assume that the bond lengths in C_2H_2 are the same as in C_2D_2, and that the moment of inertia of a molecule is the sum of the products of the masses of the atoms and the squares of their distance from the centre of mass. Hence use eqn (5.6.5) and your value of B_0 for C_2H_2 to calculate the bond lengths in ethyne.

Results

DCl—see Fig. 5.6.4.

Accuracy

Typically the band origins cannot be read from the instrument scale to better than 2 to 5 cm^{-1} (depending on wavenumber region) with a repeatability of between 1 and 3 cm^{-1}. The accuracy of B and α is determined less by absolute wavenumber accuracy than by the linearity of the grating drive. It is clear that a small departure from linearity may have a marked effect upon the measurement of α. Simple instruments in which the grating position is governed by a cam or sliding cosecant drive cannot be expected to give accurate values of these constants.

Comment

This experiment demonstrates the use of infrared rotational fine structure in the determination of the structural properties of diatomic and simple polyatomic molecules. In current research, extremely high spectral resolution and precision is achieved by the use of very long path lengths, typically 10 m, and laser interferometers to monitor grating position.[10] The direct use of interferometers (Expt 5.4) also yields equally high resolution spectra.

Technical notes

Apparatus

Infrared spectrometer,[$$] vacuum line for filling cells,[†] dessicator for storing cells. Part A: cell with NaCl windows filled with 150 mmHg each of HCl and DCl. Part B: cell with NaCl windows filled with 100 mmHg ethyne (acetylene).

References

[1] *High resolution spectroscopy*. J. M. Hollas. Butterworths, London (1982), ch. 5.
[2] *Physical chemistry*. R. S. Berry, S. A. Rice, and J. Ross. John Wiley, New York (1980), part 1, p. 256.
[3] *Molecular spectra and molecular structure, II. Infrared and Raman*

Spectra of Polyatomic Molecules. G. Herzberg. Van Nostrand, New York (1945), p. 77.
[4] Ref. 2, p. 268.
[5] Ref. 1, p. 336.
[6] *Spectroscopy*. Eds. B. P. Straughan and S. Walker. Chapman and Hall, London (1976), vol. 2, p. 140.
[7] *Symmetry and spectroscopy*. D. C. Harris and M. D. Bertolucci. Oxford University Press, New York (1978). p. 121.
[8] Ref. 3, p. 266.
[9] Ref. 1, p. 113.
[10] *High-resolution infrared spectroscopy: aspects of modern research*. K. Narahari Rao. MTP *Int. Rev. Sci., Phys. Chem*. Series 2, **3**, 317 (1976).

5.7 Heat capacity of a gas from infrared spectrum and acoustic interferometry

In this experiment we measure the heat capacity at constant volume C_V of gaseous chloromethane (methyl chloride) by two different methods.

In Part A the high-resolution infrared spectrum of chloromethane is recorded. An analysis of the spectrum, which requires a familiarity with the theory of infrared spectra of polyatomic molecules (Expt 5.6), and also with group theory, yields information about the energy levels in chloromethane. A value for the heat capacity of the gas may then be calculated by the methods of statistical thermodynamics.

Part B of the experiment involves the determination of heat capacity by measurement of the speed of sound using an acoustic interferometer. The passage of a sound wave through a gas involves rapidly alternating adiabatic compression and rarefaction, and it can be shown that the speed of the sound wave depends on the ratio of heat capacities at constant pressure C_p and constant volume C_V. C_V itself is found by extrapolating a series of speed of sound measurements to zero pressure.

Finally we compare the values of C_V obtained by the two methods, which should be in close agreement.

Theory

A(i) Vibrations of chloromethane[1,2]

The chloromethane molecule has a structure which is referred to as a *symmetric rotor* or *symmetric top*, Fig. 5.7.1. It has a three-fold axis of symmetry and two of its principal moments of inertia are equal. In group theory nomenclature[3] it belongs to point group C_{3v}, which has the character table shown in Table

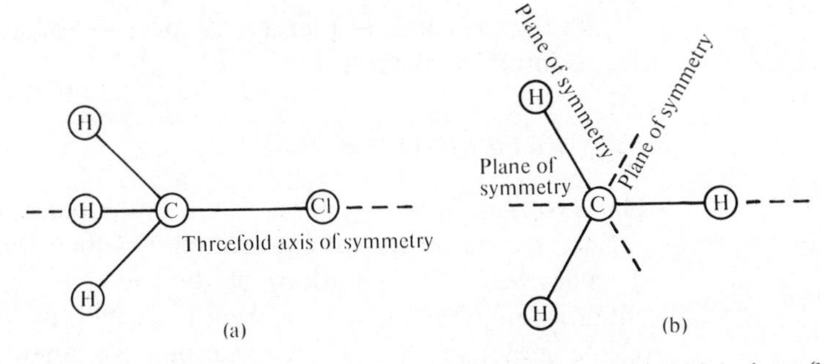

Fig. 5.7.1. Symmetry properties of chloromethane. (a) 'side' view, (b) 'top' view

5.7.1. The vibrations of a non-linear molecule containing n atoms can be rationalized in terms of $(3n-6)$ *normal modes* (p. 223). The $(3\times5-6)=9$ normal modes of vibration of chloromethane are divided into three *parallel vibrations*, of symmetry species A_1, in which the direction of dipole moment change is parallel to the symmetry axis of the top, and three degenerate pairs of *perpendicular vibrations* of symmetry species E, in which there is a change of dipole moment at right angles to the symmetry axis, Fig. 5.7.2. Both classes of vibration are

Table 5.7.1. C_{3v} character table

C_{3v}	E	$2C_3$	$3\sigma_v$	
A_1	1	1	1	z
A_2	1	1	−1	R_z
E	2	−1	0	$(x, y)(R_x, R_y)$

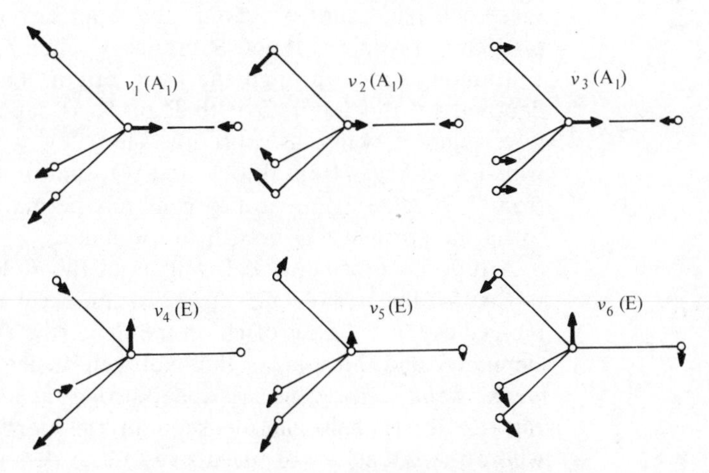

Fig. 5.7.2. Normal modes of vibration of chloromethane

infrared active, and we therefore expect to find six strong bands in the infrared spectrum.

A(ii) Rotational fine structure[5]

The two classes of vibration in chloromethane may be distinguished by the rotational fine structure of their bands. It is usual to characterize the rotations of the molecule in terms of the quantum numbers J, representing the total angular momentum of the molecule, and K, representing the angular momentum about the axis of the top. K is a component of J, and therefore K can never be greater than J.

Remember that these selection rules do not dictate the movements of the atoms in the molecule, but are simply a mnemonic guide as to which parallel vibrations are active in the infrared; $\Delta K = 0$ because rotations about the symmetry axis do not rotate the electric dipole of the molecule.

The selection rules for parallel vibrations are

$$\Delta v = \pm 1 \qquad \Delta J = 0, \pm 1 \qquad \Delta K = 0,$$

where v is the vibrational quantum number. The selection rules are the same as those for the *perpendicular* vibrations of *linear* polyatomic molecules (p. 223), and the bands exhibit the same P, Q, and R branches. The sub-bands corresponding to different values of K are almost exactly superimposable.

In perpendicular vibrations, in which a changing dipole at right angles to the symmetry axis is introduced, changes in K become possible. The selection rules are:

$$\Delta v = \pm 1 \qquad \Delta J = 0, \pm 1 \qquad \Delta K = \pm 1.$$

In this case the sub-bands are no longer superimposed, but are spread out in two series, the P type and the R type, according to the direction of the changes in K. In practice when instruments of modest resolving power are used only the Q branches of these sub-bands are observed, above an envelope of overlapping and unresolved P and R branches. The Q branches form a continuous series through the band origin. They are normally denoted by the letter Q with P or R as superscript, and the lower state K value as subscript. Thus $^{R}Q_0$ is the Q branch in which K changes from 0 to 1, and $^{P}Q_2$ that in which K changes from 3 to 2. Assignment may generally be made by taking $^{R}Q_0$ to be the strongest Q branch in the band.

A difference of statistical weights of the K levels, analogous to that for linear molecules (p. 227), can occur in symmetric top molecules. In the case of chloromethane, the three-fold axis of symmetry and the spins of the hydrogen nuclei lead to a strong, weak, weak, strong, weak, weak, strong... alternation in the ratio 2 : 1 : 1. This can be seen in the perpendicular bands where the intensities of successive Q branches reflect the populations of the successive K levels from which they originate.

Thus the strong lines in a perpendicular band are $^{R}Q_0$, $^{R}Q_3$, $^{R}Q_6$, etc., and $^{P}Q_3$, $^{P}Q_6$, etc. This is a further aid in the assignment of the band origin.

Two principal types of interaction occur to disturb the pattern described above. One is the phenomenon of *Fermi resonance.* This is a straightforward resonance interaction which occurs when two vibrational energy levels are very close to one another and are of the same symmetry. The effect of this interaction is that the levels are 'pushed apart', and there is a mixing of the wave-functions of the interacting level. This mixing can result in a greatly enhanced intensity of an overtone vibration in Fermi resonance with a fundamental. For example, in chloromethane we would expect three parallel and three perpendicular bands. In fact seven strong bands are found, one of which has been ascribed to Fermi resonance increasing the intensity of the parallel component of the overtone $2\nu_5$. The irregular appearance of the perpendicular fundamental ν_4 has also been ascribed to a Fermi resonance interaction, in this case with the level $3\nu_6$.

The second type of interaction is caused by the *Coriolis effect.* The effect arises when a body moves in a rotating framework and produces a force at right angles to both the direction of motion and to the axis of rotation. In a rotating molecule the motion of an atom in one vibrational mode may give rise to a force in a direction close to that of another mode, and an interaction between the modes may be set up. This occurs between the members of a degenerate pair of perpendicular vibrations in a symmetric top molecule and is an important factor in determining the spacing of the Q branches in perpendicular bands. It may also occur between a parallel and a perpendicular vibration, if the frequencies of these are close, and an interaction of this type is responsible for irregularities in the band ν_5 of chloromethane.

A(*iii*) *Heat capacity from spectra*[6]

The *partition function* q for a single isolated polyatomic molecule such as chloromethane may be expressed in the form:

$$q = q^{T}q^{R}q^{V}q^{E}, \tag{5.7.1}$$

where the individual partition functions are the translational, rotational, vibrational, and electronic terms respectively. The expression assumes that the energies for the separate motions are, to an adequate approximation, independent of one another.

It can be shown that the molar heat capacity at constant volume C_V is related to the molecular partition function of a molecule by the expression:

$$C_V = RT\left[\frac{\partial^2}{\partial T^2}(T \ln q)\right]_V. \tag{5.7.2}$$

The translational, rotational, vibrational, and electronic contributions to the heat capacity, C_V^T, C_V^R, C_V^V, and C_V^E respectively, may therefore be calculated, and the total heat capacity found from the sum of these terms, since:

$$C_V = C_V^T + C_V^R + C_V^V + C_V^E. \tag{5.7.3}$$

We now calculate these various heat capacities in turn.[7]

From a knowledge of the energy levels of a particle in a box it can be shown that

$$q^T = V(2\pi m k_B T/h^2)^{3/2}. \tag{5.7.4}$$

Therefore

$$\ln q^T = \tfrac{3}{2}\ln T + \text{const.}, \tag{5.7.5}$$

and, from eqn (5.7.2),

$$C_V^T = 3R/2. \tag{5.7.6}$$

This is also the classical value that follows from the theorem of *equipartition of energy*, being made up from a contribution of $R/2$ for each of the x, y, and z directions. Such agreement is expected because the translational energy levels are closely spaced.

The rotational levels are also closely spaced. Allowing an $R/2$ contribution for rotation about each of three axes by the equipartition principle:

$$C_V^R = 3R/2. \tag{5.7.7}$$

To calculate the vibrational partition function, we assume that the chloromethane molecule undergoes independent harmonic oscillations at each of its six fundamental vibrational frequencies. Then the total vibrational partition function is the product of the partition functions for each fundamental vibration $\tilde{\nu}_i$, and is given by the expression:

$$q^V = \prod_{i=1}^{6} [1 - \exp(-hc\tilde{\nu}_i/k_B T)]^{-1}. \tag{5.7.8}$$

From eqn (5.7.2), the heat capacity contribution from each fundamental vibration is therefore:

$$C_V^{V_i} = R\left(\frac{hc\tilde{\nu}_i}{k_B T}\right)^2 \frac{\exp(-hc\tilde{\nu}_i/k_B T)}{[1 - \exp(-hc\tilde{\nu}_i/k_B T)]^2}. \tag{5.7.9}$$

The total vibrational heat capacity may then be calculated from the sum of the $C_V^{V_i}$ terms.

Chloromethane is not appreciably electronically excited at room temperature, and therefore $q^E = 1$ and $C_V^E = 0$.

B. Heat capacity from speed of sound[8,9]

When sound travels through a gas, the particles of the medium move back and forth in the direction of propagation and form what is known as a *longitudinal pressure* (or *compression*) *wave.* For reasons explained below, the frequencies studied in this experiment are in the ultrasonic region at around 100 kHz. At these frequencies the time interval between the passage of a compression maximum and minimum at a particular point is very brief, of the order of a nanosecond. We can therefore assume that the pressure variation occurs rapidly enough for heat conduction to be negligible, that is, the compression and expansion in the sound wave are nearly adiabatic processes.

Adiabatic changes in perfect gases obey the expression

$$pV^{\gamma} = \text{const.}, \qquad (5.7.10)$$

where $\gamma = C_p/C_V$. In terms of density, ρ,

$$p/\rho^{\gamma} = \text{const.} \qquad (5.7.11)$$

Differentiating this expression we find that

$$\left(\frac{\partial p}{\partial \rho}\right)_S = \gamma p/\rho. \qquad (5.7.12)$$

It can be shown that the speed v of a plane longitudinal wave in a homogeneous medium of uniform density is

$$v^2 = \left(\frac{\partial p}{\partial \rho}\right)_S. \qquad (5.7.13)$$

Combining eqns (5.7.12) and (5.7.13) we see that

$$v^2 = \gamma p/\rho. \qquad (5.7.14)$$

For one mole of a perfect gas

$$p/\rho = pV/M = RT/M, \qquad (5.7.15)$$

and therefore

$$v^2 = \gamma RT/M. \qquad (5.7.16)$$

A better approximation to the behaviour of real gases is given by the virial equation of state (p. 9), which may be written as a pressure expansion of the form:

$$pV/RT = 1 + B'p, \qquad (5.7.17)$$

where the coefficient B' is a measure of the gas imperfection. Now it may be shown that eqn (5.7.13) can be expressed in the form:

$$v^2 = \left(\frac{\partial p}{\partial \rho}\right)_S = \gamma\left(\frac{\partial p}{\partial \rho}\right)_T = -\gamma\frac{V^2}{M}\left(\frac{\partial p}{\partial V}\right)_T. \qquad (5.7.18)$$

By substituting eqn (5.7.17) into this expression we obtain an equation of the form:[10]

$$v^2 = \frac{\gamma_0 RT}{M}\left[1 + 2p\left(B' + \frac{R}{C_V}\frac{d(TB')}{dT} + \frac{1}{2}\frac{R^2T}{C_VC_P}\frac{d^2(TB')}{dT^2}\right)\right], \qquad (5.7.19)$$

where γ_0 is the zero-density (or perfect gas) limit of γ. The terms in B' may be eliminated by measuring v^2 at a series of pressures and extrapolating to $p = 0$. Having found γ_0 from the value of $\gamma_0 RT/M$ at the $p = 0$ intercept, we may calculate C_V from the relation:

$$\gamma_0 = 1 + \frac{R}{C_V}. \qquad (5.7.20)$$

Apparatus

A

The principles of infrared spectrometers have been described previously (p. 224).

B

In part B of this experiment, we measure the wavelength λ of sound, of accurately known frequency ν, which is directed through the simple gas. Then the speed of the sound v may be calculated from the relationship

$$v = \lambda\nu. \qquad (5.7.21)$$

The wavelength is measured in an *acoustic interferometer*. The principle of the apparatus is straightforward. Sound of a single, accurately known frequency is generated by a diaphragm or loudspeaker at one end of a sample tube, and detected by a second diaphragm or a microphone at the other. The transmitted and received signals are fed to the 'x' and 'y' plates of an oscilloscope. The oscilloscope trace is observed as the detector is moved towards the source by a micrometer, and the positions

Fig. 5.7.3. Phase relationships in an acoustic interferometer. (a) Transmitted and received signals π radians out of phase—position noted. (b) Transmitted and received signals neither in phase nor π radians out of phase—no reading taken

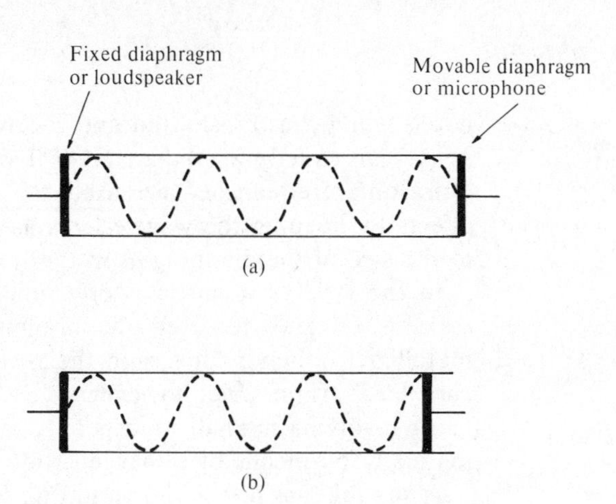

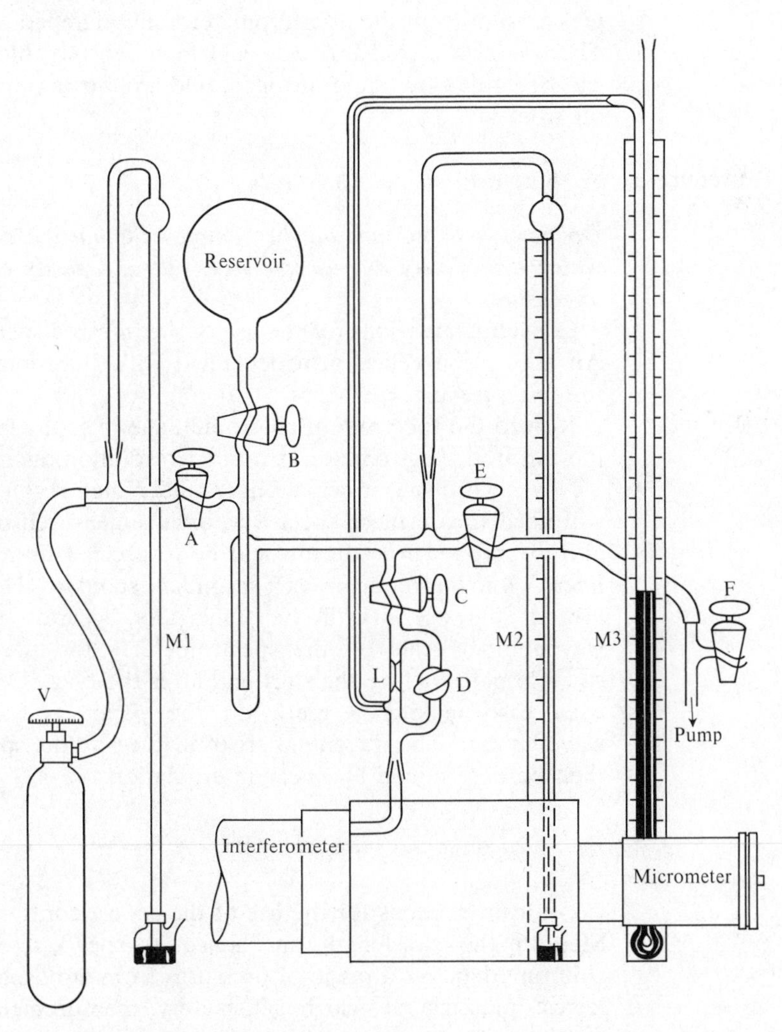

Fig. 5.7.4. Gas line and acoustic interferometer

Note that the separation of the (n)th and $(n+10)$th readings is 5λ, because the balance points represent a phase difference of $k\pi$ radians, where k is an integer, Fig. 5.7.3.

noted when the transmitted and received signals are exactly in phase, or exactly π radians (180°) out of phase, Fig. 5.7.3. Ultrasonic frequencies are used to minimize wall and edge effects by keeping the wavelengths as small as possible relative to the size of the diaphragms in the transmitter and receiver.

In the type of acoustic interferometer used in research, the transmitter and receiver diaphragms are constructed from metallized polymer films, and the receiver is moved by a micrometer. It is possible, however, to use a much simpler arrangement involving a small loudspeaker, and a microphone moved manually by means of a marked rod.[11]

A suitable gas line is shown in Fig. 5.7.4. It features a narrow leak L to prevent the interferometer being pumped out too fast. There is also a cold trap below B in which the chloromethane can be cooled by liquid nitrogen, and any air impurity pumped off from it.

Procedure

A. Spectroscopic measurements

Do not touch the taps on the sample cell, nor the cell window, which are of very hygroscopic rock salt and easily damaged by fingerprints.

Consult instructions for the use of the infrared spectrometer.* An absorption cell is provided filled with chloromethane at a suitable pressure.

Record the spectrum of chloromethane over the full range of the infrared spectrometer at normal resolution, as instructed.*

Now record the principal bands under high resolution.* You will need to use these spectra to obtain measured frequencies, and the proper procedure would be to use a calibrant gas with lines of known frequency as a secondary standard. However, for present purpose, it will be sufficiently accurate to use the wavenumber scale on the spectrometer. Stop the scan at a suitable point before the start and after the end of the band, in each case making a mark on the paper and noting the wavenumber. The wavenumbers of features in the spectrum can then be determined by linear interpolation.

B. Speed of sound measurements

Take the mean of say twenty readings by subtracting the first from the eleventh, second from the twelfth, and so on; it can be easily shown that subtracting the first from the second, second from the third, etc. only actually uses the first and last reading.

Consult instructions for the use of the acoustic interferometer.* Measure the wavelength, and hence the speed, c, of sound in chloromethane at a range of pressures from atmospheric to the lowest pressure at which satisfactory measurements can be made.

Calculation

A. C_V from infrared spectrum

Assign the bands to the various normal modes of the chloromethane molecule, using Table (5.7.2). Note the various features of the parallel (||) and perpendicular (⊥) bands, in particular the alternating intensity in the Q branches of the latter. Note also the presence of Q branches in the C—Cl stretching band ν_3 (Fig. 5.7.2) corresponding to the two isotopic forms $CH_3\,^{35}Cl$ and $CH_3\,^{37}Cl$.

Accurate vibrational frequencies for the normal modes can only be obtained by analysing the rotational fine structure of the bands. However, sufficient accuracy for this experiment can be obtained by assuming that the Q branch peaks coincide with the band centres. For the perpendicular bands, the frequency of the strongest Q branch is measured.

Calculate the heat capacity contributions for the six fundamental vibrations by means of eqn (5.7.9), not forgetting to count the degenerate vibrations twice.

Table 5.7.2. Infrared spectrum of CH_3Cl[12]

Approximate $\tilde{\nu}_{vacuum}/cm^{-1}$		Assignment
750	∥	ν_3
1000	⊥	ν_6
1350	∥	ν_2
1450	⊥	ν_5
2900	∥	$2\nu_5$
3000	∥	ν_1
3050	⊥	ν_4

Finally, add together the translational, rotational and vibrational contributions to obtain the total heat capacity.

B. C_V from speed of sound

Plot v^2 against p to obtain a value of C_V (eqns (5.7.19) and (5.7.20)). Be careful about the units.

Results

There should be close agreement between the heat capacities obtained from infrared spectroscopy and acoustic interferometry. A difference of a few per cent may arise from a lack of equilibration between energy from ultrasonic waves in the external degrees of freedom of the chloromethane molecules, and the energy in their internal degrees of freedom.[13]

Comment This experiment has introduced two interesting techniques of physical chemistry. The first is the quantitative analysis of infrared spectroscopy to give information about the internal energy of a molecule. The second, the measurement of ultrasonic absorption, is used in the study of energy transfer and relaxation processes which are an important feature of gas phase reactions.

Technical notes

Apparatus

A: Infrared spectrometer,[$$] cell containing CH_3Cl. B: acoustic interferometer (research[$†] or student type—p. 238), gas handling line (Fig. 5.7.4), CH_3Cl from cylinder.

References

[1] *Symmetry and spectroscopy*. D. C. Harris and M. D. Bertolucci. Oxford University Press, New York (1978), p. 135.
[2] *Molecular spectra and molecular structure II. Infrared and Raman spectra of polyatomic molecules*. G. Herzberg. Van Nostrand, New York (1945), p. 312.
[3] *Chemical applications of group theory*. F. A. Cotton. Interscience, New York, 2nd edn (1971), chs. 3 and 4.
[4] Ref. 2, p. 314.
[5] *Fundamentals of molecular spectroscopy*. C. N. Banwell. McGraw Hill, London, 2nd edn (1972), p. 92.
[6] *Physical chemistry*. R. S. Berry, S. A. Rice, and J. Ross. John Wiley, New York (1980), part 2, p. 771.
[7] Ref. 6, p. 763.
[8] *Fundamentals of ultrasonics*. J. Blitz. Butterworths, London, 2nd edn (1967), p. 90.
[9] Ref. 6, part 3, p. 1080.
[10] *Absorption and dispersion of ultrasonic waves*. K. F. Herzfeld and T. A. Litovitz. Academic Press, New York (1959), p. 191.
[11] *Experiments in physical chemistry*. D. P. Shoemaker, C. W. Garland, J. I. Steinfeld, and J. W. Nibler, McGraw-Hill, New York, 4th edn (1981), p. 84
[12] Ref. 2, p. 315.
[13] Ref. 10, ch. 2.

5.8 Vibrational Raman spectroscopy of simple polyatomic molecules

When electromagnetic radiation of wavelength λ_0 is passed through a gas, liquid or solid, it may be transmitted, absorbed or scattered. Most spectroscopy is concerned with the absorption or emission of radiation, and in the previous three experiments we have been concerned with the absorption of infrared radiation when molecules vibrate. In this experiment, however, we are concerned with the *scattering* of radiation by vibrating molecules.

In the *Tyndall effect*, radiation is scattered by particles which have sizes comparable to λ_0, such as those in smoke or fog. The intensity of the scattered light varies with angle and particle size, but the wavelength remains unaltered. However in the case of atoms or molecules the radiation is scattered isotropically and the effect is known as *Rayleigh scattering*. With macromolecules, the intensity of the scattering is angle dependent and may be used to elucidate their structure (Expt 6.2).

A small fraction (10^{-4}–10^{-6}) of the light incident on the sample appears with wavelengths differing slightly from λ_0. This is known as the *Raman effect*, and the shifts in wavelength are characteristic of the scattering substance. Raman spectroscopy has been revolutionized by the use of intense, monochromatic laser light sources, and is now widely employed. In this experiment we measure some molecular vibrational frequencies from the Raman shifts of the 488 nm line emitted by an Ar^+ laser and scattered by liquid tetrachloromethane (carbon tetrachloride) and carbon disulphide.

The Raman effect takes its name from the Indian scientist Sir C. V. Raman who detected the phenomenon in 1928 after its prediction by Smekal.

Theory[1,2]

(*i*) *Quantum theoretical approach*

In a Raman spectrometer the sample is irradiated by an intense source of monochromatic radiation usually in the visible part of the spectrum. Generally this radiation frequency is much higher than the vibrational frequencies of the sample molecules but is lower than their electronic frequencies. Rayleigh scattering can be regarded as the elastic scattering of incident photons of wavelength λ_0 and energy $h\nu_0$. The Raman effect, on the other hand, can be regarded as the inelastic collision of an incident photon with a molecule where as a result of the collision the vibrational or rotational energy of the molecule is changed by an amount ΔE. We can visualize the vibrational Raman effect by reference to Fig. 5.8.1. The lines designated $v = 0$ and $v = 1$ represent the vibrational levels of a diatomic molecule such as N_2. A transition directly between these two levels causes the absorption of an infrared photon of energy $\Delta E = h\nu_1$. In Rayleigh scattering, the energy of the photon $h\nu_0$ remains unchanged, whereas in Raman scattering it decreases or increases by an amount $h\nu_1$ producing *Stokes* and *anti-Stokes* lines respectively. Note that the virtual states shown in the diagram are not stationary state energy levels, but serve as convenient intermediate levels for picturing the overall scattering process, which occurs in a time less than the period of a

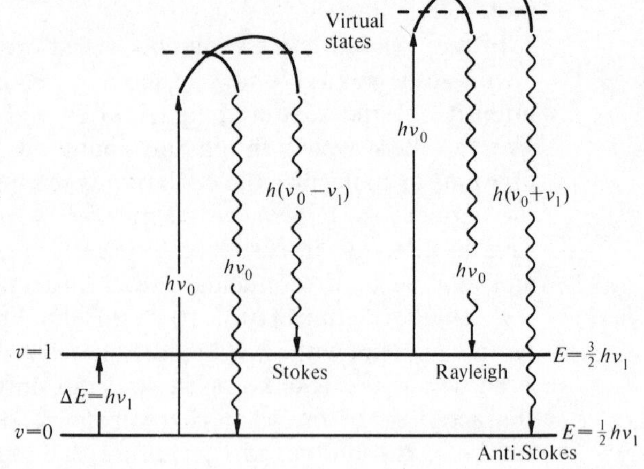

Fig. 5.8.1. Vibrational Raman transitions

molecular vibration ($\sim 10^{-13}$ s). Should the incident radiation have a frequency that causes actual absorption to a real state, the lifetime of the upper state is much longer ($\sim 10^{-8}$ s) and the resultant intense emission is termed fluorescence (Expt 5.11).

According to the Boltzmann function (p. 198) the ratio of the number of molecules in the $v = 1$ state to the number in the $v = 0$ state for a given vibration is

$$\frac{n_1}{n_0} = \frac{g_1}{g_0} e^{-h\nu_{1-0}/k_B T}, \qquad (5.8.1)$$

where ν_{1-0} is the vibration frequency of the transition between the states. At ordinary temperatures most of the molecules are in the ground state and therefore Stokes lines have greater intensities than anti-Stokes lines which originate from an excited level with a lower population.

(*ii*) *Classical theory*

The classical theory of Raman scattering was, interestingly, developed by Cabannes *after* the quantum mechanical treatment of Kramers and Heisenberg.[1]

In many respects the treatment of the scattering of light by a molecule is simplified by using the 'classical' wave picture for the light, expressed mathematically as a cosine function of frequency and time.

When a molecule is introduced into an electric field of strength E, an electric dipole moment p is induced in the molecule. If α is the polarizability of the molecule, the magnitude of the induced dipole is given by:

$$p = \alpha E. \qquad (5.8.2)$$

When electromagnetic radiation of frequency ν_0 falls on the molecule it introduces a varying electric field, which we express in terms of its amplitude E_0 and a time dependent term:

$$p = \alpha E_0 \cos(2\pi\nu_0 t). \tag{5.8.3}$$

For a vibrating molecule, the molecular polarizability α will also vary with time. For very small vibrational amplitudes the polarizability of the molecule is related to the vibrational co-ordinate Q_v by the relation:

$$\alpha = \alpha_0 + (\partial\alpha/\partial Q_v)_0 Q_v. \tag{5.8.4}$$

α_0 is the polarizability at the equilibrium configuration of the molecule, and $(\partial\alpha/\partial Q_v)_0$ refers to the rate of change of polarizability with bond length, evaluated at the equilibrium configuration.

Finally, since the molecule is vibrating with frequency ν_v, the displacement Q_v is itself a function of time:

$$Q_v = Q_0 \cos(2\pi\nu_v t), \tag{5.8.5}$$

where Q_0 is the co-ordinate of the initial position, taken as the maximum displacement from the equilibrium bond length.

By combining eqns (5.8.3), (5.8.4) and (5.8.5), and using a trigonometric identity, we obtain the expression

$$p = \alpha_0 E_0 \sin(2\pi\nu_0 t) + \left(\frac{\partial\alpha}{\partial Q_v}\right)_0 \frac{Q_0 E_0}{2}[\cos 2\pi(\nu_0 - \nu_v)t + \cos 2\pi(\nu_0 + \nu_v)t]. \tag{5.8.6}$$

It can be seen from this equation that the induced dipole moment p varies with those component frequencies ν_0, $\nu_0 - \nu_v$ and $\nu_0 + \nu_v$, which correspond to Rayleigh scattering, Stokes, and anti-Stokes Raman scattering respectively. This classical prediction for these frequencies corresponds to the quantum mechanical result for Raman transitions when $\Delta v = \pm 1$. If the vibrations cause no change in polarizability so that $\partial\alpha/\partial Q = 0$, then eqn (5.8.6) shows that the Raman component frequencies of the induced dipole moment have zero amplitudes, and therefore there is no radiation at the Raman frequencies.

(*iii*) *Raman and infrared spectrum of carbon disulphide*

The interpretation of lines in the Raman spectra of specific molecules is carried out by means of group theory.[4,5] Carbon disulphide is a linear triatomic molecule whose *normal modes* of

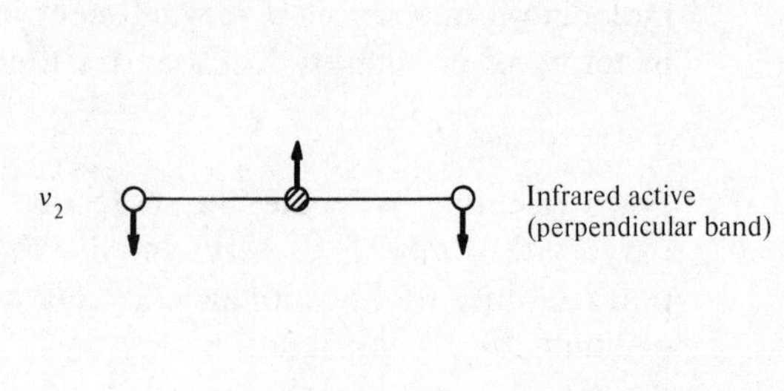

Fig. 5.8.2. Normal modes of vibration in carbon disulphide

vibration (p. 223) are shown in Fig. 5.8.2. All molecules such as carbon disulphide which possess a centre of symmetry obey the *mutual exclusion rule*[6,7] which states that its Raman active vibrations will be infrared inactive, and vice versa. If there is no centre of symmetry then some (but not necessarily all) vibrations may be both Raman and infrared active. Only the normal mode ν_1 is associated with a change in polarizability, giving rise to Raman scattering. For a vibration to be infrared active it must cause a change in dipole moment. It can be seen that this applies to ν_3, in which the change is along the molecular axis giving rise to a parallel band, and to ν_2 in which the change is at right angles to the axis, giving a perpendicular band in the infrared. As predicted, no normal mode is active both in Raman and in the infrared.

The mutual exclusion rule does not preclude transitions which are inactive both in Raman and in the infrared.

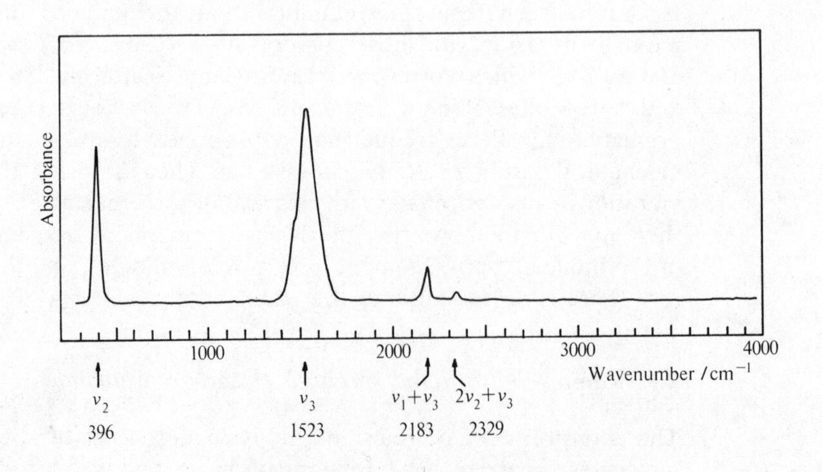

Fig. 5.8.3. Infrared spectrum of carbon disulphide (liquid)

The infrared spectrum of liquid carbon disulphide is shown in Fig. 5.8.3, and includes overtone (p. 221) and combination bands as shown.[8]

(*iv*) *Raman spectrum of tetrachloromethane*

It can be shown that the $(3n-6)$ internal degrees of freedom on a non-linear molecule of n atoms correspond to $(3n-6)$ independent normal modes of vibration.[9] Owing to the high symmetry of CCl_4, the $(3n-6)=9$ normal modes are represented by only four frequencies, since one corresponds to a degenerate pair of vibrations, and two are triply degenerate.[10] One mode corresponding to each frequency is shown in Fig. 5.8.4, together with its group theory point group. All are Raman active, but only ν_3 and ν_4 are infrared active also.

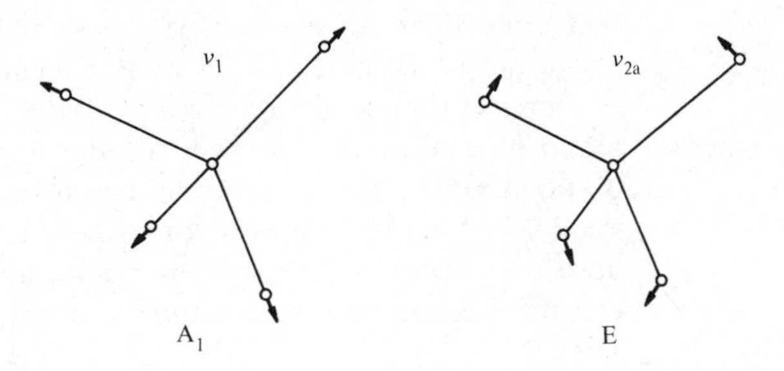

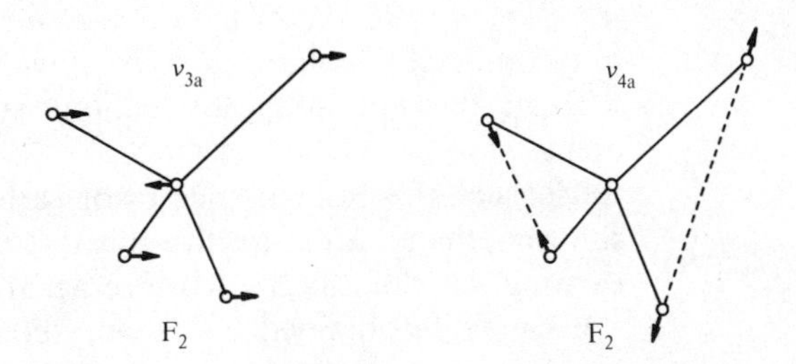

Fig. 5.8.4. Some normal vibrations of CCl_4

Apparatus In a typical Raman spectrometer, Fig. 5.8.5, light is shone onto the sample from a laser source. Lasers from a few milliwatts up to 2 W in power are used, the most popular being the Ar^+ laser which has lines at 514.5 and 488 nm with a spectral width of 0.25 cm^{-1} or less. Liquid samples are usually contained in quartz 'box' a centimetre or two in length. Solids and gases may also be studied, gases being contained in a vessel through which the laser beam passes several times. The Raman scattering is

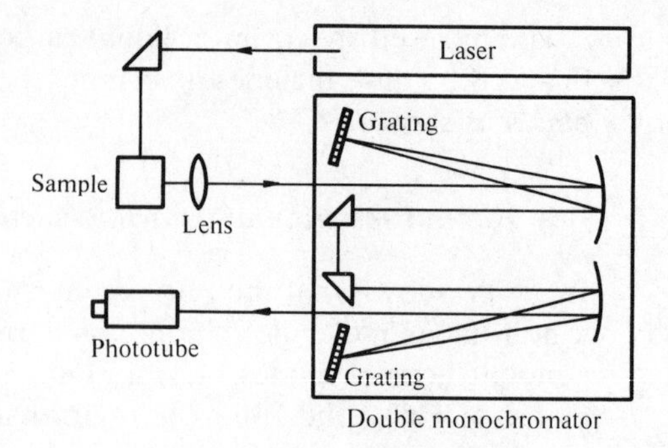

Fig. 5.8.5. Simplified optical diagram of a typical Raman spectrometer

collected by a wide lens set perpendicular (or sometimes coaxially) to the incident laser beam. It is then passed through a double grating monochromator. In modern instruments the gratings are ruled holographically, which improves throughput and stray light rejection. Both these factors are clearly very important in the detection of weak Raman lines in the presence of intense Rayleigh scattering. After leaving the second monochromator, the intensity of the beam is measured by a photomultiplier connected to an amplifier. The output from the amplifier is fed to a chart recorder, and a separate motor drives both the paper and the wavenumber scan mechanism. Modern instruments use photon counting systems interfaced to computers.[11]

Procedure

Laser radiation can cause severe eye damage if the beam, or even a reflection of it, strikes the eye. Do not carry out this experiment until you have familiarized yourself with the safety requirements.

Read the operating instructions supplied with the instrument.*

Measure the Raman spectra of carbon disulphide and tetrachloromethane. Measure the shifts of the Raman lines from the existing Ar^+ line in cm^{-1}. More accurate values of the Raman shifts may be obtained by halving the distance between corresponding pairs of Stokes and anti-Stokes lines.

Calculation

Assign the Raman shifts you have observed to vibrational transitions.[10,12] Comment on the relative intensities of the Stokes and anti-Stokes lines. Compare the Raman spectrum of carbon disulphide with the infrared spectrum shown in Fig. 5.8.3, and comment on the relationship between the two spectra.

Results[12,13] *Accuracy*

The accuracy of your result will be determined by the instrument. Typically it will be accurate to 1 cm^{-1} with a reproducibility of 0.2 cm^{-1}.

Comment This experiment illustrates Raman spectroscopy in its simplest form, in which only vibrational transitions are observed, rather than the more usual full analysis of vibration–rotation fine structure. Many current investigations of the Raman effect are concerned with much more subtle non-linear effects. An example is Coherent Anti-Stokes Raman Scattering (CARS), where two laser beams are used and resonance peaks are recorded when a Raman-active vibrational or rotational frequency is equal to the difference in frequency of the two incident beams.[14]

Technical notes

Apparatus

Laser Raman spectrometer,[$$] samples of liquid CCl_4 and CS_2.

Safety

The laser, sample chamber, and incident and scattered laser beam should be totally enclosed, and interlocks fitted to prevent operation of the laser if its radiation path is exposed at any point.

References

[1] *Raman spectroscopy in inorganic chemistry*. R. S. Tobias. *J. Chem. Educ.* **44,** 2 (1967).
[2] *Introduction to infrared and Raman spectroscopy*. N. B. Colthup, L. H. Daly, and S. E. Wiberley. Academic Press, New York, 2nd edn (1975), p. 57.
[3] Ref. 2, p. 65.
[4] *Chemical applications of group theory*. F. A. Cotton. Interscience, New York, 2nd edn (1971), chap. 10.
[5] Ref. 2, p. 155.
[6] *Fundamentals of molecular spectroscopy*. C. N. Banwell, McGraw-Hill, London, 2nd end edn (1972), pp. 137 and 148.
[7] *Symmetry and spectroscopy*. D. C. Harris and M. D. Bertolucci. Oxford University Press, New York (1978), sec. 3–6.
[8] *Infrared and Raman spectra of polyatomic molecules*. G. Herzberg. Van Nostrand, Princeton, N.J. (1945), p. 277.
[9] Ref. 2, chap. 14.
[10] Ref. 8, p. 100.
[11] *Chemical applications of Raman spectroscopy*. J. G. Grasselli, M. K. Snavely and B. J. Bulkin. John Wiley, New York (1981), chap. 2.

[12] Ref. 8, pp. 250–276.
[13] *Spectroscopy*. Eds. B. P. Straughan and S. Walker. Chapman and Hall, London (1976), vol. 2, p. 226.
[14] *Coherent Anti-Stokes Raman Scattering*. H. C. Andersen and B. C. Hudson, Specialist Periodical Report, *Molecular Spectroscopy* **5**, 142 (1978).

5.9 Formula and stability constant of a complex by spectrophotometry

When a solution containing ferric ions is mixed with a solution of salicylic acid, a violet-coloured complex is formed. Both the composition of the complex and its stability constant are found by measuring the relative absorption of various mixtures with a spectrophotometer. The free energy of formation is then calculated from the stability constant.

Theory

Since it is the anion of salicylic (2-hydroxybenzoic) acid which complexes with the ferric ion, the apparent value of the stability constant of the complex varies with pH. This experiment is carried out at a pH of about 2.5. Under these conditions the phenolic —OH group of the salicylic acid is undissociated, and the carboxylic —COOH group partly so. The hydrolysis of the ferric ions is largely suppressed, and there are no appreciable concentrations of the di- and tri-salicylate complexes.

Evidence for the existence of only one complex species in solution is gained from the presence of an *isosbestic point.*[1] Figure 5.9.1 shows a series of absorption spectra for solutions with different [Fe^{3+} : sal^-] ratios but the same total concentration, such as will be used in this experiment. The diagram shows

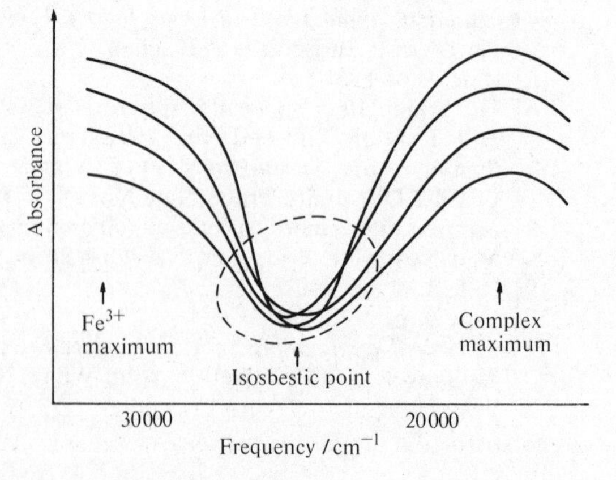

Fig. 5.9.1. Absorption spectrum of Fe^{3+}/Sal^- solutions

Isosbestic points are difficult to measure in practice, and at best manifest themselves only as a group of intersecting curves, from which a mean frequency can be estimated (Fig. 5.9.1).

the frequency where the absorption of the complex is at a maximum, and also the Fe^{3+} maximum. The salicylate ion does not absorb in these regions of the spectrum. The isosbestic point is the frequency at which the total absorbance is independent of the ratio of the concentrations of the two absorbing species. In the case of a 1 : 1 complex such as $Fe^{3+}(sal^-)$, this will clearly occur at the point where the molar absorbances of the two species are equal.

The equilibrium for the formation of a ferric salicylate complex, in the absence of any other species, may be written

$$Fe^{3+} + n(sal^-) \leftrightharpoons Fe^{3+}(sal^-)_n.$$

The *stability constant* K is defined as

$$K = \frac{[\text{complex}]}{[Fe^{3+}][sal^-]^n}, \tag{5.9.1}$$

where $[Fe^{3+}]$ and $[sal^-]$ refer to the concentrations of the free species.

The complex has an absorption maximum at a wavenumber of 18 950 cm^{-1}. Its optical absorbance a is given by the *Beer–Lambert Law* (p. 215)

$$a = \log_{10}(I_0/I) = \varepsilon l[\text{complex}], \tag{5.9.2}$$

where I_0 and I are the incident and transmitted light intensities, ε is the extinction (absorption) coefficient of the complex, and l the optical path length in the cell. Spectrophotometers record the absorbance a, rather than intensity I, so that the reading at 18950 cm^{-1} is directly proportional to the concentration of the complex since neither sal^- nor Fe^{3+} absorb at this wavenumber.

The empirical formula of the complex may be found by *Job's method*,[2] which applies to two reactants that combine to form a complex. Equimolar solutions are made up, and then mixed in volume ratios 1 : 9, 2 : 8, . . . , 9 : 1. The total reactant concentration is therefore the same in each case. The maximum amount of equilibrium product will be formed when the proportions of reactants employed correspond to the empirical formula of the product.

In practice the maximum absorbance of each solution due to the complex is plotted against the mole fraction of one component to give a *Job plot*. The maximum of this curve then indicates the empirical formula of the complex. In the present case we expect the maximum to occur for the 5 : 5 mixture, confirming the 1 : 1 formula, $Fe^{3+}(sal^-)$.

To determine the stability constant, we need to find the concentration of the complex by means of eqn (5.9.2). Since a

and l can be measured, the problem is one of finding ε, or, in practice, the product εl. There are a number of possible methods,[3] of which two of the more reliable ones are given here.

A

Let x be the concentration of Fe^{3+} added to the solution, and let y be the concentration of salicylic acid added. Then for any mixture

$$K = \frac{[\text{complex}]}{[\text{Fe}^{3+}][\text{sal}^-]} = \frac{[\text{complex}]}{(x-[\text{complex}])(y-[\text{complex}])}$$

$$= \frac{a/\varepsilon l}{(x-a/\varepsilon l)(y-a/\varepsilon l)}. \tag{5.9.3}$$

The dimensionless equilibrium constant $K^\ominus$, required later for the calculation of $\Delta G^\ominus$, is given by

$$K^\ominus = \frac{[\text{complex}]/c^\ominus}{[\text{Fe}^{3+}]/c^\ominus . [\text{sal}^-]/c^\ominus}$$

$$= Kc^\ominus, \tag{5.9.4}$$

where $c^\ominus$ is the standard concentration (1 mol dm^{-3}). Rearranging eqn (5.9.3):

$$\frac{xy}{a} = \frac{1}{\varepsilon l}\left[\frac{1}{K} + (x+y)\right] - \frac{a}{(\varepsilon l)^2}. \tag{5.9.5}$$

The total concentration $(x+y)$ is constant. It follows that a graph of xy/a against a should be linear, and that the slope will yield εl, and the intercept and this value of εl will then yield K.

B

An alternative method for obtaining εl is to measure a for a series of 1:1 mixtures at increasing dilutions. Let the equal concentrations of added Fe^{3+} and salicylic acid in the resulting solutions be c. Then for the series of mixtures the concentrations of free Fe^{3+} and sal^- are given by:

$$[\text{Fe}^{3+}] = [\text{sal}^-] = c - [\text{complex}]. \tag{5.9.6}$$

Therefore from eqns (5.9.1) and (5.9.2) it follows that

$$K = \frac{a/\varepsilon l}{(c-a/\varepsilon l)^2} \tag{5.9.7}$$

or

$$\frac{c}{a^{1/2}} = \left[\frac{1}{\varepsilon l K}\right]^{1/2} + \frac{a^{1/2}}{\varepsilon l}. \tag{5.9.8}$$

A plot of $c/a^{1/2}$ against $a^{1/2}$ should therefore be linear with slope $1/\varepsilon l$ and intercept $(1/\varepsilon l K)^{1/2}$.

Apparatus[4]

The experiment employs an ultraviolet spectrophotometer. The source of radiation in such an instrument is usually a hydrogen or deuterium lamp. The radiation is dispersed by a prism or grating, and, after passing through the sample, its intensity is usually measured by a phototube or semiconductor device. It is usual for ultraviolet spectrophotometers to have a cell carrier which can be moved to one of, say, four positions, so that several cells may be prepared and loaded together. Double-beam instruments are more common than single beam ones, and follow similar principles and layout to that of infrared double-beam instruments, p. 224.

Procedure[5]

Ferric ammonium sulphate can decompose on storing—check that the crystals or powder are pale violet, or almost colourless; the solution hydrolyses on standing and should be used immediately. The salicylic acid should also be fresh, in the form of fluffy, readily soluble crystals.

Make up 500 cm^3 of M/500 hydrochloric acid from the standard acid provided, and then make up two solutions X and Y as follows, weighing to an accuracy of ±0.003 g:
X—a solution M/400 in Fe^{3+} made by dissolving 0.603 g ferric ammonium sulphate in 500 cm^3 M/500 hydrochloric acid,
Y—a solution M/400 in salicylic acid made by dissolving 0.173 g salicylic acid in 500 cm^3 M/500 hydrochloric acid.
Fill two burettes with the solutions and make up mixtures in the ratios 1 : 9, 2 : 8, 3 : 7, etc. up to 9 : 1.

Switch on the spectrophotometer as instructed.* Use 1 cm cells (so that $l = 1$ cm), and switch to a slow scan speed. Record all the spectral curves on a single sheet of paper.

Treat the absorption cells with care. When changing solution, empty out the old solution, rinse with the new solution, and then fill with the new. Wipe only the outside of the cells with paper tissue.

First obtain the spectra of solutions X and Y separately. Note that there is little absorption at wavenumbers below 23 000 cm^{-1}.

Then obtain the spectra of the mixtures. Note the new band due to the complex which appears at a wavenumber of 18 950 cm^{-1}, with a height proportional to the complex concentration as explained previously.

Measure the spectra of a series of 5 : 5 mixtures of X and Y at increasing dilution.

Leave the cells to drain in a safe place at the end of the experiment.

Calculation

Estimate the position of the isosbestic point.

Verify the empirical formula of the complex by Job's method.

Find the extinction coefficient of the complex and its stability constant by both of the methods listed earlier. Estimate the precision of your results by drawing bounding lines on the graphs to cover the full range of slopes and intercepts compatible with your experimental results.

Show that a graph of xy/a against a (eqn (5.9.5)) would not be linear for a complex of formula $Fe^{3+}(sal^{-})_2$.

Calculate the free energy of formation of the complex, using the relation $\Delta G^{\ominus} = -RT \ln K^{\ominus}$.

Results[6]

Accuracy

It is not difficult to obtain consistent values of ε from this experiment, but K is much more sensitive to errors. Provided the experiment is carried out competently, the factor which limits its accuracy is usually the accuracy of the stock solution concentrations. If old samples are used, K may be incorrect by several orders of magnitude and there will be corresponding errors in $\Delta G^{\ominus}$.

Comment

This simple experiment introduces the use of visible (and ultraviolet) spectrophotometry, and illustrates the determination of the stoichiometry and stability constant of a complex. Stability constants can also be measured by calorimetric methods (Expt 3.5), and solubility measurements analogous to those in Expt 4.1. Most commonly, determinations are made by electrochemical methods similar to those in Expt 4.2.

Technical notes

Apparatus

Spectrophotometer,$$ 1 cm absorption cells (plastic ones are cheapest), standardized hydrochloric acid, fresh (<8 months old) ferric ammonium sulphate, fresh (<18 months old) salicylic acid.

References

[1] *Kinetics and mechanism*, J. W. Moore and R. G. Pearson. John Wiley and Sons, New York, 3rd edn (1981), p. 49.

[2] *The chemistry of the coordination compounds*. Ed. J. C. Bailar Jr, Reinhold, New York (1956), p. 569.

[3] *Molecular addition compounds of iodine. I. An absolute method for the spectroscopic determination of equilibrium constants*. N. J. Rose and R. S. Drago. *J. Amer. Chem. Soc.* **81,** 6138 (1959).

[4] *Physical methods of chemistry*. Eds. A. Weissberger and B. W. Rossiter. Wiley-Interscience, New York, (1972), Vol. 1, Part IIIB ch. 3 (by F. Grum).

[5] *Findlay's practical physical chemistry*. Rev. B. P. Levitt. Longman, London, 9th edn (1973), p. 132.

[6] *Stability constants of metal-ion complexes Supplement No.* 1, (L. G. Sillén), Special Publication No. 25, Chemical Society (1971), p. 483.

5.10 Dissociation energy of iodine from visible absorption spectrum

The electronic spectra of molecules, unlike those of atoms, are complicated by the additional possibilities that the molecule may possess energy of both rotation and vibration. As a consequence, each atomic spectrum line is replaced by a system of *bands*. The distribution of the individual bands in each system depends on vibrational energy changes, while the detailed line structure of individual bands depends on changes in rotational energy.

In this experiment we measure the positions of vibrational bands in the absorption spectrum of iodine. The spectrum extends over a wide range from red to green/blue. The energies of the vibrational transitions converge, and an extrapolation to the convergence limit yields the dissociation energy of the upper electronic state. The dissociation energy of the ground state may then be obtained by using independent information about the states of atomic iodine. The rotational fine structure of the bands is not analysed.

Theory[1]

(i) Vibration of a diatomic molecule

As mentioned in Expt 5.6, the simplest way of describing a vibrating diatomic molecule is to assume that it is a harmonic oscillator, i.e. that its potential energy depends quadratically on the change in intermolecular distance r. The allowed energy levels may be calculated by quantum mechanics[2] and are expressed as term values in wavenumber units:

$$G(v) = \omega_e(v + \tfrac{1}{2}) \qquad (5.10.1)$$

where v is the vibrational quantum number having integral values 0, 1, 2, . . . , and ω_e is the vibrational wavenumber. If we plot the *classical turning points*, i.e. maximum and minimum internuclear separations for each vibrational energy, as a function of the energy of the vibration, we obtain a simple parabolic potential energy function such as the curve shown dashed in Fig. 5.10.1. However, as can be seen, the true potential energy function for iodine has a very different shape. In particular, the outer (right-hand) wall of the potential energy curve goes asymptotically to a constant energy as $r \rightarrow \infty$. The *dissociation energy* D_0 of the molecule is the difference between this energy and the energy of the $v = 0$ ground state vibrational level.

For a non-harmonic oscillator the vibrational energy term can

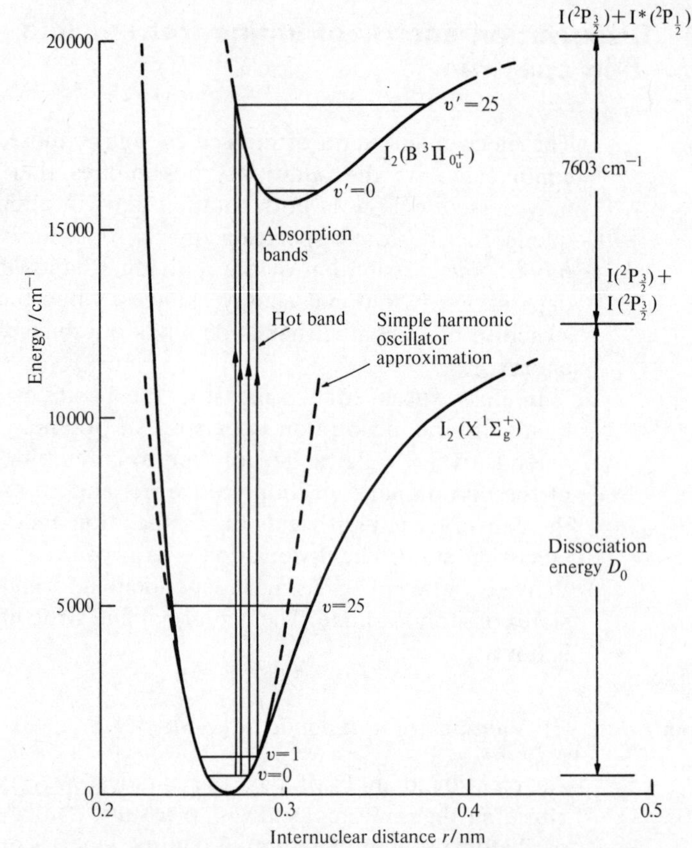

Fig. 5.10.1. Molecular potential energy curves of ground state and electronically excited state of I_2[5]. The heights of the $v=0$ and 1 levels are exaggerated, whereas those of the $v=25$ levels are correct

The curves in Fig. 5.10.1 represent potential energy only, while the horizontal lines represent the sum of potential and kinetic energy.

usually be expressed as a simple power series in terms of the vibrational quantum number v:

$$G(v)=\omega_e(v+\tfrac{1}{2})-\omega_e x_e(v+\tfrac{1}{2})^2+\omega_e y_e(v+\tfrac{1}{2})^3 \ldots \qquad (5.10.2)$$

ω_e is the hypothetical *equilibrium oscillation wavenumber* of the anharmonic system, i.e. the classical wavenumber of infinitely small vibrations about the equilibrium point. $\omega_e x_e$, $\omega_e y_e \ldots$ are the *anharmonicity constants* of the molecule. It follows that the energy of the lowest vibrational level, $v=0$, is given by the expression

$$G(0)=\tfrac{1}{2}\omega_e-\tfrac{1}{4}\omega_e x_e+\ldots \qquad (5.10.3)$$

This is known as the *zero-point energy*, and has a value of about 107 cm^{-1} for the ground state of I_2 and 63 cm^{-1} for its upper state.

How do you account for the existence of a finite zero point energy?

(*ii*) *Vibrational and electronic transition*

At room temperature, most of the iodine molecules are in the $v = 0$ ground state. On absorption of visible radiation, they undergo electronic transitions to vibrational levels in an upper electronically excited state as shown in the figure. The bands start in the red region of the spectrum and stretch into the green/blue, converging to continuous absorption at the dissociation limit. There are also *hot bands* in the red arising from transitions from the $v = 1, 2 \ldots$ states. At room temperature, however, these states are not well populated and the hot bands fade out in the yellow.

The *Born–Oppenheimer* approximation[3] is implicit in diagrams such as Fig. 5.10.1. It assumes that the electronic distribution in a molecule can be evaluated in a static nuclear framework, and that therefore the classical turning points of each vibrational energy of the molecule can be expressed solely as a function of distance, giving rise to the molecular potential energy curves shown. A further feature of the diagram is that the electronic transitions are drawn vertically, because the difference in mass between the electron which is being excited and the iodine nuclei is so great that the nuclear framework is effectively static during the time taken for the electronic transition. This is the basis of the *Franck–Condon principle*,[4] which may be used to gain information from the intensities of the vibrational bands.[5] We shall not attempt such an analysis in this experiment.

The wide range of the iodine molecular spectrum is associated with a change in structure between the lower state and the electronic excited state, which results in a displacement of the upper molecular potential energy curve to larger internuclear distance.

The measurements which we do make are of the energies of the bands. It can be seen in the diagram that the absorption spectrum at room temperature will give rise to a series of bands which converge towards the dissociation limit of the *upper* electronic state. The energy at the limit will therefore be the height of the dissociation limit of the upper state above the $v = 0$ level of the ground state. To obtain the dissociation energy of the ground state, which is the quantity we are interested in, we must subtract from the total energy the difference in energy between the dissociation products of the upper and lower states. Anticipating the result of the discussion in section (iii), we find that the upper state dissociates into $I + I^*$ with states $^2P_{3/2}$ and $^2P_{1/2}$ respectively, whereas the ground state dissociates into two iodine atoms with state $^2P_{3/2}$. The difference in energy is therefore the energy of excitation of I $^2P_{3/2} \rightarrow$ I $^2P_{1/2}$, and from atomic spectroscopy experiments we know this to be 7603 cm^{-1}.

(*iii*) *Assignment of molecular and atomic states*

The assignment of the ground and electronic excited state of molecular iodine, from which the states of the iodine atoms may be deduced, requires an understanding of spin-orbit coupling, group theory, and the formation of P, Q, and R branches in the rotational fine structure of the vibrational bands (Expt 5.6). However, the understanding of how this assignment is made is not vital to the successful performance of the experiment.

In the ground state of the I_2 molecule, which we call X, all the electrons are in closed shells and the state may be described as $\ldots\sigma_g^2\sigma_u^{*2}\pi_u^4\sigma_g^2\pi_g^{*4}$, X $^1\Sigma_g^+$ (p. 207). The nature of the upper state, B, requires some consideration. The experimental information is

(i) the transition X→B is relatively weak

(ii) the rotational structure shows that the bands contain only two branches each, R and P.

The next available orbital in the I_2 molecule is σ_u^*, and the first excited configuration is then expected to be $\ldots\pi_g^{*3}\sigma_u^*$, which gives $^1\Pi_u$, $^3\Pi_u$ molecular states. Now B cannot be the $^1\Pi_u$ state of this configuration, for the transition $^1\Pi_u \rightarrow {}^1\Sigma_g^+$ would show strong Q branches. In the iodine atom, spin-orbit coupling is large, and will be expected to be large also in I_2. The $^3\Pi_u$ state in these circumstances is split into four molecular states, characterized by $\Omega = |\Lambda + \Sigma|$, where Λ is the projection of L on the internuclear axis, and Σ that of the resultant electron spin on the internuclear axis. Thus $^3\Pi_u$ gives the states $\Omega = 2_u$, 1_u, 0_u^+ and 0_u^-. Of these only 1_u and 0_u^+ can combine with the ground state. 1_u is found to be the upper state of a near infrared system of I_2 which shows P, Q and R branches. B is therefore identified as 0_u^+, and the transition B $0_u^+ \rightarrow$ X $^1\Sigma_g^+$ yields, as it should, only R and P branches.

We must now consider in detail the relation between the molecular states of iodine and the states of the atoms into which the molecules dissociate. The ground state of the iodine atom is $\ldots p^5$ ^2P, but the separation between $^2P_{1/2}$ and $^2P_{3/2}$ is large.

Table 5.10.1. Correlation between values of molecular Ω and atomic J in iodine

Relative energy	Ω	J
0 cm^{-1}	3_u, 2_u, 2_g, $1_u(2)$, 1_g, $O_g^+(2)$, $O_u^-(2)$	3/2+3/2
7603.2 cm^{-1}	2_g, 2_u, $1_g(2)$, $1_u(2)$, O_g^+, O_u^+, O_g^-, O_u^-	1/2+3/2
15206.4 cm^{-1}	1_u, O_g^+, O_u^-	1/2+1/2

Excited states lie so high that they need not be considered here. The correlations between molecular Ω's and atomic *J*'s are shown in Table 5.10.1. Of the large number of molecular states (many of which are repulsive), there is only one state which can be identified with B, the 0_u^+ state arising from $^2P_{3/2}+{}^2P_{1/2}$.[6] Thus the dissociation limit which is observed in this experiment correlates with $I(^2P_{3/2})+I(^2P_{1/2})$, and the dissociation energy of the ground state—the quantity which is of direct chemical interest—is obtained by subtracting the excitation energy corresponding to the change of state $^2P_{1/2}$–$^2P_{3/2}$ from the measured limit (Fig. 5.10.1).

(*iv*) *Extrapolation to the dissociation limit*

The vibrational bands observed in this experiment converge towards the dissociation limit I+I*, but do not reach it. We therefore need some means of extrapolating the known energies to find the value of the limit. We may use eqn (5.10.2) to express the difference in energy between the vibrational levels (v') and ($v'+1$) in the electronic excited state in the form:

$$\Delta G(v') = G(v'+1) - G(v')$$
$$= \omega_e - x_e\omega_e(2v'+2) + y_e\omega_e[3(v')^2+6v'+3\tfrac{1}{4}]+\ldots. \tag{5.10.4}$$

If this power series converges sufficiently rapidly, it may only be necessary to use the terms in ω_e and $x_e\omega_e$. When this is the case, a plot of $\Delta G(v')$ against v' is a straight line, and the terms may be extended up to the dissociation energy; this is known as a *linear Birge–Sponer extrapolation.*[7] In practice however it is usually found that the Birge–Sponer plot has some curvature, and that a curved extrapolation gives a more accurate result. The dissociation energy D_0 is the sum of all the $\Delta G(v')$ terms, and can therefore be found from the area under the entire curve, both measured and extrapolated.

It is tempting to use the Birge–Sponer expression to calculate ω_e and $\omega_e x_e$, but, as discussed later, the extrapolation is not accurate enough to make this worthwhile.

Apparatus[8] The source of radiation for this experiment is a tungsten lamp. The light is shone through a cell, typically 20 cm long, with windows at either end. The cell contains iodine vapour, and may be heated gently by an electric winding to make the spectrum more intense, although not so much that the hot bands interfere.

The spectrum may be observed using a direct reading spectrometer (p. 200). This may be calibrated using a Zn/Cd/Hg

Table 5.10.2. Calibration using Fe/Ne lines

Calibration data					Specimen calibration		
λ_{air}/nm	Order	$\lambda_{II,air}$/nm	$\lambda_{vac}-\lambda_{air}$/nm	$\tilde{\nu}_{vac}$/cm^{-1}	Comparator x/mm	$\tilde{\nu}_{interp}$/cm^{-1}	$\Delta\tilde{\nu}$/cm^{-1}
Ne 330.97372	III	496.4606	0.1385	20 136.97	6.849	20 132.14	+4.83
Fe 248.97920	IV	497.9584	0.1389	20 076.40	12.998	20 076.40	0.00
Fe 332.37955	III	498.5693	0.1391	20 051.80	15.463	20 054.06	−2.26
Fe 250.11326	IV	500.2265	0.1395	19 985.37	22.314	19 991.96	−6.59
Ne 334.54494	III	501.8174	0.1400	19 922.01	28.864	19 932.59	−10.58
Fe 251.62630	IV	503.2526	0.1403	19 865.20	34.767	19 879.08	−13.88
Ne 336.99078	III	505.4862	0.1409	19 777.42	43.971	19 795.65	−18.23
Fe 253.36618	IV	506.7324	0.1413	19 728.77	49.095	19 749.21	−20.44
Ne 339.27974	III	508.9196	0.1419	19 643.99	58.123	19 667.38	−23.39
Ne 341.79035	III	512.6855	0.1429	19 499.70	73.664	19 526.51	−26.81
Ne 344.77028	III	517.1554	0.1439	19 331.17	92.142	19 359.02	−27.85
Ne 345.41949	III	518.1292	0.1443	19 294.83	96.171	19 322.50	−27.67
Ne 347.25711	III	520.8857	0.1450	19 192.73	107.591	19 218.99	−26.26
Ne 352.04717	III	528.0708	0.1469	18 931.59	137.414	18 948.66	−17.07
Fe 267.63716	IV	535.2743	0.1489	18 676.81	167.406	18 676.81	0.00
Ne 359.35262	III	539.0289	0.1499	18 546.72	183.068	18 534.85	+11.87

$\Delta\tilde{\nu}=\tilde{\nu}_{vac}-\tilde{\nu}_{interp}$

lamp as described on p. 205, or by use of a Fe/Ne lamp which gives rise to the lines shown in Table 5.10.2.

A much more precise result may be obtained by using a long path-length spectrometer using, for example, the Ebert arrangement shown in Fig. 5.10.2. The spectrum is recorded on a photographic plate, and emission lines from a Fe/Ne lamp are photographed adjacent to the iodine spectrum. A *comparator* (travelling microscope with vernier scale) is used to measure the positions of the iodine lines on the plate, which are then calibrated in terms of the known Fe/Ne wavelengths as described below.

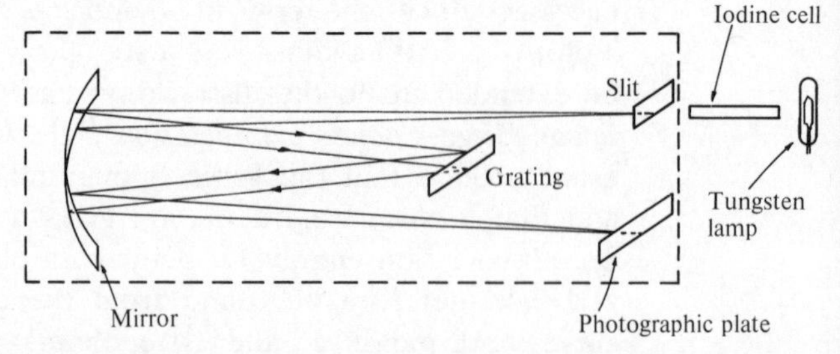

Fig. 5.10.2. The arrangement of the cell and Ebert spectrograph

Procedure If using a direct reading spectrometer, measure the bands from 540 nm as far as possible into the blue. Also measure the lines from the calibration lamp. Remember to approach each reading from the same direction, to prevent backlash errors in the drum.

If using a Hilger or Ebert spectrometer, photograph the spectrum of iodine and the calibration lamp as instructed.* Measure the positions of the two sets of lines with a comparator.

Calculation

If using a direct reading spectrometer, calibrate the instrument and correct the I_2 readings by the method described on p. 202.

If using a precision spectrograph, calibrate the instrument with the aid of Table 5.10.2. The second column shows the wavelengths of third and fourth order diffractions of lines from Fe/Ne calibration lamp. They are corrected to second order in the column labelled $\lambda_{\mathrm{II,\,air}}$. In the column headed $\tilde{\nu}_{\mathrm{vac}}$, the calibration lines have been converted from wavelengths to wavenumbers in cm^{-1}, and a small correction made to give the wavenumber in vacuum rather than in air. The next column shows a typical set of comparator readings x.

Your own comparator readings x should replace those shown in Table 5.10.2. Measure them as follows. Look at the calibrated Fe/Ne spectrum provided.* Find patterns of Fe/Ne lines which correspond to patterns of lines on the plate. You should thereby be able to locate at least one of the lines listed in the table. Set the comparator on this line, and then move it by a distance read from the table.* So confirm the positions of other lines. When they match those listed in Table 5.10.2, you are in a position to assign the correct wavenumber $\tilde{\nu}_{\mathrm{vac}}$ to a line at each end of the plate, e.g. the 20076.40 and 18676.81 cm^{-1} lines as shown.

The next task is to find a relationship between x and $\tilde{\nu}_{\mathrm{vac}}$. The most straightforward method is to start off by assuming that the relationship between the two is linear, and then to calculate the straight line from a point near each end of the plate. For example, using the lines at 20076.40 and 18676.81 cm^{-1} we might find that the linearly interpolated wavenumber $\tilde{\nu}_{\mathrm{interp}} = 20\,194.22 - 9.06423\,x$, where x is in mm and $\tilde{\nu}_{\mathrm{interp}}$ in cm^{-1}. We use this formula to calculate $\tilde{\nu}_{\mathrm{interp}}$ as shown, and calculate the differences $\Delta\tilde{\nu} = \tilde{\nu}_{\mathrm{vac}} - \tilde{\nu}_{\mathrm{interp}}$. These differences are then plotted on a graph of $\Delta\tilde{\nu}$ against $\tilde{\nu}_{\mathrm{interp}}$, and they and the straight line formula used to calculate $\tilde{\nu}_{\mathrm{vac}}$ from any value of x.

If you have a sufficiently sophisticated calculator to hand, then you could try to find a direct relationship between $\tilde{\nu}_{\mathrm{vac}}$ and your values of x.

Tabulate the wavenumbers of the band-heads you have measured. For each band-head, calculate $\Delta G(v')$, which is the energy between successive band-heads. Plot $\Delta G(v')$, against an arbitrary running integer v''. Draw a line through the points, and make a curved extrapolation from the last point down to the v'' axis. Measure the area under the extrapolated section of the curve. Add this energy to the energy of the measured bandhead

Note that the extrapolation is only made from the *last* experimental point to the dissociation limit—it would clearly be foolish to use an extrapolation for the region of the curve where experimental measurements are already available.

to obtain the dissociation limit. This corresponds to the energy X $^1\Sigma_g^+$ v = 0 → I $^2P_{3/2}$ + I $^2P_{1/2}$. Hence find the dissociation energy of ground-state molecular iodine.

Results[9]

Suppose a plate is calibrated as shown in Table 5.10.2 with the interpolation formula quoted above, and that a band-head occurs at a comparator reading x of 29.503 mm. Then $\tilde{\nu}_{\text{interp}} = 20\,194.22 - 9.06423 \times 29.503 = 19926.8\ \text{cm}^{-1}$. From graph, $\Delta\tilde{\nu}$ at this value of $\tilde{\nu}_{\text{interp}} = -10.95\ \text{cm}^{-1}$. Therefore $\tilde{\nu}_{\text{vac}} = 19\,926.8 - 10.95 = 19\,915.8\ \text{cm}^{-1}$.

Accuracy

See below.

Comment

In this experiment we have analysed the *heads* of bands caused by the rotational fine structure of each vibrational level.[10] Strictly speaking, however, we should use the energies of the band *origins*. However, because the rotational constants of the upper and lower levels are very different, the band heads are close to the band origins and very little error is introduced.

A further source of error lies in the use of the Birge–Sponer extrapolation, which is only valid if the graph of $\Delta G(v')$ against v' gives a good indication of the extrapolation to the dissociation limit. Generally this is not the case. The correct extrapolation for iodine shows positive curvature towards the dissociation limit which increases the area under the curve, and hence the estimate of the dissociation energy, by about 1 per cent. The value of v' at the dissociation limit is altered substantially (from about 65 to 87), and the use of the Birge–Sponer expression to calculate $\omega_e x_e$ and ω_e from v' is therefore very unreliable. A much more accurate *short* extrapolation (from well above $v' = 70$ in the case of iodine) employs the fact that the functional form of a molecular potential energy function at large separations is given by

$$u(r) = D_0 - C_n/r^n, \tag{5.10.5}$$

where C_n is a constant, and $n = 5$ for iodine ($^2P_{3/2} + {}^2P_{1/2}$). LeRoy and Bernstein have shown that this term is the dominant influence on the spacing of the vibrational levels very near the dissociation limit. Under these circumstances, a graph of $\Delta G(v')^{2n/(n+2)}$ against $G_{v'}$ is linear,[11] and may be used to make a *LeRoy–Bernstein* extrapolation to the dissociation limit.[12]

The calculation of better extrapolations is one of the factors which has given rise to renewed interest in the spectroscopy of simple molecules. The other is the use of tunable lasers, which yield spectra of such high precision that a complete analysis of the rotational as well as the vibrational energy levels is possible. The results may be converted, by means of the 'Rydberg–Klein–Rees' (RKR) method, into a complete molecular potential energy curve.[12]

Technical notes

Apparatus

Spectrometer, or spectrograph;$^{\$\$}$ calibration lamp; iodine cell; [Fe/Ne spectrum accurately showing the wavenumbers of about six lines]; [facilities for developing plates, comparator].

References

[1] *High resolution spectroscopy.* J. M. Hollas. Butterworths, London (1982), p. 303.
[2] *Physical chemistry.* R. S. Berry, S. A. Rice, and J. Ross. John Wiley, New York (1980), part 1, p. 256.
[3] *Quanta—a handbook of concepts.* P. W. Atkins, Clarendon Press, Oxford (1983), p. 29.
[4] Ref. 3, p. 79.
[5] *Calculation of intensity distribution in the vibrational structure of electronic transitions: The B* $^3\Pi_{O_u^+}-X\ ^1\Sigma_{O_g^+}$ *resonance series of molecular iodine.* R. N. Zare. *J. Chem. Phys.* **40,** 1934 (1964).
[6] *Dissociation energies and spectra of diatomic molecules.* A. G. Gaydon. Chapman and Hall, London, 3rd edn (1968), p. 74.
[7] Ref. 5, ch. 5.
[8] Ref. 1, p. 77.
[9] *Molecular spectra and molecular structure IV. Constants of diatomic molecules.* K. P. Huber and G. Herzberg. Van Nostrand Rheinhold, New York (1979), p. 330.
[10] *Fundamentals of molecular spectroscopy.* C. N. Banwell, McGraw-Hill, London, 2nd edn (1972), p. 214.
[11] *Energy levels of a diatomic near dissociation.* R. J. Le Roy, Specialist Periodical Report, *Molecular Spectroscopy* **1,** 113 (1973).
[12] *B* $^3\Pi_{O_u^+}-X\ ^1\Sigma_g^+$ *system of* $^{127}I_2$*: rotational analysis and long-range potential in the B* $^3\Pi_{O_u^+}$ *state.* R. F. Barrow and K. K. Yee, *J. Chem. Soc. Far. Trans. II* **69,** 684 (1973).

5.11 Fluorescence spectrum and rate of electron transfer reaction by quenching

This experiment involves a study of the fate of molecules which are in electronically excited states due to the absorption of radiation. There are two processes of emission which they may undergo—*fluorescence* and *phosphorescence.* A rough distinction between the processes is that fluorescence ceases immediately the source of exciting radiation is shut off, whereas phosphorescence does not. In this experiment we are concerned solely with fluorescence.

The experiment is in two parts. Part A involves a study of the relation between the absorption spectrum, fluorescence excitation spectrum and fluorescence spectrum of fluorescein. The fluorescence excitation spectrum is obtained by analysing the fluorescence at a single wavelength while scanning the wavelength of the excitation source.

Part B of the experiment utilizes the fact that the short lifetime of the excited state of a fluorescent species may be used as an internal clock. The fluorescence of $Ru(bipyridyl)_3^{2+}$ is made to compete with electron transfer reactions to Fe^{3+} or Cu^{2+}, and the kinetics of these reactions are deduced from the degree to which they quench the fluorescence. The theory of the experiment assumes an understanding of reaction kinetics, Section 8.

Theory

A. Absorption and fluorescence[1,2]

Figure 5.11.1 shows the molecular potential energy curve of a singlet ground state S_0 and singlet excited state S_1 of a typical organic molecule in solution. The assumptions on which this

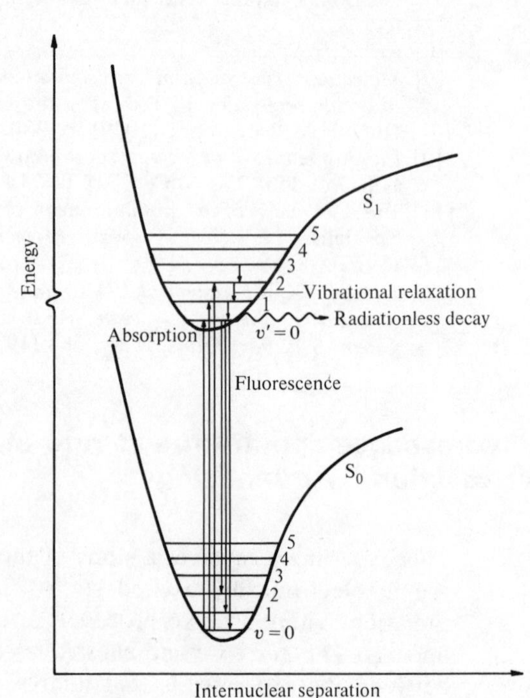

Fig. 5.11.1. Transitions in a typical organic molecule in solution. The spacing of the vibrational levels is exaggerated

type of diagram is based were explained for the case of iodine on p. 255. Note however, that the excitation of an electron in a large molecule has much less effect on the nuclear framework than occurs for a simple diatomic molecule such as iodine, and that therefore the potential energy curves of the S_0 and S_1 states are very similar. Consequently the spacing of the vibrational levels in the two states is also similar, as shown schematically in the diagram.

At room temperature, most of the unexcited molecules are in the vibrational ground state $v = 0$, and the absorption spectrum arises from transitions to vibrational levels v' of the S_1 state. In the excited state, collisions with solvent molecules cause radiationless decay down to the lowest vibrational level $v' = 0$, but do not quench the electronic excitation. The molecule therefore undergoes spontaneous emission to vibrational levels in the ground state, emitting the excess energy as fluorescense radiation.

Other important features may be explained by careful consideration of Fig. 5.11.1 and the processes involved. We might expect, for example, that the frequencies of the $v = 0 \leftrightarrow v' = 0$ transitions in the absorption and fluorescence spectrum would be the same. Sometimes they are not, a fact that may be explained in terms of the difference in solvation of the ground and excited states. The relationship between the absorption and fluorescence excitation spectrum may be explained if we know that for most complex molecules in solution, the fluorescence efficiency happens to be almost independent of the wavelength of the exciting light.

Phosphorescence occurs if, before returning to the ground state, an electronically excited molecule transfers to a triplet state T_1 by a process known as *inter-system crossing*. The transition from $T_1 \rightarrow S_0$ is then spin forbidden. In practice, because of spin-orbit coupling, the transition does take place, but only after a length of time which may be anything from 10^{-4}–1 s.

B(i) Experimental determination of electron transfer rate coefficient[3,4]

The species $(R^{2+})^*$, where $R = Ru(bipyridyl)_3$, is in a *charge transfer excited state*, so called because an electron has been partially transferred from the metal atom to the antibonding π orbitals of the liquid. The excited state has an energy of 202 kJ mol^{-1} in excess of that of the ground state, and readily undergoes electron transfer reactions of the type

$$(R^{2+})^* + X^{n+} \xrightarrow{k_q} R^{3+} + X^{(n-1)+},$$

where k_q is the quenching rate coefficient and, in this experiment, X^{n+} is Fe^{3+} or Cu^{2+}.

The excitation and fluorescence processes may be written

$$R^{2+} + h\nu \xrightarrow{\text{rate}=r} (R^{2+})^* \qquad \text{excitation (absorption)}$$

$$(R^{2+})^* \xrightarrow{k_s} R^{2+} \qquad \text{de-excitation (fluorescence and non-radiative decay).}$$

Let the concentration of $(R^{2+})^*$ be y and let that of X^{n+} be x. Then by inspection of the reaction scheme,

$$dy/dt = r - k_s y - k_q xy. \tag{5.11.1}$$

Applying the *steady-state approximation* to $(R^{2+})^*$, so that $dy/dt = 0$,

$$y = r/(k_s + k_q x). \tag{5.11.2}$$

The fluorescence intensity I is proportional to the concentration y of the excited state. Let I_0 be the fluorescence intensity when no electron acceptor ions are present and $x = 0$. I is measured for a series of steady state experiments, in which the excitation rate r is constant and always the same. Under the conditions of the experiment $x \gg y$ so that x is also effectively constant in a given experiment. Therefore

$$I_0/I = (k_s + k_q x)/k_s = 1 + x(k_q/k_s), \tag{5.11.3}$$

and a graph of I_0/I for various values of x yields a straight line of slope k_q/k_s. This is known as a *Stern–Volmer plot.* k_q may then be calculated if k_s is known.

B(ii) Calculation of rate coefficient by Marcus theory

The electron transfer process studied in this experiment is an example of an *outer-sphere redox reaction,* i.e. one in which there is little interaction between the oxidant and reductant at the moment of electron transfer.[5] Marcus has developed a theory by which the electron transfer rate constant k_q (or k_{12}) may be calculated from the equilibrium constant K_{12} of the overall reaction, and the *self-exchange rate coefficients* k_{11} and k_{22} which relate to the transfer of an electron between oxidized and reduced forms of the same species:

$$(R^{2+})^* + X^{n+} \underset{}{\overset{K_{12}}{\rightleftharpoons}} R^{3+} + X^{(n-1)+},$$

$$(R^{2+})^* + R^{3+} \xrightarrow{k_{11}} R^{3+} + (R^{2+})^*,$$

$$X^{n+} + X^{(n-1)+} \xrightarrow{k_{22}} X^{(n-1)+} + X^{n+}.$$

The required relation is then

$$k_q = k_{12} = (k_{11}k_{22}K_{12}f)^{1/2}. \tag{5.11.4}$$

K_{12} may be calculated from the electrode potentials using the relation

$$\Delta E^{\ominus} = \frac{-\Delta G^{\ominus}}{zF} = \frac{RT}{zF} \ln K_{12}. \tag{5.11.5}$$

Marcus[6] has shown that

$$\log_{10} f = \frac{(\log_{10} K_{12})^2}{4 \log_{10}(k_{11}k_{22}/Z^2)}, \tag{5.11.6}$$

where the collision frequency factor Z may be taken as $\sim 10^{11}$ dm^3 mol^{-1} s^{-1} in aqueous solution at 25 °C.

Apparatus[7] The fluorescence and fluorescence excitation spectra and intensities are measured on a fluorescence spectrophotometer. The optical system of a typical instrument is shown in Fig. 5.11.2. Light from a xenon arc is dispersed by a grating in the excitation monochromator, and radiation of the selected wavelength is

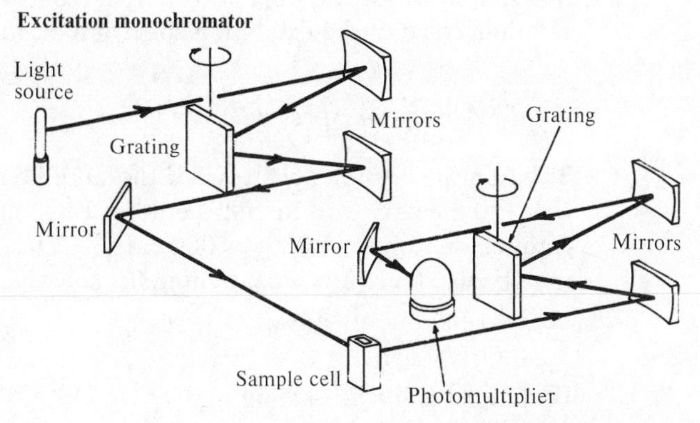

Fig. 5.11.2. The optical system of a typical fluoresence spectrophotometer

than directed onto the fluorescent sample. Fluorescence from the sample passes through the analysis monochromator, and then onto a detector which creates a signal for driving a meter or pen recorder.

Controls on the monochromators allow selection of the excitation and analysis wavelengths, and allow one or other to be driven electrically.

The absorption spectrum is measured on a ultraviolet spectrophotometer, as described on p. 251.

Procedure

Details of the stock solutions provided for this experiment are given in the Technical notes.

A(i) Fluorescence spectrum

The variation of the emitted intensity with wavelength may be recorded by scanning the analysing monochromator at a fixed wavelength of excitation. Make up 10 cm^3 of a solution of 1.0 mg dm^{-3} fluorescein in 0.1 M NaOH, and place in a fluorescence cell. Set the exciter wavelength to 440 nm and scan the analyser wavelength from 480 to 600 nm, recording the resultant fluorescence spectrum on the chart recorder of the fluorescence spectrophotometer. Adjust the sensitivity controls to obtain a reasonably large signal at the fluorescence peak.

Change the exciter wavelength to 460 nm and repeat.

A(ii) Excitation spectrum

Using the same solution, set the analysing wavelength to 540 nm, scan the exciting wavelength from 400 to 520 nm and obtain a trace of the excitation spectrum on the chart recorder.

A(iii) Absorption spectrum

Take an absorption spectrum of the undiluted fluorescein solution (10 mg dm^{-3}) on the ultraviolet spectrophotometer over the frequency range 18000–24000 cm^{-1}.* The fluorescence cells may be used and a cell containing 0.1 M NaOH should be placed in the reference beam.

B. Fluorescence quenching

Make up the solutions listed in Table 5.11.1 in 10 cm^3 volumetric flasks.

Place solutions (a) to (d) in the four silica fluorescence cells,

Table 5.11.1. Molarities of sample solutions

	$[Fe^{3+}]/10^{-3}$ M	$[Cu^{2+}]$/M
a	0	0
b	0.2	0
c	0.4	0
d	0.8	0
e	1.2	0
f	1.6	0
g	1.8	0
h	0	0.02
i	0	0.04
j	0	0.08
k	0	0.12
l	0	0.16
m	0	0.18

All of the solutions are 10^{-5} M in $Ru(bipyridyl)_3^{2+}$ and 0.5 M in sulphuric acid.

and insert the cells carefully in the numbered slots 1 to 4 in the carrier. Place the carrier in the instrument.

Set the excitation wavelength manually for 450 nm, and the analysis wavelength to 610 nm. With solution (a) in position and the shutter open, set the three sensitivity controls to give a meter reading near 100. Record the reading. Now measure the fluorescence intensities of solutions (b) to (d). Replace solutions (b)–(d) by (e)–(g) and measure the fluorescence intensities for these three solutions. In between each measurement, check the intensity from (a)—some drift is inevitable since the light intensity fluctuates slightly.

Then make a series of measurements for the Cu^{2+} solutions (h)–(m), while continuing to check the intensity with solution (a).

Finally make a correction for scattered light by measuring the intensity from 0.5 M sulphuric acid.

Calculation

A

On the same graph draw out the absorption, excitation and fluorescence spectra against a wavenumber abscissa, normalizing the highest peak of each spectrum to unity. Explain the features of the spectra as discussed in the Theory section.

B

Use Stern–Volmer plots to find k_{12} for the quenching of $Ru(bipyridyl)_3^{2+}$ by Fe^{3+} and Cu^{2+}.

Table 5.11.2. Data for quenching reactions

$[R^{2+}]^* + Cu^{2+}(aq)$		
	$[R^{2+}]^*/R^{3+}$	$k_{11} = 1 \times 10^8\ dm^3\ mol^{-1}\ s^{-1}$ $E^{\ominus} = -0.84\ V$[7]
$Cu^{2+}(aq)/Cu^+(aq)$		$k_{22} = 1 \times 10^{-5}\ dm^3\ mol^{-1}\ s^{-1}$ $E^{\ominus} = 0.16\ V$
$[R^{2+}]^* + Fe^{3+}(aq)$		
	Results for $[R^{2+}]^*$ as above.	
	$Fe^{3+}(aq)/Fe^{2+}(aq)$	$k_{22} = 4\ dm^3\ mol^{-1}\ s^{-1}$ $E^{\ominus} = 0.77\ V$

Calculate K_{12} for each system using the data given in Table 5.11.2. Hence calculate k_{12} using the Marcus relation and the self-exchange rate coefficients given in the table.

Compare the experimental and calculated values of k_{12}. The rate coefficient for quenching by Fe^{3+} by a diffusion controlled mechanism is $\sim 3.3 \times 10^{-9}\ dm^3\ mol^{-1}\ s^{-1}$. Comment.[8,9]

Results

A

The differences between the characteristics of the spectra and those anticipated from the discussion in the Theory section can be largely attributed to variations in the exciting light intensity and analyser light sensitivity with wavelength. Appropriate corrections can be made.[10]

B

The experimental and theoretical values should agree to within a factor of 2 or 3.

Comment

Spectrofluorimetry can be used to measure the excitation spectra of fluorescent compounds at concentrations far lower than would be needed to measure the absorption spectrum directly with an absorption spectrophotometer. It also has the advantage that the absorption spectrum of one component of a mixture of absorbing compounds can be picked out by tuning to the wavelength of the appropriate fluorescence emission band.

Technical notes

Apparatus

Fluorescence spectrophotometer with numbered slots for 4 sample cells;$$ u.v. spectrophotometer;$$ cells. Stock solutions for part A: 0.1M

NaOH, 10 mg fluorescein per dm^3 0.1 M NaOH. Stock solutions for part B: 0.5 M H_2SO_4, and other solutions 0.5 M in H_2SO_4: 10^{-4} M Ru(bipyridyl)$_3^{2+}$ Cl^-_2, 10^{-3} M $(NH_4)Fe(SO_4)_2$, 0.2 M $CuSO_4$.

References

[1] *Photoluminescence of solutions.* C. A. Parker. Elsevier, Amsterdam (1968), ch. 1.
[2] *Introduction to inorganic photochemistry.* G. B. Porter. *J. Chem. Educ.* **60,** 785 (1983).
[3] *Photochemistry.* R. P. Wayne. Butterworths, London (1970), p. 88.
[4] Ref. 1, p. 72.
[5] *Inorganic reaction mechanisms.* M. L. Tobe, Nelson, London (1972), ch. 9.
[6] *Mechanisms of electron transfer.* W. L. Reynolds and R. W. Lumry. Ronald Press, New York (1966), p. 133.
[7] Ref. 1, ch. 3.
[8] *Electron-transfer reactions of excited states.* N. Sutin and C. Creutz. *J. Chem. Educ.* **60,** 809 (1983).
[9] *Temperature dependence of excited-state electron–transfer reactions. Quenching of* *RuL_3^{2+} *emission by copper (II) and europium (III) in aqueous solution.* J. E. Baggott and M. J. Pilling. *J. Phys. Chem.* **84,** 3012 (1980).
[10] Ref. 1, p. 246.

5.12 Keto–enol ratios, hydrogen bonding, and functional group analysis by n.m.r. spectroscopy

This experiment illustrates some of the applications of nuclear magnetic resonance (n.m.r.) spectroscopy, and is in four parts. Part A involves some familiarization and setting up exercises. In part B we study the intermolecular hydrogen bonding in benzyl alcohol (phenyl methanol), and in part C keto–enol tautomerism in pentane-2,4-dione (acetyl acetone). Finally, in part D the use of n.m.r. in structure determination is illustrated by an analysis of the spin multiplet patterns of ethanal (acetaldehyde).

There follows a brief general introduction to the theory of the experiments, while further observations about the nature of the spectra are made in the Procedure section.

Theory[1,2]

A, B. Intensities and chemical shifts

In a proton n.m.r. spectrometer, which is the type we shall be using in this experiment, transitions occur between the two spin states allowed to the hydrogen nucleus (the proton) in an applied magnetic field B_0. These spin states are separated in energy by $2\mu B_0$, where μ is the observable component of the magnetic moment of the proton, as shown in Fig. 5.12.1. To

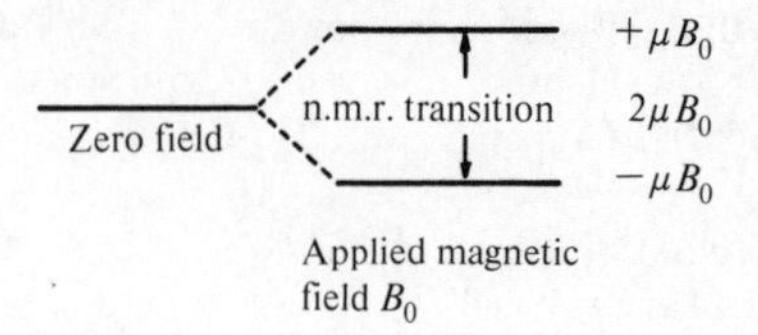

Fig. 5.12.1. Relation between a proton n.m.r. transition and applied magnetic field

cause the transition we apply radiation of frequency ν such that

$$h\nu = 2\mu B_0. \tag{5.12.1}$$

The magnetic moment of the proton originates from its spin angular momentum (of observable magnitude $\frac{1}{2} \cdot h/2\pi$), and we may write

$$\mu = \gamma \cdot \tfrac{1}{2} \cdot h/2\pi, \tag{5.12.2}$$

where γ is a proportionality constant called the *magnetogyric ratio.* Combining the two equations, we obtain the resonance condition

$$\nu = \frac{\gamma}{2\pi} \cdot B_0. \tag{5.12.3}$$

The intensity of an absorption in an n.m.r. spectrum is proportional to the number of protons giving rise to it, and this is a most useful guide in deducing the structure of the sample species.

The applied magnetic field also induces a circulation of electrons in the molecule in such a direction as to induce a magnetic field which opposes the applied one. The actual field at the nucleus is then $(1-\sigma)B_0$ where σ is the *shielding constant.* The resonance condition becomes

$$\nu = \frac{\gamma}{2\pi}(1-\sigma)B_0. \tag{5.12.4}$$

Measurements are made relative to tetramethylsilane ($(CH_3)_4Si$, TMS), which gives a proton resonance at a higher field than most other compounds. The dimensionless *chemical shift* δ of a species i is given by the relation

$$\delta_i = \sigma_{TMS} - \sigma_i.$$

δ_i is expressed in parts per million (p.p.m.) of the applied magnetic field. An alternative scale is given by

$$\tau_i = 10.00 - \delta_i,$$

and on this scale TMS is 10.00. Resonances which have positive values of δ occur to the *low field* side of TMS and are said to be *deshielded* with respect to those at higher field. One of the causes of deshielding is the reduction of electron density around the nucleus by electron withdrawing substituents, but many other factors must be taken into account. For example when

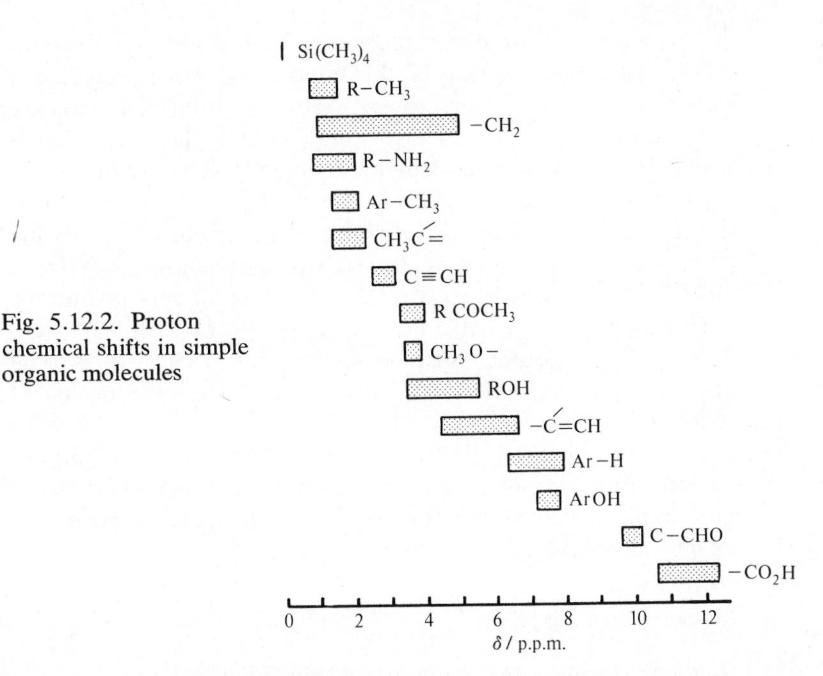

Fig. 5.12.2. Proton chemical shifts in simple organic molecules

hydrogen bonding occurs, the resonance of the proton concerned is shifted to low field. Intramolecular hydrogen bonding, such as that in keto–enol tautomerism, produces the largest effect. Some representative chemical shifts, relating to the simple molecules studied in this experiment, are shown in Fig. 5.12.2.

C. Keto–enol tautomerism[3,4]

The power of the n.m.r. method to distinguish between different chemical species in solutions is illustrated by the case of the keto–enol tautomerism of 2,4-pentanedione. The proton resonance spectrum of this substance is the superimposition of the individual spectra of the keto and enol forms, Fig. 5.12.3. Once the peaks have been correctly assigned, the ratio of the total (integrated) peak areas of the two species directly yields the equilibrium constant of the tautomerization.

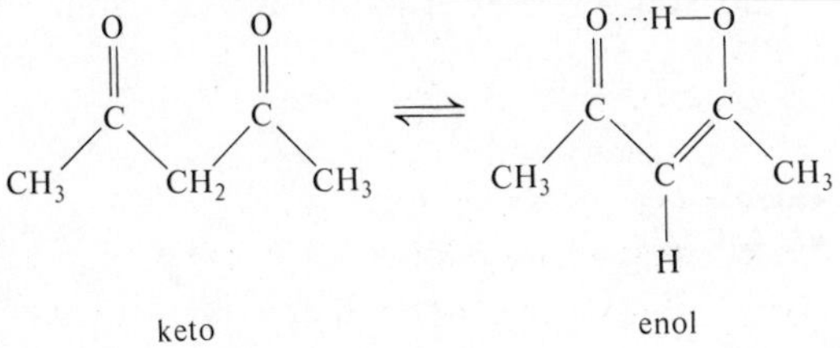

CH_3–C(=O)–CH_2–C(=O)–CH_3 ⇌ CH_3–C(=O)–CH=C(–O–H)–CH_3 (O····H–O)

keto   enol

Fig. 5.12.3. Keto and enol forms of pentanedione

The success of the n.m.r. technique for this problem depends on the fact that the rate of the tautomerization is sufficiently slow for distinct signals to be obtained from the separate species. In the presence of fast isomerizations, the n.m.r. signals would be at the weighted mean frequency. The condition for separate non-averaged signals is that the lifetime of each species should be greater than the chemical shift difference. Since the whole range of proton shifts covers only about 700 Hz at 60 MHz, this condition requires that even under very favourable circumstances (where chemical shifts of the two sites are very different) the lifetime must be longer than about 10^{-3} s. The broadening of the resonance peaks indicates a speeding up of exchange.

There is a marked difference between the chemical shifts of proton resonances in pure liquid 2,4-pentanedione and those of the sample dissolved in certain solvents, which can be attributed to intermolecular effects.

D. Spin–spin coupling[1]

In most spectra, fine structure is observed due to *spin–spin coupling*. The energy of one nucleus is modified by the coupling to a neighbouring nucleus. The magnitude of the interaction between two nuclei j and k is measured in terms of the *coupling constant* J_{jk}, which, unlike δ, is independent of field, and is therefore expressed in Hz rather than p.p.m..

If the chemical shift between two protons is of the order of, or less than, the corresponding coupling constant, the system is referred to as an AB system. An example is in 1-chloro,2-bromoethene, Fig. 5.12.4.

If the chemical shift difference between a pair of nuclei is much greater than the coupling constant between them, they are assigned letters well apart in the alphabet. For example we may have an AX_2 system such as trichloroethane, Fig. 5.12.5, in which there are two equivalent protons X, and one proton A which has a very different chemical shift. Equivalent protons do not interact with each other to produce additional fine structure in the spectrum, and may therefore be treated as a single group. The splitting pattern for an AX_2 spin system is therefore as

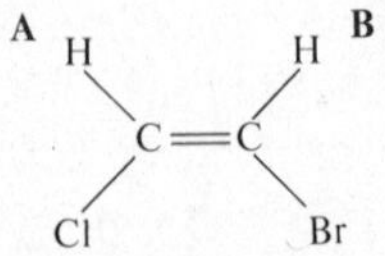

A H H B
C=C
Cl Br

Fig. 5.12.4. 1-chloro, 2-bromoethene, an AB spin system

Fig. 5.12.5. Trichloroethane, an AX_2 spin system

shown in Fig. 5.12.6. The resonance frequency ν_A of the proton A is split by X_1, and then each of these lines is split to the same degree by X_2, leading to three resonances of intensities 1 : 2 : 1 equally spaced by an amount J_{AX} as shown. The resonance at ν_X is split into two by A. Since the intensity is proportional to the number of protons present, it follows that there are two resonances of relative intensity 4 spaced an amount J_{AX} apart. Instead of drawing out the successive splittings as in the figure, the following rules may be used for magnetically equivalent nuclei: (i) the splitting of a resonance due to the coupling with p magnetically equivalent neighbour nuclei leads to the appearance of $(p+1)$ distinct lines, and (ii) the intensity distribution within the $(p+1)$ lines is that of the binomial coefficients, i.e. those of the expansion $(1+x)^p$.

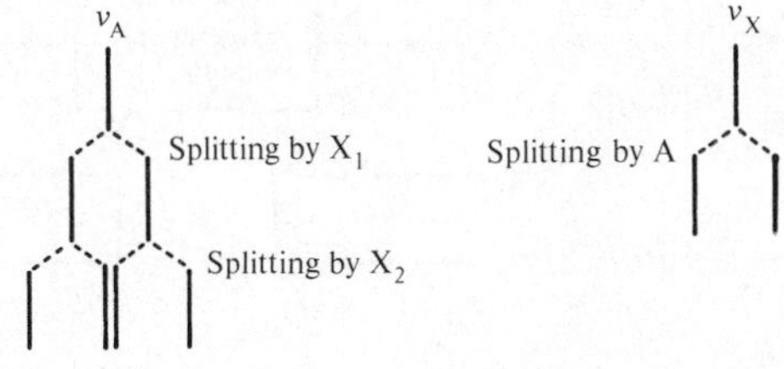

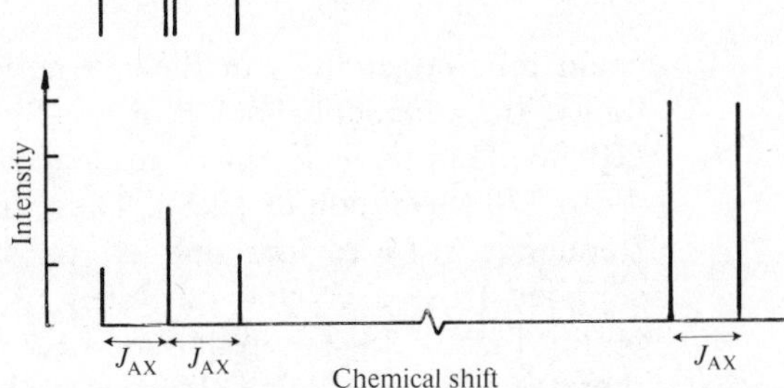

Fig. 5.12.6. Splitting in an AX_2 spin system

Apparatus[2] Like all other types of spectrometer, an n.m.r. spectrometer requires a source of radiation, a sample cell, a detector and recording device, and a means of scanning the spectrum. A magnet is also required.

Figure 5.12.7 shows the simplest arrangement for such an instrument. The sample is placed in a thin glass tube between the poles of an electromagnet or permanent magnet, and the tube is spun at about 30 Hz by an air turbine to average out

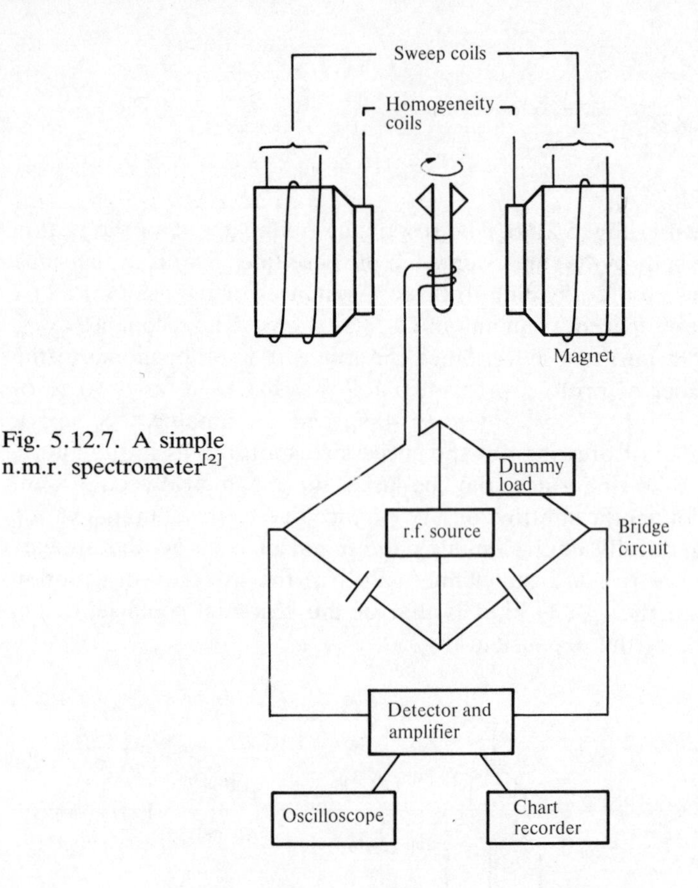

Fig. 5.12.7. A simple n.m.r. spectrometer[2]

small inhomogeneities in the magnetic field. For proton resonance the magnetic field is normally in the range 1–8 Tesla (10–80 kGauss), which gives resonance frequencies in the range 40 to 220 MHz (eqn (5.12.3)). These frequencies lie in the radio frequency (r.f.) region, and so to detect them, radiation is supplied from a crystal-controlled r.f. source. A resolution of better than 0.5 Hz is required, i.e. about 1 part in 10^8, and therefore the magnetic field must be homogeneous to this very high degree. This is achieved not only by spinning the sample, but also by the precision engineering of the magnet and the use of correction or *shim coils* would onto the pole pieces of the magnet, through which small currents are passed.

The most common method of scanning the spectrum is to keep the frequency constant and scan the magnetic field by means of sweep coils round the pole pieces. As the spectrum is scanned, resonances are produced in the sample and this unbalances the bridge circuit. The signals are picked up by a detector, amplified, and displayed on an oscilloscope and recorder. The current control to the sweep coils is usually linked to the chart recorder drive, so that the terms 'chart sweep' and 'field sweep' are generally synonymous.

Procedure

N.m.r. spectrometers are complex instruments, and it is essential that you read the instructions supplied with the spectrometer, and employ the optimum settings determined for you. Do not, however, alter the shim coils, which will have been set up for you by your instructor. Great care must be exercised in handling the sample tube, which is made of very thin glass and fits inside a delicate receiver coil. The spinner on the tube must be positioned accurately and the outside of the tube kept scrupulously clean.*

A. Familiarization experiments

Use a sample of approximately 5 per cent cyclohexane in tetrachloromethane. The spectra from experiments A(i) and A(ii) should be displayed on a single piece of chart paper, as should those from experiments A(iii), A(iv), and A(v).

A(i) Calibration

Run the spectrum, readjusting the field shift controls if necessary until a peak is observed.

The resulting peak may not be symmetrical in shape. Figure 5.12.8 shows some typical line shapes. Note the *ringing* on the

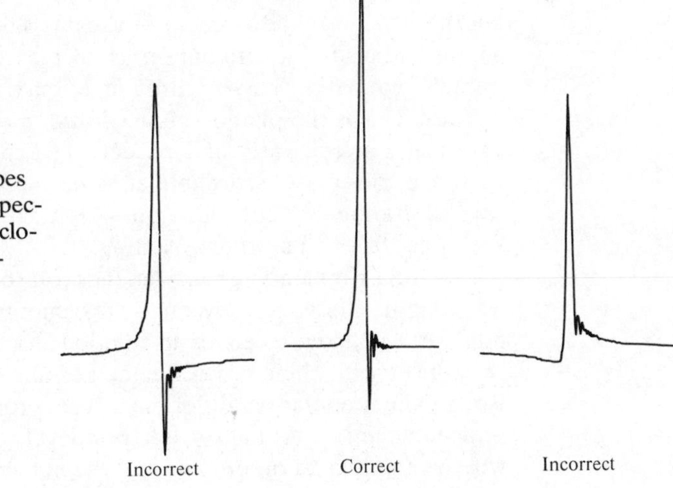

Fig. 5.12.8. Line shapes in the proton n.m.r. spectrum of 5 per cent cyclohexane in tetrachloromethane

trailing edge of the line, an artifact arising out of a transient response of the nuclear spins. This ringing will be ignored at present. Set the phase shift control for a symmetrical signal as in the figure. The effect of mis-setting the phase shift control is also illustrated. (This procedure ensures that the signal is in the absorption mode with no admixture of the dispersion mode. It is a rather critical adjustment when integrations are to be performed.)

By means of the coarse and fine shift controls position the cyclohexane peak at $\delta = 1.44$. This provides a calibration of the chart, and is equivalent to setting tetramethylsilane (T.M.S.) at $\delta = 0.00$. Note the settings of these controls: they may be changed to shift the spectrum upfield or downfield by known amounts in p.p.m., corresponding to the p.p.m. scales printed on the chart, but they should be returned to their original settings in order to read off δ or τ values from the chart.

A(ii) Resolution

Scan the cyclohexane peak as instructed, and measure its width at half its height. This measurement may be used to define the resolution of the spectrometer; check that the resolution is high enough before proceeding.*

A(iii) Saturation

With the cyclohexane peak displayed as instructed, run through the resonance using several different settings of the radiofrequency field strength control. These successive observations may either be superimposed or displayed adjacent to one another by careful adjustment of the fine shift control. Note that as the intensity of irradiation is increased, the peak height initially increases, passes through a maximum and then decreases. This is the phenomenon of *saturation*. Reduce the chart speed and sweep rate, and repeat the experiment. Saturation then becomes more noticeable at lower field strengths, that is to say, saturation is dependent on sweep rate. Return the chart speed control to its original setting.

The n.m.r. phenomenon depends on the extremely small population difference between the two energy levels in an applied field. At very low levels of irradiation the thermal coupling of the nuclei with their environment is sufficient to maintain the Boltzmann population difference. The process is called *spin-lattice relaxation*. At high irradiation levels the 'pumping' overwhelms the relaxation process and the net absorption signal falls to zero. At intermediate irradiation levels, saturation and relaxation complete, and a population difference is established which is smaller than that for Boltzmann equilibrium.

A(iv) The filter

With the peak still displayed on the same range, run the spectrum using different positions of the filter control. Select the position which gives the best signal-to-noise ratio (p. 211). Verify that this also depends on chart speed, which is ultimately set at a value which is a compromise between signal-to-noise requirements and the time needed to obtain the spectrum. The filter discriminates against high frequency random noise components from the receiver coil and early stages of amplification.

A(v) The effect of spinning

Stop the spinner and note the drastic fall in resolution when the spectrum is scanned as before. Now spin very fast. Once more the resolution deteriorates from the optimum. This is due to the liquid vortexing in the tube, producing a discontinuity in magnetic susceptibility near the r.f. coil, which distorts the magnetic field.

When the sample is spinning you may observe small peaks on either side of the main peak and symmetrically spaced. These are *spinning sidebands*. As the sample spins, the nuclei are carried around circular paths, and because of residual field gradients, they experience a modulated magnetic field which produces the sidebands. Spinning sidebands are separated from the main peak by the spinning frequency, and their intensity falls off with spinning rate.

B. Hydrogen bonding

Prepare a series of mixtures of benzyl alcohol and CCl_4; suitable ratios by volume are 2 : 1, 1 : 1, 1 : 2 and 1 : 5. Measure the changes with composition of the shift of the aromatic, methylene and hydroxyl protons.

C. Keto–enol ratios

Prepare the following four samples in n.m.r. sample tubes, using equal volumes of each component for samples (b), (c) and (d):

(a) Pure 2,4-pentanedione (acetyl acetone),
(b) 2,4-pentanedione + CCl_4,

(c) 2,4-pentanedione + ethanoic (acetic) acid,
(d) 2,4-pentanedione + triethylamine.

Allow the samples to equilibrate in the pre-heater, and measure their spectra as instructed. Record the δ chemical shift values of the five peaks and assign them to the appropriate protons. One is a broad peak of low intensity which is easily overlooked.

The intensities of the peaks are measured by their areas, and, as mentioned previously, are proportional to the number of protons causing them. N.m.r. spectrometers usually incorporate an automatic integrator for measuring the areas.

D. Spin–spin coupling

Take a sealed sample of ethanal and display the n.m.r. spectrum on a single chart. Measure the chemical shift between the methyl and aldehyde protons. Record the two spin multiplets on expanded scales and measure the spin–spin coupling constant J_{AX}, verifying that all the splittings are equal.

Calculation

B. Hydrogen bonding

Consider whether the breaking of hydrogen bonds between OH groups of benzyl alcohol leads to a high-field or low-field shift.

C. Keto–enol ratios[3,4]

Obtain the keto–enol ratios by comparing the intensities of the CH and CH_2 peaks and of the two CH_3 peaks.

The occurrence of marked differences between the chemical shifts of resonances in the pure liquid and its solutions indicates intermolecular effects. Which of the solvents investigated show evidence of such solvent–solute interaction? Does the nature of the solute–solvent interaction, if any, correlate with the accompanying change in the position of the keto–enol ratio? If there is no solute–solvent interaction, what other factors are there to account for change in this ratio?

In view of the changes in linewidths of certain resonance peaks, indicating a speeding up of exchange, what conclusions can be drawn about OH exchange between ethanoic acid and *enol* pentane-2,4-dione, and about the loss of H by the CH_2 group in the keto form?

D. Spin–spin coupling

Ethanal provides a good example of an AX_3 type spin system. Assign the resonances on the spectrum, explaining your reasoning. What changes would you expect to see in the ethanal spectrum if it were to be recorded on a spectrometer which operated at a field twice as intense as that of the present spectrometer? If two of the methyl protons in ethanol were to be replaced by chlorine atoms, what kind of spin multiplet structure would you expect to see, assuming that the chemical shift remains large compared with the spin–spin coupling? Is it permissible to neglect the magnetic effects of the chlorine nuclei? Give your reasons.

Results

Some specimen n.m.r. lines are shown in Fig. 5.12.8. The quality of the spectra you obtain in this experiment will depend on how well the instrument is set up and used; do not hesitate to ask the advice of an instructor about this.

Comment

This experiment introduces the basic instrumentation and a few simple applications of proton n.m.r.. In recent years there have been very great advances in both areas. Modern instruments increasingly use Fourier transform (F.T.) techniques, which were introduced in Expt 5.4. A short, powerful burst of radiation (typically 1 kW for 1 μs) is applied to the sample over a wide frequency range. The sample then gives a decaying, oscillating signal known as free induction decay which contains all the information about the allowed transition frequencies. Fourier transformation is the mathematical process, performed by a computer built in to the spectrometer, which converts the decay signal into a frequency spectrum.

A simple extension of the present experiment would be to study proton exchange processes at different temperatures, and thus obtain the exchange activation energy.[4] Other applications include the study of molecular conformation, the properties of solids, and the spectra of other nuclei such as ^{11}B, ^{13}C, ^{14}N, ^{17}O, ^{29}Si, ^{31}P, ^{33}S, and the halogens.

Whole-body n.m.r. scanners, currently being installed in major hospitals, detect ^{31}P in ATP, an energy-storage molecule present in healthy living cells.

Technical notes

Apparatus

N.m.r. spectrometer;$$ sample tubes; spec. grade CCl_4, cyclohexane, 2,4-pentanedione (acetyl acetone), ethanoic (acetic) acid, triethylamine, benzyl alcohol (phenyl methanol), sealed sample of ethanol (acetaldehyde).

Note

The ethanal sample will gradually decompose, and should be checked by looking for extraneous peaks between the low field and high field peaks.

References

[1] *Nuclear magnetic resonance spectroscopy*. R. K. Harris. Pitman, London (1983).

[2] *Magnetic resonance*. K. A. McLauchlan. Clarendon Press, Oxford (1972).

[3] *High resolution nuclear magnetic resonance*. J. A. Pople, W. G. Schneider and H. J. Bernstein, McGraw-Hill, New York (1959), ch. 17.

[4] *Proton exchange behavior in some hydrogen-bonded systems*. W. G. Schneider and L. W. Reeves. *Ann. New York Acad. Sci.* **70,** 858 (1957–8).

5.13 Conformational energies by molecular quantum mechanics

This experiment involves the quantum mechanical calculation of the energies of various molecules as a function of changes in bond lengths, bond angles and torsion angles. The calculations are carried out by means of a computer program based on the *extended Hückel theory*. The approach involves relatively crude approximations, but nevertheless the results are in agreement with experiment for a number of properties.

The variety of molecules to which the method can be applied is only limited by the size and speed of the computer. This experiment illustrates two simple applications. Part A involves the calculation of the energy of the carbon dioxide molecule as a function of bond angle, the fitting of a parabola to the calculated energies, and the calculation of the bending force constant from the coefficients of this curve. In Part B the barrier to internal rotation in ethane is calculated, and a determination is made of how this barrier varies with the C—C bond length.

Theory

(*i*) *Molecular orbitals*

Much of molecular quantum chemistry is devoted to solving the *Schrödinger equation* for an isolated molecule. This beautifully compact equation,

$$H\psi = E\psi, \qquad (5.13.1)$$

in which ψ is the wave-function, H the Hamiltonian operator, and E the energy, cannot be solved for many-body systems such as molecules. Therefore a range of approximations must be made.

Firstly we make the *orbital approximation*. The molecular wavefunction is written as a product of one-electron functions or orbitals,

$$\psi = \phi_1\phi_2 \ldots \phi_n. \tag{5.13.2}$$

This approximation is familiar for atoms, which we are accustomed to describing in terms such as 'the electronic structure of lithium is $1s^2 2s$.' Each electron has its own wave function, which in the case of an atom is of 1s, 2s, 2p, . . . symmetry. The same is true in molecules but the symmetries will depend on the point group of the molecule.

In order to know ψ, we need to know the individual functions ϕ_i. A further approximation is to represent each ϕ_i in terms of known atomic wavefunctions multiplied by numerical coefficients. This *linear combination of atomic orbitals* (*LCAO*) *approximation* allows us to express any molecular orbital ϕ_i as an expansion of known atomic orbitals (χ_k) multiplied by coefficients C_{ik}:

$$\phi_i = \sum_k C_{ik}\chi_k. \tag{5.13.3}$$

Thus for H_2 the most tightly bound molecular orbital is called $1\sigma_g$ for symmetry reasons (p. 207), and expressed as

$$\phi_{1\sigma_g} = C_1 1s_A + C_2 1s_B. \tag{5.13.4}$$

To determine the ϕ_i and hence ψ for a molecule we need to determine the coefficients C_{ik}.

(*ii*) *Secular equations and determinants*

To obtain the wavefunction we start with a one-electron wave equation which looks rather like the Schrödinger equation but which refers to only one electron at a time, i.e.

$$H\phi_i = \varepsilon_i\phi_i, \tag{5.13.5}$$

where H is the one-electron Hamiltonian and ε_i the orbital energy of electron i, which by *Koopmans' theorem*[1] is approximately equal to its ionization potential.

All molecular orbital methods start with this one-electron equation and then make the LCAO substitution (eqn (5.13.3) to give the expression

$$H\sum_k C_{ik}\chi_k = \varepsilon_i \sum_k C_{ik}\chi_k. \tag{5.13.6}$$

If we now multiply both sides of the equation by one of the set of atomic orbitals, say χ_1, and integrate over all positions of the electron, we obtain the crucial *secular equations*

$$\int \chi_1 H \sum_k C_{ik}\chi_k \, d\tau = \int \chi_1 \varepsilon_i \sum_k C_{ik}\chi_k \, d\tau. \qquad (5.13.7)$$

These may be tidied up by introducing shorthand notations for the integrals,

$$\int \chi_1 H \chi_k \, d\tau \equiv H_{1k}$$

and

$$\int \chi_1 \chi_k \, d\tau \equiv S_{1k}$$

Thus the secular equation may now be written as

$$\sum_k C_{ik}(H_{1k} - \varepsilon_i S_{1k}) = 0. \qquad (5.13.8)$$

If we know all the H, ε, and S terms we would have a set of simultaneous equations for the coefficients C_{ik} only soluble if

$$\det |H_{1k} - \varepsilon_i S_{1k}| = 0. \qquad (5.13.9)$$

This is the *secular determinant* which is the starting point for all molecular orbital calculations, the simplest and crudest being the extended Hückel theory.

(*iii*) *Extended Hückel theory*

Extended Hückel theory was introduced by Raoul Hofmann, who won the Nobel prize in 1982 and who is best known for his part in the famous Woodward–Hofmann rules.

In extended Hückel theory (EHT) the nature of the one-electron Hamiltonian is never explored, and may difficulties are avoided by replacing difficult integrals by experimental parameters such as ionization potentials.

Every hydrogen atom in the molecules we shall examine has a 1s atomic orbital, χ_{1s}, and every main (non-hydrogen) atom has χ_{2s} and χ_{2p} orbitals. The integrals S_{1k} are computed properly by putting in actual atomic orbitals appropriate to the atoms in the particular molecule, at the correct points in space corresponding to a defined molecular geometry. All the integrals H_{11} are replaced by ionization potentials, and the off-diagonal H_{1k} terms are approximated by

$$H_{1k} = 0.5KS_{1k}(H_{11} + H_{kk}), \qquad (5.13.10)$$

where K is a parameter introduced to make the answers close to experiment for some simple model systems.

With these parameters the secular determinant is now a polynomial in ε whose roots—the series of orbital energies—are found by matrix diagonalization. For each ε_i we can then substitute back into the secular equations to get the coefficients, hence the functions ϕ_i and the desired function ψ.

(*iv*) *The computer program*

The computer program is in the form of a main calling routine, NEWEHT, which calls up a series of subroutines in the order shown in Fig. 5.13.1.

From NEWEHT, control first passes to BUILDZ which constructs the molecule in co-ordinate terms from the input data. Within BUILDZ itself the subroutines VECTOR and VPROD assist in the process of building the molecule. Before emerging from BUILDZ back into the main program it is possible to print out the co-ordinates calculated in this subroutine for main atoms and hydrogens.

The subroutine MAIN is then called. This uses the Cartesian co-ordinates calculated in BUILDZ, and Slater exponents written into the program, to calculate the separation between each pair of main atoms and hence the overlap between their orbitals.

In BETAH both the hydrogen–hydrogen overlaps (where relevant) are calculated from their separation, and Slater orbitals and the hydrogen-main atom overlaps are calculated to give the complete overlap matrix, S_{ij}. BETAH then calculates Hamiltonian matrix elements, H_{ij}, H_{ii}, from the valence state ionization potentials, then the remaining elements A_{ij} $(i \neq j)$.

H (containing the Hamiltonian matrix) and S (containing the overlap matrix) are carried into EIGEN.

EIGEN itself calls the subroutine JNH and hence TRY, VAL, and VEC consecutively: these carry out diagonalization using the Householder method. S_{ij} is not a unitary matrix at the

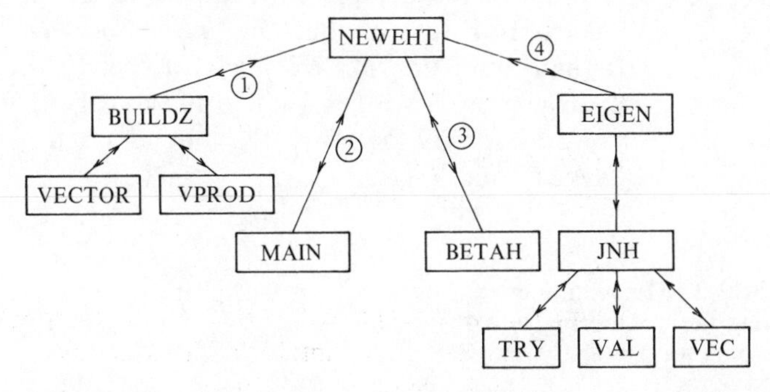

Fig. 5.13.1. Structure of the NEWEHT computer program

beginning of EIGEN (individual orbitals are non-orthogonal), so two diagonalizations must be performed to convert $(H-ES)V=0$ from a generalized eigenvalue problem to an ordinary one in order that the diagonalization yields the energies and coefficients of the molecular system. These are carried back into EIGEN in arrays E and H respectively.

Control passes back to NEWEHT where the energy of the molecule is calculated as a sum of the individual orbital energies in the array E, multiplied by the occupancy. If a potential energy scan is being undertaken, the process is repeated for the new co-ordinates, and the results printed at every angle of rotation.

(v) *The Z-matrix*

Information about the molecule to be studied is input to the computer program in the form of a *Z-matrix*.[2] The drawing up of a *Z*-matrix is one of the chief tasks in this experiment, and we shall now go through the procedure for the case of methylamine, Fig 5.13.2. The rows of the *Z*-matrix, Table 5.13.1, correspond to the numbering of the atoms in the figure. The molecule has the following geometry: bond lengths C—N = 0.1474 nm, N—H = 0.1014 nm, C—H = 0.1093 nm; angle H—N—H = 105.8°, angle H—N—C = 112.2° and angle H—C—N = 109.5°.

First we choose an atom which lies on a convenient reference axis, which we call the Z-axis. In the case of methylamine, we choose the nitrogen atom, and write down its atomic number as shown in Table 5.13.1. We then choose another main atom which lies on the Z-axis. For methylamine, this is the carbon atom of atomic number 6. We continue row 2 by entering the bond length XA, where X = C = atom 2 and A = N = atom 1. As the bond lies along the Z-axis, this is sufficient to define the position of the C atom with respect to the nitrogen atom.

Next in row 3 of the table we choose one of the H atoms (atomic number = 1). We need to specify the position of the hydrogen atom with respect to the nitrogen, and to do this we not only specify the H—C (XA) bond length, but also the HCN (XAB) bond angle.

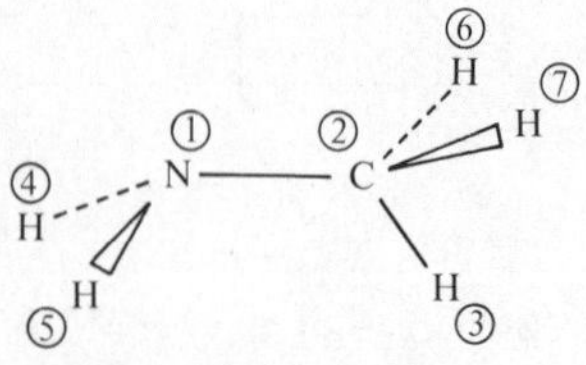

Fig. 5.13.2. Methylamine, showing the numbering of the atoms for the Z-matrix

Table 5.13.1. Z-matrix for methylamine

Atomic number of atom X	Label of atom A	Bond length XA/nm	Label of atom B	Angle XAB/°	Label of atom C	Dihedral angle XAC/°
7						
6	1	0.1474				
1	2	0.1093	1	109.5		
1	1	0.1014	2	112.2	3	300.5
1	1	0.1014	2	112.2	3	59.5
1	2	0.1093	1	109.5	3	120.0
1	2	0.1093	1	109.5	3	240.0

Row 4 refers to a hydrogen atom bonded to the nitrogen. The dihedral angle is found from a *Newman projection* of the molecule, as shown in Fig 5.13.3. All the dihedral angles must be taken in the same sense—in this case clockwise.

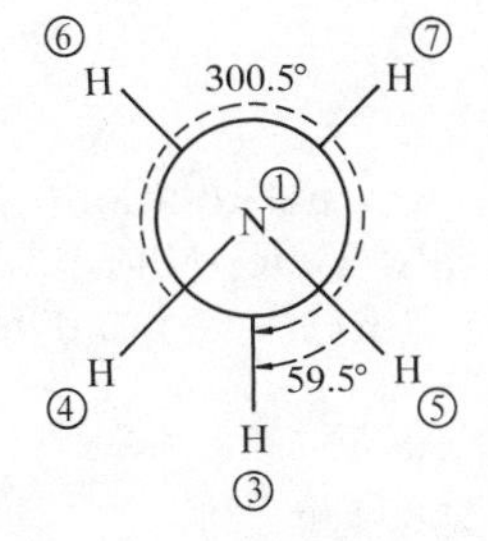

Fig. 5.13.3. Newman projection of the methylamine molecule

(*vi*) *Graphics*

The co-ordinates of the atoms are already available from BUILDZ and, if a graphics facility is available, can be displayed as a two dimensional stick-model of the nuclear skeleton. If the program is being used to calculate the variation in energy with conformation, the nuclear skeleton can be seen to rotate as the calculation proceeds. For example, for a 30° rotation about the C–C axis in ethane, the display will change as shown in Fig. 5.13.4.

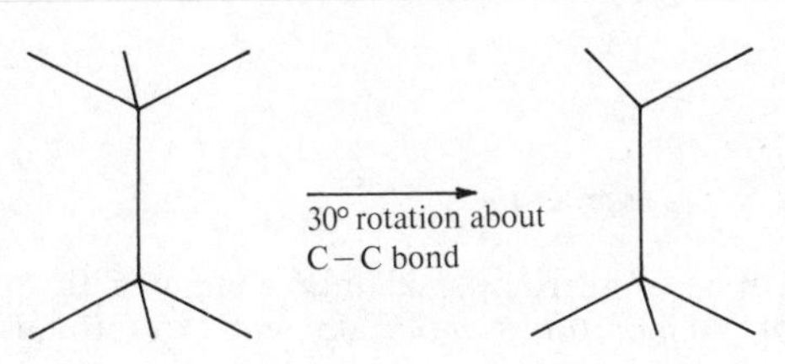

Fig. 5.13.4. Graphics display of a 30° rotation about the C—C bond in ethane

Apparatus The experiment requires a main-frame, mini- or fast micro-computer, linked to a printer and, preferably, a graphics terminal.

Procedure Calculate the Z-matrices of the carbon dioxide and ethane molecules (and any others you have been asked to study*). The C—O bond length in carbon dioxide is 0.162 nm. In ethane, the C—H bond length is 0.109 nm, the C—C bond length is 0.154 nm, and the HCH bond angle is 109.5°.

Run the program as instructed.*

Calculation *A*

Calculate the energy of the carbon dioxide molecule as a function of bond angle θ. Calculate the energies for three different values of θ, such as 180°, 170° and 160°. Fit the energies to a parabola of the form $E = a\theta^2 + b\theta + c$. Calculate the force constant, $d^2E/d\theta^2$, of the bending vibration.

B

Calculate the barrier to internal rotation about the C—C bond in ethane at 30° intervals. Determine how the energy of the rotational barrier varies for C—C bond lengths in the range 0.12–0.18 nm.

Results Force constant for CO_2: 1680 N m^{-1}.

Accuracy

Theoretical calculations are often much less reliable than experiment. This technique gives an accurate result for the barrier to internal rotation in ethane. You will discover, however, that it does not give an accurate answer for the force constant in CO_2.

Extended Hückel theory predicts that the water molecule is linear!

Comment The extended Hückel theory is one of the most widely used quantum mechanical methods for calculating conformational energies, geometries, and orbital energies and ionization potentials. Inorganic chemists refer to it as the Wolsfberg–Helmholtz method.

Technical notes

Computer program

On a fast micro or small mini computer, the program takes of the order of 20 min for a molecule with 15 orbitals, and about 50 min for molecule with 50 orbitals. Program in FORTRAN available from the author.

References

[1] *Ab initio molecular orbital calculations for chemists.* W. G. Richards and D. L. Cooper. Clarendon Press, Oxford, 2nd edn (1983), p. 52.
[2] *Quantum pharmacology.* W. G. Richards. Butterworths, London, 2nd edn (1983), p. 165.

Section 6
Structure of macromolecules

Physical chemists have traditionally shown an interest in the structure of macromolecules, and this section contains three experiments which illustrate how such structures may be determined. The purpose of the experiments is not to dwell on the associated theory, which is somewhat involved, but to give a feel for the techniques used and the magnitude of the parameters obtained.

Experiment 6.1 illustrates the simplest technique for determining the structure of a polymer, which is to measure the viscosity of dilute solutions. The results are used to find the viscosity-average molar mass of the sample. The experiment also illustrates how a range of other parameters may be found by use of additional information about the sample.

In Expt 6.2, the amount of light scattered from polymer solutions and gels is used to find further structural constants. The method is an absolute one, unlike the measurement of viscosity. The skills demanded of the student are perhaps the greatest of all in this book, for the samples must be prepared entirely free of dust. Fortunately, however, the analysis of the results by means of a Zimm plot is very robust, in that a considerable number of measurements are extrapolated down to a single intercept, and even poor results yield sensible parameters. Part A of the experiment requires a commercial light-scattering photometer, whereas the second part, involving the study of gel-sol transitions, could be performed on a much simpler instrument.

Finally, in Expt 6.3, we study solid polymers, and investigate their transition temperatures. The technique of the experiment is very easy, only requiring the operation of an audiofrequency bridge. The theory is more complex, but again, the purpose of the experiment is to illustrate the nature of the results, which in this case give important information about the temperature range over which a solid polymer is strong without being brittle.

Although osmometry is a common technique in structure determinations, it is not featured here because we have found that the maintenance of the membrane is very difficult in a teaching laboratory.

6.1 Conformation of macromolecules by dilute-solution viscometry

The viscous drag created by the presence of random-coil polymers in a flowing solvent is a measure of the size, not the mass, of the polymer molecules. Measurement of the dilute-solution viscosity of polymer solutions provides one of the most easily obtained and widely used items of information about the molecular structure of the samples. The theory employed in this experiment is complicated, but only because many parameters are obtained from a few simple measurements. If only the viscosity-average molar mass $\langle M_V \rangle$ is required, as defined below, then the theory is simple.

Pointed brackets are used to indicate the mean value of the term inside them.

In this experiment we measure the viscosity of polystyrene in cyclohexane and toluene at 34°C and concentrations in the range 1–5 g dm^{-3}. From these measurements a large number of properties of polystyrene are determined.

Theory[1]

(i) The conformation of polymer molecules in solution

Let us consider simply the backbone of a long-chain polymer molecule such as polystyrene. Since the backbone is flexible, there are an infinite number of shapes of *conformations* which this backbone can assume. When a polymer is dissolved in a *good solvent,* e.g. polystyrene in toluene, interactions between the solvent and segments of the backbone are energetically more favourable than segment–segment interactions. The macromolecules will accordingly expand to maximize solvent–segment interactions (consistent with their attainment of minimum Gibbs free energy), as shown in Fig. 6.1.1.

In a *poor solvent,* segment–segment interactions are more favourable than segment–solvent interactions, and the molecule correspondingly assumes a much more compact shape as shown. Now it happens that the non-zero volume of the segments of a polymer molecule cause the molecule to extend more than its *unperturbed dimensions* (indicated below by a subscript 0), which describe the shape of the backbone one would predict simply on the basis of bond lengths, bond angles and restrictions

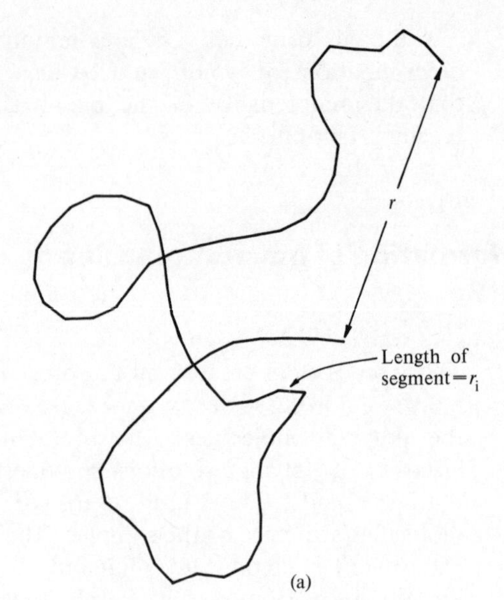

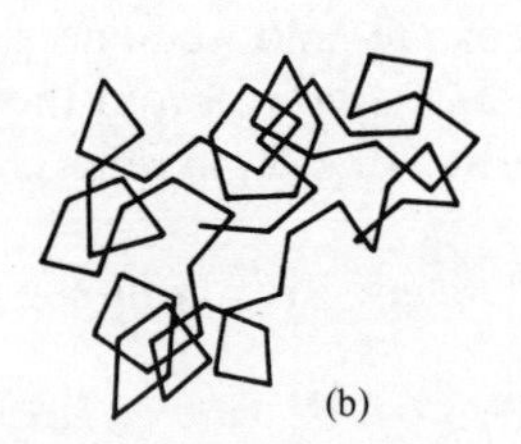

Fig. 6.1.1. Conformation of a long-chain polymer (a) in a good solvent and (b) in a poor solvent

to free rotation. When the dimensional contraction resulting from poor solvency exactly cancels the increase in size due to non-zero segmental volumes, the solvent is termed a *theta* (θ) *solvent.* An example is polystyrene in cyclohexane at 34 °C. As we expect, the viscosity of polystyrene in the good solvent toluene is greater than its viscosity in the theta solvent cyclohexane, because of the difference in conformation.

We must now attempt a mathematical description of the conformation of a polymer. Polymers such as polystyrene which have unbranched chains are often characterized by their root-mean-square (r.m.s.) end-to-end distance defined by

$$\langle r^2 \rangle^{1/2} = \left(\frac{\sum_i n_i r_i^2}{\sum_i n_i} \right)^{1/2}, \tag{6.1.1}$$

where n_i is the number of molecules each having an end-to-end length r_i (Fig. 6.1.1). This definition is clearly unsatisfactory for branched polymers, which are multi-ended, and so resort is also made to the r.m.s. radius of gyration given by

$$\langle \rho^2 \rangle^{1/2} = \left(\frac{\sum_i n_i \rho_i^2}{\sum_i n_i} \right)^{1/2}. \qquad (6.1.2)$$

Here ρ_i is the radius of gyration of the molecule in the ith conformation:

$$\rho_i^2 = \frac{\sum_j m_{ij} x_{ij}^2}{\sum_j m_{ij}}, \qquad (6.1.3)$$

where x_{ij} is the distance of the mass element m_{ij} from the centre of mass of the ith conformation.

For a linear random coil macromolecule such as polystyrene, the unperturbed r.m.s. radius of gyration $\langle \rho^2 \rangle_0^{1/2}$ is related to $\langle r^2 \rangle_0^{1/2}$ through

$$\langle \rho^2 \rangle_0^{1/2} = (\langle r^2 \rangle_0 / 6)^{1/2}. \qquad (6.1.4)$$

A simple model of a linear macromolecule is of a volumeless, freely jointed chain with many links (total n) each of length l. The r.m.s. end-to-end length $\langle r^2 \rangle_{0f\theta}^{1/2}$ of such a random flight chain is given by

$$\langle r^2 \rangle_{0f\theta}^{1/2} = n^{1/2} l. \qquad (6.1.5)$$

This prediction can be made by the theory used to describe a two-dimensional random walk in Brownian Motion, Expt 7.1.

The three subscripts refer respectively to unperturbed dimensions (explained above), free rotation (i.e. any value of the bond rotation angle ϕ), and any valence angle θ between one segment and the next ($\theta = 180°$ for a straight chain).

For a real macromolecule, allowance must be made for at least three additional properties of the polymer backbone: for the bond angles, for restrictions to free rotation about bonds, and for the non-zero volume of the chain segments.

First, if the valence angle θ has a constant value along the backbone in the range 0°–180°, then the r.m.s. end-to-end length becomes

$$\langle r^2 \rangle_{0f}^{1/2} = n^{1/2} l \left(\frac{1 - \cos\theta}{1 + \cos\theta} \right)^{1/2}. \qquad (6.1.6)$$

The influence of restricted valence angle is illustrated in Fig. 6.1.2.

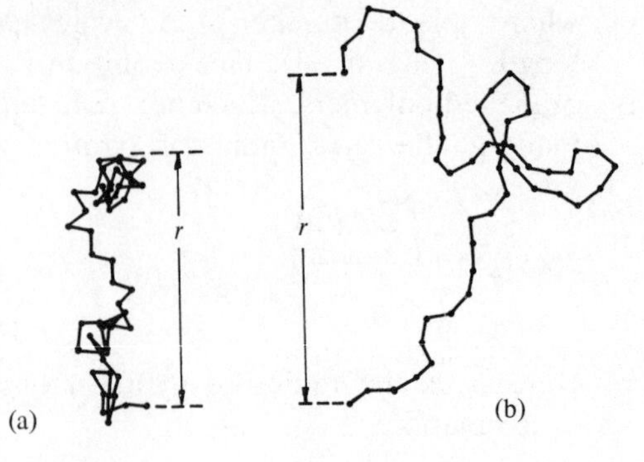

Fig. 6.1.2. The end-to-end length r of a fifty link chain in a two-dimensional random walk for (a) a freely jointed chain and (b) a chain with the valence angle restricted to $90° \leq \theta \leq 180°$

Second, limitations to free rotation can be expressed by a steric factor σ, giving

$$\langle r^2 \rangle_0^{1/2} = \sigma \langle r^2 \rangle_{0f}^{1/2}. \tag{6.1.7}$$

If the bond rotational potentials were truly independent (i.e. the conformation of bond i was independent of those of bonds $(i-1)$ and $(i+1)$), then it can be shown that $\sigma = [(1+\langle \cos\phi \rangle)/(1-\langle \cos\phi \rangle)]^{1/2}$ where the statistical averages are taken over all the angles ϕ to which the bond rotations are restricted. In reality, the bond rotational potentials of a polymer chain are strongly interdependent. The *ab initio* calculation of σ is difficult and so σ is usually determined experimentally. Its value often lies in the range 1.5 to 3.

Finally, as mentioned earlier the non-zero segmental volumes of real molecules expand their sizes to larger than that calculated from $\langle r^2 \rangle_0^{1/2}$. One index of this long-range intramolecular expansion is α, where

$$\langle r^2 \rangle^{1/2} = \alpha \langle r^2 \rangle_0^{1/2}. \tag{6.1.8}$$

What will be the values of α for a theta solvent and for a good solvent?

(ii) *Viscosity of dilute polymer solutions*

In this section we shall use the parameters η_0 = viscosity of the solvent, η = viscosity of the solution, and c_2 = polymer concentration. The viscosity functions derived from these are listed in Table 6.1.1. It can be seen that the intrinsic viscosity $[\eta]$ has units of reciprocal concentration ($dm^3\ g^{-1}$).

The presence of dissolved macromolecules in a solution increases the energy dissipation during laminar flow, relative to that for the pure solvent. This is a consequence of the perturbation of the solvent flow by the polymer segments. Flexible linear

Table 6.1.1. Nomenclature of solution viscosity

Common name	Recommended name	Symbol and defining equation
Relative viscosity	Viscosity ratio	$\eta_r = \eta/\eta_0 = t/t_0$
Specific viscosity	—	$\eta_{sp} = \eta_r - 1 = (\eta - \eta_0)/\eta_0 = (t - t_0)/t_0$
Reduced viscosity	Viscosity number	$\eta_{red} = \eta_{sp}/c$
Inherent viscosity	Logarithmic viscosity number	$\eta_{inh} = (\ln \eta_{red})/c$
Intrinsic viscosity	Limiting viscosity number	$[\eta] = (\eta_{sp}/c)_{c=0} = [(\ln \eta_{red})/c]_{c=0}$

macromolecules in solution behave hydrodynamically as though they were solvent-impermeable cores with peripheral solvent-permeable segments.

The Kirkwood–Riseman theory for the intrinsic viscosity of such macromolecules, of molar mass M_2, gives the expression

$$[\eta] = \Phi\langle r^2\rangle^{3/2}/M_2, \tag{6.1.9}$$

where Φ is a universal constant. The best value of Φ predicted theoretically for a polymer of homogeneous molar mass is 2.84×10^{26} g mol^{-1}, where $[\eta]$ is expressed in dm^3 g^{-1} and $\langle r^2\rangle^{3/2}$ is in m^3. Experimentally Φ has been found to lie in the range $(2.0\text{–}2.5) \times 10^{26}$ for unfractionated and partially fractionated polymers, with perhaps the best value to date being $2.1(\pm 0.2) \times 10^{26}$. Equation (6.1.9) is transformed into the so-called Flory–Fox formation by inserting eqn (6.1.8):

$$[\eta] = \Phi\alpha^3\langle r^2\rangle_0^{3/2}/M_2 = K\alpha^3(M_2)^{1/2}, \tag{6.1.10}$$

where $K = (\Phi\langle r^2\rangle_0^{3/2}/M_2^{3/2})$ is a constant for a given polymer.

By measuring the intrinsic viscosity $[\eta]_\theta$ of a macromolecule in a θ-solvent ($\alpha = 1$), its unperturbed r.m.s. end-to-end length can be calculated from

$$\langle r^2\rangle_0^{1/2} = ([\eta]_\theta M_2/\Phi)^{1/3}. \tag{6.1.11}$$

The expansion factor in good solvents may be determined from viscosity measurements using the relation

$$\alpha = ([\eta]/[\eta]_\theta)^{1/3}. \tag{6.1.12}$$

Equation (6.1.10) suggests, and experiments confirm, that the molar mass can be estimated in a θ-solvent from an expression of the form

$$[\eta]_\theta = K^*_\theta (M_2)^{1/2}. \tag{6.1.13}$$

The value of the constant K^*_θ for polystyrene in cyclohexane at 34 °C is found experimentally to be $8.2(\pm 0.2)\times 10^{-5}\ \mathrm{dm^3\,g^{-3/2}\,mol^{1/2}}$ for polymers in the molar mass range of $(2\text{–}400)\times 10^4\ \mathrm{g\,mol^{-1}}$. The more general form of eqn (6.1.10) can also be recast into the empirical Mark–Houwink relationship,

$$[\eta] = K^* M_2^a, \tag{6.1.14}$$

where normally $0.5 \le a < 1$ and both a and K^* are constants for a given molar mass range of a particular polymer. Results for linear polystyrene in the molar mass range $(1\text{–}5)\times 10^5\ \mathrm{g\,mol^{-1}}$ in toluene at 34 °C yield $K^* = 1.15\times 10^{-5}$, and $\alpha = 0.72$.

Rigorously *mono-disperse* polymers have molecules which are all of the same mass; in practice, the polymerization processes used in their manufacture yield *poly-disperse* polymer molecules with a range of molar masses.

These theories are best applied to rigorously monodisperse polymers. When applied to polydisperse systems, rather complicated averages result. For example, the viscosity average molar mass $\langle M_V\rangle$ derived from the application of the Mark–Houwink equation is

$$\langle M_V\rangle = \left(\frac{\sum_i n_i M_i^{1+a}}{\sum_i n_i M_i}\right)^{1/a}, \tag{6.1.15}$$

where n_i is the number of macromolecules of molar mass M_i. $\langle M_V\rangle$ obviously varies with the exponent a but since $0.5 \le a < 1$, it lies somewhere between the number average ($\langle M_N\rangle = \sum_i n_i M_i / \sum_i n_i$) and the mass average ($\langle M_M\rangle = \sum_i n_i M_i^2 / \sum_i n_i M_i$) molar mass. The value of $\langle M_V\rangle$ measured in good solvents is often within 20 per cent of the value of $\langle M_M\rangle$.

The reduced viscosity at non-zero polymer concentrations c may be expanded in the binomial form

$$\eta_{\mathrm{red}} = [\eta] + k'[\eta]^2 c, \tag{6.1.16}$$

where the so-called Huggins constant k' is commonly ~0.35–0.40 for uncharged flexible polymers in good solvents.

Apparatus

Two types of glass capillary viscometer are in common use for the characterization of dilute polymer solutions. The simpler Ostwald or Cannon–Fenske viscometer, Fig. 6.1.3, is a constant volume device. It comprises a fine capillary about 10 cm long with a bulb above. There are calibration marks at positions (c), (b), and (f). Enough liquid is introduced into the wider limb at

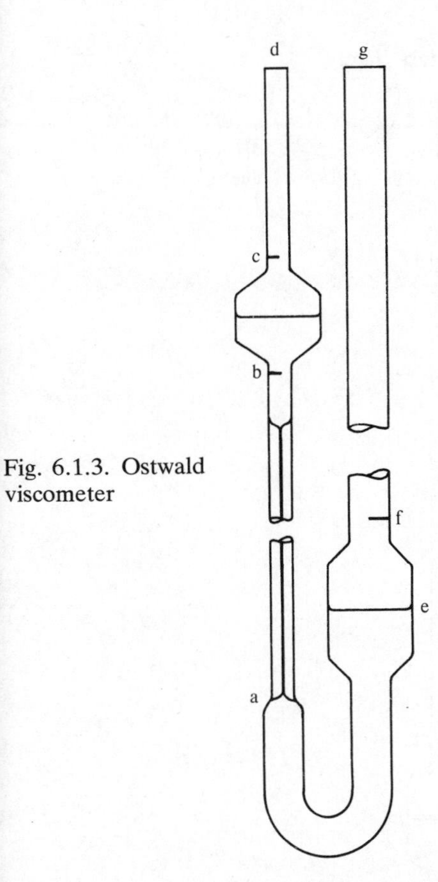

Fig. 6.1.3. Ostwald viscometer

(g), using a pipette, to fill the viscometer to point (f). By means of a pipette filler attached at (g), the liquid is blown up so that its upper surface is above (c). The liquid is kept in this position by placing a finger over the tube at point (d). The liquid is then released, and the time t is measured for it to flow back from mark (c) to (b) under a pressure caused by its own weight.

In an Ubbelohde viscometer, Fig. 6.1.4, there is a side-arm (b) from the lower bulb (a) which keeps the bulb at atmospheric pressure during a run. This ensures that the effluent from the capillary tube flows down the walls of the bulb (a) in a manner which is independent of the liquid level in the main reservoir, and thus of the total volume of solution. The mode of action is said to employ a *suspended level.* It is convenient in that it allows the polymer solution to be diluted simply by adding more solvent. To reset the viscometer, the 3-way taps shown in the figure are repositioned and air, dried by silica gel, is blown through both (b) and (c). By contrast the Cannon–Fenske vis-

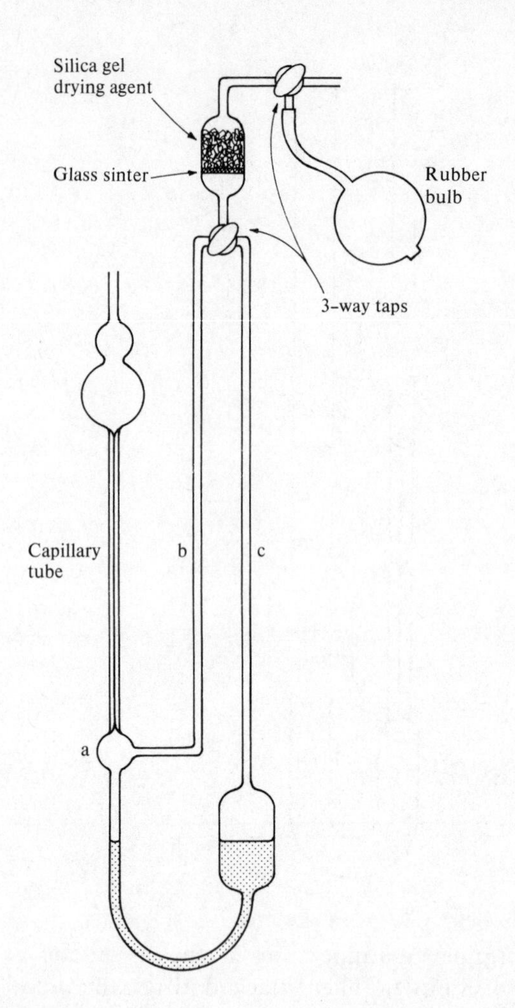

Fig. 6.1.4. Ubbelohde viscometer

cometer has to be emptied of solution, cleaned, and refilled with the same volume of a more dilute solution.

Ubbelohde viscometers are often observed visually, although automated devices are available which measure the efflux time by means of photocells at the upper and lower marks, dilute the solution by addition of solvent from an automatic titrator, blow the solution up through the capillary tube again, and repeat the cycle as many times as required.

For both types of viscometer,

$$\eta = At\boldsymbol{\rho} - B\boldsymbol{\rho}/t^2, \tag{6.1.17}$$

where η is the viscosity of the solution of density $\boldsymbol{\rho}$, t the efflux time, and A, B are constants, with appropriate units, which refer to the particular viscometer used. Similarly for the standard liquid (water):

$$\eta' = At'\rho' - B\rho'/t'^2. \quad (6.1.18)$$

Viscometers are designed so that, for efflux times greater than 100 s or so, the second terms are negligible. Therefore:

$$\eta/t\rho = \eta'/t'\rho', \quad (6.1.19)$$

from which η may be calculated. η_0, the viscosity of the particular solvent, may be found in a similar way.

The viscometer must be mounted in a water bath at 34 °C, and maintained at a constant temperature to within ±0.02 K.

Procedure[2]

The dissolution of polystyrene in cyclohexane is facilitated by a short exposure to low intensity ultrasonic radiation.

Make up solutions of accurately known concentrations of about 5 g dm^{-3} of polystyrene in toluene and polystyrene in cyclohexane.

Rinse the viscometer with one of the solvents and let it drain. Then place the viscometer in the constant temperature bath, securely fastened and with the upright tubes accurately vertical.

With a pipette or syringe, pipette the required volume* of solvent (e.g. 10 cm^3) into the viscometer. Wait at least 10 min until thermal equilibrium has been achieved.

Now bring the liquid level in the viscometer above the upper graduation mark. This can be done by attaching a dust-free tube to the non-capillary arm (tubes (b) and (c) in the case of the Ubbelohde viscometer) and applying pressure with a rubber bulb or, ideally, with low-pressure filtered nitrogen from a supply. Alternatively, in the case of the Cannon–Fenske viscometer, attach a dust-free tube to the capillary tube and apply suction with a pipette filler.

Allow the liquid to drain down the capillary tube. Start the time as the meniscus passes the upper graduation mark, and stop it as the meniscus passes the lower mark.

Make repeated determinations of the efflux time for each solution. Obtain two readings which agree to within 0.1 s or 1 per cent of their mean, whichever is larger, and use their mean value for the calculations. If agreement cannot be obtained, it is likely that the variation results from foreign material in the capillary tube or inadequate temperature control.

If using an Ostwald or Cannon–Fenske viscometer, remove it from the bath, empty it, and allow it to drain. If a supply of filtered air or nitrogen is available, blow it dry. Measure the efflux times of the 5 g dm^{-3} stock solutions. Also measure the times for at least two other solutions for each solvent in the concentration range 1–5 g dm^{-3}, prepared by quantitative dilution of the stock solutions with the appropriate solvent. Between each run dismantle and clean the viscometer as above, rinsing it with solvent to ensure that no solution remains.

Experiments using an Ubbelohde viscometer are carried out over this same concentration range. For a series of solutions with the same solvent, the dilution can be carried out by pipetting appropriate solvent volumes directly into the viscometer. Mix each solution well by closing the tube above (a) with your finger, and applying pressure alternately through tubes (b) and (c).

Leave the viscometer clean and dry at the end of the experiment.

Calculation

(i) Plot a graph of reduced viscosity (Table 6.1.1) against polymer concentration, showing the results for the two sets of solutions on the same graph. Draw straight lines through the points for each solvent and extrapolate them to zero concentration to obtain the intrinsic viscosity of polystyrene in toluene ($[\eta]$) and of polystyrene in cyclohexane ($[\eta]_\theta$).

(ii) Calculate the viscosity average molar mass $\langle M_V \rangle$ from both equations (6.1.13) and (6.1.14), by assuming that $\langle M_V \rangle$ measured in cyclohexane $= M_2$.

(iii) Calculate an average value for $\langle r^2 \rangle_0^{1/2}$ from eqn (6.1.11). The best value of M_2 to insert into this equation is $\langle M_N \rangle$ but since it is usually unknown, the value of $\langle M_V \rangle$ measured in cyclohexane suffices. The average of $\langle r^2 \rangle_0^{1/2}$ so calculated differs by less than 20 per cent from that derived from $\langle M_N \rangle$.

(iv) Calculate an average value for $\langle \rho^2 \rangle_0^{1/2}$ from eqn (6.1.4).

(v) Estimate an average value for $\langle r^2 \rangle^{1/2}$ in toluene using eqn (6.1.9). Again the value of $\langle M_V \rangle$ measured in cyclohexane is used for M_2.

(vi) Calculate an average value of α in toluene at 34 °C from eqn (6.1.12).

(vii) Calculate the average number $\langle n \rangle$ of styrene monomer units ($C_6H_5.CH{=}CH_2$) per polystyrene molecule, using the value of $\langle M_V \rangle$ in cyclohexane.

(viii) Calculate a value for $\langle r^2 \rangle_{0f}^{1/2}$ from eqn (6.1.6), taking the backbone C—C bond length contributed by each monomer unit as 0.154 nm and $\theta = 109°$. Carry out the calculation by using either an appropriate geometric formula, or a scale drawing.

(ix) Estimate σ from eqn (6.1.7).

(x) Calculate an average value for the fully-stretched, planar zig-zag length of the macromolecule using $\langle n \rangle$. What percentage of this contour length is $\langle r^2 \rangle_0^{1/2}$?

(xi) Determine the least-squares value of k' in toluene using eqn (6.1.16).

Results[1]

For a sample with $\langle M_V \rangle \sim 250\,000$ g mol^{-1}, $\langle \rho^2 \rangle_0^{1/2}$ is ~35 nm. Literature values of α (for a well-fractionated sample) and σ are 1.25 and 2.4 respectively.

Accuracy

No reliance can in general be placed on student values of k' determined in θ-solvents.

There are many factors which can effect the accuracy of viscosity determinations, including the solvation or entanglement of the polymer chains.[3] One of the main difficulties is that the viscosities are dependent on the shear rate, and more accurate determinations use devices such as slowly rotating Cartesian divers.[4]

Comment

It is clear that viscosity measurements can yield a great deal of information about the structure of polymer molecules in solution. They can also be used to measure the rate of polymerization.[5] However, the method is a relative rather than an absolute one. To obtain the comprehensive results in this experiment much additional information has had to be employed, including various empirical constants and the fact that cyclohexane is a theta solvent for polystyrene at 34 °C. To obtain absolute information directly it is necessary to employ a method such as light scattering, as described in the next experiment.

Technical notes

Apparatus

Ostwald, Cannon–Fenske or Ubbelohde viscometer with solvent flow time between 1 and 3 min; water bath controlled at 34 ± 0.01 °C; stopwatch; broad molar mass sample of polystyrene, either commercial, or prepared from styrene;[1,5] toluene; cyclohexane; [supply of low-positive-pressure filtered air or nitrogen]; [source of low intensity ultra-sonic radiation].

References

[1] *Conformation of macromolecules—a physical chemistry experiment.* D. H. Napper. *J. Chem. Educ.* **46,** 305 (1969).
[2] *Experiments in polymer science.* E. A. Collins, J. Bareš and F. W. Billmeyer Jr. John Wiley, New York (1973), pp. 147 and 394.
[3] *Experimental methods in polymer chemistry.* J. F. Rabek. Wiley Interscience, Chichester (1980), p. 130.
[4] *A rotating Cartesian–Diver viscometer.* S. J. Gill and D. S. Thompson, *Proc. Nat. Acad. Sci.* **57,** 562 (1967).
[5] *Polymerization kinetics and viscometric characterization of polystyrene.* J. H. Bradbury. *J. Chem. Educ.* **40,** 465 (1963).

6.2 Properties of polymer solutions by light scattering photometry[1]

Many characteristics of polymers in solutions and gels may be determined by measuring the amount of light which is scattered by their molecules. Primarily, light scattering studies provide an absolute method of measuring molar masses. However, because polymers are created by the successive addition of smaller units to form a chain, the molar masses of molecules in a particular sample always vary.[2] The term which is measured is then some average over the molar mass distribution. Light scattering studies yield a quantity known as the *mass average molar mass* M_M:

$$M_M = \sum_i n_i M_i^2 \Big/ \sum_i n_i M_i \tag{6.2.1}$$

where n_i is the number of molecules of molar mass M_i, and $\sum_i$ represents the sum over the mass distribution.

Other properties which may be studied by light scattering concern the various levels of structure of polymers. The *primary structure* of a polymer is concerned with the type and number of atoms in a polymer chain. If, as in polystyrene, there is only one chain conformation, then we may refer to the *secondary structure* as the mode and degree of coiling of each chain. *Tertiary structure* concerns interactions between different parts of the same molecule. No information on the tertiary structure of polystyrene is obtained from this experiment, and in order to derive its secondary structure, we need to use the results of other experiments which show that the molecules are in the form of random, unkinked coils. *Quaternary structure* arises from the interactions of neighbouring molecules.

This experiment is in two parts. Part A involves the measurement of the mass average molar mass of polystyrene molecules in toluene, and also their degree of coiling. Part B is an investigation of the changes in quaternary structure which occur during the gel/sol transition ('melting') of gelatin.

Theory

A(i) Light scattering[3]

The classical theory of light scattering developed by Rayleigh, based upon the electromagnetic theory of Maxwell, adequately describes most light scattering encountered in polymer science. Rayleigh assumed a scattering volume, small compared to the wavelength of light, filled with a homogeneous continuum. The oscillating electric field of the incident radiation induces in the

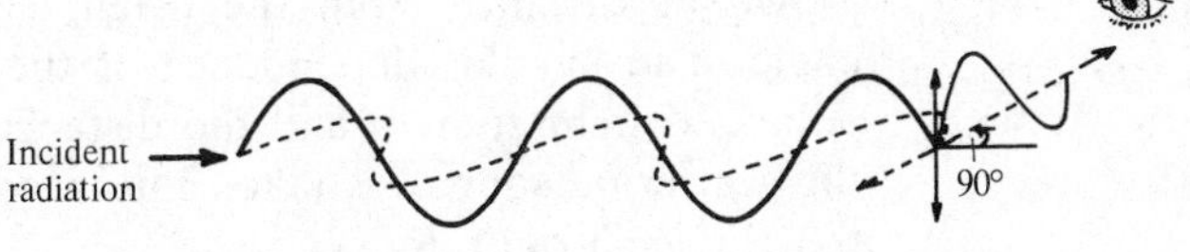

Fig. 6.2.1. Scattering of vertically and horizontally polarized light

continuum an oscillating dipole by polarizing charges in the continuum. This induced dipole, oscillating in phase with the electric field of the incident light, sets up its own oscillating field and becomes a secondary source of low intensity radiation. If the scattering volume is small compared to the amplitude and wavelength of the electric field of the incident light, then to a good approximation the electric field in the scattering volume is uniform at any instant. Then the dipole induced in the scattering volume is simply the product of the polarizability of the continuum and the field strength of the incident field at the time. The polarizability can be expressed in the limit of high frequency in terms of the refractive index of the continuum (p. 186). Since the amplitude of the scattered light is proportional to the polarizability of the scattering particle, and consequently to its mass, the contributions of the larger particles are greater than those of the smaller, and it is the mass average relative molar mass which is obtained from light scattering studies.

Consider the case of vertically and horizontally polarized light being scattered by a molecule and observed at 90° to the incident beam, as shown in Fig. 6.2.1.

The vertically polarized electric field of the incident light causes vertical polarization of the charge distribution in the sample molecule which will re-emit light at 90°, as well as at other angles. The horizontally polarized component, however, causes excitation parallel to the line of observation which will not scatter light in this direction. These considerations may be extended to other angles, from which it follows directly that the angular dependence is as shown in Fig. 6.2.2.

The diagram is a graph in polar co-ordinates, intensity being

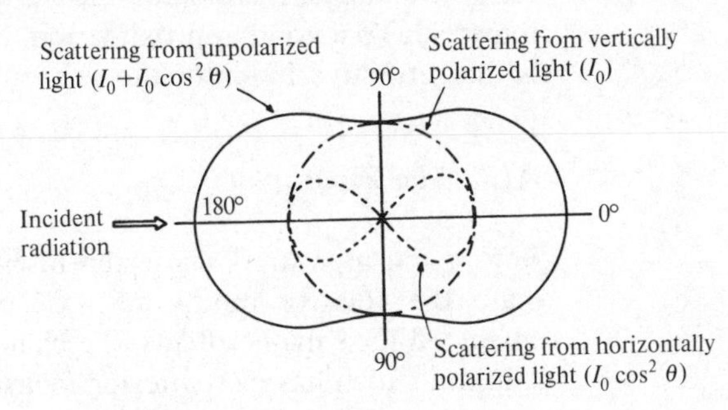

Fig. 6.2.2. Variation of light scattering intensity with angle of observation

shown as distance from the origin at a particular scattering angle. The angular dependence of the scattering from an unpolarized light source, and the distance r of the detector from the scattering source, is taken into account in a term known as the *Rayleigh ratio*, R_θ:

$$R_\theta = \frac{r^2 i_\theta}{I_0(1+\cos^2\theta)}. \tag{6.2.2}$$

Note that the units of the Rayleigh ratio are length^{-1}; it represents the fraction of light scattered at an angle θ per unit path length through the system.

In this equation i_θ is the intensity of the scattered light per unit volume of solvent when observed at a distance r from the solvent and at an angle θ between the incident and scattering rays, and I_0 is the intensity of the unpolarized incident beam.

For small isotropic particles, R_θ is related to the mass average relative molar mass, M_M, of the solute by the Debye equation:

$$\frac{Kc}{R_\theta} = \frac{1}{M_M P(\theta)} + 2A_2 c + \ldots \tag{6.2.3}$$

A_2 is the second virial coefficient of the scattering equation, which characterizes deviations from ideality due to interactions between solute and solvent. It is identical with a corresponding term obtained from osmotic pressure measurements. Terms with higher powers of c are ordinarily negligible. $P(\theta)$ is a term which corrects for interference from scattering centres located in different parts of molecules large with respect to the wavelength of the incident light.

The constant K is a measure of the amount of light scattering which arises from the characteristics of the sample itself and the wavelength of the incident beam:

The enormous increase in K with decreasing λ indicates that the blue end of the visible spectrum is scattered much more efficiently than other wavelengths, and explains the blue colouration of the sky.

$$K = \frac{2\pi^2 n_0^2}{L\lambda^4}\left(\frac{\mathrm{d}n}{\mathrm{d}c}\right)^2. \tag{6.2.4}$$

Here n is the refractive index of the solution, λ is the wavelength in the medium of the incident beam, c is the concentration of the polymer solution and $\mathrm{d}n/\mathrm{d}c$ is the specific refractive index increment (measured in a differential refractometer). To a good approximation, $\mathrm{d}n/\mathrm{d}c = (n - n_0)/c$, where n_0 is the refractive index of the solvent.

A(*ii*) *The Zimm plot*[3]

Before the measured scattering intensities can be used to estimate the relative molar mass of polystyrene, they must be corrected for solvent effects, the change with θ of the volume of solution which has its scattering measured, and the angle depen-

Table 6.2.1. Angle functions for calculation of Zimm plot

	30°	37.5°	45°	60°	75°	90°
$\sin\theta/(1+\cos^2\theta)$	0.286	0.374	0.471	0.693	0.905	1.000
$\sin^2\theta/2$	0.067	0.103	0.146	0.250	0.371	0.500
	105°	120°	135°	142.5°	150°	
$\sin\theta/(1+\cos^2\theta)$	0.905	0.693	0.471	0.374	0.286	
$\sin^2\theta/2$	0.629	0.750	0.854	0.897	0.933	

dence of the scattering intensity discussed earlier. These may be calculated respectively by: (i) subtracting the solvent intensity i° from the solution intensity i at each angle, (ii) multiplying the intensities by $\sin\theta$ and (iii) dividing them by $(1+\cos^2\theta)$. The resulting intensities are termed α.

In terms of experimental variables, eqn (6.2.3) may be expressed as

$$\frac{Kc}{R_\theta}=\frac{K\,i^\circ_{90}}{R^\circ_{90}}\left(\frac{c}{\alpha}\right)^{\theta=0}_{c=0}=\frac{1}{M_\mathrm{M}P(\theta)}+2A_2c+\ldots \tag{6.2.5}$$

where i°_{90} is the scattering intensity, and R°_{90} the Rayleigh ratio, for the solvent at 90°.

$(c/\alpha)^{\theta=0}_{c=0}$ is the intercept of c/α obtained by independently extrapolating θ and the concentration c to zero. The quantity may be determined by use of a *Zimm plot* in which c/α is plotted against $(\sin^2\theta/2+kc)$. Figure 6.2.3 shows a Zimm plot for polystyrene in benzene. k is an arbitrary parameter chosen to spread the points along the abscissa, and a correct choice of k (by trial and error) is essential if sensible extrapolations are to be made. Note that $k\neq K$.

The polystyrene solutions studied in this experiment contain linear coiling chain molecules with a distribution of degrees of polymerization. It may be shown that for such systems the ratio of the initial slope of the solvent line to the intercept is given by:[4]

$$\frac{\text{initial slope }(c=0)}{\text{intercept}}=\frac{8\pi^2}{3\lambda^2}\langle r^2\rangle_z. \tag{6.2.6}$$

$\langle r^2\rangle_z$ is the *z-average* mean square distance between the ends of the chain, defined by the equation

$$\langle r^2\rangle_z=\sum_i n_iM_i^2\langle r^2\rangle_i\Big/\sum_i n_iM_i^2, \tag{6.2.7}$$

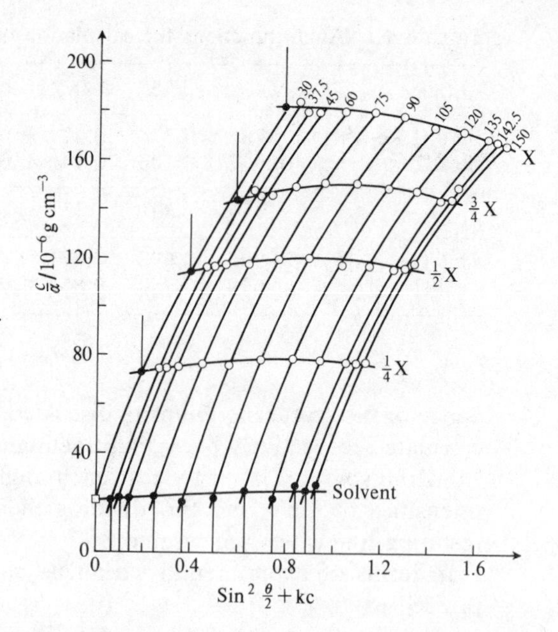

Fig. 6.2.3. Zimm plot for polystyrene in benzene[1]

where $(r^2)_i$ is the square of the end-to-end chain length of a molecule of mass i. The pointed brackets are used to indicate the mean value of the term inside them.

More commonly, polymer molecules are characterized in terms of their *radius of gyration* ρ (p. 291). For flexible coiled molecules such as those of polystyrene, it may be shown that:[5]

$$\langle\rho^2\rangle = \langle r^2\rangle_z/6. \qquad (6.2.8)$$

B. Gelatin and its properties

Gelatin is a *polyelectrolyte*, i.e. a macromolecular compound which contains many ionizable groups, such as —COOH and —NH_2, within the same molecule. The molecules are capable of being either positively or negatively charged according to whether positive ions (including H^+) or negative ions (including OH^-) are predominantly adsorbed. They also possess an *isoelectric point* at which the net charge on the particles is zero—positive and negative ions being adsorbed to an equal extent. In distilled water, large electrostatic effects between neighbouring polyelectrolyte chains occur. At the isoelectric point, which for gelatin occurs at pH 5.1, the maximum number of charges are available for these interactions, which also provide light scattering centres. The formation of aggregates in gelatin, as discussed below, can only be observed in solutions containing salt, which prevents these interactions masking the more subtle structural effects.

Light scattering experiments have shown that gelatin molecules are in the form of random coils when dissolved in water. Under certain conditions, such as a lowering of temperature, the molecules intertwine to enclose all the surrounding water molecules, and they form an easily deformable pseudo-solid known as a *gel.* A gel may be converted back to its liquid, or *sol,* form on heating, and in the case of gelatin this process is reversible.

We might expect that systems containing gel linkages would exhibit an increase in the amount of light they scatter with increasing concentration, as a result of the increasing size or number of the density fluctuations in the gel which form the scattering centres. However, an alternative structure for gelatin gels has been proposed.[6] It has been suggested that under certain conditions gelatin solutions contain aggregates which are cross-linked by very small crystallites. These aggregates show the characteristic of being increasingly ordered at high concentrations, whereupon the amount of light scattered is reduced and may actually decrease with concentration.

Apparatus

In a light scattering photometer a collimated light beam of known wavelength shines on to a sample cell. The light scattered by the sample, which is of low intensity, is measured by a sensitive photomultiplier which can either be rotated round the axis of the cell (Part A) or which is fixed at 90° to the incident beam (Part B). There is a mask in the apparatus which prevents the photomultiplier being damaged by exposure to the incident beam, and a shutter which must be used to prevent exposure to daylight when the sample chamber is opened.

The optical layout of a typical instrument[4,7] is shown in Fig. 6.2.4. The light source (a) is a water-cooled mercury vapour lamp. The beam passes through a heat filter (b), prism (c), lenses (d, e) and a slit (f). A green filter (g) absorbs all but the 546 nm green line of the mercury vapour emission (ring position Gr). A polarizer (h) may also be incorporated, but is not used in this experiment. The beam then passes through an adjustable slit (i). A flint-glass cylinder (j) in the beam scatters light to a standardizing photomultiplier (k) which compensates for changes in the intensity of the source by adjusting the high voltage supply to the measuring photomultiplier (s).

The beam passes through a polished window (l) into a vat (m) containing dust-free toluene. This has a similar refractive index to that of the glass sample cell (n) and almost entirely eliminates

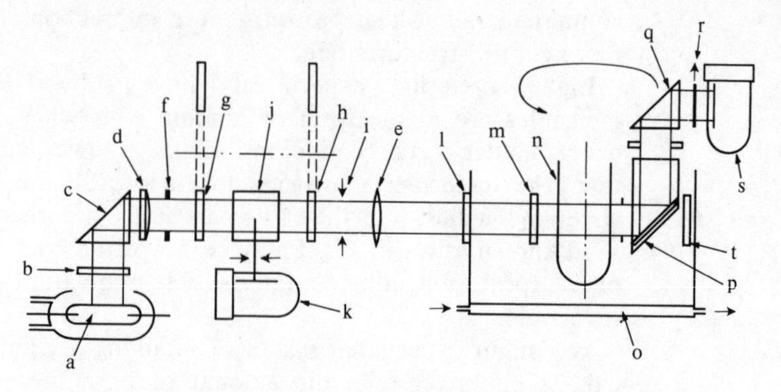

Fig. 6.2.4. Optical layout of a typical light scattering photometer

reflections and refraction from the face of the cell. It also allows the use of cells which are not optically perfect. The vat also contains an internal heater or a heat-exchanger (o) through which water is circulated from an external thermostatically controlled water bath. An air-gap reflection prism (p) reflects the scattered light via another prism (q) into the measuring photomultiplier (s) which has a manually operated shutter (r). (Do not confuse the slit and the shutter.) Unscattered light passes into a light trap (t)—an important component of the instrument.

The output from the photomultiplier is displayed on a meter with a range switch, and output to a chart recorder for reasons explained below.

Dust, the enemy of light scatterers, manifests itself as tiny flashes of light in the illuminated sample.

A periscope is incorporated in the instrument. This may be used for checking the position of the cell and for detecting dust in the sample.*

Procedure

Chromic acid is highly corrosive. Wear a laboratory coat and safety glasses, and use a filler when pipetting.

There are two special features of the experiment of which you should be mindful.

(i) The smallest amount of dust in the sample solution or on the walls of the cell will invalidate your results. To remove dust from the cell walls, they must have been soaked in chromic acid for at least twelve hours. The side walls of the cells must not be touched; they should be lifted by their base and holders. Dust is removed from the sample solution by a pressure filtration unit incorporating a 0.22 μm filter.

(ii) The properties of the solutions for both parts of the experiment depend on their history. For meaningful results all the solutions should be stirred and left to stand for about the same length of time and at the same temperature. The properties of polystyrene solutions (part A) vary with time due to the settling of the polymer, and samples should always be pipetted from the top half of each stock solution. The dependence on thermal history of the gelatin solutions (part B) precludes the repeated use of the same solution—you must be right first go!

Preparation of cells

The cells should be standing in a chromic acid bath. One or more cells are required for Part A, and five cells for Part B. Very carefully pour out the chromic acid from the centre of each into a beaker, and wash the discarded acid down a sink. Remove a cell from the bath and wash it with pure demineralized water over a sink. Then wash the cell inside and out with water from a high-pressure tap, taking care to hold it firmly by its holder and base. Do *not* touch the side walls of the cell. Wash again with demineralized water and inspect the cell to ensure that all impurities have been removed. Invert the cell in a drying oven, switch on the oven (70 °C) and leave to dry. Repeat the procedure with the other cells if required.

Place the perforated disc in the lower portion of the filtration unit by the apparatus.* Place a 0.22 μm Millipore filter onto the disc. (The filters are packaged between blue wrapping papers.) Place the PTFE ring over the filter and reassemble the filtration unit. Carefully place the complete unit in the oven and leave to warm.

A. *Polystyrene*

In a 500 cm^3 conical flask make up 250 cm^3 of a solution of polystyrene in toluene of concentration X, where $X = 4 \times 10^{-3}$ g cm^{-3}. Stir without heat for at least 1 h. Using a pipette to draw off the top of the solution, make up 40 cm^3 total volumes in 100 cm^3 conical flasks, of concentrations X, $3X/4$, $X/2$, $3X/8$, and $X/4$, by diluting with appropriate volumes of toluene.

Switch on the photometer following the instructions provided.*

Remove the cells and filtration unit from the oven and allow to cool.

Adjust the chart recorder to a fairly slow speed* and switch it on. The use of the chart recorder in the quantitative measurements on polystyrene allows the accurate estimation of the true reading, which is the minimum recorded. Dust adds noise to the signal, and if excessive will shift the baseline and so invalidate the results.

A standard glass rod is available. Place this in the beam and adjust the beam slit to give the recommended intensity.* Record the exact reading. The slit control must then not be altered during part A.

Rinse out the filter unit with toluene, and then filter a sample of toluene into a clean cell. Pressure may be applied from the low pressure compressed air line*, but do not filter faster than two or three drops per second. Take readings for natural light at all the preset angles from 30° to 150°. At each angle, use the maximum sensitivity possible without the reading going off scale. Write the settings on the chart paper. Then

$$\text{scattering intensity} = i = \frac{\dfrac{\text{chart recorder reading}}{100} \times \text{range setting}}{\text{standard}}. \quad (6.2.10)$$

Twist the cell through 180° to check the measurements. Take the minimum value at each angle if some dust is present, or re-filter and repeat the readings if the signal noise is excessive. Since the molecules of polystyrene are small in comparison to the wavelength of light, the readings should be symmetrical about 90°.

Take similar sets of readings for the five polystyrene solutions. The intensity of scattered light should increase steadily with polymer concentration. If it does not, re-filter the solutions and repeat the determinations as necessary. Be particularly careful with the solvent determinations, especially the value at 90° (i°_{90}) which fixes the absolute value of all the other readings taken.

B. Gelatin

Make up the following five solutions in 250 ml conical flasks (0.15 M NaCl solution is provided):

(1) 0.9 g gelatin in 100 cm^3 0.15 M NaCl solution;
(2) 3.0 g gelatin in 100 cm^3 0.15 M NaCl solution;
(3) 6.0 g gelatin in 100 cm^3 0.15 M NaCl solution;
(4) and (5) 0.9 g gelatin in 100 cm^3 distilled water.

Stir solutions (4) and (5) for 1 h at a heat setting of about 4, depending on the hot plate. When stirred, find the pH of the solutions by dipping in a strip of narrow range pH paper. Adjust the pH of solution (4) to pH 5, and of solution (5) to pH 8 by adding 1 M HCl or 1 M NaOH solution dropwise from a fine dropping pipette. Remove the indicator paper. Set solutions (1), (2), and (3) to stir for about 1 h at heat setting 4.

Remove the filtration unit from the drying oven (it will be hot—use a cloth). Mount it on its stand. Support one of the dry cells underneath it. Fill the filtration unit to a specified height* with solution (4), apply pressure from the low pressure compressed air line and filter the solution into the cell. Cover the cell with a metal lid and stand it in an ice–salt freezing mixture. Quickly wash out the top of the unit with solution (5), then filter the correct amount of this solution into another cell. Cover this also and leave to stand in the freezing mixture. Renew the filter in the unit and leave the unit to warm in the oven.

When solutions (1), (2), and (3) have been stirred, adjust their pH to 5, and filter them into three cells (solution 3 last). The solutions should be clear and golden. Stand the cells in freezing mixture for at least 10 min until solutions (2) and (3) set, and solution (1) at least partially sets. Thoroughly clean the filtration unit, replace the filter, and leave the unit to dry.

Switch on the photometer and adjust the chart recorder to a slow speed as instructed.*

Remove the cell containing the solution (1) gel from the freezing mixture, wipe the walls of the cell carefully with a very clean cloth and insert it into the sample vat. Open the shutter and quickly adjust the slit to give a recorder deflection of about 60. Follow the change in scattered light intensity as the gel melts, for half an hour or until the trace ceases to change.

Repeat for solutions (2) and (3) without altering the slit adjustment. Meanwhile allow samples (4) and (5) to melt by dipping them into the thermostat bath. Then take readings at 90°. Since the solutions have the same thermal history, the results may be compared with the final readings for solution (1). (Take glass standard readings if the slit is altered for solutions (4) and (5)—see eqn. (6.2.10).)

Shut down

Dismantle and clean the filtration unit. Empty and clean out all the cells. Then dip them into the chromic acid bath and very carefully fill each with the acid using a pipette and pipette filler. Do not allow chromic acid to touch the holders or the metal top of the vat.

Switch off the photometer, thermostat bath and oven. Turn off the water supply

Calculation *A. Polystyrene*

Use a computer program for these calculations if one is available.* You will require light scattering intensities at all the marked angles for the solvent and each solution, and the concentration of each solution. Tabulate values of $\sin^2\theta/2$, i, $(i - i_0)$, $\alpha = (i - i_0)/(\sin\theta/(1 - \cos^2\theta))$, and $(\sin^2\theta/2 + kc)$.

A value of k must be chosen to give a Zimm plot with a similar shape to that shown in Fig. 6.2.3. Try $k = 100\ \text{cm}^3\ \text{g}^{-1}$ initially, and then halve or double its value if necessary. If k is too low, it may prove impossible to extrapolate the points to zero concentration. If k is too large, the curves for small angles may intersect.

Once a suitable value of k has been chosen, plot the Zimm plot on graph paper. Draw lines through each set of points for a particular angle or concentration, as in the example. (See also the discussion in the Results section). The points for pure solvent occur where the extrapolations at constant angle intersect a straight line at $\sin^2\theta/2$ (i.e. $(\sin^2\theta/2 + kc)$ at $c = 0$). The points at zero angle occur at kc for each line ($(\sin^2\theta/2 + kc)$ at $\sin^2\theta/2 = 0$). Finally draw lines through these extrapolation points. The two lines should intersect on the vertical axis at the required value of $(c/\alpha)^{\theta=0}_{c=0}$. $P(\theta) = 1$ at $\theta = 0$, and $2A_2c = 0$ when $c = 0$. So at the intercept, from eqn (6.2.5),

$$\frac{K\,i^\circ_{90}}{R^\circ_{90}}\left(\frac{c}{\alpha}\right)^{\theta=0}_{c=0} = \frac{1}{M_\text{M}}. \qquad (6.2.11)$$

The absolute scattering power of toluene may be measured on the photometer by comparing the intensity at 90° with that from the glass standard. The glass standard has in turn been calibrated with respect to benzene,* for which R_{90} is $1.3 \times 10^{-5}\ \text{cm}^{-1}$.[8] This allows the intercept from the Zimm plot to be converted to an absolute value.

The other parameters which are required for toluene are $\text{d}n/\text{d}c = 0.11\ \text{cm}^3\ \text{g}^{-1}$ and $n = 1.49$,[9] which from eqn (6.2.4) give $K = 9.91 \times 10^{-8}\ \text{cm}^2\ \text{mol g}^{-2}$ for $\lambda = 546$ nm at 25 °C. Use these data, and the absolute value of the intercept to calculate M_M, the relative molar mass of polystyrene, by means of eqn (6.2.5).

Calculate the z-average r.m.s. distance between the ends of the chain, $\langle r^2\rangle_z^{1/2}$, and the r.m.s. radius of gyration $\langle\rho^2\rangle^{1/2}$ (eqns (6.2.6) and (6.2.9)).

Calculate the mass M_mon of the styrene monomer. Using also your value of M_M, the mass average relative molar mass of the polystyrene sample, calculate the quantity M_M/M_mon, which is the mass-average degree of polymerization. Estimate the length l of the extended polystyrene chain, given that each monomer unit adds 0.155 nm. Thus calculate the degree of coiling $= l/\langle r^2\rangle_z^{1/2}$.

B. Gelatin

Interpret your results for gelatin solutions (1), (2) and (3) in the light of the possible structures for gelatin discussed earlier. Deduce the cause of the fundamental difference in shape and gradient of the three curves.

Interpret your results for solutions (1), (4) and (5) in terms of electrostatic interactions between chains and the isoelectric point of gelatin at pH 5.1.

Results[5,6]

A. Polystyrene

See Fig. 6.2.3.

Accuracy

As mentioned previously excessive dust in the sample will shift the baseline and cause errors in both parts of the experiment. If the photometer is not properly aligned for part A, noise will occur at some angles near 0° or 180°. 'Force-fitting' to the experimental results of lines of the correct form for the Zimm plot can remove many of the errors caused by these factors. A complete set of anomalous readings may be obtained if one of the solutions is accidentally stirred, and these may be ignored.

B. Gelatin

See ref. 1.

Comment

The experiment which you have carried out illustrates the elucidation of the structure of a polymer by use of a light scattering photometer. Despite the complexities of experimental technique the method provides a powerful means of characterizing macro-molecular compounds. A particularly useful feature is that the accuracy of the measurements increases with molar mass, in contrast with osmotic pressure measurements where the accuracy diminishes. The use of laser light sources has greatly increased the sensitivity of the method and has allowed also the study of the structures of liquid mixtures such as benzene/cyclohexane.[10]

Technical notes

Apparatus

Photometer thermostatted at 30 °C, either commercial,[$$] or laboratory-constructed for Part B only;[†] pipette rack; 2 wooden racks for cells; chromic acid vat (e.g. glass pneumatic trough) with plastic lid drilled to support cells in acid, on plastic tray; brass filtration unit with spare perforated discs to support filter;[† or $] pressure line to give low pressure of compressed air to suit filtration unit;[†] 0.22 μm Millipore filter papers, toluene, chromic acid on plastic tray. A: stirrer unit; glass standard previously calibrated relative to benzene; polystyrene, toluene [pressure line]; [Zimm plot computer program (program in BASIC available from author)]. B: pH 4–6 indicator paper, pH 6–8 indicator paper; stirrer/heater unit; clean cloth; gelatin, 0.15 M sodium chloride solution, ~1 M HCl, ~1 M NaOH.

Maintenance

Apparatus must be kept clean, and sample vat topped up with dedusted toluene. A: Alignment of apparatus must be checked periodically.

References

[1] *Light scattering by polymers. Two experiments for advanced undergraduates.* G. P. Matthews. *J. Chem. Educ.* **61,** 552 (1984).

[2] *Molecular Weight and Molecular Weight Distributions in Synthetic Polymers,* T. C. Ward, *J. Chem. Educ.,* **58,** 867, (1981).

[3] *Physical chemistry of macromolecules.* C. Tanford. John Wiley, New York (1961), ch. 5.

[4] *Apparatus and methods for measurement and interpretation of the angular variation of light scattering. The scattering of light and the radial distribution function of high polymer solutions,* B. H. Zimm. *J. Chem. Phys.* **16,** 1099 and 1093 (1948).

[5] Ref 3, p. 165.

[6] *A study of gelatin molecules, aggregates and gels,* H. Boedtker and P. Doty. *J. Phys. Chem.* **58,** 968 (1954).

[7] *Description d'un appareil pour l'étude de la diffusion de la lumière, J. Chim. Phys.* **51,** 201 (1954).

[8] Ref. 4, p. 1116.

[9] *Polymer handbook,* Eds. J. Brandrup and E. H. Immergut, John Wiley and Sons, New York, 2nd edn (1975).

[10] *Light scattering studies of molecular liquids.* D. Kivelson and P. A. Madden. *Ann. Rev. Phys. Chem.* **31,** 523 (1980).

6.3 Properties of a solid polymer by dielectric relaxation

The *relaxation* of a system can be defined as its time-dependent return to equilibrium after it has experienced a change in the influences acting upon it. The changing influence in this experiment is an alternating electric field. In the case of charged species which are mobile, as in metals or electrolyte solutions, the alternating field will simply give rise to conductivity effects. In a polymer sample, however, the movement of molecules is constrained and the sample exhibits *dielectric relaxation.* The relaxation is affected by the internal motion of molecules, and the variation of the relativity permittivity (dielectric constant) of a polymer sample with temperature and frequency can therefore give information about structure.

In this experiment we study the properties of a solid sample of polyethylmethacrylate (PEMA) over a frequency range from 10 to 10^5 Hz at temperatures from 10–65 °C. From the results we identify the *β-transition* at which the side groups of the polymer chains start to rotate, and also the *glass* or *α-transition* in which the polymer chains themselves begin to rotate. At the α-transition the chains may even begin to diffuse by the process of *reptation*, i.e. by slithering snake-like movements in hypothetical tubes defined by the molecules' own dimensions.

Confusingly, α, β, and γ refer simply to the order in which transitions are observed as the temperature of the sample is reduced, not to corresponding processes in different polymers.

Theory[1,2]

The first part of the Theory section of Expt 5.2, p. 183, provides an introduction to dipole moments.

Note that in Expt 5.2, relative permittivity is given the symbol ε_r, whereas here, to simplify the notation, we refer to it as ε.

(*i*) *Dielectric relaxation*

Several mechanisms contribute to dielectric relaxation in a polymer. There are two essentially instantaneous processes ($\sim 10^{-13}$ s), namely the displacement of the electron cloud with respect to the nuclei, and minor distortions of bond angles. These are equivalent to the electronic and distortion polarization mentioned on p. 186. They contribute an amount ε_∞ to the total equilibrium relative permittivity ε_s.

Meanwhile, over a much wider range of lower rates ($\sim 10^{-12}$ to 10^4 s), movements in the polymer chains take place in such a way as to reduce the internal field still more. This *orientation polarization* contributes the remainder of the relative permittivity $\Delta\varepsilon$, defined as:

$$\Delta\varepsilon = \varepsilon_s - \varepsilon_\infty, \tag{6.3.1}$$

where $\Delta\varepsilon$ is a real quantity. The rates of the processes may be expressed in terms of a single relaxation time τ, such that

$$\varepsilon(t) = \varepsilon_\infty + \Delta\varepsilon(1 - e^{-t/\tau}). \tag{6.3.2}$$

τ is the characteristic relaxation time which represents the time variation of the polarization (P) in the medium, so that if the electric field is instantaneously removed when $t=0$, the step-response function $dP/dt = -P/\tau$, or $P = P_0 \cdot e^{-t/\tau}$. τ is the time taken for the polarization to drop to 1/e of its value on removal of the field. It is a considerable oversimplification to use a single macroscopic relaxation time to describe the many relaxation processes occurring in different parts of the polymer molecules, and one of the problems of the experiment is to understand why relaxation at the glass (α) transition does not obey eqn (6.3.2) save in a broadly qualitative way.

To understand the way in which the relative permittivity of a polymer can be separated into components, we must first consider a field which is oscillating so slowly that the molecules follow the oscillations of the imposed potential exactly. Under these circumstances the displacement current in the dielectric will be exactly 90° out of phase with the voltage, with the former leading the latter. In simple terms this arises from the fact that the rotation of the dipoles will be most rapid when the voltage changes most rapidly, i.e. when it passes through the instantaneous value of zero (Fig. 6.3.1). In a graph showing the relationship between the current $I(t)$ and voltage $V(t)$, it can be shown that for this case there is no component of the current in step with the e.m.f., Figs 6.3.1 and 6.3.2(a).

If we then increase the frequency to an extent that the motions of the molecules begin to lag the stimulus, there is a fall off, or *dispersion*, of the relative permittivity from ε_0 towards ε_∞, the value for a non-polar material arising solely from electronic and distortion polarization. The phase displacement is shown in Fig. 6.3.2(b).

The behaviour of the relative permittivity can also be represented by a phase graph, Fig. 6.3.2(c), in which the total

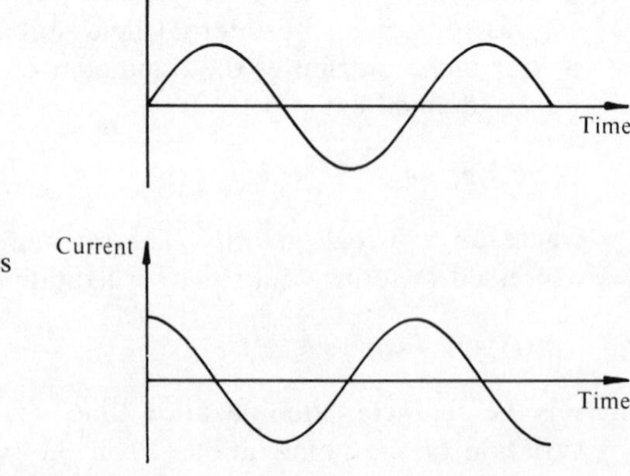

Fig. 6.3.1. Relation between applied e.m.f. and induced current in a dielectric at low frequencies

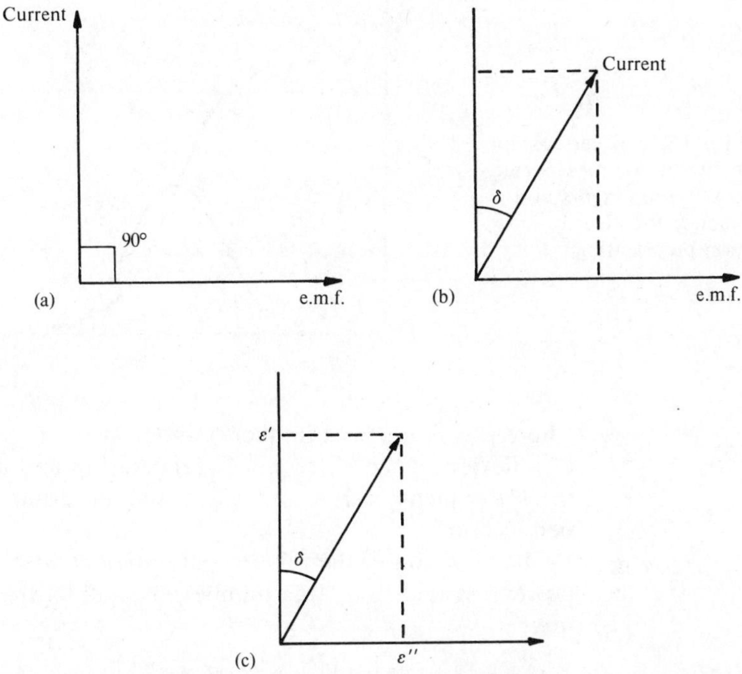

Fig. 6.3.2. Ways of expressing the phase relationships in a dielectric: (a) current and e.m.f. 90° out of phase, as in Fig. 6.3.1; (b) dielectric with loss factor δ; (c) corresponding graph of ε' and ε''

complex relative permittivity ε^* has two components at those frequencies for which loss occurs in the medium. This is represented by

$$\varepsilon^* = \varepsilon' - i\varepsilon'', \tag{6.3.3}$$

where i, the anticlockwise rotation operator equal to $\sqrt{-1}$, generates the imaginary axis of the Argand diagram, ε' is the real relative permittivity, and ε'' is the imaginary component. ε'' represents loss-processes and is a measure of the conductance of the medium. It can be seen from the graph that

$$\varepsilon''/\varepsilon' = \tan\delta, \tag{6.3.4}$$

where $\tan\delta$ is known as the *dissipation factor* or *loss tangent.*

Non-polar polymers such as polythene show remarkably low loss factors, and their use as low loss v.h.f. dielectrics has been a vital factor in the development of radar.

Figure 6.3.3 shows the frequency dependence of ε' and ε'' in the β-relaxation region of polyvinyl ethanoate (acetate). As the frequency is increased, ε' drops from ε_s to ε_∞, and ε'' goes through a maximum. The relationship between field frequency and dipole rotation rates in the ideal case is

$$2\pi f_{\max}\tau = \omega_{\max}\tau = 1, \tag{6.3.5}$$

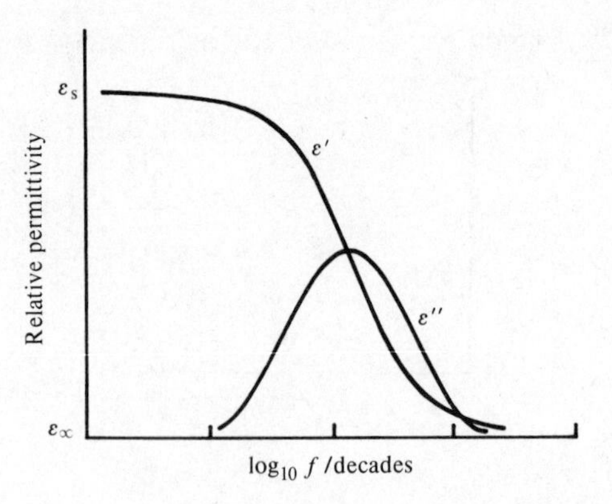

Fig. 6.3.3. Relationship between complex permittivity components and frequency for ideal orientation polarization[4]

where f_{max} is the field frequency in Hz at which there is a point of inflexion in the ε'/frequency relationship and a maximum in the ε''/frequency curve, and ω_{max} is the same quantity in radians per second.

The variation of the relative permittivities with frequency can be represented in a large number of cases by the *Debye equations*

$$\varepsilon' = \varepsilon_\infty + [(\varepsilon_s - \varepsilon_\infty)/(1+\omega^2\tau^2)], \tag{6.3.6}$$

and

$$\varepsilon'' = (\varepsilon_s - \varepsilon_\infty)\omega\tau/(1+\omega^2\tau^2). \tag{6.3.7}$$

(*ii*) *Measurement of relative permittivity*

We must now relate the relative permittivities to properties which can be measured in an electrical circuit. The arrangement is shown schematically in Fig. 6.3.4, where a polymer sample between parallel capacitor plates is shown to be equivalent to a capacitance C_x in parallel with a resistance R_x. We define the relationship

$$\varepsilon^* = C^*/C_0, \tag{6.3.8}$$

where C^* = complex capacitance, and C_0 is the capacitance *in*

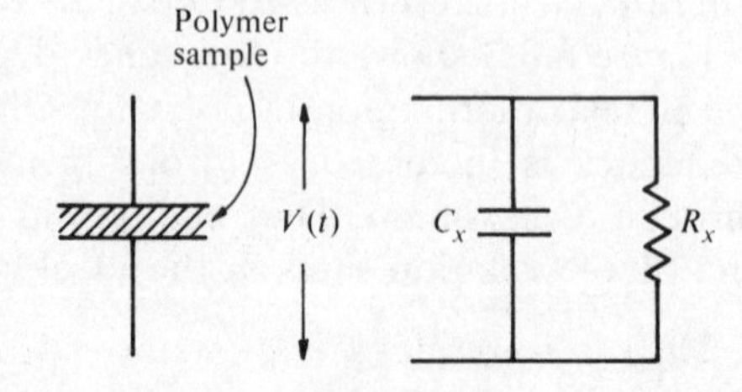

Fig. 6.3.4. The polymer sample between the plates of the capacitance cell, and the equivalent electrical circuit

vacuo. Suppose that we apply a voltage $V(t)$ to the sample such that

$$V(t) = V_0 \cdot e^{i\omega t}. \tag{6.3.9}$$

By definition the current $I(t)$ is the rate of change of charge $q(t)$ with time:

$$I(t) = d[q(t)]/dt = d[C^* V(t)]/dt. \tag{6.3.10}$$

It follows from the last three equations that

$$I(t) = C_0 \varepsilon^* i\omega V_0 e^{i\omega t} = C_0 \varepsilon^* i\omega V(t). \tag{6.3.11}$$

Therefore from eqn (6.3.3)

$$I(t) = C_0(\varepsilon' - i\varepsilon'') i\omega V(t) = C_0(i\omega\varepsilon' + \omega\varepsilon'') V(t). \tag{6.3.12}$$

Now the theory for the electrical circuit shown in the figure tells us that

$$I(t) = \frac{V(t)}{\text{impedance}} = V(t)\left(\frac{1}{R_x} + i\omega C_x\right). \tag{6.3.13}$$

Comparing eqns (6.3.12) and (6.3.13), we see that

$$\varepsilon' = C_x/C_0, \tag{6.3.14}$$

and

$$\varepsilon'' = G_x/\omega C_0, \tag{6.3.15}$$

where G_x = conductance = $1/R_x$. Therefore, from eqns (6.3.4) and (6.3.5),

$$\tan\delta = \varepsilon''/\varepsilon' = G_x/\omega C_x = G_x/2\pi f C_x. \tag{6.3.16}$$

(*iii*) *Effect of temperature*

Figure 6.3.5 shows the variation in ε with temperature for a typical polymeric solid. At very low temperatures, the chains in a polymer molecule undergo only low-amplitude vibrations about fixed positions. As the temperature is increased, there first occurs a β-transition during which the side groups of the polymer chain start to rotate. At a higher temperature T_g, a *glass- or α-transition* occurs, in which the polymer chains themselves begin to rotate and the polymer changes from a glassy to a rubbery state. Note that the β and α transitions are not always fully separable, especially when the dipolar side-groups are bulky. At a higher temperature still, T_{fus}, the sample softens and the molecules flow.

The glass transition temperature for rubber is below room temperature, whereas for polystyrene it is above room temperature; thus heated polystyrene ($\geq$116 °C) is rubbery, but if rubber itself is cooled in liquid nitrogen it becomes glassy and can be smashed with a hammer.

The transition temperatures are identified by maxima in the

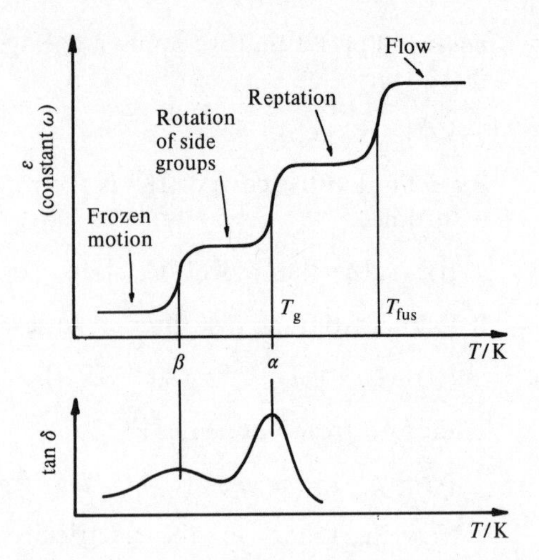

Fig. 6.3.5. Schematic diagram showing variation in relative permittivity and loss factor of a polar polymer with temperature

loss tangent $\tan\delta$ and hence in the measured quantity $G_x/2\pi fC_x$ (eqn (6.3.16)).

Apparatus

The apparatus comprises an audio frequency bridge analogous to the r.f. bridge shown on p. 274. The bridge is connected to a dielectric cell, the latter being immersed in a water bath containing a stirrer, heater, and thermometer.

Details of the dielectric cell are shown in Fig. 6.3.6. The PEMA sample is in the form of a disc, typically 1 mm thick and 15 mm in diameter, which is placed between the two condenser plates. The plates are accurately parallel to each other so that the capacitance *in vacuo*, C_0, is equal to $A\varepsilon_0/d$, where A is the area of each plate, d is the distance between the plates, and ε_0 is the relative permittivity of free space. A guard ring eliminates edge effects.

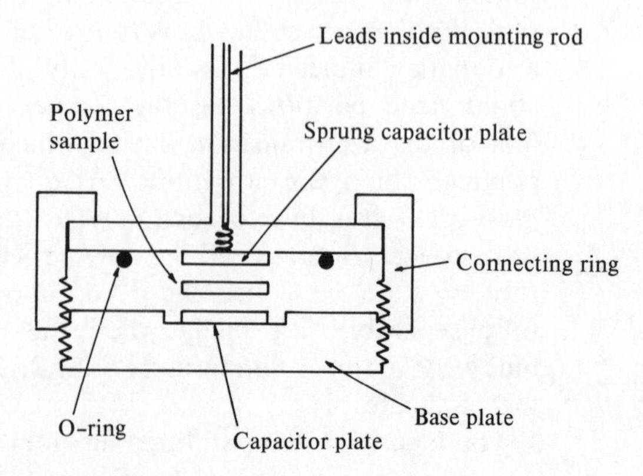

Fig. 6.3.6. Dielectric cell for measurement of relative permittivity of a solid disc of polymer

Procedure Mount the sample in the cell as instructed.* Connect the cell to the audio frequency bridge, switch on the bridge, and allow the electronics to warm up. Immerse the cell in the water bath, which has been pre-cooled to about 10 °C by the addition of ice.

Set the frequency to 17.0 Hz, and measure the capacitance C and conductance G at at least 15 temperatures in the range 10–70 °C. Tabulate the results ready for the calculations described in the next section.

In a detailed examination of this polymer in which a complete map of ε' and ε'' against ω and T is required, it would normally be necessary to repeat the entire experiment at frequencies of about 25, 35, 65, 300, 400, 800, 1000, 5000, 10 000, 30 000, 60 000, and 90 000 Hz, but for survey purposes it is adequate to work at one low, one medium and one high frequency.

An alternative approach which brings out the nature of the glass transition is to maintain the water bath at a constant temperature of 55 °C, and to measure C and G at frequencies in the range 10–80 000 Hz, increasing the frequency by a factor of between 1.5 and 2 between each reading. The whole procedure is then repeated for temperatures of 60, 65, and 70 °C.

Calculation At each fixed frequency f, tabulate values of T, C, G, G/ω, and $\tan \delta$ (eqn (6.3.16)).

Similarly, at each fixed temperature T, tabulate f, C, G, G/ω, and $\tan \delta$.

Plot the following graphs. Interpret each in terms of the β and glass transitions in PEMA and summarize the results qualitatively.

(i) Capacitance against temperature, at various frequencies,
(ii) G/ω, and hence ε'', against temperature, again plotting all the different frequency lines on one graph,
(iii) G/ω against $\ln \omega$,
(iv) C against $\ln \omega$,
(v) $\tan \delta$ against $\ln \omega$,
(vi) $\tan \delta$ against T,
(vii) G/ω against C at four constant temperatures just above the glass transition (55, 60, 65, 70 °C).

Results[3] See Fig. 6.3.7.

Comment The technical significance of polymeric materials lies in their use as plastics, rubbers, or fibres. Experiments to determine the conditions of β and glass transitions are an important part of the characterization of a particular polymer, since its flexibility

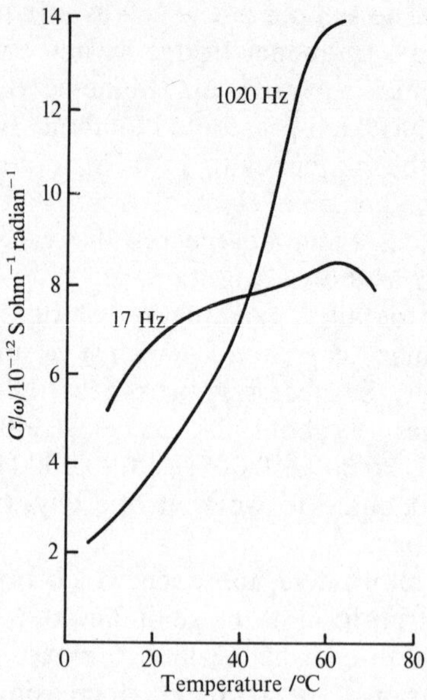

Fig. 6.3.7. Variation of G/ω with temperature for solid poly-ethyl methacrylate

or brittleness is determined by the extent of molecular motion within the sample. It often proves useful to use a polymer at temperatures between the two transitions because in this region the polymer is solid, but nevertheless pliable because rotations of the side-groups of the polymer molecules can take up energy.

Experiments to measure the dielectric relaxation of dilute polymer solutions are also much used, since inter-chain phenomena are minimized, and intra- and intermolecular effects may be distinguished.[4]

Technical notes

Apparatus

Wayne–Kerr A.C. bridge covering range 10–10^5 Hz;[$ or $$] dielectric cell;[††] thermostat bath, stirrer, thermometer, ice.

Preparation of samples

Samples can be prepared by evaporation of solvent (butanone) from a flat Petri dish of PEMA solution.[5] Avoid ingress of water vapour which on absorption by the polymer can lower the glass transition temperature and accentuate the dielectric loss. Errors arise if there is any air gap between the sample and plates, and electrical contact may be ensured by evaporating a metal film onto the faces of the sample. In practice, however, the sample rapidly conforms to the electrodes once the glass transition has been traversed.

References

[1] *Molecular motion in high polymers.* R. T. Bailey, A. M. North, and R. A. Pethrick. Clarendon Press, Oxford (1981) ch. 1 and 5.
[2] *Dielectric absorption.* M. Davies, *Quart. Revs.* **8,** 250 (1954).
[3] *Anelastic and dielectric effects in polymeric solids.* N. G. McCrum, B. E. Read and G. Williams. John Wiley, London (1967) p. 258.
[4] *Dielectric relaxation in polymer solutions.* A. M. North. *Chem. Soc. Revs.* **1,** 49 (1972).
[5] *Experiments in polymer science.* E. A. Collins, J. Bareš, and F. W. Billmeyer, Jr. John Wiley, New York (1973) p. 472.

Part III Change

Section 7
Particles in motion

The nine experiments in this section involve the study of particles in motion. In the first three experiments the particles are uncharged species, and in the remainder they are ions.

The subject is introduced in Expt 7.1 by the direct observation, under a microscope, of the Brownian motion of latex particles in water. The theory of the experiment not only yields an equation from which Avogadro's constant may be determined, but also introduces important considerations about the statistics and nature of random processes.

In Expt 7.2 the particle motion is simulated by a molecular dynamics computer program, and the results allow the structure of solid, liquid and gaseous phases to be compared. Although the molecular dynamics technique itself is sophisticated, the experiment requires only rudimentary computing skills to operate, and the results can be readily interpreted.

The viscosity of a gas, arising from the transfer of momentum during collisions, is measured in Expt 7.3 by means of a capillary flow viscometer. This device comprises an easily operated gas line by means of which the flow times of a gas through a capillary may be measured. The temperature variation of viscosity is compared with the predictions of the simple kinetic theory of gases, and the deviations explained in terms of a more rigorous kinetic theory which accounts for the effects of inelastic collisions.

The mobilities of ions, which are the underlying concern of the remaining experiments, may be measured by finding the conductivities of ionic solutions. Experiment 7.4 introduces the basic theory and experimental techniques associated with such measurements. It also illustrates their most common use, which is as a measure of the relative number of ions in a solution, and hence of such properties as dissociation constants and solubilities.

Conductivities are not only affected by the number of ions in a solution, but also by the mobilities of the individual ions. Experiment 7.5 demonstrates that the viscosity of the solutions is a major factor in determining the magnitude of ionic mobilities. The procedure involves the determination of solution viscosities at various temperatures by means of a simple glass viscometer, as well as the measurement of conductivities by the techniques introduced in Expt 7.4. The cations and anions do not carry the charge equally, and the subsequent two experiments, 7.6 and 7.7, involve the measurements of the fraction of the charge carried by each. In Expt 7.6 this is achieved by a simple technique in which a moving boundary is set up in a capillary tube, and in Expt 7.7 by measuring the e.m.f. of an electrochemical cell. Although the techniques in Expt 7.7 are those associated with equilibrium electrochemistry, Section 4, the process is a kinetic one and the theory of the experiment is based on Fick's first law of diffusion.

Finally there are two experiments in which the reactions of ions are governed by transport processes. Experiment 7.8 involves the measurement of the kinetics of an ion exchange reaction by titration. The results are shown to agree with equations based on Fick's first law of diffusion, and not on the mass-action kinetic equations encountered in Section 8. The last experiment involves the polarographic study of the reduction of substituted nitrobenzene compounds. The apparatus employs a dropping mercury electrode, of which the theory is too complicated to be included in full. Nevertheless the experiment demonstrates how polarographic measurements of the electrode processes may be correlated with the structure of the reacting species by means of the so-called Hammett σ parameter.

7.1 Avogadro's constant from Brownian motion[1]

In the summer of 1827 the botanist Robert Brown observed that the particles in an aqueous suspension of pollen were in rapid oscillatory motion.[2] He then examined a large number of other suspensions of both organic and inorganic matter, and to his surprise observed similar motion in all of them. Nearly a century elapsed before Einstein, unaware of previous attempts to explain the phenomenon, made conclusive mathematical predictions of the rate of diffusion of suspended particles caused by the random impacts of solvent molecules.[3]

For the sake of thoroughness, Brown included in his range of suspensions a powdered fragment of the Sphinx.

In this experiment we observe an aqueous suspension of latex particles under a microscope and measure the extent of this Brownian motion. We then use the equation derived by Einstein to calculate Avogadro's constant.

Theory

(*i*) *General considerations*

It is instructive to consider two of the factors which militated for so long against a kinetic explanation of Brownian motion.[3,4] The first involves a simple calculation of conservation of momentum, which shows that Brownian motion is two orders of magnitude greater than would be expected from a typical collision between a solvent molecule and a macroscopic Brownian particle. Early workers could not explain this discrepancy in terms of multiple collisions, because they assumed that if the collisions were random their net effect on the Brownian particle would be zero. However, such a sequence would in fact be a highly *ordered* one. It can be shown that if N *random* collisions occur, the net displacement will be proportional to $\sqrt{N}$, or since $N \propto \tau$, the time interval between observations, the net displacement is proportional to $\sqrt{\tau}$.

If, however, the extent of Brownian motion is calculated in terms of the conservation of energy, the speed $\dot{s} = \mathrm{d}s/\mathrm{d}t$ of the particle is given by the *principle of equipartition of energy* as

$$\overline{m\dot{s}^2}/2 = 3kT/2, \qquad (7.1.1)$$

which yields a value much *greater* than that observed. The explanation lies in a closer examination of the term $\overline{\dot{s}^2}$. Figure 7.1.1 shows the two-dimensional track of a Brownian particle

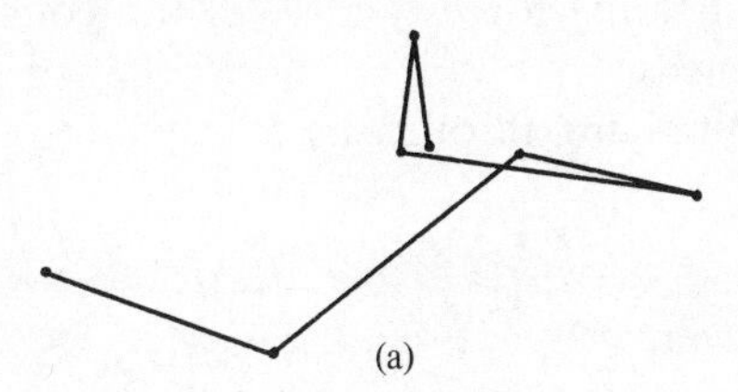

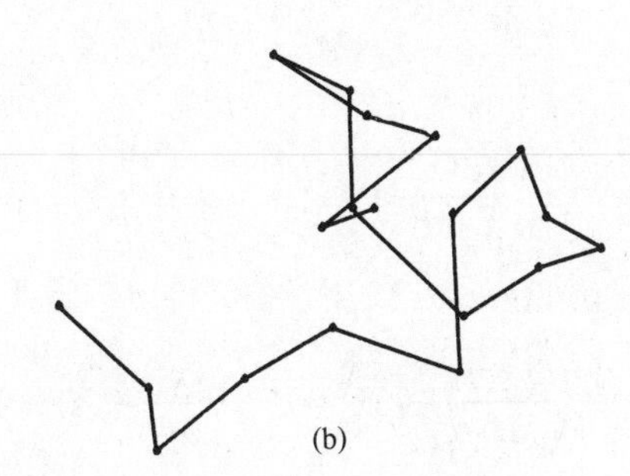

Fig. 7.1.1. Two-dimensional track of the same Brownian particle observed (a) every 30 s and (b) every 10 s

observed every 30 s for 3 min. However, if the *same* random walk had been observed every 10 s, there would have been three times as many points, as shown in the lower diagram. Every third point is the same as in the first diagram, but the overall impression is that the particle is moving $\sqrt{3}$ times as fast. In fact during the time τ the particle undergoes millions of collisions. So as the time between observations is decreased still further, the apparent velocity continues to increase, and only when τ is of the order of the time between the collisions will eqn (7.1.1) hold true. If this important factor is taken into account, as in the derivation below, there is good agreement between theory and observation.

(*ii*) *Derivation of Einstein's expression*[5]

Suppose we make a somewhat artificial division of the forces on a suspended particle of mass m into those of ordinary continuous-fluid hydrodynamics and those dealt with from a statistical-atomic view point; in the first category is an ordinary 'frictional' force proportional to velocity, while all other external influences of the fluid are combined in the second. For the motion of a particle in any specified but arbitrary direction, which we take as the x direction,

$$m\ddot{x} = -f\dot{x} + X. \tag{7.1.2}$$

The dot and double-dot represent the first and second derivatives with respect to time. $(-f\dot{x})$ designates the x-component of the frictional forces and X is the combined effect of all other influences.

Multiplying through by x:

$$m\ddot{x}x = -f\dot{x}x + Xx. \tag{7.1.3}$$

However,

$$\dot{x}x = \frac{1}{2}\frac{\mathrm{d}(x^2)}{\mathrm{d}t},$$

and

$$\ddot{x}x = \frac{1}{2}\frac{\mathrm{d}}{\mathrm{d}t}\left(\frac{\mathrm{d}(x^2)}{\mathrm{d}t}\right) - \dot{x}^2.$$

Substituting:

$$\frac{m}{2}\frac{\mathrm{d}}{\mathrm{d}t}\left(\frac{\mathrm{d}(x^2)}{\mathrm{d}t}\right) - m\dot{x}^2 = -\frac{f}{2}\frac{\mathrm{d}(x^2)}{\mathrm{d}t} + Xx. \tag{7.1.4}$$

We form such an equation for each particle which is suspended in the fluid, and take the mean of these expressions for all particles:

$$\frac{m}{2}\frac{\mathrm{d}}{\mathrm{d}t}\overline{\left(\frac{\mathrm{d}(x^2)}{\mathrm{d}t}\right)}-\overline{m\dot{x}^2}=-\frac{f}{2}\frac{\overline{\mathrm{d}(x^2)}}{\mathrm{d}t}+\overline{Xx}. \tag{7.1.5}$$

Let us now assume (strictly, a special proof is required) that the mean value $\overline{Xx}$ vanishes, because the force X varies in a completely irregular manner. Further, according to the equipartition principle, the kinetic energy of the particles is

$$\overline{m\dot{x}^2}/2=k_{\mathrm{B}}T/2, \tag{7.1.6}$$

because we are now dealing with motion in one dimension only. Since

$$\overline{\frac{\mathrm{d}}{\mathrm{d}t}\left(\frac{\mathrm{d}(x^2)}{\mathrm{d}t}\right)}=\frac{\mathrm{d}}{\mathrm{d}t}\overline{\left(\frac{\mathrm{d}(x^2)}{\mathrm{d}t}\right)},$$

then

$$\frac{m}{2}\frac{\mathrm{d}}{\mathrm{d}t}\overline{\left(\frac{\mathrm{d}(x^2)}{\mathrm{d}t}\right)}+\frac{f}{2}\frac{\overline{\mathrm{d}(x^2)}}{\mathrm{d}t}=k_{\mathrm{B}}T. \tag{7.1.7}$$

For brevity, let $\overline{\mathrm{d}(x^2)/\mathrm{d}t}=u$. Then eqn (7.1.7) can be written

$$\frac{m}{2}\frac{\mathrm{d}u}{\mathrm{d}t}+\tfrac{1}{2}fu=k_{\mathrm{B}}T. \tag{7.1.8}$$

This is a differential equation in u, for which the general solution is

$$u=2k_{\mathrm{B}}T/f+C\exp\left(\frac{-ft}{m}\right), \tag{7.1.9}$$

where C is an integration constant. Now because of the small value of m the quotient f/m is a very large number, so that the exponential term has no influence after the first extremely small time interval, and

$$u=\frac{\overline{\mathrm{d}(x^2)}}{\mathrm{d}t}=2k_{\mathrm{B}}T/f. \tag{7.1.10}$$

After integration from $t=0$ to $t=\tau$, this gives

$$\overline{x^2}-\overline{x_0^2}=2k_{\mathrm{B}}T\tau/f. \tag{7.1.11}$$

If we now set $x_0=0$ when $t=0$, and, to indicate its small value, write $\overline{\Delta x^2}$ instead of $\overline{x^2}$, then

$$\overline{\Delta x^2}=2k_{\mathrm{B}}T\tau/f. \tag{7.1.12}$$

The quantity $\overline{\Delta x^2}$ has the following meaning. A particle is observed at time 0 and time τ. During this time interval it has undergone a displacement Δs, whose projection on the x-axis is

Δx. The same particle is observed at later time, always separated by the same interval τ, i.e. at times $2\tau, 3\tau, \ldots$, and Δx is determined for each interval. These values are squared and their mean is computed, the result being $\overline{\Delta x^2}$. As illustrated previously, the displacements thus observed are in no sense the actual path of the particle, nor is $\Delta x/\tau$ the x-component of its velocity.

An expression for the factor f is now introduced from hydrodynamics. According to *Stokes's Law* the viscous force F acting on a sphere of radius r, moving with velocity v in a fluid of viscosity η, is given by

$$F = 6\pi\eta r v. \tag{7.1.13}$$

Therefore f, the force per unit of velocity, is

$$f = F/v = 6\pi\eta r. \tag{7.1.14}$$

Combining this with eqn (7.1.12) yields *Einstein's expression* for the root mean square displacement $(\overline{\Delta x^2})^{1/2}$:

$$(\overline{\Delta x^2})^{1/2} = (k_B T\tau/3\pi\eta r)^{1/2}, \tag{7.1.15}$$

or

$$(\overline{\Delta x^2})^{1/2} = (RT\tau/3\pi\eta r L)^{1/2}, \tag{7.1.16}$$

where R is the gas constant and $L = R/k_B$ = Avogadro's constant. Thus L may be found by measuring $(\overline{\Delta x^2})^{1/2}$ in a solvent of known viscosity.

(*iii*) *The Normal distribution*

The distribution would become closer to Normal if the number of observations was increased, or if the 'class width', in this case half a graticule spacing, was decreased to give an almost continuous rather than discrete distribution.

Since the Brownian motion of the particle is random, the values of Δx approximate to a *Normal distribution*, Fig. 7.1.2. The Normal distribution may be plotted in the form of a cumulative probability curve, Fig. 7.1.3. This shows, for example, that if 100 measurements are made, 84 of them will occur below the

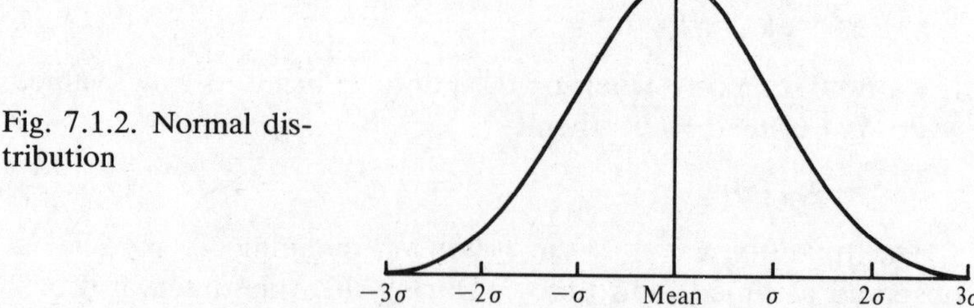

Fig. 7.1.2. Normal distribution

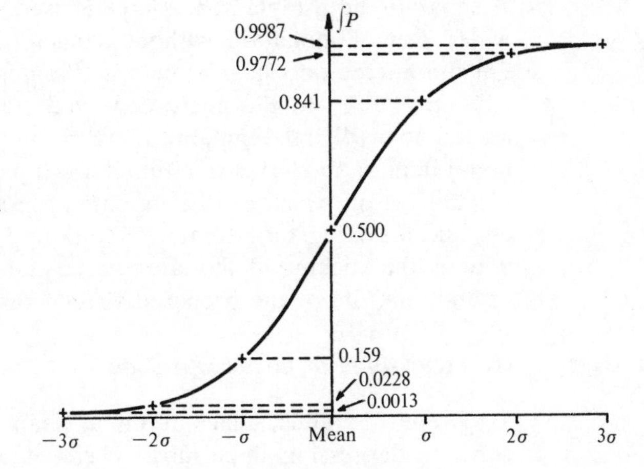

Fig. 7.1.3. Normal distribution expressed as a cumulative probability curve

mean and up to 1 standard deviation above the mean. In this experiment the results are plotted on *probability graph paper*, which has its ordinates plotted so that this cumulative probability curve becomes a straight line.

Apparatus

The availability of suspensions of particles of accurately known diameter has greatly facilitated the quantitative study of Brownian Motion.

The sample solution is a very dilute aqueous suspension of latex particles of diameter within 1 per cent of an accurately known value around 10^{-6} m (1 μm or 1 micron).

The apparatus comprises a standard microscope fitted with a $\times 40$ objective and a widefield $\times 15$ Huygens eyepiece. Into the eyepiece is mounted a graticule with a line spacing approximately twice the diameter of the particles when viewed through the microscope. The central portion is illustrated in Fig. 7.1.4.

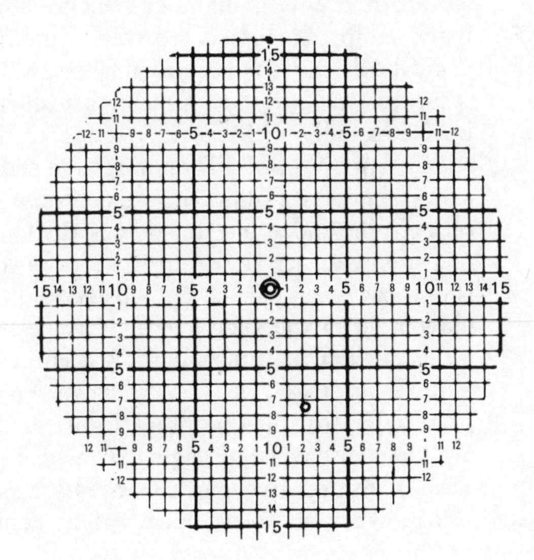

Fig. 7.1.4. Central portion of the eyepiece graticule

A separate, adjustable light source is used, defocussed to give a wide, even illumination without danger of convection effects. On the microscope stage is mounted a spirit level, and also, in close proximity to the microscope slide, a thermocouple connected to a digital temperature read-out. An additional, optional item of apparatus is a 'tumbler'—a platform rotating once per minute on which can be mounted a microscope slide, edge on, and a boiling tube of stock solution. The tumbling motion inhibits the sticking of the latex particles to the glass tube and extends the life of the prepared slides from hours to days.

Procedure

(*i*) *Preparation of microscope slide*

To prepare a slide, place it on a clean lens tissue with its concave depression uppermost. Clean it with a lens tissue if necessary. Fill the depression to overflowing with solution using a dropping pipette. Slide a cover slip sideways over the depression, centre it, and ensure there is no air bubble trapped. While holding the edge of the cover slip with a finger-nail, absorb the excess solution on the slide with a lens tissue. Having ensured that the cover slip is still central, seal it to the microscope slide by painting Canada balsam around its edge with the brush provided. Return the brush to the solvent bottle.

(*ii*) *Observation of the slide*

Mount the slide on the slide carrier of the microscope. Switch on the microscope lamp, and ensure that the entire liquid surface is evenly illuminated because convection currents occur if the lamp is focussed too well. *Carefully* rack down the lens using both the coarse and fine focus (large and small knobs respectively) until the ×40 lens just touches the cover slip. Rack up fractionally with the fine focus. Centre the slide using the slide carrier controls. Then move the slide gently back and forth using one of the slide carrier controls. As you are doing this, observe the slide, and slowly rack the lens upwards with the fine focus. Never rack downwards with the coarse control. The slide is in focus when particles move across the field of vision with the movement of the slide carrier. Centre the slide again.

A useful hint is not to press your eye to the lens, but to support it a centimetre or two away with the edge of your hand; don't strain your eye—relax it to focus at infinity and don't have the illumination too bright.

Check that the microscope is accurately vertical—*why*?

Refer to Fig. 7.1.5 which shows how to recognize a latex particle in Brownian motion. Adjust the focus by small amounts until you discover the jigging Brownian particles. What you see are not the opaque particles themselves but their projections, which, when in focus, have bright centres due to diffraction

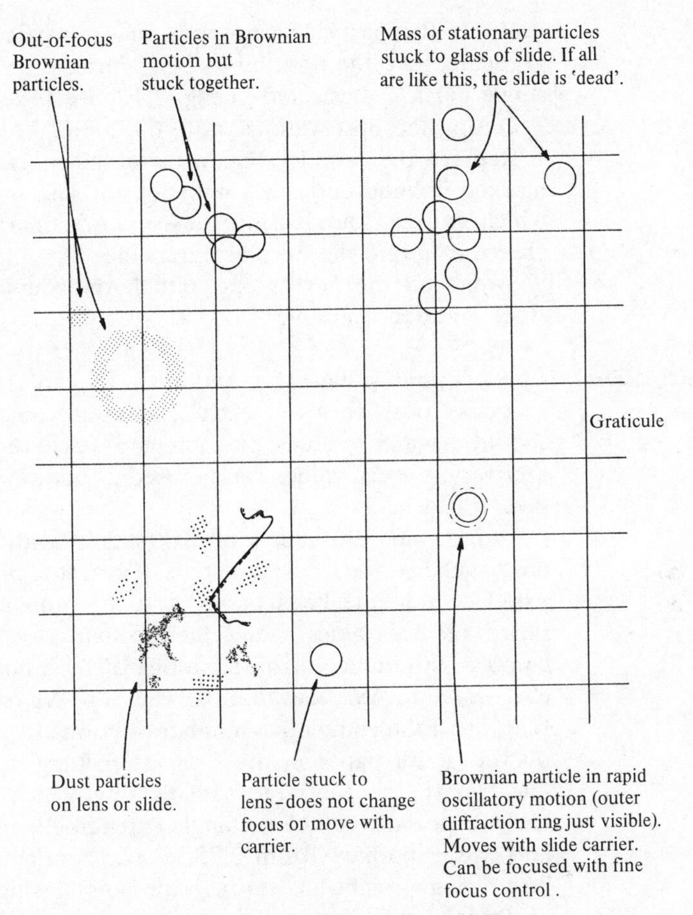
Out-of-focus
Brownian
particles.
Particles in Brownian
motion but
stuck together.
Mass of stationary particles
stuck to glass of slide. If all
are like this, the slide is 'dead'.
Graticule
Dust particles
on lens or slide.
Particle stuck to
lens – does not change
focus or move with
carrier.
Brownian particle in rapid
oscillatory motion (outer
diffraction ring just visible).
Moves with slide carrier.
Can be focused with fine
focus control.

Fig. 7.1.5. Identification of Brownian particles

effects. It may take your eye some time to adjust to observations through the eyepiece graticule, especially if you are not accustomed to using a microscope.

(*iii*) *Measurement of Brownian motion*

For the experiment, you and your partner will each need to observe a chosen particle for 25 minutes, reporting its x and y co-ordinates at 30 s intervals. You may have to refocus slightly to keep the particle in view. If you lose a particle before 25 min have elapsed, choose another one and continue the observations. You will need a total of $(49+P)/2$ min viewing time, where P is the number of different particles you observe. Practice for a minute or two before carrying out the experiment itself. If the particle is not touching a particular line on the graticule, take the position as 0.5. Thus the co-ordinates of the single particle illustrated in Fig. 7.1.4 are $(+2, -7.5)$.

During the observations, note the temperature of the slide.

Replace the sample slide with the stage graticule, which is marked in hundredths of a millimetre. Focus onto the graticule, which can be found using the centering marks on the slide. Hence calibrate the eyepiece graticule.

Switch off the microscope and thermocouple, but leave the stock solution tumbling.

Calculation

Convert your values of x and y to Δx and Δy, the change in graticule position over each 30-second time interval. You should have 50 values of each and since the directions are arbitrary, the Δy values can be used as additional Δx values to give 100 in all.

Order your 100 values of Δx (still in graticule units) from most negative to most positive. Turn the probability graph paper so that the linear scale is at the bottom, and scale the range of Δx values along the bottom axis. Work out the cumulative number of observations of Δx's equal or less than each value of Δx. Note that the values of Δx are only specified to $\pm\frac{1}{4}$, and the cumulative number of points to $\pm\frac{1}{2}$. Plot the data on the graph paper in the form of rectangles illustrating this uncertainty; e.g. the hundredth reading, which cannot be plotted as a point, will be a rectangle extending from 99.5 upwards, and width, perhaps, from 2.75 to 3.25 graticule units. Draw a best fitting straight line through the points, which, as the extent of the Brownian motion is symmetrical about the mean $= 0$, must pass through the point $(-\frac{1}{4}, 50)$. (*Why not the point* $(0, 50)$*?*). Also draw two other lines through the point $(-\frac{1}{4}, 50)$ which just include all the experimental results.

Using the data given in Fig. 7.1.3, measure on the graph paper the absolute length or ordinate which corresponds to one standard deviation σ on the abscissa. Hence, from the absolute gradient of your best-fitting straight line, calculate σ. Then use your graticule calibration to obtain an absolute value of σ in metres. σ is the root mean square deviation of x from the mean. Thus since the mean is zero, $\sigma = (\overline{\Delta x^2})^{1/2}$.

Interpolate the data in Table 7.1.1 to find the viscosity η of water at the temperature of the experiment.

Hence use Einstein's eqn (7.1.16) to find Avogadro's constant. Use the gradients of the bounding lines on your graph to estimate the statistical uncertainty of your results, and hence estimate its total error limits.

Table 7.1.1. The viscosity of water near room temperature

T/°C	$\eta/10^{-3}$ N s m^{-2}	T/°C	$\eta/10^{-3}$ N s m^{-2}
15	1.139	23	0.933
16	1.109	24	0.911
17	1.081	25	0.890
18	1.053	26	0.871
19	1.027	27	0.851
20	1.002	28	0.833
21	0.978	29	0.815
22	0.955	30	0.798

Results[1]

Accuracy

If carried out carefully, this experiment should yield Avogadro's constant to within 10 per cent of the accepted value. The most common source of error is convection currents caused by a light source which is too well focussed. Usually these can be recognized because they yield a graph which does not pass through the central point ($-\frac{1}{4}$, 50).

Comment

This experiment provides an introduction to random processes, which are often misunderstood. Brownian motion represents the mechanism for the random movement of particles in a liquid, or indeed a gas. The movement may manifest itself as diffusion, say of particles in a non-uniform colloidal suspension, or, if we examine a volume of fluid fixed in space, as fluctuations in concentration. An understanding of Brownian motion is thus fundamental to our knowledge of the structure of liquids and gases, and gives insights into such apparently unrelated problems as noise in electronic circuits[4] and the distribution of heat in a solid body.[7]

The historic importance of the experiment is demonstrated by the fact that Perrin received a Nobel prize for the work which he carried out on it in 1908.[6]

Technical notes

Apparatus[1]

Microscope, eyepiece,† [tumbler†], as described in Apparatus section; microscope slides with concave indent, cover slips, Canada balsam, brush kept in stoppered bottle of xylene solvent, stage graticule marked in hundredths of a millimetre. 2.5 ml bottle of ~1 μm diameter Dow uniform latex particles in water, probability graph paper.

Design notes

Design of tumbler is not critical. Eyepiece graticule is an image on reversal film of an original diagram, photographed down so that its line spacing is approximately twice the diameter of the particles when viewed through the microscope. Insert shims in microscope assembly so that the lens can just touch microscope slide, but cannot be racked through it. Keep bottle of suspension in refrigerator, and dissolve one small drop in a boiling tube of purified water to make stock solution. Stock solution must be almost transparent—a milky solution is too concentrated. Replace stock solution every week.

References

[1] *Brownian motion. An undergraduate laboratory experiment.* G. P. Matthews. *J. Chem. Educ.* **59,** 246 (1982).

[2] *A brief account of microscopical observations made in the months of June, July and August,* 1827, *on the particles contained in the pollen of plants; and on the general existence of active molecules in organic and inorganic bodies.* R. Brown. *Phil. Mag.* **4,** 161 (1828).

[3] *Investigations on the theory of the Brownian movement.* A. Einstein, tr. A. D. Cowper. Methuen, London (1926), ch. II, IV and V.

[4] *Noise and fluctuations: an introduction.* D. K. C. MacDonald. Wiley, New York (1962), p. 8.

[5] *Statistical thermodynamics.* J. F. Lee, F. W. Sears, and D. L. Turcotte. Addison-Wesley, Massachusetts, 2nd edn (1973) p. 299.

[6] *Crucial experiments in modern physics.* G. L. Trigg. Von Nostrand Reinhold, New York (1971) ch. 4.

[7] *Brownian motion and potential theory.* R. Hersh and R. J. Griego. *Sci. Am.* **220** (3), 67 (1969).

7.2 Computer simulation of phases by molecular dynamics

Over the past two decades, computer experiments have become an invaluable tool in the study of physical properties, especially those of liquids. Such studies employ computer programs which generate the positions and trajectories of a small number of particles, calculate the effects of collisions between them, and monitor various properties of the system such as temperature and pressure.

An important part of these programs is an equation known as a *potential function*, which describes the energy of interaction between the particles as a function of distance. An obvious goal of computer simulation experiments would be to mimic the properties of a real fluid by using an accurate potential function. In practice this is very difficult, either because the true interactions are unknown, or because they are expressed by very cumbersome equations. Nevertheless, although the interactions in the simulated system may not be entirely realistic, the fact that they are totally unambiguous means that precise, quasi-experimental data about a system may be obtained. Theoretical models may be tested and their properties compared with experimental results, and information may be obtained on properties of theoretical importance which cannot be easily measured in the laboratory.

In this experiment we carry out a *molecular dynamics* calculation, in which the particles are assigned starting positions, and initial, random velocities such that the net momentum of the system is zero. The subsequent trajectories of the particles are then calculated step by step on the assumption that they obey the classical equations of motion. Typically each step corresponds to a time interval of the order of 10^{-14} s, and between 10^3 and 10^5 steps are followed, depending on the property under investigation. As the simulation proceeds, running averages, and totals are calculated, and from these are obtained the values of various properties of the system such as its temperature, pressure, energy, and the structure of the system in terms of the radial distribution function explained below.

We specify the forces between the particles by an equation known as the *Lennard–Jones* (12–6) *pair potential energy function* or, more simply, as the *LJ potential.* The trajectories, velocities and distribution of the particles are followed at various temperatures and densities, and a number of properties of the solid, liquid and gas are calculated. The energies and distances involved in the calculations are scaled using parameters which relate to argon, and a comparison is made between the theoretical results and experimental measurements on the real system.

Theory[1]

(*i*) *Interaction potential*

The Lennard–Jones (12–6) potential function, Fig. 7.2.1, provides a very convenient approximation to the interaction energy between two non-polar particles, such as the molecules of a noble gas:

$$U(r) = 4\varepsilon[(\sigma/r)^{12} - (\sigma/r)^6]. \tag{7.2.1}$$

ε is the equilibrium energy, usually referred to as the *well depth*, and the *collision diameter* σ is the separation at which $U(r) = 0$. Another useful parameter is the *equilibrium separation*, r_m, at which $U(r) = -\varepsilon$, $\mathrm{d}U(r)/\mathrm{d}r$ is zero, and there is no force between the molecules. For the LJ potential, $r_m = 2^{1/6}\sigma$.

The negative r^{-6} term arises from the attractive, long range interactions known as *London* or *dispersion* forces. These can be thought of as arising from an instantaneous dipole moment in one particle inducing a dipole in a nearby particle, which then causes an attractive force.

When the particles approach each other more closely, there is a reduction in the electron density between the two nuclei, and the particles repel strongly. It is convenient to express this

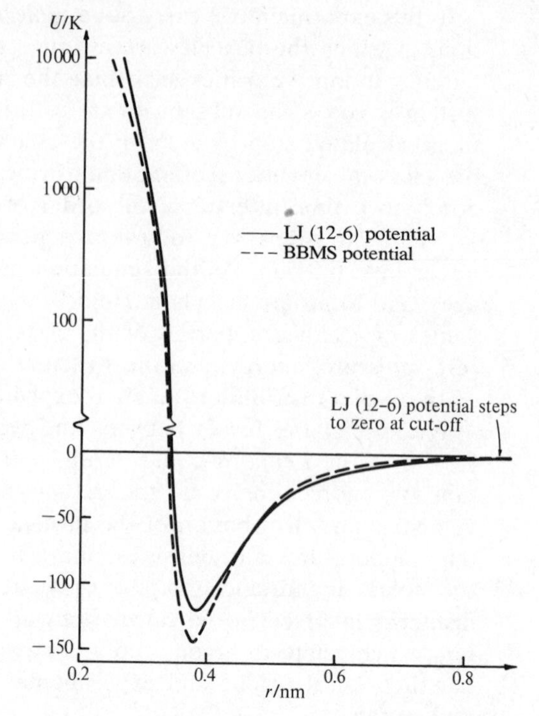

Fig. 7.2.1. Pair potential energy functions for argon. Note that the upper energy scale is logarithmic

repulsive interaction as an r^{-12} term, although a negative exponential term would have more theoretical justification.

It can be seen in the figure that the LJ potential predicts a small attractive interaction at long distances. For the sake of economy in computing time, it is customary to truncate the interaction at a separation r_c, where r_c is at most half the cube length L. In the present experiment, we employ a cut-off at a maximum value of r_c of 2.5σ, as shown in the figure. The contribution of more distant particles to properties such as energy and pressure is approximated by integration over a uniform distribution of particles beyond r_c.

Kelvins are unusual units for energy, but are convenient because then the well-depth is measured in the same units as the temperature of the sample, which is a measure of the kinetic energy of the particles.

Although as explained previously the LJ (12–6) potential does not accurately describe the interactions of any known chemical system, it is instructive to use the values of the well depth ε and collision diameter σ which give as near a reproduction as possible of the properties of a real system. Commonly used LJ parameters for argon are $\varepsilon/k_B = 120$ K and $\sigma = 0.3405$ nm.

Also shown in the figure is the Barker–Bobetic–Maitland–Smith or *BBMS potential* for argon. This is a realistic function containing ten complicated mathematical terms,[2] with $\varepsilon/k_B =$

142.5 K and $\sigma = 0.3355$ nm. It can be seen that its shape is different from that of the LJ function.

(ii) The virial term

Not only do we need to calculate the energy of a particle as a function of distance between it and its neighbours, but also the forces acting on it. Consider two particles a distance r apart, as shown in Fig. 7.2.2. If we take repulsive forces as being positive, then the force F_x acting in the x direction is

$$F_x = -\frac{\partial U}{\partial x} = -\frac{\partial U}{\partial r}\left(\frac{\partial r}{\partial x}\right). \tag{7.2.2}$$

By Pythagoras' theorem in three dimensons, (Fig. 7.2.2),

$$r = (x^2 + y^2 + z^2)^{1/2}, \tag{7.2.3}$$

and therefore

$$\frac{\partial r}{\partial x} = x/r. \tag{7.2.4}$$

Thus

$$F_x = -\frac{1}{r}\left(\frac{\partial U}{\partial r}\right)x, \tag{7.2.5}$$

and the force along r, represented by the vector $\mathbf{F}$, is

$$\mathbf{F} = -\frac{1}{r}\left(\frac{\partial U}{\partial r}\right)\mathbf{r}, \tag{7.2.6}$$

where $\mathbf{r}$ is a vector along r in the same direction as $\mathbf{F}$. The term $-1/r\ (\partial U/\partial r)$ is known as the *virial.*

(iii) The basic cell

A major limitation of any computer simulation is the number of particles which can be included. There are 6×10^{23} particles in a

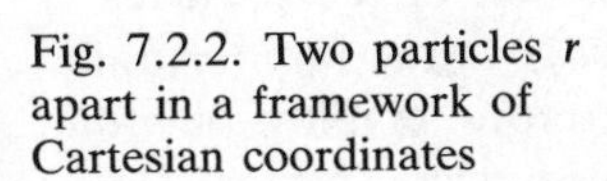

Fig. 7.2.2. Two particles r apart in a framework of Cartesian coordinates

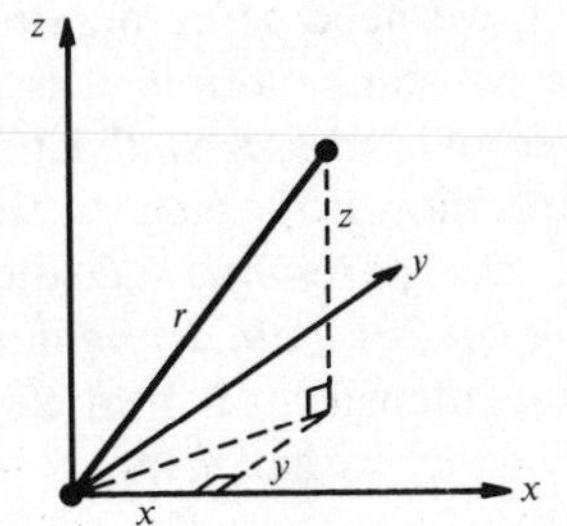

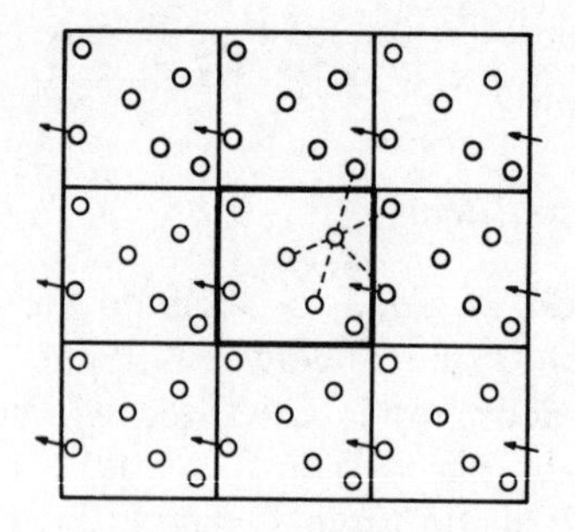

Fig. 7.2.3. The basic cell (in the centre), and effect of periodic boundary conditions

mole, but a computer would take many centuries to calculate even one time step for this number of particles. The size N of the sample which is studied in practice is usually less than 1000 particles, and in the present experiment N is only 32. To minimize surface effects, and simulate more closely the properties of an infinite system, a *periodic boundary condition* is imposed. This is illustrated for a two-dimensional system in Fig. 7.2.3. The particles under investigation lie in the central cell. When one particle leaves the cell, as shown by an arrow, the move is balanced by another particle entering the cell with the same velocity through the opposite face. The only interactions which are calculated are those between a chosen particle and *either* every other particle *or* its image in a neighbouring cell if that is closer. These interactions are shown by dashed lines in the figure. The procedure is known as *minimum imaging*, and is a corollary of periodic boundary conditions.

In this experiment we use cubic cells, and arrange the particles initially in a face-centred cubic lattice which corresponds to the arrangement in solid argon. Such a distribution requires $N=4n^3$ particles per cell, where n is an integer, and our 32 particle calculation clearly satisfies this condition.

(*iv*) *Radial distribution function*[3]

In studying the properties of the various phases of a Lennard–Jones (LJ) fluid, we need some measure of structure. Since the particles are in constant motion, it is not possible to judge the overall structure of a phase by the position of the particles at any one instant, nor is it easy to judge structure from the trajectories of the particles. We therefore employ a *radial distribution function*, or *pair distribution function* $g(r)$. $g(r)$ is a measure of the density of particles in a limitingly small volume element $d\tau$, situated at a distance r from a given

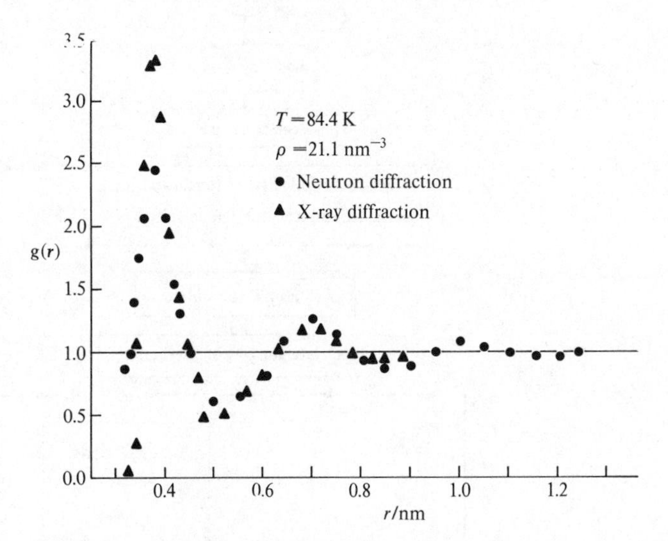

Fig. 7.2.4. Experimental determination of $g(r)$ for liquid argon near the triple point[4]

particle, relative to the average density ρ. Thus $\rho g(r)\,\mathrm{d}\tau$ is the average number of particles in the volume element. Note that this average number is not normalized to unity, but that

$$\int_0^\infty \rho g(r) 4\pi r^2\,\mathrm{d}r = N - 1 \approx N, \tag{7.2.7}$$

so that $g(r)$ is normalized to unity as $r \to \infty$.

For a perfect solid, $g(r)$ is a periodic array of spikes which reflect the certainty of finding other particles in the lattice at fixed distances from the selected particle. The spikes continue indefinitely with increasing r, although in practice $g(r)$ is never calculated beyond half the cube length or about 5σ, whichever is the smaller. A solid is said to have *long range order.*

Liquids have *short range order,* characterized by a hump in the pair distribution function at a distance of about r_m. However, at greater separations, $g(r)$ approaches the value of 1 as the structure around the chosen particle disappears, and the density approaches its average value for the liquid. The pair distribution function for liquid argon, obtained from neutron scattering studies and X-ray diffraction measurements, is given in Fig. 7.2.4.

Think carefully about the meaning of $g(r)$; why, for example, does $g(r) \to 0$ as $r \to 0$, and what do you think $g(r)$ for a gas will look like?

(*v*) *The computer program*

The flow diagram for the molecular dynamics program is shown in Fig. 7.2.5. At the beginning of every run, the particles are

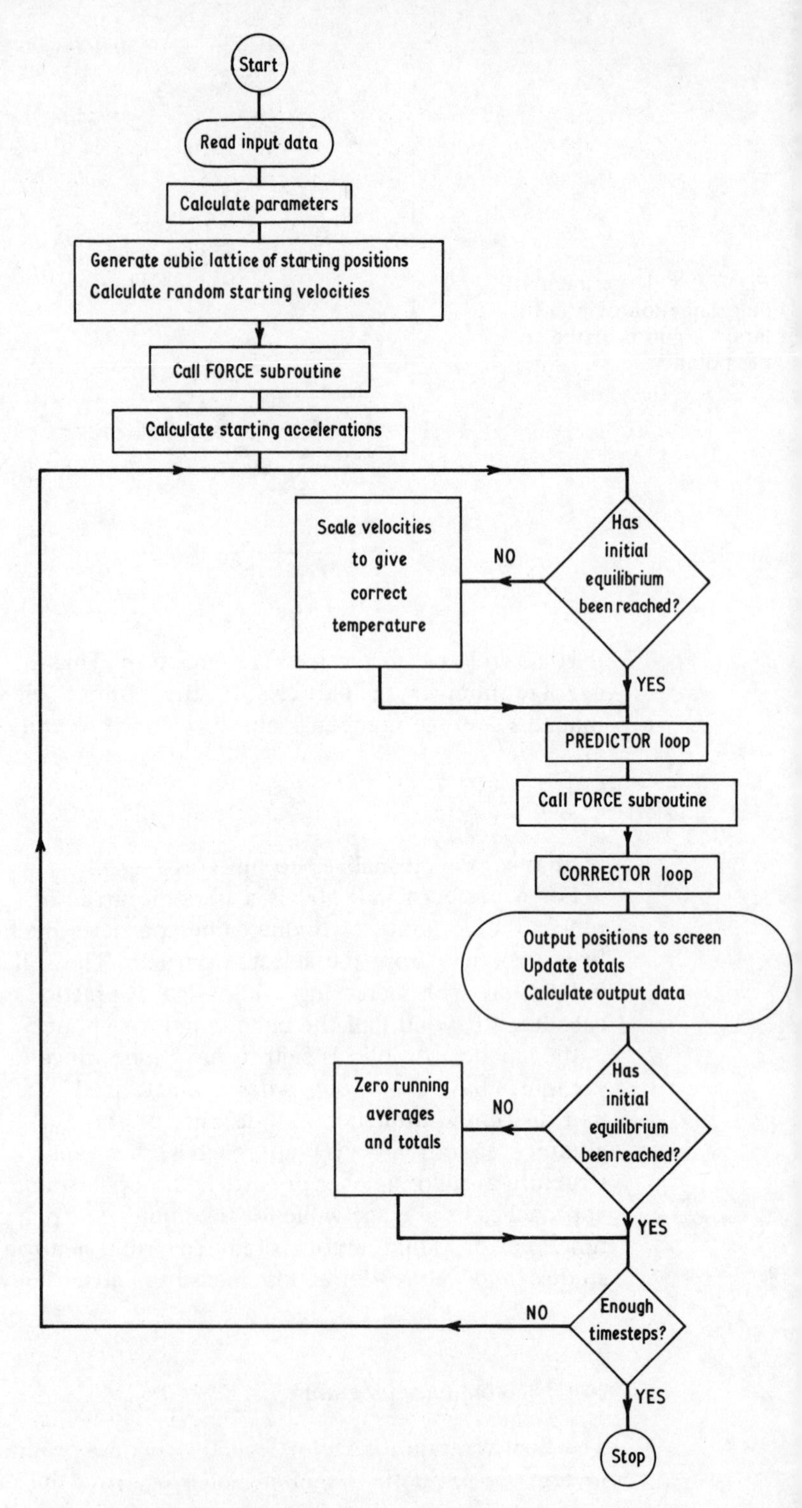

Fig. 7.2.5. Flow diagram for the molecular dynamics program

positioned in a lattice structure. This prevents the particles starting unnaturally close to each other, and, incidentally, makes the development of the structure of the phase very obvious.

After the initial co-ordinates and velocities of the particles have been assigned, a subroutine named FORCE is called. This subroutine calculates the distances between every particle and its neighbours. It then employs the LJ potential to find the force between any particles which are closer than the cut-off distance.

In the main part of the program, the trajectories are first calculated by the PREDICTOR loop on the assumption that there are no changes in the interactions between the particles. The methods used for solving Newton's equations of motion are well known,[1] but will not be detailed here. The FORCE subroutine is then called, and the CORRECTOR loop corrects the trajectories of any interacting particles. This predictor/corrector technique is an efficient method for molecular dynamics calculations.

The calculations continue until an initial, arbitrary equilibrium is reached, at which the memory of the initial lattice configuration has been lost. In the current program, the initial equilibrium is defined as being achieved when the particles have moved a specified total mean square distance, or when 500 timesteps have been calculated, whichever is the sooner.

After the initial equilibrium has been reached, the image on the screen, which shows the trajectories of the particles, is erased, and the running averages and totals are zeroed, including those used to calculate $g(r)$. The calculations then proceed for a specified number of timesteps. The running totals and averages are printed out periodically so that the progress of the calculations can be monitored, and at the end of the program the pair distribution function $g(r)$ is calculated. After the program has finished, print-outs of the running totals, $g(r)$ and the trajectories are obtained for analysis.

(*vi*) *Reduced quantities*

The molecular dynamics calculations in this experiment are calculated in terms of dimensionless *reduced quantities*, denoted by an asterisk, which may then be converted to real units for any chemical species containing non-polar, spherical particles by means of the appropriate values of ε and σ. Thus the reduced temperature T^* ('tee star') is $k_B T/\varepsilon$, where T is the absolute temperature and ε/k_B the well depth, both in kelvins. The reduced variables employed may be summarized as follows:

distance	$r^* = r/\sigma$,
energy	$U^*(r^*) = U(r)/\varepsilon$,
temperature	$T^* = k_B T/\varepsilon$,
pressure	$p^* = p\sigma^3/\varepsilon$,
volume	$V^* = V/\sigma^3$,
number density	$\rho^* = \rho\sigma^3 = N\sigma^3/V$,
timestep	$\Delta t^* = \Delta t/\tau_0$,
velocity	$v^* = dr^*/dt^* = r^*\tau_0/t$.

For the (12–6) function the time interval τ_0 is conventionally given the value:[6]

$$\tau_0 = (m\sigma^2)/48\varepsilon)^{1/2}, \tag{7.2.8}$$

where m is the molecular mass.

In reduced variables, the LJ potential becomes:

$$U^*(r^*) = 4[(r^*)^{-12} - (r^*)^{-6}]. \tag{7.2.9}$$

(*vii*) *Phase diagram for an LJ fluid*[5,6]

When carrying out the molecular dynamics simulation, we must be mindful of the phase diagram for a LJ fluid shown in Fig. 7.2.6. The boundary between the two-phase gas + liquid region and the fluid region is known as the *coexistence curve*. C is the

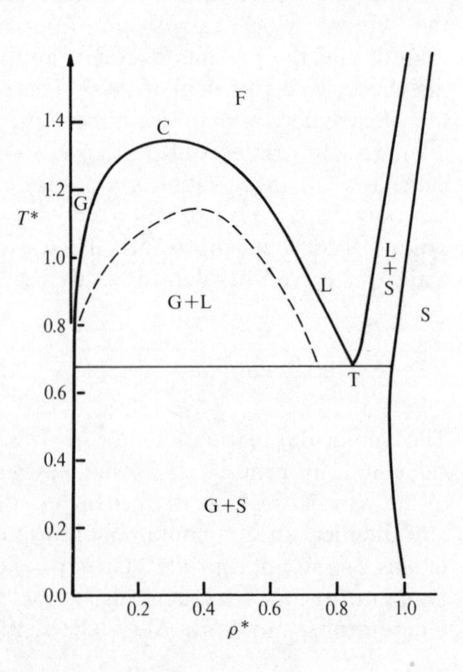

Fig. 7.2.6. Phase diagram of argon, showing the boundaries between solid (S), liquid (L), and vapour (G) or fluid (F). C is the critical point and T the triple point. —— full LJ (12–6) potential, ---- LJ potential truncated at 2.5σ, as in this experiment (the corresponding triple point is not known).

critical point and T the *triple point* of the system. If we were to run a simulation in one of the two-phase regions, e.g. $T^* = 1.0$ and $\rho^* = 0.4$, the pressure of the system would fluctuate over large numbers of time-steps.

Apparatus The molecular dynamics program used in this experiment is written in FORTRAN 77 or BASIC, and may be run on a main-frame, mini-, or microcomputer with printer.

Procedure (*i*) *Programming*

Your first task is to supply two functions which are missing from the molecular dynamics program already in the computer. When calculating the interaction energy of the particles, the FORCE subroutine calls a function eng(rdsq). This calculates the reduced potential for a value of rdsq $= (r^*)^2$. To calculate the force, a second function is called, namely pot(rdsq). This provides the virial term $-1/r^* \cdot (\partial U^*/\partial r^*)$.

Note that a great deal of computing time is saved by working in r^2 space rather than r space, because the distances between the particles are calculated by Pythagoras' theorem in three dimensions, and finding the square-root of a number is a very inefficient process.

Suppose that the interaction between two particles was given by the function $U^*(r^*) = 1/(r^*)^{18}$. Then your energy subroutine in FORTRAN might look like this:

(6 spaces before each line)

```
function eng(rdsq)
eng = 1./rdsq **9.
return
end
```

The corresponding virial term would be given by

$$-\frac{1}{r^*}\left(\frac{\partial U^*}{\partial r^*}\right) = -\frac{1}{r^*}\left(\frac{-18}{(r^*)^{19}}\right) = \frac{18}{(r^*)^{20}},$$

and therefore the function might read:

(6 spaces before each line)

```
function pot(rdsq)
rdsqi = 1./rdsq
r4 = rdsqi*rdsqi
r8 = r4*r4
pot = 18.*r8*r8*r4
return
end
```

The computer program will call the functions which you write a great many times, and if they are inefficiently written the program will take much longer to run. The fastest way of calculating several power terms with a common factor is to carry out sequential multiplications of the type shown in the second example above. Notice also that the numbers have decimal points after them, to show that they are real numbers rather than integers. This is because it is dangerous to carry out *mixed-mode arithmetic* in which real numbers and integers are mixed together. Note also that the FORTRAN language assumes that any variables beginning with i, j, k, l, m, or n are integers.

Write function routines for the potential and virial terms of the Lennard–Jones (12–6) potential (eqn (7.2.9)). Type these into the computer, and compile and load them with the molecular dynamics program as instructed.*

(*ii*) *Trying the program*

The use of too coarse a timestep may cause two particles to take up positions unnaturally close to each other; if this happens the system will 'blow up', with both the running averages and the trajectories taking on seemingly random and illogical values.

All the runs in this experiment should be carried out using a timestep of 10^{-14} s.

The program and your functions should first be tried out by specifying an artificially short initial equilibrium period at a total mean square displacement of 2σ, and a total run of only 10 timesteps. Study a gas (e.g. $T^* = 1.2$ and $\rho^* = 0.05$). Run the program as instructed.*

The system should run to initial equilibrium with T^* constant. This condition is imposed on the system by a continual rescaling of velocities. After the initial equilibrium, the total energy should be constant. The constancy of the energy is a check that the equations of motion are being solved correctly. Strictly speaking, the energy should remain absolutely steady (to within 1 part in 10^6), but with an artificially short initial equilibrium period, it may be found that it drifts somewhat.

Having ascertained that your function routines work correctly, print out copies of them for your account of the experiment.

At the end of this and every run of the program, print out the map of the particle trajectories, the running averages and the (cumulative) pair distribution function $g(r)$. Carefully label all the output. Indicate the magnitude of σ on the trajectory diagrams.

(*iii*) *Running the program*

To obtain $g(r)$ for a LJ solid, run the program for only 100 timesteps with a brief initial equilibrium (total mean square displacement 0.05σ) somewhere in the solid region on the phase diagram.

Then carry out a run for a LJ liquid corresponding to the conditions shown in Fig. 7.2.4, using an initial equilibrium condition of 100σ and a total of 2000 timesteps of 10^{-14} s.

Finally run the program for LJ gas (e.g. $T^* = 1.2$, $\rho^* = 0.05$) with the other conditions as above.

You may also wish to try other conditions of your own choosing. For example, you could test the effect of using only the repulsive r^{-12} term of the Lennard–Jones (12–6) function, or investigate the two-phase region. You could also run the program with parameters for krypton instead of argon ($\varepsilon/k_B = 200$ K, $\sigma = 0.358$ nm), remembering that after the initial calculation of the absolute quantities, the run will be identical to an argon run with the same *reduced* parameters, and should therefore not be duplicated.

(*iv*) *Logging off*

Before finally logging off from the computer, delete the function routines you have written, the input file, and all the output files.

Calculation For the liquid argon run, display the velocities of the thirty-two particles as histograms as shown in Fig. 7.2.7. Find out approximately how many encounters at $r < r_c$ are required per particle before the velocities take up the Gaussian distribution shown in the figure.

Comment on the difference in the pair distribution function $g(r)$ for solid, liquid and gaseous argon. Explain how $g(r)$ illustrates the structure of the three phases. Compare your results for liquid argon with the experimental results shown in Fig. 7.2.4 by plotting both sets of points on the same graph, and explain any differences.

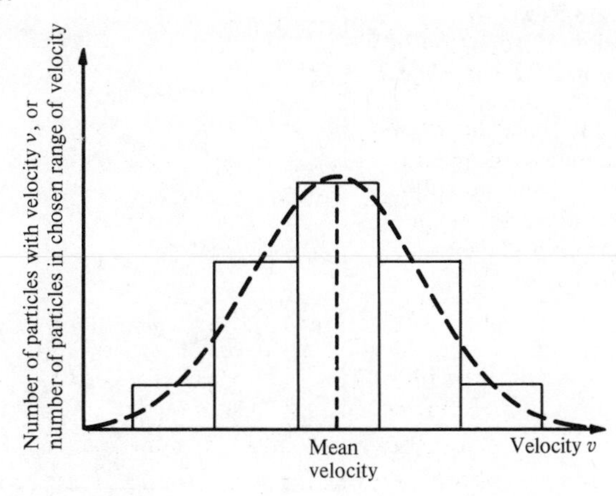

Fig. 7.2.7. Gaussian distribution of velocities, and a histogram corresponding to this distribution

Note that the concept of mean free path length is only strictly defined for hard spheres; real particles whose interaction potential extends to $r = \infty$ never travel in exactly straight lines.

From the tracks of the argon particles in the gas, estimate the mean path length between encounters. Note that the projections of the tracks in two-dimensions are, on average, a factor of $(3/2)^{1/2}$ shorter than the actual three-dimensional trajectories. Compare the mean path length with the kinetic theory expression for the mean free path length λ:

$$\lambda = 1/(2^{1/2}\pi\rho\sigma^2). \tag{7.2.10}$$

Results

Figure 7.2.7 shows a histogram of the velocities of the particles when they have assumed a Gaussian distribution. Tracks for a LJ gas at $T^* = 0.8$, $\rho^* = 0.01$ are shown in Fig. 7.2.8.

Accuracy

For such a small number of particles and with a limited number of timesteps, one cannot be sure that the system has reached a good enough equilibrium to give reliable results. If a few hundred particles are simulated, a run of 1000 steps is usually sufficient to give the internal energy and the non-perfect contribution to the pressure to an accuracy of the order of 1 per cent. Other problems in molecular dynamics systems with a small number of particles include the fact that periodic phenomena with a wavelength greater than the cell length are suppressed (these include critical point phenomena).

Comment

Computer simulations may also be carried out by *Monte Carlo* methods. As the name implies, these calculations involve a

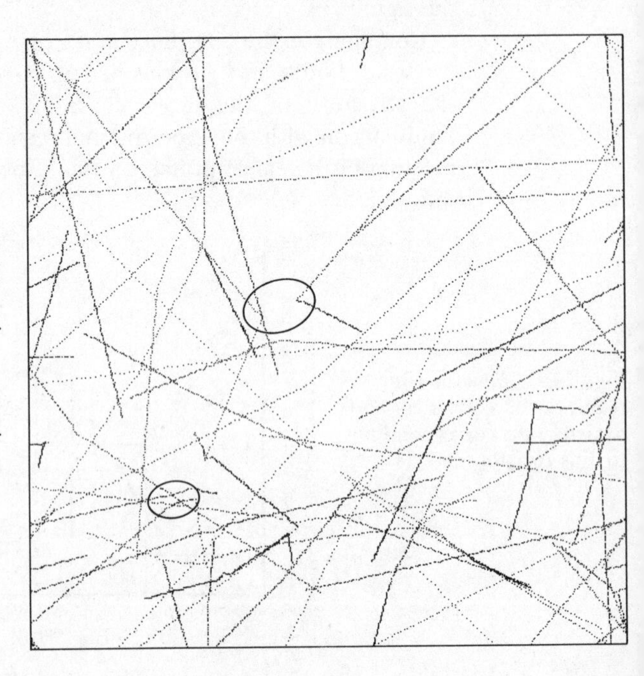

Fig. 7.2.8. Two-dimensional projection of tracks for a 32-particle LJ fluid at $T^* = 0.8$ and $\rho^* = 0.01$. Each dot represents a time-step of 10^{-14} s. A head-on collision has occurred at the centre, and a glancing collision at bottom left

random, gambling element.[7] A system of N particles interacting through some known potential is allocated a set of arbitrarily chosen initial coordinates. A sequence of configurations of the particles is then generated by successive random displacements. However, some of the configurations which occur are rejected to ensure that the overall selection is characteristic of a thermal equilibrium system. The order in which the configurations occur is arbitrary, since no time scale is involved.

If only static properties are required, Monte Carlo methods, in which the number of particles N, the volume V, and temperature T are fixed, often give more useful results than molecular dynamics calculations in which N, V, and total energy U are fixed. However, molecular dynamics calculations have the great advantage that time-dependent phenomena may be studied. An example is the calculation of velocity auto-correlation functions which show how the velocities of the particles change with time.[8] Inhomogeneous systems, such as the liquid–gas interface, have also been studied by molecular dynamics methods, and the surface tension of liquid droplets calculated.[9]

An 'ensemble' in thermodynamics is characterized by the fixing of three properties; the Monte Carlo method is a simulation of the canonical ensemble (constant N, V, T), and the molecular dynamics method is a simulation of the microcanonical ensemble (constant N, V, U).

Technical notes

Apparatus

See Apparatus section. The runs take about 15 min on a large mainframe computer, about 15 h on a fast mini-computer, and up to 4 min per timestep on a Commodore 64. Programs in FORTRAN 77 and Commodore 64 BASIC available from author.

References

[1] *Theory of simple liquids.* J. P. Hansen and I. R. McDonald. Academic Press, London (1976), ch. 3.
[2] *The intermolecular pair potential of argon.* G. C. Maitland and E. B. Smith. *Mol. Phys.* **22,** 861 (1971).
[3] *Statistical mechanics.* D. A. McQuarrie. Harper and Row, New York (1976), p. 258.
[4] *Radial distribution functions of fluid argon.* A. A. Khan. *Phys. Rev.*, 2nd series, **134A,** 367 (1964).
[5] Ref. 1, p. 2.
[6] *Equation of state for the Lennard–Jones fluid.* J. J. Nicolas, K. E. Gubbins, W. B. Streett, and D. J. Tildesley. *Mol. Phys.* **37,** 1429 (1979).
[7] *The application of Monte Carlo methods to physicochemical problems.* M. A. D. Fluendy and E. B. Smith. *Quart. Rev.* **XVI,** 241 (1962).
[8] *Computer 'experiments' on classical fluids. III Time-dependent self-correlation functions.* D. Levesque and L. Verlet. *Phys. Rev.*, 3rd series, **2A,** 2514 (1970).
[9] *Molecular theory of capillarity.* J. S. Rowlinson and B. Widom. Clarendon Press, Oxford (1982), ch. 6.

7.3 Gas viscosities by capillary flow viscometer

In this experiment we determine the viscosity of a gas by measuring the time which the gas takes to flow through a capillary tube under a known pressure drop. The capillary tube is immersed in an oil bath so that a range of temperatures may be studied. Capillary flow experiments are more suited to relative rather than absolute measurements, for reasons explained below. In this experiment, therefore, the relative viscosities of argon and air are measured by determining their relative flow times over the temperature range 298–500 K. The absolute viscosity of argon is then calculated from published measurements of the viscosity of air, and the results compared with the predictions of a rigorous kinetic theory of gases.

Theory

(*i*) *Laminar flow*

The gas pressure and dimensions of the capillary tube used in this experiment are carefully chosen so that the gas passes through the tube by *laminar* or *Poiseuille flow*. This is streamline viscous flow in which the fluid velocities are symmetrically distributed about the axis of the capillary tube and the velocity at the wall of the tube is zero.

First let us consider the laminar flow of an incompressible fluid through a capillary tube, Fig. 7.3.1. The viscous shearing stress acting over the surface of a short cylinder is, by definition, equal to $\eta A(\mathrm{d}v/\mathrm{d}a)$, where A is the surface area of the cylinder, $(\mathrm{d}v/\mathrm{d}a)$ is the velocity gradient across the tube, and η is the *shear viscosity coefficient* or simply the *viscosity*. For a short cylinder of the fluid of radius a and length $\mathrm{d}x$, the surface area

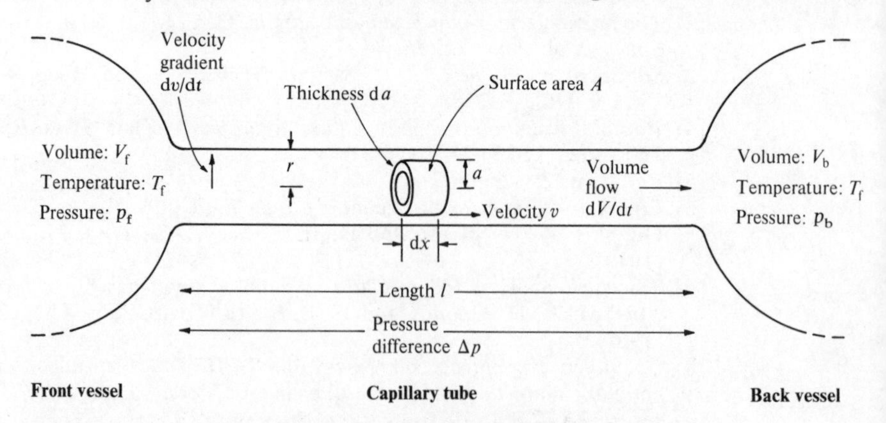

Fig. 7.3.1. Parameters describing the laminar flow of a fluid through a capillary tube

A is $2\pi a\,\mathrm{d}x$. We can equate the viscous shearing stress with the force $\pi a^2\,\mathrm{d}p$ arising from the pressure difference $\mathrm{d}p$ between the ends of the cylinder:

$$\pi a^2\,\mathrm{d}p = \eta 2\pi a\,\mathrm{d}x(\mathrm{d}v/\mathrm{d}a). \tag{7.3.1}$$

We now choose any particular position along the tube, and integrate the velocity v of the gas over the radius a of the tube. If we assume that $v=0$ at the walls, where $a=r$, then

$$v = -\frac{1}{4\eta}\frac{\mathrm{d}p}{\mathrm{d}x}(r^2-a^2). \tag{7.3.2}$$

To calculate the volume flow rate through the capillary tube, we consider a series of concentric cylinders of circumference $2\pi a$ and thickness $\mathrm{d}a$. Each of these cylinders travels at its own velocity v, and the total volume flow rate $\mathrm{d}V/\mathrm{d}t$ is therefore calculated by integrating over all these cylinders:

$$\begin{aligned}\frac{\mathrm{d}V}{\mathrm{d}t} &= \int_0^r 2\pi a v\,\mathrm{d}a = \int_0^r \frac{2\pi a}{4\eta}\frac{\mathrm{d}p}{\mathrm{d}x}(a^2-r^2)\,\mathrm{d}a\\ &= -\frac{\pi r^4}{8\eta}\frac{\mathrm{d}p}{\mathrm{d}x}.\end{aligned} \tag{7.3.3}$$

On integrating over the length l of the tube, we find that

$$\frac{\mathrm{d}V}{\mathrm{d}t} = \frac{\Delta p\,\pi r^4}{8\eta l}, \tag{7.3.4}$$

where Δp is the difference in pressure between the ends of the capillary tube.

However, since a gas is compressible, the volume flow varies along the tube. Assuming the gas to be perfect:

$$\frac{\mathrm{d}V}{\mathrm{d}t} = \frac{RT}{p}\frac{\mathrm{d}n}{\mathrm{d}t}, \tag{7.3.5}$$

where p is the mean pressure of a small volume $\mathrm{d}V$ containing $\mathrm{d}n$ moles of gas. Since $\mathrm{d}n/\mathrm{d}t$ is independent of position along the tube, combining eqns (7.3.3) and (7.3.5) and then integrating over the length l of the capillary gives *Poiseuille's equation* for a compressible fluid:

$$\frac{\mathrm{d}n}{\mathrm{d}t} = -\frac{\pi r^4 p}{8\eta RT}\frac{\mathrm{d}p}{\mathrm{d}x} = \frac{\pi r^4}{16\eta RTl}(p_\mathrm{f}^2 - p_\mathrm{b}^2), \tag{7.3.6}$$

where p_f and p_b are the pressures in the front and back vessels, Fig. 7.3.1.

Assuming that the gas in the front vessel behaves perfectly,

$$p_\mathrm{f} V_\mathrm{f}/T_\mathrm{f} = n_\mathrm{f} R. \tag{7.3.7}$$

If V_f varies so slowly that it may be regarded as constant for the time dt, we may relate dn/dt to the rate of pressure drop in the front vessel by:

$$-\frac{\mathrm{d}p_f}{\mathrm{d}t}\left(\frac{V_f}{T_f}\right)=-\frac{\mathrm{d}n_f}{\mathrm{d}t}R. \qquad (7.3.8)$$

Therefore from eqn (7.3.6)

$$-\frac{\mathrm{d}p_f}{\mathrm{d}t}=\frac{\pi r^4}{16\eta l}\frac{T_f}{TV_f}(p_f^2-p_b^2). \qquad (7.3.9)$$

For a pressure drop in the front vessel from p_f^0 to p_f^t, integration gives

$$-\int_{p_f^0}^{p_f^t}\frac{V_f\,\mathrm{d}p_f}{(p_f^2-p_b^2)}=\frac{\pi r^4}{16\eta l}\frac{T_f}{T}t, \qquad (7.3.10)$$

where t is the time of flow.

Equation (7.3.10) may in principle be used to determine the viscosity of a gas from measurements of flow times t under known pressure conditions. However, as can be seen, the flow time is dependent on the fourth power of the capillary tube radius r, and it is impracticable to measure this radius to an accuracy four times greater than the required accuracy of the viscosity measurements (p. xxii). Capillary viscometers are therefore always used for measurements relative to a standard gas of known viscosity.

There are two relative methods. Method A is to determine the flow times t_a and t_b for two gases (viscosities η_a and η_b), using the same capillary tube under identical conditions of temperature and pressure. Then from eqn (7.3.10):

$$t_a/t_b=\eta_a/\eta_b. \qquad (7.3.11)$$

Method B is to measure flow times t_1 and t_2 for the same gas at different capillary temperatures T_1 and T_2. Under the same pressure conditions, eqn (7.3.10) shows that

$$\frac{t_1}{t_2}=\frac{\eta_1}{\eta_2}\frac{l_1r_2^4}{l_2r_1^4}\frac{T_1}{T_2}. \qquad (7.3.12)$$

Since the expansion of the capillary tube is negligible in the temperature range of the present experiments,

$$\frac{t_1}{t_2}=\frac{\eta_1T_1}{\eta_2T_2}. \qquad (7.3.13)$$

(*ii*) *Other types of flow*

There are in fact five different types of flow which a gas may undergo as it passes through a capillary tube. One may define a *Knudsen number* $K = \lambda/d$, where λ is the mean free path and d the tube diameter, and for the laminar flow which we have just discussed, K is $\sim 10^{-5}$.

At pressures such that the mean free path is of the order of 1–10 per cent of the capillary diameter, ($K \approx 0.01$ to 0.1), *slip flow* occurs, i.e. the layer of gas next to the wall of the tube instead of being stationary, slips along it. The reason for this can be simply visualized as follows. The mean velocity of the gas infinitesimally close to the wall is not zero, for although half of the molecules suffered their last collision at the wall, the other half last collided with a molecule at a distance from the wall of the order of the mean free path. The flow times through the capillary tube must therefore be corrected by means of the expression

$$t_{\text{corrected}} = t_{\text{measured}}\left(1 + \frac{4\lambda}{r}\right), \tag{7.3.14}$$

where λ is the mean free path at the mean pressure. By a simple form of kinetic theory:

$$\lambda \approx \frac{\eta(3RT/M)^{1/2}}{[(p_f + p_b)/2]}. \tag{7.3.15}$$

Turbulent flow occurs when the fluid velocity v increases to such an extent that the flow becomes full of irregular eddying motions. A safe criterion for laminar flow is that the Reynolds number $N_{Re} = 2r\rho v/\eta$ is less than 1000. Above this value, especially for a slightly curved capillary, turbulence may occur and the above analysis would be inapplicable.

At low pressures where the mean free path is comparable to the capillary diameter, *Knudsen* or *free molecular flow* occurs, and molecule-wall collisions govern the flow rate. Alternatively, if the mass flow rate through the capillary tube is sufficiently great or the pressure at the pump sufficiently small, the gas flow velocity may become comparable to the speed of sound at some point along the capillary tube. The resulting *sonic flow* causes the pressure to remain constant, and then to drop suddenly to p_0 at the outlet when the gas expands laterally. Sonic flow is prevented in this experiment by maintaining the downstream pressure at 10 mm Hg, rather than connecting the end of the capillary tube directly to a vacuum pump.

(*iii*) *Comparison with kinetic theory*

It is instructive to compare the viscosity measurements obtained in this experiment with values predicted by the kinetic theory of gases. The simplest kinetic theory expression for viscosity is derived from a model of a gas made up of rigid, elastic, non-interacting spheres of diameter σ moving with mean velocity $\bar{c}$. The net result of the random thermal motions of the molecules is to transport momentum from faster to slower moving layers; it is this momentum transport which tends to counteract the velocity gradient set up by the shear forces acting on the gas. The resulting expression is:

$$\begin{aligned}\eta &= \tfrac{1}{3}\rho\bar{c}\lambda \\ &= \tfrac{1}{3}Nm(8k_BT/\pi m)^{1/2}\cdot\frac{1}{\sqrt{2}\pi N\sigma^2} \\ &= \tfrac{2}{3}\frac{(mk_BT)^{1/2}}{\pi^{3/2}\sigma^2},\end{aligned} \tag{7.3.16}$$

where N is the number of molecules per unit volume and m the molecular mass.

This simple model predicts firstly that the viscosity of a gas should be independent of density or pressure, and secondly that it should be proportional to $T^{1/2}$. At pressures of about 1 atm the viscosity is indeed found to be independent of pressure; however, both at high pressures (where $\lambda \sim \sigma$) and low pressures (where $\lambda \gg \sigma$ and the molecules collide with the vessel walls more frequently than with each other) the viscosity coefficient is found to vary with pressure. The temperature dependence is found to be nearer $T^{0.7}$ than $T^{0.5}$, which is a consequence of the forces which real molecules exert on each other.

To obtain realistic results, it is necessary to evaluate the velocity-distribution function of the molecules and to consider the dynamics of molecular collisions.[1] We then find that

$$\eta = \frac{5(\pi mk_BT)^{1/2}}{16\pi\sigma^2}. \tag{7.3.17}$$

Of course, real gas molecules are not rigid elastic spheres, but interact with each other at a distance with a pair potential energy $U(r)$. This intermolecular potential may be approximated by the Lennard–Jones (12–6) potential, described on p. 337 and in Fig. 7.2.1. The intermolecular potential has the effect of making the diameter of a molecule which is undergoing a collision appear temperature-dependent. Equation (7.3.17) must be modified for a real gas to

$$\eta = \frac{5(\pi m k_B T)^{1/2}}{16\pi\sigma_{RS}^2 \Omega^{(2,2)*}}. \tag{7.3.18}$$

The correction factor $\Omega^{(2,2)*}$ is known as an *omega integral*, and is a function of $T^* = k_B T/\varepsilon$ only (p. 343). It is of the order of unity and accounts for the apparent temperature dependence of the molecular diameter. Its precise value can be calculated from integrations over the pair potential energy function.

Apparatus

The layout of the capillary viscometer is shown in Fig. 7.3.2. The front vessel is a one litre bulb, and the pressure of gas in this bulb is measured with a cathetometer (p. 68). The capillary tube, and preheater in the form of a coiled tube of suitable length, are immersed in an oil bath. The pressure of the back vessel is measured with a McLeod gauge, a modified mercury manometer for measuring low pressures.

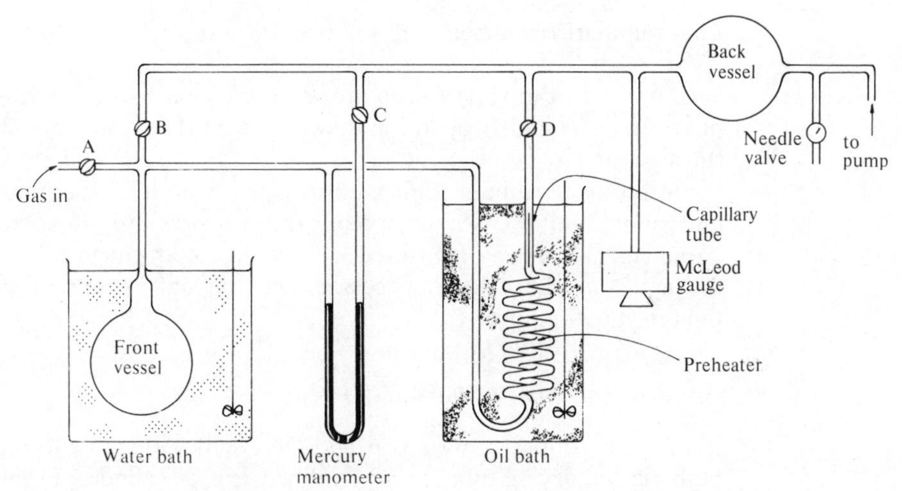

Fig. 7.3.2. The capillary viscometer

Procedure

Keep well clear of the hot oil bath. Mercury splashes must be avoided: read the following instructions carefully, and always operate the taps slowly keeping an eye on the mercury manometer.

(*i*) *Measurements to be made*

Measurements of the relative viscosities of argon and air should be made at several temperatures in the range 298 to 500 K. A suitable scheme is given below. Each of the runs in (a) and (b) takes about half an hour.

(a) A calibration experiment at 298 K using air at an initial pressure of about 500 mm Hg, making measurements of the time required for the reservoir pressure to drop to about 450, 400, and 350 mm Hg. Repeat this experiment to test the results for reproducibility. If the results differ by more than a few percent, seek advice.
(b) Two runs at 298 K for argon under the same conditions as in (a). Calculate the viscosity of argon from your measurements by method A.
(c) If your 298 K measurements give consistent and reproducible results, it is permissible to carry out the elevated temperature experiments using one pressure drop only—say 500 to 50 mm Hg. Record the flow times for air and argon for three more temperatures in the range 298–500 K.

(*ii*) *Set up*

First familiarize yourself with the operation of a McLeod gauge ('Vacustat').*

Check that the rotary vacuum pump is open to the atmosphere (taps A and B open), and switch it on. Leave it running throughout the experiment.

Flush out the tubing connected to tap A with air (or argon), and pump out the whole apparatus to a pressure of about 1 mm Hg (1 Torr) as measured on the McLeod gauge. Close down the needle valve if necessary, but on no account overtighten it.

(*iii*) *Viscosity measurement*

Isolate the sample bulb (taps B and D closed). Admit air to the bulb via the drying tube, or argon direct from a cylinder, to the required initial pressure (500 mm Hg for the first experiment).

Adjust the cathetometer so that the base and telescope are horizontal. Record the levels in the manometer. Measure the pressure with the McLeod gauge, and add this to the manometer reading to obtain the true bulb pressure. Close off the manometer side-arm (tap C).

Open the needle valve sufficiently to give a downstream pressure of about 10 mm Hg.

Adjust the position of the cathetometer to a level on the non-evacuated manometer limb corresponding to the first 'final pressure' value (450 mm Hg in the first experiment). Note the reading.

Allow gas to flow through the capillary tube by opening tap

D, and at the same time start a stopwatch. Observe the mercury level through the cathetometer and note the time as the meniscus passes the horizontal cross-wire. A piece of paper placed with its edge level with the horizontal cross-wire may help to define better the point at which the meniscus passes it. Readjust the cathetometer to a higher position corresponding to the next pressure drop (40 mm Hg in the first experiment) and level the telescope. Note the total time from the beginning of the run for the mercury to reach this level. Repeat this procedure for each pressure drop. Monitor the bulb and capillary tube temperature throughout the run. Check periodically that the base of the cathetometer is still horizontal.

In all subsequent experiments, it is essential to fill the bulb to the same initial pressure, and record flow times for the same pressure drops, as those used in the reference run. The reference will be air at the same temperature for method A, and either argon or air at 25 °C for method B.

Calculation

(i) Calculation of viscosities

Experimental measurements of the viscosity of air have been fitted to a smoothing curve of the type

$$\ln \eta = A \ln T + B/T + C/T^2 + D, \tag{7.3.19}$$

where T is the temperature in kelvins, and η the viscosity in units of $10^{-7}\,\mathrm{kg\,m^{-1}\,s^{-1}}$. Calculate the viscosity of the air standard at each temperature, given that $A = 0.669878$, $B = 42.96133$, $C = -12680.57$ and $D = 1.408$.[2]

Now calculate the approximate viscosity of argon at each temperature using method A, eqn (7.3.11). Then use your values of η for air and argon to correct all the flow times for slip flow, eqns (7.3.14) and (7.3.15).

Check that all the other deviations from Poiseuille flow are negligible for the conditions of your experiments.

Calculate the true viscosity of argon at each temperature using method A. Also calculate the viscosities of argon and air at each of the elevated temperatures using method B.

Estimate the experimental error associated with each measurement and hence with your quoted viscosities. Compare your results graphically with published values for air[2] and argon,[3] remembering to associate error bars with both your results and the literature values before commenting on the agreement.

(ii) Comparison with kinetic theory

Using your results for the viscosity of argon, calculate the rigid sphere diameter σ of an argon atom from eqn (7.3.17) at each

Table 7.3.1. The omega integrals $\Omega^{(2,2)*}$ for calculating viscosities for the Lennard–Jones (12-6) potential[4]

T^*	$\Omega^{(2,2)*}$	T^*	$\Omega^{(2,2)*}$
1.90	1.197	2.70	1.069
1.95	1.185	2.80	1.058
2.00	1.175	2.90	1.048
2.10	1.156	3.00	1.039
2.20	1.138	3.10	1.030
2.30	1.122	3.20	1.022
2.40	1.107	3.30	1.014
2.50	1.093	3.40	1.007
2.60	1.081	3.50	1.000

experimental temperature, and plot a graph of T against σ. Comment on the temperature dependence and compare your values with σ_{LJ} given below.

Values of ε/k_B and σ have been calculated for argon on the assumption that the properties of all the noble gases differ only by the change in certain scaling parameters.[5] For the LJ (12–6) potential they are $\varepsilon/k_B = 152.8$ K and $\sigma_{LJ} = 0.329$ nm. Using these approximate values for the potential parameters, together with eqn (7.3.18) and the tabulated correction factors $\Omega^{(2,2)*}$ (Table 7.3.1), calculate the viscosity of argon at several temperatures within your experimental range. In order to save time, you will find it best to evaluate the viscosity at temperatures such that the reduced temperature $T^* = k_B T/\varepsilon$ has a value included in the tabulations.

Note how these parameters for argon differ from the two sets given on p. 338, and in general how the estimates of well-depth and collision diameter vary according to the potential used and property studied.

Plot these calculated viscosities on your graph and compare them with your experimental results.

Results

Accuracy

This experiment is capable of yielding results accurate to 5 per cent, better than those of early research workers (see Comment).

Comment[1]

Considerable controversy has surrounded the results of viscosity determinations of even the most common gases. Experiments in the 1930s and 1940s produced results which agreed closely with the predictions of the Lennard–Jones (12–6) potential. In the early 1970s, however, these were shown to be in error by as much as ten per cent at temperatures above 1000 K. Also at this time, *inversion methods* were developed, based on the kinetic theory employed in this experiment, which could convert experimental measurements into numerical potential functions.

The application of these methods to the new results has given potential functions for the noble gases and their mixtures which agree well with functions derived from molecular beam experiments, the spectra of noble gas dimers, and measurements of gas imperfection (Expt 1.2).

Technical notes

Apparatus

As shown in Fig. 7.3.2,[†] including pyrex glass capillary tube radius 0.1 mm, length 20 cm;[††] stopwatch; instructions for McLeod gauge; [hood and fan over oil bath to remove fumes].

Design note

The pre-heater and capillary tube must be arranged so that convection effects do not oppose the gas flow; errors of ~10 per cent may otherwise result. The entry and exit joins to the capillary tube should be made as smooth as is possible by ordinary glass-blowing techniques. If a capillary tube of different dimensions is used, check that all the conditions for laminar flow are satisfied. Helical capillaries require corrections for curved pipe flow.[6]

References

[1] *Intermolecular forces.* G. C. Maitland, M. Rigby, E. B. Smith, and W. A. Wakeham. Clarendon Press, Oxford (1981) p. 299 and ch. 1.
[2] *Viscosities of oxygen and air over a wide range of temperatures.* G. P. Matthews, C. M. S. R. Thomas, A. N. Dufty, and E. B. Smith. *J. Chem. Soc. Far. Trans I* **72,** 238 (1976).
[3] *Critical reassessment of viscosities of* 11 *common gases.* G. C. Maitland and E. B. Smith. *J. Chem. Eng. Data* **17,** 150 (1972).
[4] *Molecular theory of gases and liquids.* J. O. Hirschfelder, C. F. Curtiss and R. B. Bird. John Wiley, New York, and Chapman and Hall, London (1954) p. 1126.
[5] *Proc. 4th Symposium on thermophysical properties.* ASME, New York (1968), article by R. Dipippo and J. Kestin, p. 204.
[6] *Viscosities of the inert gases at high temperatures.* R. A. Dawe and E. B. Smith. *J. Chem. Phys.* **52,** 693 (1970).

7.4 Dissociation constants and solubility from conductance of electrolyte solutions

Many compounds wholly or partly dissociate into ions when dissolved in water. Aqueous ions conduct electricity, and so conductance measurements can be used to measure the degree of dissociation. In this experiment we use such measurements to determine the acidity constant of ethanoic (acetic) acid, the acidity constant of anilinium chloride (aniline hydrochloride) and the solubility of potassium periodate KIO_4.

Theory[1]

(i) Measuring conductivity

The *resistivity* ρ of a material is the resistance of a unit cube to a current flowing normally between two opposite faces, and is equal to (potential gradient)/(current density). If, instead of being a unit cube, the conductor is of length l and cross-sectional area a, then

$$\rho = \frac{Ra}{l}, \tag{7.4.1}$$

where R is the resistance. The *electric conductivity* κ is the reciprocal of the resistivity:

$$\kappa = 1/\rho = l/Ra. \tag{7.4.2}$$

The SI units of κ are $\mathrm{S\,m^{-1}}$, where S (= siemens) is a reciprocal ohm (Ω^{-1}, or 'mho'). However, electric conductivities are usually quoted in units of $\mathrm{S\,cm^{-1}}$ or $\Omega^{-1}\,\mathrm{cm^{-1}}$.

A more useful quantity when dealing with electrolyte solutions is the *molar conductivity*, since it enables one to compare the conductivities of equivalent numbers of ions under similar circumstances. The molar conductivity Λ is defined as the conductivity per mole of electrolyte, i.e.

$$\Lambda = \kappa/c, \tag{7.4.3}$$

where c is the electrolyte concentration. The units of Λ usually quoted are $\mathrm{S\,cm^2\,mol^{-1}}$; in this case the units of c must be $\mathrm{mol\,cm^{-3}}$.

The molar conductivity of an electrolyte solution can be measured by immersing a *conductivity cell* in it. A simple

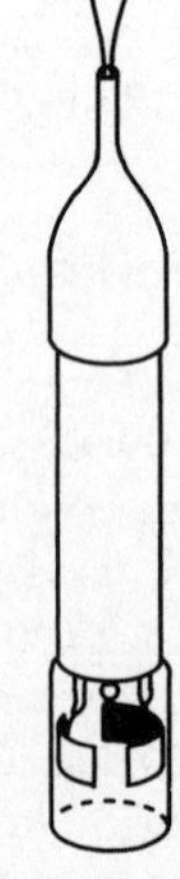

Fig. 7.4.1. A commercial conductivity cell

commercial conductivity cell is shown in Fig. 7.4.1, and comprises two platinum electrodes mounted inside an open glass tube. For any particular conductivity cell the ratio l/a is a constant known as the *cell constant C*. From eqns (7.4.2) and (7.4.3),

$$\Lambda = l/Rac = C/Rc = GC/c, \tag{7.4.4}$$

Some conductivity meters have a control which is set to the cell constant of the cell in use; they then display the conductivity κ (usually in μS cm^{-1}) from which the molar conductivity can be calculated using eqn (7.4.3).

where G is the observed conductance (S or Ω^{-1}). To avoid having to construct cells of uniform and accurately known dimensions, the cell which is to be used is filled with a standard solution of concentration c and known molar conductivity Λ. The conductance G of the solution is then measured, and the cell constant C calculated from the expression above.

(*ii*) *The conductivity of strong and weak electrolytes*

The variation of molar conductivity with concentration, Fig. 7.4.2, or dilution, Fig. 7.4.3, reveals two classes of behaviour, and we call the two types of electrolytes *weak* and *strong*.

Strong electrolytes such as potassium chloride dissociate fully into ions when dissoved in water, and therefore have relatively high molar conductivities. At low concentrations this molar conductivity has been found to obey an expression of the type

$$\Lambda = \Lambda_0 - A'c^{1/2}, \tag{7.4.5}$$

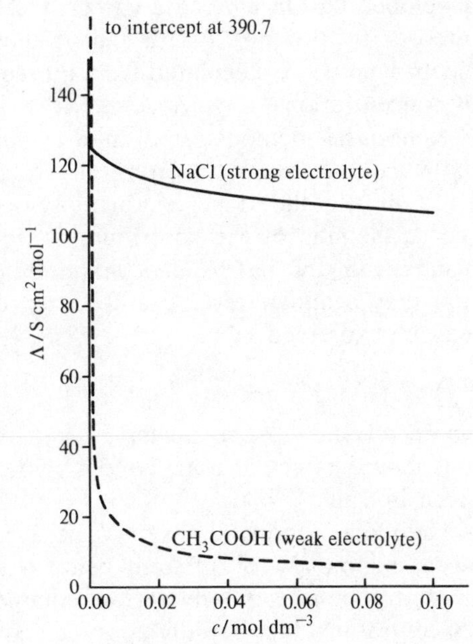

Fig. 7.4.2. Molar conductivities of a strong and a weak electrolyte as a function of concentration c

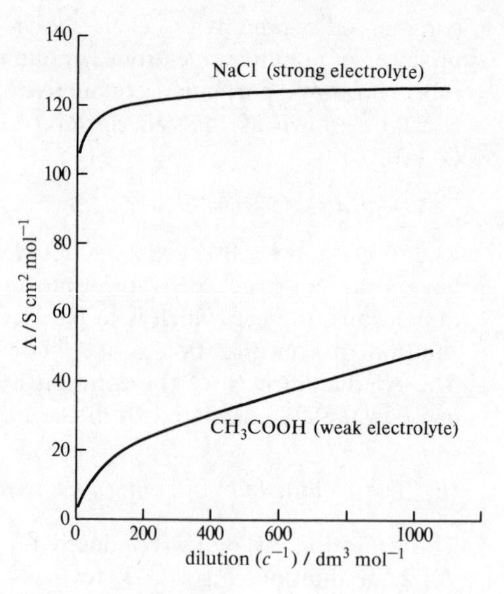

Fig. 7.4.3. Molar conductivities of a strong and weak electrolyte as a function of dilution c^{-1}

where Λ_0 is the molar conductivity at infinite dilution ($c = 0$), and A' is a constant which depends on the solvent, the temperature, the nature of the electrolyte (whether it dissociates into singly or multiply charged cations and anions), and the value of Λ_0. The variation in Λ with concentration is caused by interactions between the ions; a theoretical expression for A' was developed by *Onsager*, and agrees well with experiment.[2] The importance of eqn (7.4.5) is that it allows Λ_0 values for strong electrolytes to be calculated from measurements of Λ at a series of concentrations.

Kohlrausch demonstrated another important property of the molar conductivities of strong electrolytes, namely that at infinite dilution the molar conductivity of any salt can be expressed as the sum of the contributions from individual ions. This results from the *independent migration of ions* (at infinite dilution they cannot interact), and for an electrolyte $\nu_+ A^{z+} \nu_- B^{z-}$ may be expressed

$$\Lambda_0 = \nu_+ \lambda_+^0 + \nu_- \lambda_-^0, \tag{7.4.6}$$

where λ_+^0 and λ_-^0 are the *ionic conductivities* of the individual ions. Some values of ionic conductivities at infinite dilution are given in Table 7.4.1.

Verify from Table 7.4.1 that Λ^0 ($BaCl_2$) is 280.1 S cm^2 mol^{-1} at 25 °C.

The molar conductivities of weak electrolytes, such as organic acids, exhibit a very different behaviour. At higher concentrations they have a relatively low conductivity, but at very low concentrations they rapidly approach conductivities comparable

Table 7.4.1. Ionic conductivities at infinite dilution and 25 °C[3]

Cation	λ^0/S cm^2 mol^{-1}	Anion	λ^0/S cm^2 mol^{-1}
H^+	349.8	OH^-	197.8
Li^+	38.7	Cl^-	76.4
Na^+	50.1	Br^-	78.2
K^+	73.5	IO_4^-	54.5
NH_4^+	73.4		
Ba^{2+}	127.3		

to those of strong electrolytes. This behaviour may be explained by the dominant effect of the degree of dissociation α of the electrolyte on its molar conductivity. α is related to the observed molar conductivity Λ by the *Arrhenius expression*

$$\alpha = \Lambda/\Lambda_i \approx \Lambda/\Lambda_0, \tag{7.4.7}$$

where Λ_i is the hypothetical molar conductivity which the solution would have at a concentration αc, if it were fully dissociated. As α is small, the approximation $\Lambda_i \approx \Lambda_0$ is quite a good one.

In this experiment we wish to determine the dissociation (acidity) constant of ethanoic acid, for which we clearly need to find a value of α and hence of Λ_0. However, it is evident from Fig. 7.4.2 that there is no possibility of finding Λ_0 with any accuracy from extrapolation of Λ to $c = 0$.

Thus we must employ one of two other methods of calculating α. The first method is to employ Kohlrausch's expression (7.4.6), using ionic conductivities which have been calculated from the Λ_0 value for strong electrolytes derived from eqn (7.4.5).

The second method employs a simple equilibrium expression, often referred to as *Ostwald's dilution law*. Consider the ionization of a weak acid HA

$$\begin{array}{ccc} H_2O + HA & \leftrightharpoons & H_3O^+ + A^-. \\ c(1-\alpha) & & c\alpha \quad\quad c\alpha \end{array}$$

The dissociation constant or *acidity constant* K_a, expressed in terms of concentrations rather than activities, is

$$K_a = \frac{c\alpha \cdot c\alpha}{c(1-\alpha)} = \frac{c\alpha^2}{1-\alpha}. \tag{7.4.8}$$

From eqn (7.4.7) it follows that

$$K_a = \frac{c(\Lambda/\Lambda_0)^2}{1-\Lambda/\Lambda_0} = \frac{c\Lambda^2}{\Lambda_0(\Lambda_0 - \Lambda)}. \tag{7.4.9}$$

Rearranging:

$$c\Lambda = K_a(\Lambda_0)^2(1/\Lambda) - K_a\Lambda_0. \qquad (7.4.10)$$

The linearity of this plot is partly fortuitous; the approximation $\alpha = \Lambda/\Lambda_0$ and the use of ionic concentrations rather than activities both produce small errors, but these largely cancel each other.[4]

A graph of $c\Lambda$ against $1/\Lambda$ should therefore be linear, and K_a and Λ_0 may be found from the slope and intercept.

(iii) The hydrolysis of salts

Aqueous solutions of salts of weak bases and strong acids, such as anilinium chloride which we shall refer to as BHCl can be considered to be fully dissociated:

$$\begin{array}{lcll} \mathrm{BHCl} & \rightarrow & \mathrm{BH^+} + & \mathrm{Cl^-} \\ & & c & c \end{array}$$

The species BH^+ then undergoes partial hydrolysis:

$$\begin{array}{llll} \mathrm{BH^+} & + \mathrm{H_2O} \leftrightharpoons & \mathrm{B} + & \mathrm{H_3O^+}, \\ (1-\alpha)c & & \alpha c & \alpha c \end{array}$$

where α the degree of hydrolysis. Assuming that the total molar conductivity Λ is the sum of the molar conductivities of the charged species, then

$$\Lambda = (1-\alpha)\lambda_{\mathrm{BH^+}} + \alpha\lambda_{\mathrm{H_3O^+}} + \lambda_{\mathrm{Cl^-}}. \qquad (7.4.11)$$

An excess quantity of the free base B (aniline) is then introduced into the solution. The base suppresses the hydrolysis so that $\alpha \rightarrow 0$ and the molar conductivity becomes

$$\Lambda' = \lambda_{\mathrm{BH^+}} + \lambda_{\mathrm{Cl^-}}. \qquad (7.4.12)$$

It follows directly from eqns (7.4.11) and (7.4.12) that

$$\alpha = \frac{\Lambda - \Lambda'}{\lambda_{\mathrm{H_3O^+}} - \lambda_{\mathrm{BH^+}}} = \frac{\Lambda - \Lambda'}{\lambda_{\mathrm{HCl}} - \lambda_{\mathrm{BH^+Cl^-}}}. \qquad (7.4.13)$$

Therefore

$$\alpha \approx \frac{\Lambda - \Lambda'}{\lambda^0_{\mathrm{H_3O^+}} + \lambda^0_{\mathrm{Cl^-}} - \Lambda'}. \qquad (7.4.14)$$

Thus the dissociation or *hydrolysis constant* K_h may be evaluated, since

$$K_h = \frac{c\alpha^2}{1-\alpha}. \qquad (7.4.15)$$

The acidity constant K_a of the species BH^+ is related to K_h by the expression

$$K_a = K_w/K_h, \qquad (7.4.16)$$

where K_w is the ionic product for water.

(iv) Solubility

Derive an approximate equation to calculate the solubility of potassium periodate from a measurement of the conductivity of a saturated solution at 25 °C.

Apparatus

A typical conductivity cell used to measure the conductivity of an electrolyte solution was shown in Fig. 7.4.1. The platinum electrodes in the cell may have been *blacked* (given a finely divided surface) to promote rapid equilibrium between the electrodes and the solution. The measurement of conductance in the cell is not simply a question of applying a voltage across the electrodes and measuring the resulting direct current, since this would give rise to polarization at the electrodes and electrolysis of the solution. It is necessary to apply an alternating current to prevent this, and to construct an electronic circuit not only to measure the resistance but also to compensate for capacitance effects caused by the alternating current. The circuitry of a typical modern instrument is shown in Fig. 7.4.4. Other instruments employ bridge circuits in which both the resistance and capacitance of the cell is balanced.[5,6]

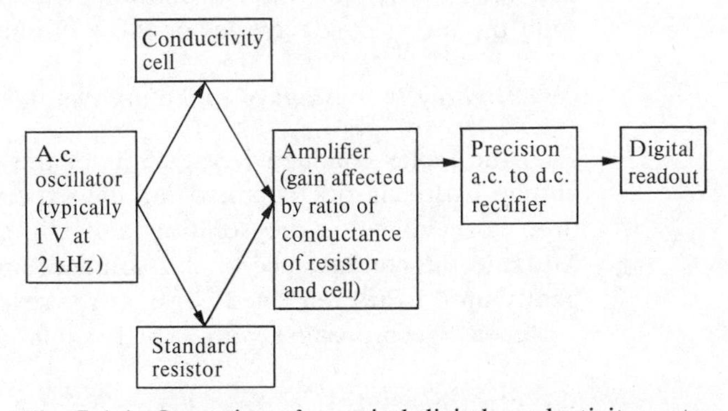

Fig. 7.4.4. Operation of a typical digital conductivity meter

It is important to mount the conductivity cell and the sample solution in a thermostat tank, because the conductance of electrolyte solutions changes by as much as 2 per cent per kelvin. It is also important to use purified water from which CO_2 has been removed by boiling in a flask connected to an aspirator.

Procedure

(i) Set up

Set up and switch on the conductivity cell and conductivity meter or bridge as directed.*

Check with an independent thermometer that the water is near 25 °C and measure its temperature accurately.

Boil some distilled water under vacuum, using the flask and aspirator provided. Lightly stopper the flask while the water is still hot.

Before each conductivity measurement, wash the conductivity probe with a little of the purified water. Then wash it with a little of the sample solution. Wash out the sample tubes in a similar manner. Allow the samples at least 10 min to attain thermal equilibrium before taking any readings.

(ii) Calibration of conductivity meter

Calibrate the conductivity meter,* or measure the cell constant given that the electric conductivity of 0.02 M KCl is 2.777×10^{-3} S cm^{-1}.[2]

(iii) Acidity constant of ethanoic acid

Check the stock solution of ethanoic acid by titration.* Make up five solutions of concentration 0.05 M, 0.025 M, 0.02 M, 0.01 M, and 0.005 M by quantitative dilution with the purified water. Find the molar conductivities of the solutions.

(iv) Hydrolysis constant of anilinium chloride

Find the molar conductivity at 25 °C of the 0.05 M solution of aniline hydrochloride supplied for the experiment. Add a few drops of aniline to the solution and stir gently for 15 min. Measure the conductance of this solution, and continue to stir gently until it has remained constant for several minutes.

Repeat the procedure with a 0.01 M solution.

(v) Solubility of potassium periodate

Add excess potassium periodate to some of the purified water in a sample tube, and stir for 5 min. Leave the tube in the thermostat tank for 15 min and then find the electric conductivity of the solution.

Calculation Subtract the conductance of the purified water from all the measurements you have made, except those for ethanoic acid. The exception arises from the fact that the major contribution to the conductivity of the pure water comes from the dissociation of carbonic acid formed from dissolved carbon dioxide, and the presence of ethanoic acid suppresses this dissociation.

By Kohlrausch's method calculate the acidity constant of ethanoic acid in all the solutions studied. If your values of K_a show a trend with increasing c, rather than a random scatter, plot $\ln K_a$ against αc and extrapolate to $\alpha c = 0$. If the results simply show a random scatter, average them. So obtain a value for K_a at infinite dilution.

Also calculate Λ^0 and K_a by means of eqn (7.4.10).

Calculate the acidity constant of anilinium chloride, making any reasonable assumptions necessary. Find the solubility (not the solubility product) of potassium periodate at 25 °C.

Results

Accuracy

Your discussion of the results for ethanoic acid should include a critical comparison of the methods of finding K_a. The literature value of K_a is 1.75×10^{-5} mol kg^{-1} $\simeq 1.75 \times 10^{-8}$ mol cm^{-3}. *Do conductivity measurements provide an accurate method of measuring acidity constants and solubilities?*

Comment

With greater care over measurement of the conductance, thermostatting, and purification of both solute and solvent, results of very high precision can be obtained and can be a very useful tool in the determination of equilibrium constants. Conductance measurements are particularly suited to studies on non-aqueous solvents and on systems at high temperatures and pressures.

Technical notes

Apparatus

Conductivity meter,[$] conductivity cell, thermostat tank, thermometer, round-bottomed flask connected to aspirator and mounted over gauze and bunsen burner, sample tubes, volumetric solutions of 1 M ethanoic (acetic) acid and M/20 anilinium chloride, research grade KCl, aniline, potassium periodate.

References

[1] *Ions in solution 2, introduction to electrochemistry.* J. Robbins, Clarendon Press, Oxford (1972), ch. 2.
[2] *The physical chemistry of electrolytic solutions.* H. S. Harned and B. B. Owen, Reinhold, New York, 3rd edn (1958), ch. 6 and appendices.
[3] Ref. 2, p. 231.
[4] Ref. 2, p. 310.
[5] *Electrolyte solutions.* R. A. Robinson and R. H. Stokes. Butterworths, London, 2nd edn (1970), p. 87.
[6] *A simple conductivity bridge for student use.* G. M. Muha. *J. Chem. Educ.* **54,** 677 (1977).

7.5 Temperature variation of conductivity and viscosity

Early this century Walden noticed that the product of the molar conductivity of an electrolyte at infinite dilution, Λ_0, and its viscosity η is not greatly dependent on the solvent and almost independent of temperature:

$$\Lambda_0 \eta \approx \text{const.}$$

In this experiment we test this observation, known as *Walden's rule*, and see what insight it gives into the nature of the movement of ions in an electrolyte solution.

Theory

(*i*) *Walden's rule*[1]

Let us assume that the ions in an electrolyte solution are spheres of radius a moving in a viscous, structureless fluid. We then apply hydrodynamic considerations to the motion, and obtain the *Stokes equation*

$$F = 6\pi a \eta v, \tag{7.5.1}$$

The precise magnitude of the coefficient 6π depends on the assumptions made about the behaviour at the surface of the sphere in the velocity field created in the fluid.[2]

where v is the drift velocity of an ion under an applied force F in a fluid of viscosity η.

If the charge on an ion is ze, where z is positive for cations and negative for anions, and the ion is in an electric field E, then it experiences a force

$$F = zeE. \tag{7.5.2}$$

It follows from eqns (7.5.1) and (7.5.2) that for a particular ion,

$$E \propto \eta v. \tag{7.5.3}$$

By definition, the mobility u of an ion is its velocity under a unit applied field. Therefore for an electric field of strength E:

$$u = v/E. \tag{7.5.4}$$

The molar conductivity Λ_0 of an ion at infinite dilution is directly proportional to its mobility u. Combining the various relations, we conclude that

$$\Lambda_0 \propto u = v/E \propto 1/\eta, \tag{7.5.5}$$

and that therefore

$$\Lambda_0 \eta = \text{const}, \tag{7.5.6}$$

which is Walden's rule. $\Lambda_0\eta$ is known as the *Walden product.* For dilute solutions we may assume that the observed conductivity Λ is very similar to Λ_0 and also proportional to u, and that therefore $\Lambda\eta$ is constant.

(ii) Temperature dependence of viscosity[3] *and conductivity*

While testing Walden's rule, it is of interest to determine the actual form of the temperature dependence of η at constant pressure. To a good approximation, we find that

$$\eta = A \exp(B/RT), \tag{7.5.7}$$

where A and B are constants. The theory of liquids is very complex, and there has been much debate as to the reason for this relation. Undoubtedly the dominant factor is the change in the density of the liquid with temperature. If density changes are prevented by measuring the viscosities of liquids at constant volume, rather than at constant pressure, η is found to vary much less. It nevertheless still decreases exponentially with temperature, following an equation of the type shown, but with a smaller value of B. Some workers have explained this by comparing eqn (7.5.7) with an Arrhenius-type expression (eqn (8.2.5), p. 401), and have concluded that viscosity is an activated process with some determinable activation energy B.

In general it is often useful to determine a constant of this type when some property of a system varies exponentially with temperature, even when the exact nature of the underlying process is unknown. In this experiment we determine the activation energy B for the variation of the viscosity η with temperature, and a corresponding parameter C for the temperature dependence of conductivity:

$$\Lambda = A' \exp(C/RT). \tag{7.5.8}$$

Apparatus

The conductivities of the solutions are measured with a conductivity bridge or meter connected to a conductivity cell, as described on p. 365.

The liquid viscosities are measured by means of an *Ostwald viscometer*, which provides a convenient method of making fairly accurate relative determinations. The operation of this device has been described on p. 297 and in Fig. 6.1.3.

The radius r of the viscometer's capillary tube is such that the flow of liquid through it is laminar (non-turbulent). The flow may therefore be described by *Poiseuille's equation* for a non-compressible fluid (eqn (7.3.4), p. 351), which we write in the form:

If the radius of the capillary tube is too large the flow becomes turbulent; if t is too small, dust and grease may affect the flow times.

$$\eta = \frac{\Delta p\, \pi r^4}{8l} \frac{t}{V}. \tag{7.5.9}$$

V is the volume of liquid flowing in a time t through the capillary tube of length l. r, V, and l are the same for two runs. The pressure difference Δp across the capillary tube depends on the head of liquid and its density ρ. Assuming these terms remain constant with temperature (as discussed in the Results section), then

$$\eta \propto t. \tag{7.5.10}$$

Both the conductivity cell and the viscometer are mounted in a glass-fronted thermostat tank.

Procedure

The experiment involves the measurement of the viscosity and conductivity of a 10^{-2} M solution of lithium chloride at five temperatures in the range 25–45 °C.

Check with a thermometer that the temperature of the water in the thermostat tank is 25.0 °C.

Obtain about 300 cm^3 of distilled water. Connect the flask provided to an aspirator, and boil the water under vacuum. *Lightly* stopper the flask while the water is still hot. Use the water to make up a lithium chloride solution of concentration 10^{-2} M$\pm$5 per cent in a 100 cm^3 volumetric flask. (The concentration need not be measured more precisely than this.)

Thoroughly clean the viscometer and check that there are no particles in its capillary tube. Rinse the viscometer with distilled water, followed by propanone (acetone), and finally dry it by sucking air through it with the aspirator provided.

Calibrate the conductivity meter,* or measure the cell constant (p. 361) given that the electric conductivity of 0.02 M KCl is 2.777×10^{-3} S cm^{-1}.[4]

Mount in the thermostat tank the conductivity cell in its sample tube, and also the viscometer. Leave them there for the duration of the experiment. Fill the sample tube with the lithium chloride solution, and add the required amount of solution into the viscometer from a pipette.

Leave the solution for 10 min to reach thermal equilibrium. Measure the conductivity of the solution. Then measure the flow time of the liquid by the method described in the previous section. Make two further measurements of flow time, and another of conductivity.

Repeat the procedure at 5° intervals up to 45 °C.

At the end of the experiment, turn the thermostat tank back to 25 °C ready for other users, because the water takes a long time to cool.

Calculation Draw a graph of your results to determine whether they obey Walden's rule over the temperature range studied. Since LiCl dissociates into ions which are highly hydrated, we might expect deviations from Walden's rule because the ions are not behaving as spheres in a homogeneous viscous fluid. Is this what is found?

Draw a second graph to show how well the temperature dependence of the viscosity of the liquid obeys eqn (7.5.7), and find the parameter B in units of kJ mol^{-1}.

By a similar method, find the corresponding parameter C for the variation of conductivity with temperature (eqn (7.5.8)).

Results[4] *Accuracy*

Both the density and the volume of the sample of water change with temperature. The head of water is kept very nearly constant by the fact that the lower liquid level is in a wide bulb. The change in density over the temperature range is only ~0.5 per cent, and can be ignored.

Comment The important conclusion which we should draw from this experiment is that the dominant effect on conductivity is the viscosity of the solvent, and that the effects of ionic interactions mentioned on p. 361 are merely perturbations on this.

The Walden product $\Lambda_0\eta$ is a useful parameter in the study of electrolyte solutions, and continues to be the subject of research.[5] For lithium chloride, the Walden product changes with solvent, which suggests that the Li^+ ions (and possibly the Cl^- ions) are hydrated in water. Further evidence for this is given by the unexpectedly low conductivity of Li^+ in relation to its small size (Table 7.4.1, p. 363).

Technical notes

Apparatus

Conductivity bridge or meter;$ conductivity cell; Ostwald viscometer; glass-fronted thermostat tank maintained to ±0.05 K or better; distilled, or preferably double-distilled, water; flask with bunsen burner and aspirator; stopclock; lithium chloride.

Design note

Choice of capillary radius in viscometer: a liquid of density ρ flowing through a tube of radius r with a velocity v will exhibit laminar flow if the 'Reynolds number' $2r\rho v/\eta$ is less than 1000 (p. 353). Minimum flow times must be at least 100 s to give reasonable precision. A nominal viscometer constant of 0.01 cSt $s^{-1} = 10^{-4}$ St s^{-1} satisfies these conditions.

References

[1] *Physical chemistry.* P. W. Atkins. Oxford University Press, Oxford, 3rd edn (1986), sect. 27.1.
[2] *Theory of simple liquids.* J. P. Hansen and I. R. McDonald. Academic Press, London (1967), p. 277.
[3] *The liquid state.* J. A. Pryde. Hutchinson University Library, London (1969), ch. 9.
[4] *The physical chemistry of electrolytic solutions.* H. S. Harned and B. B. Owen. Reinhold, New York, 3rd edn (1958), p. 285 and refs. cited therein.
[5] *Electrochemistry. Specialist Periodical Reports,* **1,** 92 (1970), (by G. J. Hills).

7.6 Transport numbers by moving boundary method

When two electrodes are placed in an electrolyte solution and a potential is applied across them, a current passes. The current is carried through the solution by the movement of the solvated ions, positive cations moving towards the negative cathode, and negative anions towards the positive anode. However, as demonstrated on p. 362, the mobilities, and hence conductivities, of the cations and anions in an electrolyte solution are unlikely to be the same. Consequently different fractions of the total current are carried by the different ions, a phenomenon perfectly compatible with the maintenance of the overall electrical neutrality of the solutions.[1] The fractions are known as the *transport numbers* or *transference numbers* of the cations and anions.

Extreme values of transport numbers can occur in certain solid and fused electrolytes; in silver chloride and bromide crystals all the current is carried by the cations, whereas in barium and lead chloride it is carried entirely by anions.

The most accurate direct method of measuring transport numbers, and often the simplest, is the *moving boundary method.* Two electrolyte solutions are used, with a common anion or cation. They are placed in a tube in such a manner that there is a distinct, visible boundary between them. An electrode is positioned at each end. A known quantity of electricity is passed and, under correctly chosen conditions, the boundary remains sharp. From a knowledge of the amount of electricity passed, and the extent of movement of the boundary, the transport number of the faster non-common ion may be measured.

In this experiment the formation of the boundary is achieved very simply. Dilute hydrochloric acid is used as one electrolyte solution. The second solution, cadmium chloride, is then formed from the dissolution of a cadmium anode as the current is passed, giving rise to what is known as an *autogenic boundary.* The boundary is made visible by the addition of an indicator, and its speed of movement yields the transport number of the aqueous H^+ ion relative to Cl^-.

Theory

(*i*) *Derivation of the transport number equation*[2]

The total current passing through a solution under a given potential is proportional to the sum of the ionic mobilities of the cations, u_+, and anions, u_-, (p. 368). The cations carry a fraction $u_+/(u_+ + u_-)$ and the anions a fraction $u_-/(u_+ + u_-)$, and it is these fractions which are the transport numbers t_+ and t_-. Clearly $(t_+ + t_-)$ is always unity.

Figure 7.6.1 shows, in schematic form, the system which we shall study. It represents a vertical section through an upright tube in which the boundary at x_0 is formed by the juxtaposition of two solutions of electrolytes, NX and MX, with a common anion X. A steady current i is passed through the tube for a time t, corresponding to the passage of a charge $i \cdot t$. During the time t, the boundary moves from the position x_0 to the position x_t. All the cations M in the volume V of solution between x_0 and x_t must, during this time, have crossed the plane at x_t. If M is in the form of univalent cations of concentration c, then the amount of electricity carried by them is cVF, where F is the Faraday constant. Then the transport number of the cations M is given by the relation:

$$t_{M^+} = \frac{\text{current carried by M}^+ \text{ ions}}{\text{total current}} = \frac{cVF}{it}. \tag{7.6.1}$$

(*ii*) *Complications*[3]

The success of the moving boundary method depends crucially on the maintenance of a sharp boundary between the electrolyte solutions. We now consider briefly the various disturbing effects on the boundary, and how they are minimized or overcome.

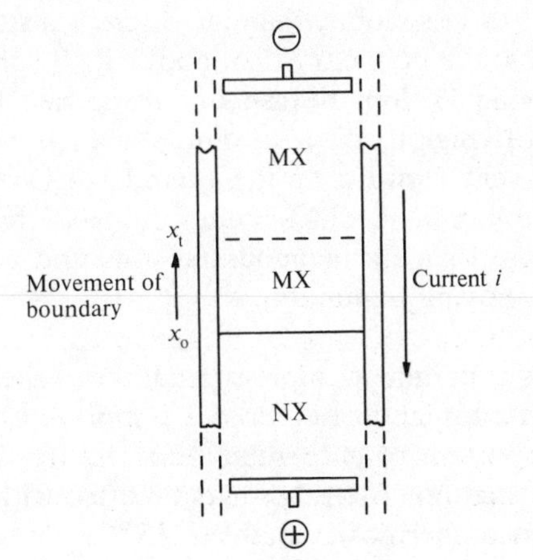

Fig. 7.6.1. Schematic representation of the tube and boundary

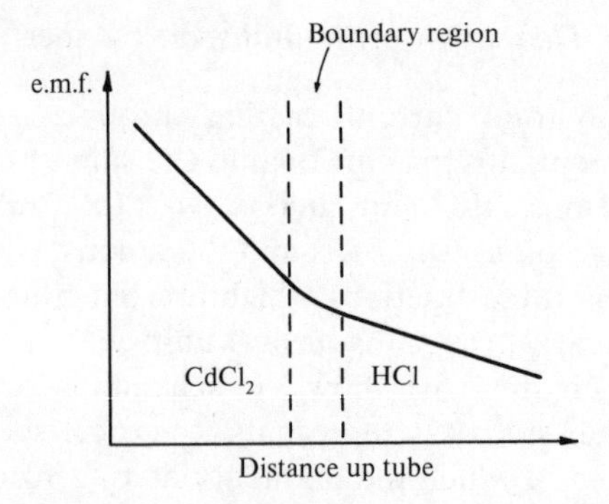

Fig. 7.6.2. E.m.f. gradient in the region of the moving boundary

Clearly if the tube is irregular, the boundary will be adversely affected, and so as smooth a tube as possible is used. It is also necessary to maintain a constant current. *Why does the current tend to vary*? The boundary is also influenced by thermal effects. Convection currents caused by radial temperature gradients are minimized by the use of a fairly narrow bore tube, and, for very precise work, the heating effect of the electric current is compensated by a thermostatted water jacket round the tube.

However the major effect on the boundary is the inter-diffusion of the two electrolyte solutions. This is minimized by careful choice of the species involved. Referring to Fig. 7.6.1, we see that for the present experiment with a common anion and ascending boundary, we need to choose a cation N which is both less mobile and denser than M. This condition is well satisfied with $N = Cd^{2+}$ and $M = H^+$. *What would be a suitable choice of N for a descending boundary*? We must also choose a common anion which is not evolved as gas at the anode. *Why is chlorine not evolved in this experiment*?

Figure 7.6.2 illustrates the way in which inter-diffusion is minimized by the difference in mobility of the cations. Because Cd^{2+} is less mobile than H^+, there is a steeper potential gradient below the boundary than above it. If some of the relatively fast moving H^+ ions diffuse or are carried by convection into the $CdCl_2$ region, they encounter a higher potential gradient and are sent forward to the boundary. On the other hand, if the cadmium ions diffuse into the HCl region they move more slowly than the aqueous H^+ ions and are finally overtaken by the moving boundary.

The self-restoring effect on the boundary is so strong that it will still reform and continue travelling at the same rate even after the current has been interrupted for many minutes.

Apparatus

This experiment employs graduated tubes about 20 cm long and with internal diameters of 2–3 mm. A Cd electrode is fixed into the bottom of each tube, and a silver/silver chloride electrode into the top. The tubes may be fitted with water jackets supplied from a thermostat bath at 25 °C. Current is supplied from a

200 V d.c. power unit. The unit is fitted with a constant-current control circuit, or with a variable resistor and ammeter so that the current may be kept constant by manual adjustment.

Procedure

The 200 V power unit can give quite a severe shock; do not touch bare conductors.

Clean and freshly prepare the silver,silver chloride electrodes as described on p. 134.

Fit a moving boundary tube with its Cd electrode. Add just enough methyl violet indicator to some 0.10 M HCl for its colour to be clearly visible when the solution is in the narrow part of the tube. Rinse and then fill the tube with this solution, taking care that no bubbles are trapped on the Cd electrode or in the narrow tube; a hypodermic syringe with a very long needle is useful for this operation. Then fit the silver, silver chloride electrode into the top of the tube, and mount the tube vertically in a clamp stand.

Connect the electrodes, with the correct polarity, to one of the current control units, and connect this to one of the pairs of output terminals on the 200 V d.c. power supply, or as directed.*

Adjust the power supply controls to the minimum current position.* Switch on the unit, and adjust the controls to obtain a current of 2.0 mA. Do not let the acid stand in prolonged contact with the electrode before starting.

It may be useful to turn up the current to bring the boundary more quickly into the narrow part of the tube. Otherwise keep the current at 2.0 mA throughout the experiment either by setting the control of a current-regulated supply, or by continual manual adjustment.

When the boundary has moved up about 1 cm start the stop clock as the boundary passes one of the calibration marks. Record the times taken for the boundary to move through successive volumes of 0.05 cm^3, until it has passed through a total volume of 0.40 cm^3.

Also carry out the experiment with 0.10 M HCl with 0.050 M HCl at a current of 1.0 mA. Perform the experiments concurrently if the apparatus allows.

Repeat each determination at least once.

Calculation

For each run, find the average time for the boundary to sweep out a volume of 0.05 cm^3. Calculate the transport number of the aqueous H^+ ion relative to Cl^- under each set of conditions, either taking average values or using the results which you judge to be most reliable.

Results

Precision

Discuss the main factors limiting the precision of your results, and compare their estimated magnitude with the deviations between successive readings.

Accuracy

In making an estimate of the accuracy of the results, consider whether any minor effects have been omitted from the preceding discussion. Is the solvent stationary, for example, and is the volume of the solution constant?[2,3] If you have already carried out the next experiment, compare the accuracy of the two methods.

Comment

Most systems which have been studied by the moving boundary method are not as convenient as the one used here. If it is impossible to make an autogenic boundary by the dissolution of one of the electrodes, then a boundary may be made by physically shearing one solution away and replacing it with another.[3] Boundaries which are not marked by a colour change may be observed by an optical system which detects the change in refractive index. This is particularly necessary in the study of proteins, for which the moving boundary method is often used. In recent studies the position of the boundary has been detected by electrodes on either side of it.[4]

Various other methods are available for measuring transport numbers.[5] The most common is *Hittorf's method*, in which the electrolysis cell is divided into compartments, and the change in concentration of the electrolyte found by chemical analysis. The next experiment illustrates another method, which is to measure the e.m.f. of a cell with a liquid junction.

Transport numbers give information about hydration and complex formation in solution; an interesting example is that of fairly concentrated cadmium iodide solutions, in which the transport number of cadmium is negative due to the formation of species such as CdI_3^- and CdI_4^{2-}.[7]

Transport numbers vary with both temperature and concentration. As the temperature of an electrolyte solution increases, the values approach 0.5, implying that the mobilities of the ions become more equal. Transport numbers exhibit a $c^{1/2}$ concentration dependence in dilute solution, as described by the Onsager equation (p. 361).[6]

Technical notes

Apparatus

200 V d.c. power supply, preferably with more than one output, and with manual or automatic current control in the range 0.5–4 mA;†† moving boundary tubes;†† cadmium electrodes;† silver electrodes; inert (platinum) electrodes; conc. NH_4OH; 0.1 M HCl; methyl violet indicator solution; stopwatch; [hypodermic syringe with very long needle].

Design note

Moving boundary tubes can be made from 1 cm^3 graduated pipettes.

References

[1] *Elements of physical chemistry*. S. Glasstone and D. Lewis. Macmillan, London, 2nd edn (1970), p. 420.
[2] *Physical chemistry*. W. J. Moore. Longman, London, 5th edn (1972), p. 433.
[3] *The principles of electrochemistry*. D. A. MacInnes. Reinhold, New York (1939), ch. 4.
[4] *Electrochemistry. Specialist Periodical Reports* **1,** 76 (1970).
[5] *Electrolyte solutions*. R. A. Robinson and R. H. Stokes. Butterworths, London, 2nd edn (1970) p. 102.
[6] Ref. 2, p. 221.
[7] Ref. 1, p. 428.

7.7 Liquid junction potentials and transport numbers from cell e.m.f.s

If we wish to make equilibrium measurements of the e.m.f. of a cell we usually try to avoid liquid junctions by careful choice of the electrodes or by use of a salt bridge (Expt 4.2). Suppose, however, that the following cell is set up:

$$\mathrm{Ag, AgCl} \mid 10^{-2}\,\mathrm{M\,HCl} \mid 10^{-3}\,\mathrm{M\,HCl} \mid \mathrm{AgCl, Ag}$$

where the junction between the two concentrations of acid is in the form of a sintered glass plug, or some other device which restrains, but does not prevent, diffusion. There is a concentration gradient across the junction, as shown in Fig. 7.7.1, and so there will be a tendency for ions to diffuse from left to right. However, the H^+ ions are more mobile than the Cl^- ions, and so will diffuse faster. The result is a net positive charge on the lower concentration side of the boundary, and a corresponding negative charge on the other. This charge separation will continue to increase until the potential gradient which it sets up

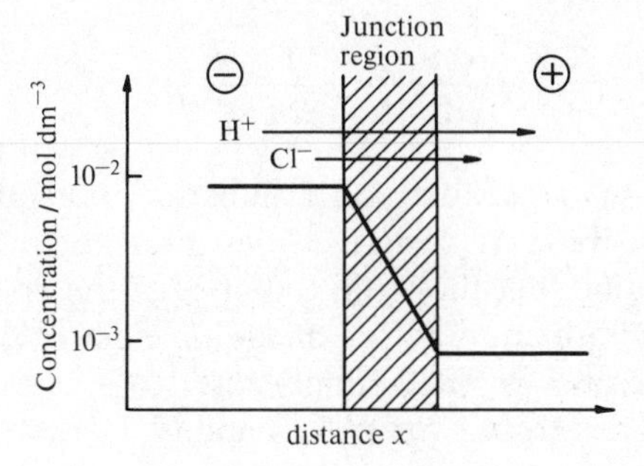

Fig. 7.7.1. Variation of concentration and charge across the liquid junction of an HCl concentration cell

slows the H^+ ions and speeds the Cl^- ions to the extent that they diffuse at the same rate. The resulting steady potential difference is known as the *liquid junction potential.* The potential arises from the difference in charge carried by the two ions, and thus yields their *transport numbers* or *transference numbers*, defined as the fraction of the total current carried by each ion.

The junction potential of a cell may be measured by finding its e.m.f. and subtracting the calculated e.m.f. of the corresponding cell without a liquid junction.

In this experiment we measure the liquid junction potential of the cell just described, and also of cells containing LiCl, NaCl, and KCl instead of HCl. From them we obtain the transport number of aqueous H^+, Li^+, Na^+, and K^+ relative to Cl^-.

Theory

(*i*) *Liquid junction potential*[1,2]

Let us suppose that the liquid junction of the cell is contained in a tube of unit cross-section. The motion of a neutral species i down the concentration gradient at the junction is described by *Fick's first law of diffusion*, which states that the flux J_i is proportional to the concentration gradient $\mathrm{d}_i/\mathrm{d}x$, or more strictly $\mathrm{d}a_i/\mathrm{d}x$, where a is the activity:

$$J_i = -D_i \, \mathrm{d}a_i/\mathrm{d}x. \tag{7.7.1}$$

The flux is the amount of substance passing through the unit area in unit time. The proportionality constant D_i is known as the *diffusion coefficient.*

Note that the flux in Fig. 7.7.1 is positive with respect to x, i.e. the ions move from left to right, that the potential gradient is positive, but that the concentration gradient is negative.

If the species is an ion, then its flux is also affected by the potential gradient $\mathrm{d}E/\mathrm{d}x$ across the boundary. We find that for any position across the junction (any value of x)

$$J_+ = -D_+\left(\frac{\mathrm{d}a_+}{\mathrm{d}x} + a_+\frac{F}{RT}\frac{\mathrm{d}E}{\mathrm{d}x}\right), \tag{7.7.2}$$

and

$$J_- = -D_-\left(\frac{\mathrm{d}a_-}{\mathrm{d}x} - a_-\frac{F}{RT}\frac{\mathrm{d}E}{\mathrm{d}x}\right). \tag{7.7.3}$$

For a steady liquid junction potential, the fluxes of the positive and negative ions must be equal, i.e. $J_+ = J_-$. We assume that the ionic activities of the positive and negative ions are both equal to the mean ionic activity, $a_+ = a_- = a_\pm$, which in this case we shall simply refer to as a.

Thus from eqns (7.7.2) and (7.7.3), we find that for any point in the junction,

$$(D_- - D_+)\frac{da}{dx} = (D_+ + D_-)a\frac{F}{RT}\frac{dE}{dx}. \tag{7.7.4}$$

Integrating across the region of the liquid junction, we find that

$$\frac{(D_- - D_+)}{(D_+ + D_-)}\ln\left(\frac{a_2}{a_1}\right) = \frac{F}{RT}(E_2 - E_1), \tag{7.7.5}$$

where E_1 and E_2 are the potentials, and a_1 and a_2 the mean ionic activities, on either side of the junction.

The transport numbers of the univalent ions, t_+ and t_-, are directly proportional to their diffusion coefficients. Since by definition $t_+ + t_- = 1$, it follows that

$$t_+ = D_+/(D_+ + D_-), \tag{7.7.6}$$

and

$$t_- = D_-/(D_+ + D_-). \tag{7.7.7}$$

The liquid junction potential E_{Jn} is the difference$(E_2 - E_1)$, and hence from eqns (7.7.5), (7.7.6), and (7.7.7):

$$E_{Jn} = E_2 - E_1 = \frac{RT}{F}(t_- - t_+)\ln\left(\frac{a_2}{a_1}\right). \tag{7.7.8}$$

(*ii*) *Cell e.m.f.*

Consider the cell

Ag, AgCl | XCl(a_1) | XCl(a_2) | AgCl, Ag

There are three contributions to the observed e.m.f. E_{obs}, namely the liquid junction potential and the two potential differences at the Ag, AgCl electrodes. The e.m.f. E at each electrode can be calculated from the appropriate *Nernst equation*, p. 123:

$$E = E^{\ominus}_{Ag,AgCl} - \frac{RT}{F}\ln a_{Cl^-}. \tag{7.7.9}$$

The overall contribution of the electrodes is the e.m.f. of the right-hand electrode minus that of the left:

$$E_R - E_L = -\frac{RT}{F}\ln a_2 + \frac{RT}{F}\ln a_1. \tag{7.7.10}$$

Thus the cell e.m.f. E_{obs} is

$$E_{obs} = \frac{RT}{F} \ln(a_1/a_2) + E_{Jn}, \qquad (7.7.11)$$
$$= \frac{RT}{F} \ln[(\gamma_{Cl^-(c_1)} \cdot c_1)/(\gamma_{Cl^-(c_2)} \cdot c_2)] + E_{Jn},$$

where $\gamma_{Cl^-(c_1)}$ represents the activity coefficient of the Cl^- ion at concentration c_1. E_{Jn} may therefore be found from a measurement of E_{obs} with known c_1 and c_2, provided that the value of the ratio of the activity coefficients, $\gamma_{Cl^-(c_1)}/\gamma_{Cl^-(c_2)}$, is known.

Substituting E_{Jn} from eqn (7.7.8) we find that

Equation (7.7.12) shows that we could obtain work from the cells used in this experiment; where does the energy come from?

$$E_{obs} = (t_- - t_+ - 1)\frac{RT}{F} \ln \frac{a_2}{a_1} \qquad (7.7.12)$$
$$= 2t_+ \frac{RT}{F} \ln \frac{\gamma_{Cl^-(c_1)} \cdot c_1}{\gamma_{Cl^-(c_2)} \cdot c_2}$$

since by definition $t_+ + t_- = 1$. This expression allows the calculation of t_+ from a measurement of E_{obs}.

Finally let us consider some extreme values of t_+ and t_-. We notice from eqn (7.7.12) that if $t_+ = 0$, signifying that there is no transport of current by cations, then $E_{obs} = 0$. If, on the other hand, the current is carried entirely by cations, $t_+ = 1$, and not at all by anions, there *is* an e.m.f., of value $2RT/F \ln[(\gamma_{Cl^-(c_1)} \cdot c_1)/(\gamma_{Cl^-(c_2)} \cdot c_2)]$. *Why is this?*

The glass electrode used for pH measurements (p. 152) can be thought of as a half-cell incorporating an ion-selective glass membrane which forms a liquid junction with the test solution at which $t_+ \approx 1$.

We could in fact achieve the situation of $t_+ = 1$ by replacing the liquid junction with two cation-reversible electrodes. Thus for the case of $X^+ = H^+$, two Pt | H_2 electrodes could be used, giving the cell combination:

$$\text{Ag, AgCl} \mid \text{HCl}(c_1) \mid \text{H}_2 \mid \text{Pt} - \text{Pt} \mid \text{H}_2 \mid \text{HCl}(c_2) \mid \text{AgCl, Ag}$$

Apparatus

The cell can either be arranged as shown in Fig. 7.7.2, or using a variation of one of the arrangements shown on p. 127. The salt bridge is filled with the more dilute solution of whichever chloride is being studied.

The e.m.f.s in this experiment must be measured with a precision potentiometer (p. 131); digital voltmeters draw too much current and are not sufficiently precise.

Procedure

Prepare two Ag, AgCl electrodes as described on p. 134. Dip them both into the same solution of 10^{-2} M HCl and check that the e.m.f. between them is 3 mV or less.

Set up the cell

$$\text{Ag, AgCl} \mid \text{XCl}(10^{-2}\,\text{M}) \mid \text{XCl}(10^{-3}\,\text{M}) \mid \text{AgCl, Ag}$$

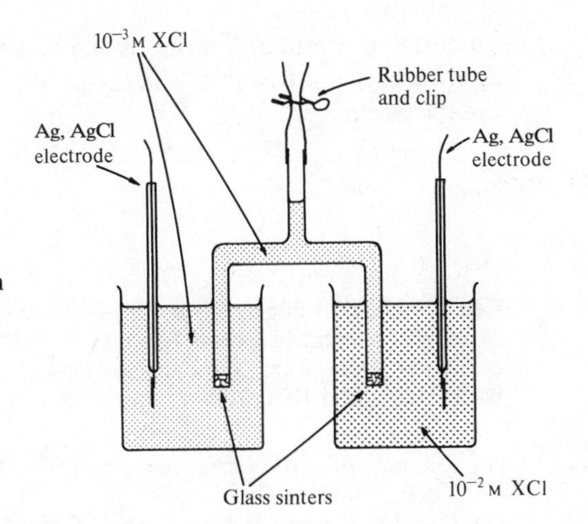

Fig. 7.7.2. Arrangement of cell with liquid junction

with $X^+ = K^+$, by the method described above. Check that the levels in the two sides of the cell are equal so that no siphoning occurs. A hypodermic syringe may prove useful for filling the bridge, which must not contain bubbles.

Deduce the true e.m.f. of the cell from readings taken every 2 min for at least 10 min.

Wash out the cell vessel very thoroughly with de-ionized water, and repeat the experiment for $X^+ = Li^+$, Na^+, and K^+.

Calculation Calculate the liquid junction potential E_{Jn} and transport number t_+ for the aqueous cations H^+, Li^+, Na^+, and K^+ with respect to Cl^-. You are given that the ratio of the chloride ion activity coefficients $\gamma_{Cl^-}(10^{-2}\ \text{M})/\gamma_{Cl^-}(10^{-3}\ \text{M})$ is approximately 0.94 (assuming that $\gamma_- = \gamma_\pm$) for any of the species X^+ at temperatures between about 10 and 30 °C.[3]

Explain the relative magnitudes of t_+. What is the significance of the values of E_{Jn} and t_+ for the potassium ion?

Results *Accuracy*

Explain why this experiment yields quite accurate values for t_{Li^+}, t_{Na^+}, and t_{K^+}, but is considerably less accurate for t_{H^+}. If you have also carried out Expt 7.6, compare the accuracy of the moving boundary method with the accuracy of the present experiment.

Comment The moving boundary method described in the previous experiment is generally the preferred method for measuring transport numbers. The importance of the present experiment lies in its introduction of the effects of the diffusion of ions on e.m.f. Attempts have been made to describe the processes at the liquid junction in term of thermodynamic quantities,[4] but such an approach is unsatisfactory for a process which is essentially of kinetic origin.[5]

Technical notes

Apparatus

Cell vessel and bridge etc. (e.g. Fig. 7.7.2);† precision potentiometer;$ silver wire electrodes; conc NH_4OH, 0.1 M HCl and power supply for electrode preparation; standard solution of HCl; standard solutions of LiCl, NaCl, and KCl.

References

[1] *The principles of electrochemistry*. D. A. MacInnes. Reinhold, New York (1939), p. 461.
[2] *Physical chemistry*. P. W. Atkins. Oxford University Press, Oxford, 3rd edn (1986), sect. 27.2.
[3] *The physical chemistry of electrolytic solutions*. H. S. Harned and B. B. Owen. Reinhold, New York, 3rd edn (1958), pp. 716, 736.
[4] Ref. 1, p. 220.
[5] *Electrochemistry. Specialist Periodical Reports* **1,** 72 (1970), (by G. J. Hills).

7.8 Kinetics of a diffusion controlled heterogeneous reaction by titration

The rates of heterogeneous chemical reactions frequently depend not on the rate of the chemical reaction itself but rather on the rate at which the reacting species can be brought together.

In this experiment we investigate the kinetics of the exchange of K^+ ions in solution with H^+ ions on a suspension of ion exchange resin. Under normal conditions of stirring, perfect transport between the bulk of the solution and the resin surface cannot be achieved because the layer of liquid in contact with the surface is stationary. Effectively there is stagnant film of liquid around the solid particles, and the only mechanism of transport through the film is by diffusion. Since this is a slower process than the exchange reaction itself, the kinetics of the overall reaction are *diffusion controlled.*

Theory[1,2]

The ion exchange reaction may be written

$$K^+ + H_R^+ \leftrightharpoons K_R^+ + H^+,$$

where K^+ and H^+ are the solvated species, and H_R^+ and K_R^+ the ions on the resin.

The mass-action equilibrium constant for the exchange of the ions on the resin with the ions in the solution is

$$K = \frac{[K_R^+][H^+]_r}{[K^+]_r[H_R^+]}, \tag{7.8.1}$$

where the subscript r indicates the concentration in the solution at the surface of the resin.

To obtain equations which describe the reaction, we must consider a simplified model of the resin. We assume that it is a solid with a flat surface of area A, and that adhering to the surface is a thin, stationary layer or *Nernst film* of thickness δ. If the difference in concentration of a species in the solution at the surface of the resin and in the stirred solution outside the Nernst film is Δc, then we assume that there is a concentration gradient $\Delta c/\delta$ in the Nernst film, normal to the resin surface.

By *Fick's first law of diffusion*,

$$J = -D\frac{\Delta c}{\delta}, \tag{7.8.2}$$

where D is the diffusion coefficient. The flux J is the amount of substance n passing through unit area in unit time, and we may therefore express this equation in the form

$$\frac{1}{A}\frac{dn}{dt} = -D\frac{\Delta c}{\delta}. \tag{7.8.3}$$

Applying this relation to the diffusion of hydrogen ions from the resin to the solution we obtain the expression:

$$\frac{dn}{dt} = -\frac{A \cdot D_{H^+}}{\delta}([H^+]_s - [H^+]_r), \tag{7.8.4}$$

where D_{K^+} is the diffusion coefficient of the hydrogen ions, and the subscript s refers to the concentration in the stirred bulk of the solution. Note that the concentration difference is negative, because there is a higher concentration at the resin than in the solution, and that the corresponding positive dn/dt refers to diffusion away from the surface of the resin.

During the present experiment, the solution is always kept at a high pH. Therefore $[H^+]_s - [H^+]_r \approx -[H^+]_r$, and to a good approximation

$$\frac{dn}{dt} = \frac{A \cdot D_{H^+}}{\delta}[H^+]_r. \tag{7.8.5}$$

Similarly for the diffusion of potassium ions

Why is the magnitude of dn/dt the same for H^+ and for K^+?

$$\frac{dn}{dt} = -\frac{A \cdot D_{K^+}}{\delta}([K^+]_s - [K^+]_r). \tag{7.8.6}$$

In this case $[K^+]_s > [K^+]_r$, and dn/dt is negative indicating that potassium ions diffuse from the solution to the resin.

It follows from an inspection of the equilibrium reaction that the amount of substance which has diffused at time t, relative to the amount still to diffuse, is given by

$$\frac{n_t}{n_\infty - n_t} = \frac{[K_R^+]}{[H_R^+]}. \tag{7.8.7}$$

On combining eqns (7.8.1), (7.8.5), (7.8.6), and (7.8.7), we find that

$$\frac{dn}{dt}\left(\frac{D_{K^+}}{D_{H^+} \cdot K} \cdot \frac{n_t}{n_\infty - n_t} - 1\right) = \frac{[K^+]_s \cdot A \cdot D_{K^+}}{\delta}. \tag{7.8.8}$$

Integrating this expression, and using the limit $n = 0$ at $t = 0$, we obtain the kinetic equation

$$\frac{n_\infty}{n_t} \cdot \ln\left(\frac{n_\infty}{n_\infty - n_t}\right) = \frac{AD_{H^+}[K^+]_s K}{\delta}\left(\frac{t}{n_t}\right) + \left(1 + \frac{D_{H^+}K}{D_{K^+}}\right). \tag{7.8.9}$$

n_t and n_∞ are directly proportional to the experimental titres a_t and a_∞, these being the amounts of potassium hydroxide solution necessary to neutralize the protons which have been exchanged. The diffusion control of the exchange reaction according to eqn (7.8.9) may therefore be tested by plotting graphs of $a_\infty/a_t \cdot \ln[a_\infty/(a_\infty - a_t)]$ against t/a_t. These should be straight lines of slope proportional to $[K^+]_s$ and thus to the initial concentration of KCl in the solution, and should all have the same intercept at $t = 0$.

Apparatus

The reaction is carried out in a 600 cm^3 beaker on a magnetic stirrer, and the potassium hydroxide solution added from a 10 cm^3 semi-micro burette. The indicator used is bromocresol green.

Procedure

First determine the colour of bromocresol green in both acid and alkali solutions.

Weigh 1 g of resin into a 600 cm^3 beaker. Add 100 cm^3 of deionized water and ten drops of the indicator. Arrange a magnetic stirrer to agitate the resin suspension vigorously, and try to keep the stirring rate constant throughout the experiment. Set up a 10 cm^3 semi-micro burette to deliver 0.1 M KOH solution into the beaker.

Semi-micro burettes are best filled by suction.

Note the temperature of the solutions.

Into a second beaker pipette 5 cm^3 of 0.01 M KCl and 100 cm^3 water.

To start the exchange reaction, pour the contents of the

second beaker into the first, note the time on a stopwatch, and *quickly* run in about 1 cm^3 of KOH solution. Note the titre.

Note the time when the indicator changes back to the acid form, and rapidly add a further smaller amount of KOH. Again note the time when the indicator changes.

Continue in this way until the reaction becomes too slow for convenient observation. Obtain about twenty-five readings of time t for the colour of the solution to change after a total of a_t cm^3 of KOH has been added to the solution.

After the final colour change, add about 1 g of solid KCl to displace the remaining acidity still present in the resin. Allow the beaker and contents to stand for an hour, and titrate the contents with the 0.1 M KOH. The final titre gives the total exchange capacity a_∞.

Repeat the experiment using 2 cm^3 of 0.1 M KCl in the second beaker and then again using 10 cm^3 of 0.01 M KCl. Dilute each with 100 cm^3 water as before.

Calculation

Plot a_t/a_∞ against t. What does the shape of this curve reveal about the kinetics of the reaction?

Draw graphs, in the manner explained previously, to ascertain whether the exchange reaction is diffusion controlled.

What mechanism can you propose for the process at the resin?

Results

Although it should be clear that the reaction does not obey either 1st or 2nd order kinetics, conclusive evidence of a diffusion controlled reaction will depend on the particular resin used.

Comment

The rates of heterogeneous reactions are also affected by the rate of stirring and by temperature. Faster stirring decreases the thickness δ of the liquid layer adhering to the solid surface, and so increases the rate of the reaction. A change in temperature will alter both the rate of diffusion and the rate of the reaction itself, and in extreme cases may result in a change from diffusion control to chemical control.[2]

Many fast homogeneous reactions are also diffusion controlled. These include the gas phase recombination reactions of atoms and radicals, acid-base reactions in solution, and the quenching of electronically-excited molecules (Expt 5.11).[3]

Technical notes

Apparatus

600 cm^3 beaker, 10 cm^3 semi-micro burette with fairly fast delivery, magnetic stirrer, an H-form resin with working range of pH 1–14,

bromocresol green (blue), solid KCl, 0.1 M KOH solution, 0.1 M KCl, 0.1 M HCl, [0.01 M KCl].

References

[1] *The diffusion process for organolite exchangers.* J. J. Grossman and A. W. Adamson. *J. Phys. Chem.* **56,** 97 (1952).
[2] *Transport control in heterogeneous reactions.* L. L. Bircumshaw and A. C. Riddiford. *Q. Revs.* **6,** 157 (1952).
[3] *Reaction kinetics.* M. J. Pilling. Clarendon Press, Oxford (1975), p. 71.

7.9 Kinetics of substituted nitrobenzene reduction at a dropping mercury electrode

In this experiment we study the kinetics of the electrochemical reduction of a series of substituted nitrobenzenes, by measuring the current-voltage curves or *polarograms* in an appropriate cell.

The current in the cell is determined by two factors. The first is the rate of transfer of electrons to and from the electrode at which the reaction is taking place, and the second factor is the rate of transport, by diffusion, of the electroactive species to the electrode. Thus although we shall concentrate on the kinetics at the electrode, it is nevertheless essential that the transport is well defined and calculable.

A static electrode in a beaker of solution does not give rise to the desired transport characteristics, and we therefore employ a *dropping mercury electrode* (DME).[1] This comprises a thin column of mercury contained in a capillary tube which dips into the test solution. The mercury is allowed to fall slowly down the tube and gradually forms a droplet at the end. The expansion of the droplet ensures a supply of fresh solution at the electrode surface. Then the drop falls off, stirring the solution as it falls, and leaving fresh solution as the next drop starts to form. By this means both the surface of the electrode and the solution around it are continually renewed.

Theory

(i) Processes at the DME

Let us consider the reduction of an electroactive species X by z electrons to form a species Y:

$$X + ze^- \underset{k_{-1}}{\overset{k_1}{\rightleftharpoons}} Y$$

The charges on X and Y differ by z, and k_1 and k_{-1} are the heterogeneous rate coefficients.

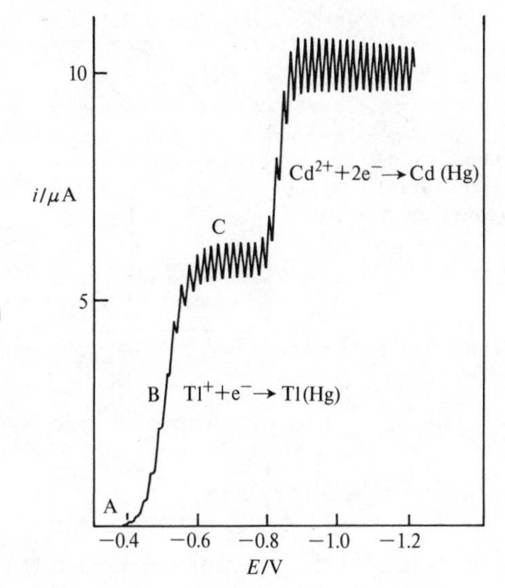

Fig. 7.9.1. Typical polarogram at a DME for the reduction of Tl^+ and Cd^{2+}[1]

Figure 7.9.1 shows a typical polarogram for a solution containing two electroactive species, in this case Tl^+ and Cd^{2+}. These ions are reduced to their elements, which form amalgams with the mercury of the electrode. The two reduction processes may be identified with the two steps in the polarogram, and three regions, labelled A, B, and C for the Tl^+ reduction, may be identified for each step. At A the potential is insufficient to drive the electrode reaction and no current is seen. In region B the current rises as the potential becomes sufficiently negative to reduce the electroactive species Tl^+. Eventually the current reaches a limiting value i_{LIM} in region C. Here the potential is such that all the X reaching the electrode is reduced and the current is therefore limited by the rate at which X can diffuse to the electrode. The limiting current shows a saw-tooth pattern corresponding to the cyclic formation of successive droplets. The current at the electrode rises as a drop expands, and then falls to a minimum again as the drop separates from the electrode.

We must now express these effects quantitatively. Figure 7.9.2 shows the concentration profiles for three different times in the growth of a droplet. t_D is the time for a drop to form, and x is the distance from the surface of the electrode. When the droplet is starting to form, the concentration rises steeply with x from zero to c_∞, its value in the bulk solution. As the droplet expands, the profile becomes less steep, as shown in the figure.

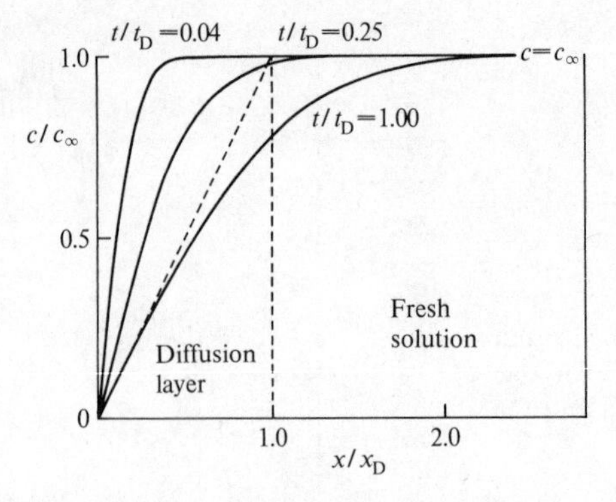

Fig. 7.9.2. Concentration profiles for three different times in the growth of a droplet[1]

The current to the drop electrode is

$$i = zFAD\left[\frac{\partial c}{\partial x}\right]_0, \tag{7.9.1}$$

where A is the area of the electrode (which we express in units of m^2) and D is the diffusion coefficient of X ($m^2\,s^{-1}$). $(\partial c/\partial x)_0$ is the limiting concentration gradient at $x = 0$. This term has the value c_∞/x_D, where x_D is the thickness of the diffusion layer, as shown in the diagram for the case where the droplet is fully formed.

The limiting current i_{LIM} at the maximum of the saw-tooth curve is expressed in terms of D, t_D, m (the rate of flow of mercury in $g\,s^{-1}$), and [X] (in mol dm^{-3}), by the *Ilkovic equation*

$$i_{LIM} = 0.063FzD^{1/2}m^{2/3}t^{1/6}[X]. \tag{7.9.2}$$

The polarogram can be characterized by a half-wave potential where the current reaches half of the limiting value, i.e. $i = i_{LIM}/2$. Two different types of polarogram are encountered and the significance of the half-wave potential is different in the two cases, as described below.

(*ii*) *Reversible waves*

In the case of reversible waves, the rates of reduction of X and the oxidation of Y are fast and so there is an equilibrium set up between X and Y at the surface of the electrode. In other words, the surface concentrations $[X]_0$ and $[Y]_0$ will be given by the Nernst equation (p. 128)

$$E = E^{\ominus} - \frac{RT}{zF}\ln\frac{[Y]_0}{[X]_0}, \tag{7.9.3}$$

where E is the potential of the electrode and $E^{\ominus}$ is the standard electrode potential, both measured with respect to a common reference electrode (an Ag, AgCl electrode here). At the half-wave potential $E_{1/2}$

$$[\mathrm{X}]_0 = [\mathrm{Y}]_0, \tag{7.9.4}$$

and so from the Nernst equation,

$$E_{1/2} = E^{\ominus}. \tag{7.9.5}$$

Thus for reversible waves the half-wave potential is identical to the standard electrode potential.

(*iii*) *Irreversible waves*

If the electron transfer reactions are slow, no equilibrium is set up. In this case $|E_{1/2}| > |E^{\ominus}|$ since a potential above the half-wave potential has to be applied to cause the electrode reactions to occur at a measurable rate. This extra potential is the *overpotential* $\eta = E - E^{\ominus}$.

It may be shown that

$$k_1 = k_1^0 \exp\left[-\frac{\alpha zF}{RT}\eta\right], \tag{7.9.6}$$

and that

$$k_{-1} = k_{-1}^0 \exp\left[(1-\alpha)\frac{zF}{RT}\eta\right], \tag{7.9.7}$$

where k_1 and k_{-1} are the rate coefficients introduced earlier. k_1^0 and k_{-1}^0 are the corresponding standard rate coefficients and α is the transfer coefficient,[2] which is often approximately equal to 1/2. The total current is given by

$$i = zFA(k_1[\mathrm{X}]_0 - k_{-1}[\mathrm{Y}]_0). \tag{7.9.8}$$

At potentials where k_1 is significant, k_{-1} is so small that the second term in eqn (7.9.8) can be neglected. $[\mathrm{X}]_0$ will then be governed by the balance between the rate at which X is consumed at the electrode and the rate at which X can diffuse to the electrode. This is accounted for in the following expression valid at any point on the wave:

$$i = i_{\mathrm{LIM}} \Big/ \left[1 + \frac{D}{k_1^0 x_{\mathrm{D}}} \exp\left(\frac{\alpha zF}{RT}\eta\right)\right], \tag{7.9.9}$$

or

$$\ln\left(\frac{i_{\mathrm{LIM}}}{i} - 1\right) = \frac{\alpha zF(E - E^{\ominus})}{RT} + \ln\frac{D}{k_1^0 x_{\mathrm{D}}}. \tag{7.9.10}$$

At the half-wave potential $i = i_{LIM}/2$ and so from eqn (7.9.9)

$$k_1^0 = \frac{D}{x_D}\exp\left(\frac{\alpha zF}{RT}(E_{1/2} - E^{\ominus})\right), \tag{7.9.11}$$

or

$$E_{1/2} = E^{\ominus} + \frac{RT}{\alpha zF}\ln k_1^0 + \frac{RT}{\alpha zF}\ln x_D/D. \tag{7.9.12}$$

Thus for irreversible waves $E_{1/2}$ is related logarithmically to the standard rate coefficient k_1^0 for the reduction of X.

(*iv*) *Reduction of nitrobenzenes*

What are the products of the electrochemical reduction of the nitrobenzenes?

In this experiment $E_{1/2}$ is measured for a range of substituted nitrobenzenes which show irreversible waves. We expect the rate of reduction to be determined by the inductive and mesomeric effects of the substituents in the molecule. The results of these effects on any reaction of a substituted benzene derivative can be estimated quantitatively by means of the *Hammett σ parameter.* In general we find that

$$\log_{10} k_1^0(\mathrm{X}) = \log_{10} k_1^0(\mathrm{H}) + \rho\sigma, \tag{7.9.13}$$

where $k_1^0(\mathrm{H})$ is the rate coefficient for the case where X = H, ρ is a constant whose value depends on the reaction in question, and σ is a value for the particular substituent obtained from tables.

To illustrate the use of this equation, let us consider a different series of reactions, namely the hydrolysis of substituted ethyl benzoates:

$$\mathrm{X{-}C_6H_4{-}CO_2Et} \xrightarrow{k} \mathrm{X{-}C_6H_4{-}CO_2H}.$$

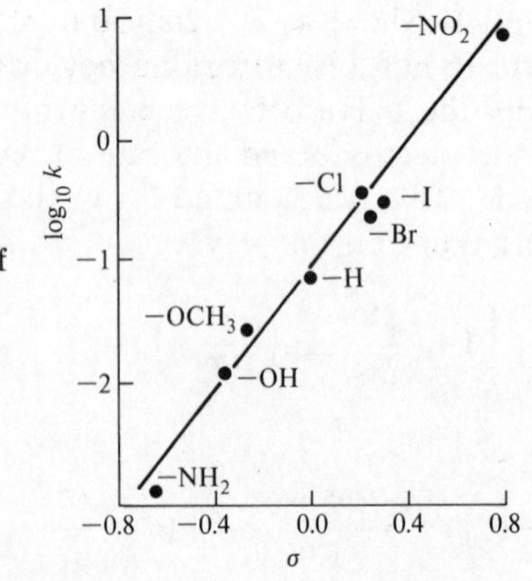

Fig. 7.9.3. Hydrolysis of substituted ethyl benzoates ($R = ArCO_2Et$)

Table 7.9.1. Hammett σ parameters for substituted nitrobenzenes

X	meta	para	
	σ	σ	σ^-
—COOH	0.35	0.44	0.75
—CHO	0.36–0.41	0.43–0.47	1.0
—OMe	0.12	−0.27	
—Cl	0.37	0.24	
—$SO_3^-K^+$	0.05	0.35	
—OH	0.12	−0.37	

Figure 7.9.3 shows that the logarithm of rate coefficient k for various substituents X correlates closely with σ.

In Table 7.9.1 are shown values of σ for substituted nitrobenzenes, as studied in this experiment. Notice that σ has different values according to whether X is meta or para to $S = -NO_2$. If S is directly conjugated to X then a parameter σ is used. This is because in addition to inductive effects and indirect resonance effects, σ^- deals with resonance effects due to quinonoid structures such as that shown in Fig. 7.9.4, which are produced if S can denote electrons.

In this experiment we measure the half-wave potentials of a range of substituted nitrobenzenes. If we combine eqns (7.9.12) and (7.9.13), and assume that $E^\ominus$ is independent of σ,[3] then we may conclude that

$$E_{1/2} = \frac{2.303RT\rho\sigma}{\alpha zF} + \text{const.} \tag{7.9.14}$$

Apparatus

The apparatus is shown schematically in Fig. 7.9.5. The potentiostat fixes a defined potential E between the dropping mercury *working electrode* and the reference electrode. The applied potential gives rise to a current i, and the operational amplifier causes all of this to pass through the Pt gauze counter electrode. Nitrogen is passed over the solution to remove oxygen which is electroactive.

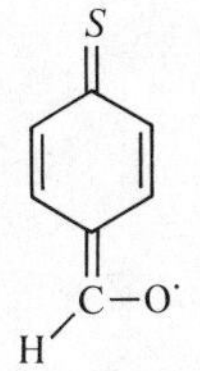

Fig. 7.9.4. Example of a quinonoid structure

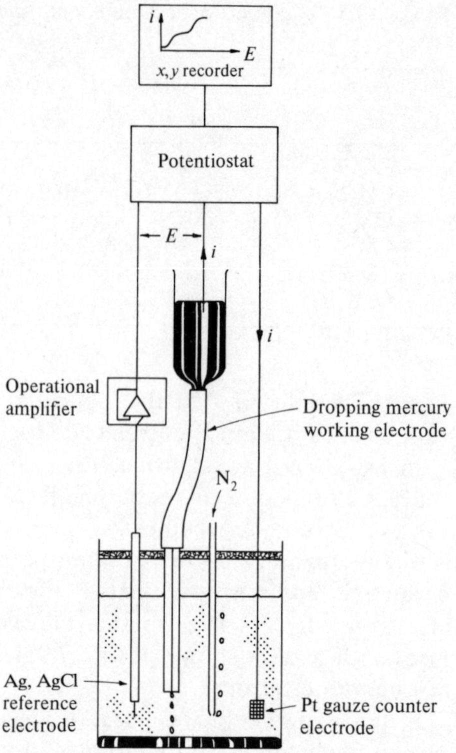

Fig. 7.9.5. Polarography apparatus incorporating a dropping mercury electrode

Procedure

Why is KCl present in the solutions?

You are provided with six solutions of various nitrobenzenes in a mixture of 1 M KCl and 10^{-2} M KCl. The solutions also contain dissolved oxygen which is electroactive at potentials of interest. This is removed by purging with nitrogen. Fill one sample vessel with the first solution using sufficient to cover the electrodes (the exact volume is not important). Attach the vessel to the head carrying the electrodes and N_2 bubbler, and bubble nitrogen gently through the solution for approximately 10 min. By the end of this period oxygen should have been eliminated (otherwise a wave with $E_{1/2} \simeq -0.05$ V will be seen). The nitrogen should be turned off when polarograms are being recorded.

Ensure that the Ag, AgCl reference electrode, the Pt gauze counter electrode and the mercury column are connected to the appropriately labelled sockets in the rear of the potentiostat.*

Start mercury drops falling and switch on the polarograph as instructed.*

Run polarograms for each of the 6 samples provided taking

care to write the potential and current scales onto each chart. Take care that each sample is degassed before a proper polarogram is run. When you have completed the polarograms:

(i) Turn off the nitrogen.
(ii) Close the tap from the mercury column.
(iii) Switch off the potentiostat.
(iv) Clean the vessels. Put all residues including any mercury in the waste bottle provided and also all washings from cleaning the vessels.

Calculation

For each wave measure the limiting current and hence find the half-wave potential $E_{1/2}$, i.e. the point where the current $i = i_{LIM}/2$. A *polarographic maximum* will be observed during one run. In this case measure i_{LIM} beyond the maximum when it has reached a steady value independent of potential. In other cases a second wave will be seen after the first, whereupon only the first wave should be considered.

Tabulate your values of half-wave potential together with the appropriate Hammett σ value. Plot a graph of $E_{1/2}$ against σ (eqn (7.9.14)), and draw a straight line through the results. Hence find the value of ρ for the reaction.

For the case of *meta* $NO_2—C_6H_4—SO_3^-$ plot a graph to show the validity of eqn (7.9.10).

Results

Accuracy

Some potentiostats only read to 10 mV, and therefore the accuracy of the results may be limited by their precision. However, all the graphs should yield good straight lines.

Comment

Polarography is a convenient and reasonably sensitive analytical technique for organic species and heavy metals. The sensitivity of the method can be increased by modulating the applied voltage with alternating current or with pulses, and using a lock-in amplifier which detects only the modulated signal. Such methods allow polarography to be used for the detection of concentrations as low as 10^{-7} M.

Other devices which give well-defined transport characteristics are tube electrodes, which may be used as flow-through detectors at the end of chromatography columns, and rotating-disc and ring-disc electrodes.[4]

Technical notes.

DME and polarograph (Fig. 7.9.5);[$] approx. 10^{-3} M sample solutions (Table 7.9.1) in 1 M KCl + 10^{-2} M HCl.

References

[1] *Electrode kinetics*. W. J. Albery. Clarendon Press, Oxford (1975), p. 71–8.
[2] Ref. 1, p. 5.
[3] *The elucidation of organic electrode processes*. P. Zuman, Academic Press, New York (1969), p. 118.
[4] Ref. 1, p. 49.

Section 8
Reaction kinetics

We have already studied the thermodynamics of various chemical reactions in Section 3. In Section 5 we encountered the use of fluorescence for the measurement of reaction rates, and Section 7 included a reaction in which the rate was determined by the transport of the reacting species. In this section we examine the kinetics of ten reactions in which the rate is governed by mass action and is therefore directly affected by the concentration of reactants. The first six experiments involve reactions in solution, and the other four concern the gas phase.

In terms of practical skills, the first six experiments require an ability to handle solutions quantitatively. In addition they give practice at using various items of analytical equipment. Experiment 8.3 employs a polarimeter, Expt 8.4 a conductivity meter, and Expt 8.5 an ultra-violet spectrophotometer. Experiment 8.6 introduces the techniques for measuring the radioactivity of solutions.

The theory of the solution experiments becomes gradually more involved as we work through the section. Experiment 8.1 is a basic introduction to the study of reaction kinetics in terms of the rate equation and order of a reaction. In Expt 8.2 we study a 'clock reaction' by the initial rates method. Experiment 8.3 introduces general acid and base catalysis. In this investigation we also encounter a method of analysing data which avoids taking a reading when the reaction has gone to completion, and this is employed again in the following experiment. Experiment 8.4 includes the measurement of activation enthalpy and entropy, and Expt 8.5 introduces the problems of analysing consecutive and parallel reactions. The kinetics of radioactive decay studied in Expt 8.6 are straightforward, but we also need to understand the method by which radiation is detected, and the statistics of the random disintegration processes which we are monitoring.

The gas phase experiments, 8.7–8.10, all assume a good

understanding of the basic theory of reaction kinetics and the steady-state approximation, and an ability to apply these to the reactions being studied. They also require considerable competence in using gas lines. Mistakes can not only spoil an apparatus, but may also lead to accidents, expecially in the last experiment. It is therefore strongly recommended that these experiments only be attempted after at least one simpler gas line experiment, such as one from Sections 1 or 3, has been completed. Otherwise, these experiments require few extra practical skills other than in the use of a gas chromatograph in Expt 8.7.

The gas-phase reactions are described in order of increasing rate, so as to highlight the different experimental techniques which are required for their study. Experiment 8.7 concerns a reaction which takes tens of minutes to complete, and which can therefore be investigated by a combination of pressure measurements and gas chromatography. In Expt 8.8 we use a flow tube method to study reactions which take only about 1 s. The iodine recombination reaction examined in Expt 8.9 is over in about 20 ms, and we therefore employ flash photolysis to initiate the reaction, and examine the progress of the reaction with a simple spectrometer connected to a storage oscilloscope or transient recorder. The hydrogen/oxygen explosion of Expt 8.10 is too fast to monitor, but we can obtain information about the reaction by finding the conditions under which an explosion takes place.

The fact that the first experiment in this book involved gently raising the pressure of a gas until it balanced a density balance, and the last involves an explosion, gives an indication of the path we have travelled in this, our investigation of experimental physical chemistry.

8.1 Kinetics of propanone iodination by titration

In studying a chemical reaction one of the first things to investigate is how its rate varies with the concentrations of the reactants. The observations may then be expressed in the form of an empirical *rate equation*, and mechanisms proposed which are compatible with this expression.[1–3]

In this experiment, we study the iodination of propanone (acetone) in an aqueous solution of sulphuric acid. The *stoichiometric equation* for the reaction is

$$CH_3COCH_3 + I_2 \rightarrow CH_2ICOCH_3 + H^+ + I^-.$$

A mixture is made up containing known concentrations of propanone, sulphuric acid and water. The reaction is started at a known time by adding a measured quantity of iodine. Aliquots are withdrawn from the reaction vessel at intervals, quenched with sodium ethanoate (acetate), and the unreacted iodine titrated against sodium thiosulphate. The experiment is then repeated with different concentrations of reactants.

Theory

The rate equation of a reaction can only be determined experimentally; it does not necessarily bear any relationship to the stoichiometric equation for the reaction.

The kinetics of a reaction between reactants A, B, C... may be expressed by a *rate equation* of the form

$$\text{rate} = k[\text{A}]^{\text{a}}[\text{B}]^{\text{b}}[\text{C}]^{\text{c}}\ldots, \tag{8.1.1}$$

where k is the *rate coefficient*, [A] is the concentration of A, [B] the concentration of B, and so on. The *order* of the reaction with respect to A is a, with respect to B is b, etc. According to the values of a, b, c..., the reaction is termed zeroth, first, second... order with respect to A, B, C.... The *overall order* of the reaction is a+b+c....

In this experiment we measure the rate of the reaction by finding how the concentration [A] of one of the reactants (iodine) varies with time. If the reaction is observed over a time during which the concentration [B], [C]... are reasonably constant, then eqn (8.1.1) may be written

$$\text{rate} = -\frac{\text{d}[\text{A}]}{\text{d}t} \approx k'[\text{A}]^{\text{a}}, \tag{8.1.2}$$

where k' is the *effective rate coefficient:*

$$k' = k[\text{B}]^{\text{b}}[\text{C}]^{\text{c}}\ldots. \tag{8.1.3}$$

On integrating eqn (8.1.2) we find that:

$$[\text{A}] \approx [\text{A}]_0 - k't \quad \text{for} \quad \text{a} = 0, \tag{8.1.4}$$

$$\ln[\text{A}] \approx \ln[\text{A}]_0 - k't \quad \text{for} \quad \text{a} = 1, \tag{8.1.5}$$

$$\frac{1}{[\text{A}]} \approx \frac{1}{[\text{A}]_0} + k't \quad \text{for} \quad \text{a} = 2, \tag{8.1.6}$$

where $[\text{A}]_0$ is the concentration of A at the start of the reaction ($t = 0$). It can be seen that if the reactant concentration [A] decreases linearly with time, then the reaction is zeroth order with respect to the reactant (eqn (8.1.4)), if ln[A] decreases linearly with time the reaction is first order (eqn (8.1.5)), and if 1/[A] increases linearly with time then the reaction is second order (eqn (8.1.6)). By fitting the experimental data to whichever equation is appropriate, we can find a value of the effective rate coefficient k' from the slope.

To be able to write the full rate equation (8.1.1), we also need to find the order of the reaction b, c ... with respect to other possible reactants B, C .... The most direct method would be to observe the rate of change of B, C ... with respect to time, and use equations analogous to eqn (8.1.2). However for this reaction, as well as many others, it is either impossible or inconvenient to follow the concentrations of the other species. Instead we vary the initial concentrations $[B]_0$, $[C]_0$ ..., and measure the rate $-d[A]/dt$ as before. For each run, we then find k' using eqn (8.1.4), (8.1.5), or (8.1.6). Then from eqn (8.1.3)

$$\ln k' = \ln k + b \ln[B] + c \ln[C] + \ldots, \tag{8.1.7}$$

or, since we are looking at the initial rate of the reaction

$$\ln k' = \ln k + b \ln[B]_0 + c \ln[C]_0 + \ldots. \tag{8.1.8}$$

Note that the effective rate coefficient k' varies with concentration, whereas by definition the actual rate coefficient k does not.

So a graph of $\ln k'$ against $\ln[B]_0$, at constant $[A]_0$, $[C]_0$, ..., will have slope b, and a graph of $\ln k'$ against $\ln[C]_0$ at constant $[A]_0$, $[B]_0$ ... will yield c.

Procedure

Make up the solutions listed in Table 8.1.1 in stoppered 250 cm^3 conical flasks, and place them in a thermostat bath set at 25 °C. Also place in the thermostat bath a separate flask containing about 120 cm^3 of a 0.1 M solution of iodine. (The iodine is dissolved in KI to increase its solubility). Leave the solutions in the bath for at least ten minutes.

Remove the first flask from the thermostat bath. Add 20 cm^3 of iodine and mix by swirling the solution. Note the time accurately, or start a stop-clock.

Why does sodium ethanoate quench the reaction?

Every 5 min withdraw 20 cm^3 of solution with a pipette, add it to about 10 cm^3 of the 1 M sodium ethanoate solution to quench the reaction, and titrate the remaining iodine against the 0.01 M sodium thiosulphate.

Repeat the experiment for the other solutions, some of which may be studied concurrently.

Table 8.1.1. Solution compositions/cm^3

Run	Propanone	1 M H_2SO_4	H_2O
1	20	10	150
2	15	10	155
3	10	10	160
4	20	15	145
5	20	5	155

Calculation

For each run, plot the concentration of iodine against time. Find which of eqns (8.1.4) to (8.1.6) describes the shape of the graphs. Hence find the order of the reaction with respect to iodine as explained in the Theory section.

From the results of runs 1 to 3 find out how k' varies with the concentration of propanone, and hence the order with respect to propanone. Similarly use the results of runs 1, 4, and 5 to find the order with respect to $[H^+]$.

On the basis of your results, answer the following questions:

(i) Given the order of the reaction with respect to the three reactants studied, what species are involved in the rate determining step?

(ii) Given that there is an equilibrium between propanone and its enol,

$$CH_3{-}\overset{\overset{\displaystyle O}{\|}}{C}{-}CH_3 \rightleftharpoons CH_2{=}\overset{\overset{\displaystyle OH}{|}}{C}{-}CH_3,$$

and that halogens rapidly attack carbon-to-carbon double bonds, devise a plausible mechanism for the reaction.

(iii) Under your reaction scheme, would the reaction of bromine with propanone be more or less rapid than the reaction of iodine with propanone?

Results

Accuracy

The graphs of the results from this experiment should show very little scatter, yield the rate equation unambiguously and give a rate coefficient within 50 per cent of the true value.

Comment

This experiment demonstrates one of the simplest methods by which the kinetics and mechanism of a reaction may be elucidated. The halogenation of propanone may also be studied spectrophotometrically.[4]

Technical notes

Apparatus

Six 250 cm^3 conical flasks with stoppers; thermostat tank to take flasks, set at 25 °C; burette and pipettes. Propanone (acetone); standardized solutions of 1 M sulphuric acid, 0.01 M sodium thiosulphate, and 0.1 M iodine in KI; approx. 1 M sodium ethanoate (acetate) solution.

References

[1] *Physical chemistry*, P. W. Atkins. Oxford University Press, Oxford, 3rd edn (1986), secs. 28.1 and 28.2.

[2] *Reaction kinetics*. M. J. Pilling. Clarendon Press, Oxford (1975), ch. 1.

[3] *Kinetics and mechanism*. J. W. Moore and R. G. Pearson. John Wiley and Sons, New York, 3rd edn (1981), chaps 1 and 2.

[4] *Experimental physical chemistry*. F. Daniels *et al.* McGraw-Hill, New York, 7th edn (1970), expt. 25.

8.2 Persulphate–iodide clock reaction by initial rates method

In this experiment we investigate the kinetics of the reaction between persulphate and iodide ions, for which the stoichiometric equation is

$$\underset{\text{persulphate}}{S_2O_8^{2-}} + 2I^- \rightarrow 2SO_4^{2-} + I_2.$$

The rate of the reaction may be measured by adding a small, known concentration of thiosulphate ions. The iodine produced from the persulphate reaction is then very rapidly reduced back to iodide by the thiosulphate ions:

$$\underset{\text{thiosulphate}}{2S_2O_3^{2-}} + I_2 \rightarrow 2I^- + \underset{\text{tetrathionate}}{S_4O_6^{2-}}.$$

This continues until all the thiosulphate has been converted to tetrathionate, whereupon free iodine is formed in the solution. The colour of the iodine is enhanced by the addition of starch solution. The time interval between the start of the reaction and the change in colour of the solution is a measure of the initial rate of the reaction. Self-indicating reactions of this type are known as *clock reactions.*

In Part A of this experiment we investigate the effect of the concentrations of the reactants on the rate of the persulphate-iodide reaction, and thereby obtain the *rate equation.* The effect of temperature is studied in Part B, and the results used to find the *activation energy* of the reaction. Finally in Part C we measure the effect of the ionic strength of the solution, from which a possible structure of the *reaction intermediate* may be deduced.

Theory

A. Rate equation

At constant temperature and ionic strength, the *rate equation* for the reaction (p. 397) may be written

$$\text{reaction rate} = -\frac{d[S_2O_8^{2-}]}{dt} = k[S_2O_8^{2-}]^m[I^-]^n, \qquad (8.2.1)$$

where the integer m is the order of the reaction with respect to $S_2O_8^{2-}$, and the integer n is the order with respect to I^-.

In this experiment we use the *initial rate method* to find m and n. The basis of the method is to measure the rate of the reaction over a period which is short enough for the reaction not to have proceeded significantly, but long enough to be unaffected by the time which the solutions take to mix at the beginning of the reaction. *What complication would arise if this reaction were monitored for a longer period of time?*

For a chosen set of initial conditions, we measure the time interval Δt between the start of the reaction and the appearance of the blue colour, corresponding to the production of an amount of iodine $\Delta[I_2]$. The rate of iodine production is equal to the rate of persulphate consumption, and thus to the rate of the reaction. If the amount of thiosulphate used satisfies the conditions for the measurement of initial rate, we may write

$$\text{rate of reaction} = -\left(\frac{d[S_2O_8^{2-}]}{dt}\right)_{\text{initial}} = \left(\frac{d[I_2]}{dt}\right)_{\text{initial}} = \frac{\Delta[I_2]}{\Delta t}. \qquad (8.2.2)$$

We then carry out further investigations with different initial conditions but the same amount of thiosulphate. Thus $\Delta[I_2]$ is also the same for each run, and it follows that

$$-\left(\frac{d[S_2O_8^{2-}]}{dt}\right)_{\text{initial}} = \frac{\text{const.}}{\Delta t}. \qquad (8.2.3)$$

Substituting back into eqn (8.2.1), and taking logarithms,

$$\ln(1/\Delta t) = \ln k + m \ln[S_2O_8^{2-}] + n \ln[I^-] - \text{const.} \qquad (8.2.4)$$

Thus a plot of $\ln(1/\Delta t)$ against $\ln[S_2O_8^{2-}]$ at constant $[I^-]$ will yield a straight line of slope m, and a graph of $\ln(1/\Delta t)$ against $\ln[I^-]$ at constant $[S_2O_8^{2-}]$ will yield n.

B. Temperature

The rate coefficients k of many simple reactions are found to vary with temperature T according to the *Arrhenius equation*

$$k = A e^{-E_a/RT}, \qquad (8.2.5)$$

where A is the *pre-exponential factor* and E_a the *activation energy* of the reaction. This expression was originally proposed by Arrhenius and van't Hoff on the basis of the van't Hoff isochore (p. 91).[1] A better understanding of the equation may be gained, however, by the interpretation of A as a measure of the frequency of encounters between reactants in the solution, and $e^{-E_a/RT}$ as the Boltzmann factor which gives the fraction of the molecules which have the required energy E_a to react.[2]

For reactions such as the present one which show Arrhenius-type behaviour

$$\ln k = \ln A - E_a/RT. \tag{8.2.6}$$

Thus both A and E_a may be obtained from a graph of $\ln k$ against $1/T$ provided that E_a is independent of temperature, which is usually a good approximation.

C(*i*) *Effect of ionic strength*

The rate coefficient of an ionic reaction depends on the *ionic strength* of the solution (p. 116). This *primary kinetic salt effect* may be understood qualitatively in terms of the favourable interactions between the reactants and activated complex and the *ionic atmospheres* of oppositely charged ions which surround them in solution. Three cases may be identified:

(a) If the charges on the reactants have the same sign, the activated complex will be more highly charged than the reactants. Increasing the ionic strength of the solution will therefore favour the interactions between the activated complex and the ionic atmosphere relative to those between the reactants and the ionic atmosphere, and will thus increase the rate coefficient.

(b) If the charges on the reactants have different signs, the charge on the activated complex will be lower than the charges on the reactants, and the rate coefficient will decrease with ionic strength.

(c) If one of the reactants is uncharged, there will be no change in the rate coefficient with ionic strength.

This qualitative explanation of the primary kinetic salt effect is all that is needed for the present experiment. However for the more advanced student a quantitative description, incorporating the Debye–Hückel theory of ionic solutions, is given in the section which follows.

C(*ii*) *Derivation of the Brønsted equation*[3]

Let us consider a simple reaction between two ions A and B which proceeds through an activated complex $(AB)^{\ddagger}$:

$$A + B \xrightarrow{k_2} AB^{\ddagger} \xrightarrow{k^{\ddagger}} \text{products.}$$

Then if z_A and z_B are the charges on the ions A and B, the charge on the activated complex will be $(z_A + z_B)$.

The rate equation for this second order reaction will be

$$\text{rate} = k_2[\text{A}][\text{B}] = k^{\ddagger}[\text{AB}^{\ddagger}], \tag{8.2.7}$$

where k_2 is the rate coefficient. If we assume that the activated complex is in equilibrium with the reactants,

$$K^{\ddagger} = \frac{a_{\text{AB}^{\ddagger}}}{a_{\text{A}}a_{\text{B}}} = \frac{[\text{AB}^{\ddagger}]\gamma^{\ddagger}}{[\text{A}][\text{B}]\gamma_{\text{A}}\gamma_{\text{B}}}. \tag{8.2.8}$$

In this equation the equilibrium constant $K^{\ddagger}$ is expressed as a quotient of the activity of the activated complex $a_{\text{AB}^{\ddagger}}$ and the activities of the reactants a_{A} and a_{B}. These activities are then expressed as products of concentrations and activity coefficients γ.

Combining eqns (8.2.7) and (8.2.8), and taking logarithms,

$$\log_{10} k_2 = \log_{10}(k^{\ddagger}K^{\ddagger}) + \log_{10}(\gamma_{\text{A}}\gamma_{\text{B}}/\gamma^{\ddagger}). \tag{8.2.9}$$

We now use the Debye–Hückel limiting law (p. 117), which may be written in the form

$$\log_{10}\gamma_i = -Az_i^2 I^{1/2},$$

to find the activity coefficient for each species i. Then, after making the substitution

$$z^{\ddagger} = z_{\text{A}} + z_{\text{B}},$$

we obtain the *Brønsted equation*

$$\log_{10}k_2 = \text{const} + 2Az_{\text{A}}z_{\text{B}}I^{1/2}. \tag{8.2.10}$$

The ionic strength I is defined by

$$I = \tfrac{1}{2}\sum_i c_i z_i^2, \tag{8.2.11}$$

where c_i is the concentration (strictly molality) of each ionic species, and z_i the charge of each species. The summation must include *all* the ionic species present in the solution, not merely the reactant ions.

Does the Brønsted equation predict the three types of behaviour discussed in Section C(i)?

Apparatus

A thermostat bath is required for Part B; it is set initially to room temperature. It is convenient to use three burettes to dispense the stock solutions.

Procedure

You are provided with the following solutions:

(a) 0.1 M $(NH_4)_2S_2O_8$,
(b) 0.1 M $(NH_4)_2SO_4$,
(c) 0.01 M $Na_2S_2O_3$.

Verify that the total ionic strength of solution (d) is the same as that of solution (a) and of solution (b).

Fill two of the burettes with solutions (a) and (b).

Make up a solution (d) by filling a 500 cm^3 volumetric flask with 333 cm^3 of solution (b), dissolving in it 8.30 g (0.050 mole) of solid KI, and making up to 500 cm^3 with de-ionized water.

A. *Rate equation*

Place 10 cm^3 of 0.1 M $S_2O_8^{2-}$ solution (a) and one drop of starch solution in a beaker. In a 100 cm^3 conical flask place 10 cm^3 of 0.1 M KI solution (d) and 5 cm^3 0.01 M thiosulphate solution (c). Mix and time to the appearance of the blue colour. Repeat the experiment with different volumes of the $S_2O_8^{2-}$ and I^- solutions as indicated in Table 8.2.1, keeping the ionic strength constant in each case by making up the volume with the ammonium sulphate solution (b).

Table 8.2.1. Composition of solutions/cm^3

Run	(a) ($S_2O_8^{2-}$)	(b) ($(NH_4)_2SO_4$)	(c) ($S_2O_3^{2-}$)	(d) (I^-)
1	10	0	5	10
2	10	2	5	8
3	10	4	5	6
4	10	6	5	4
5	8	2	5	10
6	6	4	5	10
7	4	6	5	10

B. *Effect of temperature*

Make up a mixture of 10 cm^3 of solution (a) with one drop of starch solution, and a second solution containing 10 cm^3 of (c) and 10 cm^3 of (d). Place the persulphate/starch and iodine/thiosulphate solutions in the thermostat bath for 5 min, and note the temperature of the bath. Mix and note the time to the appearance of the blue colour.

Repeat the procedure at 5 temperatures up to 50 °C.

C. *Effect of ionic strength*

Compare the initial rates (at room temperature) for the two reaction mixtures shown in Table 8.2.2. Note that the second solution has the same concentration of reactants but lower ionic strength.

Table 8.2.2. Mixture compositions in cm^3 for testing effect of ionic strength on reaction rate

Run	(a) ($S_2O_8^{2-}$)	(c) ($S_2O_3^{2-}$)	(d) (I^-)	
1	10	5	10	+ 10 cm^3 0.1 M $(NH_4)_2SO_4$ solution
2	10	5	10	+ 10 cm^3 water

Calculation A. Calculate the values of m and n, the order of the reaction with respect to persulphate and iodide ions.

B. From the effect of temperature on the rate coefficient, calculate the value of the activation energy E_a and the pre-exponential factor A. Also find the value of the rate coefficient at 298 K, if no experiment has been performed at this temperature.

C. Comment on the effect of the reduction in ionic strength on the rate of the reaction.

Results *Accuracy*

The iodide/persulphate reaction follows a simple rate equation, and m and n should be integral within experimental error. The reaction obeys the primary kinetic salt effect in general (but see below), and there should be no ambiguity about the qualitative effect of change in ionic strength on reaction rate.

Comment Clock reactions form an elegant, although obviously not generally applicable, method of studying reaction rates. With regard to the persulphate/iodide reaction, there is good agreement with the Brønsted equation provided that the persulphate concentration is kept low. However, it has also been shown that the kinetics of this reaction are strongly dependent on the nature of the added cations,[4] which underlines the danger of assuming that 'inert' salts added to increase the ionic strength of a solution take no part in the reaction.

Technical notes

Apparatus

Thermostat tank, burettes etc., 0.1 M ammonium persulphate, 0.1 M ammonium sulphate, and 0.01 M sodium thiosulphate.

References

[1] *Chemical kinetics*. K. J. Laidler. McGraw-Hill, New York, 2nd edn (1965), p. 50.

[2] *Physical chemistry*. P. W. Atkins. Oxford University Press, Oxford, 3rd edn (1986), sect. 28.3(b).

[3] Ref. 1, chap. 5.
[4] *Salt effects in solution kinetics.* C. W. Davies. *Prog. React. Kinetics* **1,** 161 (1961).

8.3 Mutarotation of dextrose by polarimeter

It is found that organic compounds which have structures which are not superimposable on their mirror images are *optically active.* If plane polarized light is passed through an aqueous solution of an optically active substance, the plane of polarization is rotated by an extent which can be measured with a *polarimeter.*

If in a reaction the angle of rotation caused by the reactants and their products is different, then the angle will change as the reaction proceeds, and the kinetics of the reactions may be studied by polarimetry. The *mutarotation* of dextrose (α-D-glucose) is a reaction of this type.[1] A freshly prepared solution of this compound does not retain its initial optical activity, but the rotation falls with time to a constant value. This is due to the first order isomeric change of α-D-glucose to an equilibrium mixture of both α-D- and β-D-glucose, as shown in Fig. 8.3.1.

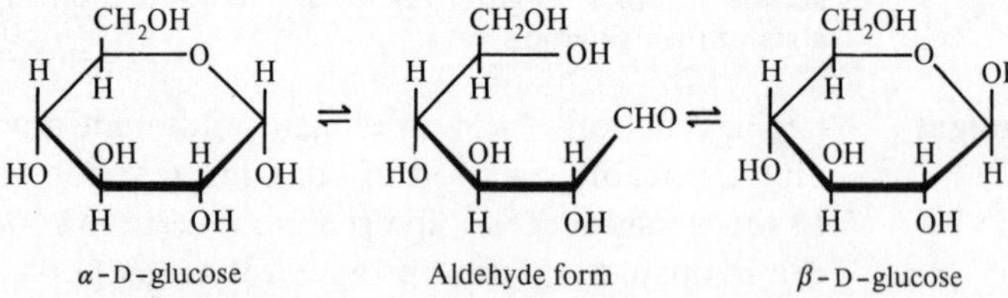

Fig. 8.3.1. Equilibrium in solution between α–D–glucose and β–D–glucose

The mutarotation of dextrose was one of the first reactions shown to undergo *general acid and base catalysis.* In this experiment we investigate this effect and measure the rate constants of the reactions for catalysis by hydroxonium ions H_3O^+, and water.

Theory

(*i*) *Specific rotation*

The specific rotation $[\alpha]$ of a substance in solution is defined as the observed rotation α of the plane of polarization of polarized light passing through a cell of unit length containing a solution of the substance at unit mass concentration. The quantity depends on temperature and the wavelength of light used, which are written as a superscript and subscript respectively. Thus the specific rotation of light from the sodium D line at 298 K would be written $[\alpha]_D^{298}$. The identity of the solvent should also be quoted, since the specific rotation is dependent upon it. Under some conditions the specific rotation may also vary with concentration, although it is justifiably assumed not to in this experiment.

For a solution of concentration c of a homogeneous optically active substance in an inactive solvent such as water

$$[\alpha] = \alpha / lc, \tag{8.3.1}$$

where l is the length of the cell. Thus for a given substance in a particular cell, the observed rotation α is directly proportional to its concentration.

(*ii*) *Mutarotation of dextrose*

The first order reactions which cause the mutarotation of dextrose may be written

$$\alpha\text{-D-glucose} \underset{k_2}{\overset{k_1}{\rightleftharpoons}} \beta\text{-D-glucose}. \tag{8.3.2}$$

concn: [A] [B]

At the start of the reaction there is no β-glucose present, i.e. $[B]=0$ at $t=0$. The reaction proceeds to the equilibrium position where the rates of the forward and backward reactions become equal, i.e. $k_1[A]=k_2[B]$. We define a parameter y which is the displacement of the reaction from the ultimate equilibrium position:

$$y = \frac{d[B]}{dt} = \frac{-d[A]}{dt} = k_1[A] - k_2[B]. \tag{8.3.3}$$

Then

$$\frac{dy}{dt} = k_1 \frac{d[A]}{dt} - k_2 \frac{d[B]}{dt}$$

$$= -y(k_1 + k_2). \tag{8.3.4}$$

Integration of this differential equation gives

$$y = y_0 e^{-(k_1+k_2)t} = y_0 e^{-kt} \tag{8.3.5}$$

This relation is a general one and is one of a family of such equations found, for example, in relaxation techniques such as the temperature jump method.

where $y = y_0$ at $t=0$. This expression shows that the displacement from equilibrium declines exponentially with a rate constant given by the sum of k_1 and k_2, which we call the *mutarotation coefficient k*.

It follows from the definition in section (i) that we may express the observed optical rotation α at a time t in the form

$$\alpha = a[\mathrm{A}] + b[\mathrm{B}], \tag{8.3.6}$$

where a and b are constants for an experiment in a particular apparatus at constant temperature. When the system has reached equilibrium at $t = \infty$,

$$\alpha_\infty = a[\mathrm{A}]_\infty + b[\mathrm{B}]_\infty. \tag{8.3.7}$$

Since the total concentration of glucose does not change, it follows that $[\mathrm{A}]+[\mathrm{B}]=[\mathrm{A}]_\infty+[B]_\infty$, and therefore from eqns (8.3.6) and (8.3.7)

$$\alpha - \alpha_\infty = (a-b)([\mathrm{A}]-[\mathrm{A}]_\infty). \tag{8.3.8}$$

In a similar way it may be shown from eqn (8.3.3) that

$$y - y_\infty = k([\mathrm{A}]-[\mathrm{A}]_\infty). \tag{8.3.9}$$

However $y_\infty = 0$ and therefore from eqn (8.3.8)

$$y = k(\alpha-\alpha_\infty)/(a-b). \tag{8.3.10}$$

Substituting this expression into eqn (8.3.5) we find that

$$\ln(\alpha-\alpha_\infty) = \ln[y_0(a-b)/k] - kt. \tag{8.3.11}$$

If $\alpha = \alpha_0$ at $t = 0$, it follows that

$$\ln(\alpha-\alpha_\infty) = \ln(\alpha_0-\alpha_\infty) - kt. \tag{8.3.12}$$

From this equation we may obtain a value of the mutarotation coefficient k from the slope of a plot of $\ln(\alpha-\alpha_\infty)$ against t.

However, α_∞ is the rotation when the reaction has reached equilibrium, and this takes many hours to achieve. Also, if the single reading of α_∞ is wrong, it will cause an error in k even if every other measurement of α is correct. We therefore avoid the use of α_∞ by employing a method of analysis devised by Guggenheim and Swinbourne,[2,3] as follows.

$2n$ readings are taken at regular intervals of time, and then each of the first n readings is paired in chronological order with one of the remaining n readings; for instance, if n is 20, the first reading is paired with the twenty-first, the second with the twenty-second and so on. Thus each pair of readings α_t and $\alpha_{t+\Delta t}$ will be separated by the same time interval, Δt. Ideally Δt should be several half-lives.

From eqn (8.3.12):

$$\alpha_t = \alpha_\infty + (\alpha_0-\alpha_\infty)\cdot \mathrm{e}^{-kt}. \tag{8.3.13}$$

It also follows from eqn (8.3.12) that

$$\ln(\alpha_{t+\Delta t}-\alpha_\infty)=\ln(\alpha_0-\alpha_\infty)-k(t+\Delta t), \tag{8.3.14}$$

and therefore that

$$\alpha_{t+\Delta t}=\alpha_\infty+(\alpha_0-\alpha_\infty)\cdot \mathrm{e}^{-kt}\cdot \mathrm{e}^{-k\Delta t}. \tag{8.3.15}$$

Multiplying eqn (8.3.15) by $\mathrm{e}^{k\Delta t}$,

$$\alpha_{t+\Delta t}\cdot \mathrm{e}^{k\Delta t}=\alpha_\infty\cdot \mathrm{e}^{k\Delta t}+(\alpha_0-\alpha_\infty)\cdot \mathrm{e}^{-kt}. \tag{8.3.16}$$

Subtracting eqn (8.3.16) from eqn (8.3.13) we find that

$$\begin{aligned}\alpha_t &= \alpha_{t+\Delta t}\cdot \mathrm{e}^{k\Delta t}+\alpha_\infty-\alpha_\infty\cdot \mathrm{e}^{k\Delta t}\\ &= \alpha_{t+\Delta t}\cdot \mathrm{e}^{k\Delta t}+\text{const.}\end{aligned} \tag{8.3.17}$$

Thus k may be determined from the slope of a graph of α_t against $\alpha_{t+\Delta t}$.

The constant term, containing α_∞, k, and Δt, need not be known. Moreover the result will not be affected by any impurities present which only contribute to α_∞. One further point about the plot is that as the time t tends to infinity, so $\alpha_t \rightarrow a_{t+\Delta t}\rightarrow \alpha_\infty$. It follows that the infinity reading may be found by extrapolating the experimental plot until it crosses the line $\alpha_t=\alpha_{t+\Delta t}$, because at $t=\infty$ there is no difference between consecutive measurements of α.

This method can be applied to any experimental property which varies exponentially with time, and we employ it again in the next experiment.

(iii) *General acid and base catalysis*[4]

The mutarotation of tetra-O-methyl-α-D-glucopyranose is slow in dry pyridine or in dry cresol, but fast in a mixture of the two solvents or in either solvent when moist; how may this be explained?

As mentioned earlier, the mutarotation of dextrose is subject to general acid and base catalysis, i.e. the reaction is catalysed in both the forward and backward directions by acids, bases and amphoteric solvents such as water. Let us consider the mutarotation of an aqueous solution of dextrose in the presence of hydroxonium ions and hydroxyl ions. Under these circumstances the mutarotation is catalysed by H_2O, H_3O^+, and OH^-, and the mutarotation coefficient obeys a rate equation of the form

$$k=k_1+k_2=k_{H_2O}[H_2O]+k_{(H_3O^+)}[H_3O^+]+k_{(OH^-)}[OH^-]. \tag{8.3.18}$$

The terms in square brackets are the concentrations of the various species. The concentration of water varies only slightly from 55.6 mol dm^{-3} in aqueous systems, so we may include this

k_{H_2O} represents the combined catalytic action of the water molecule as an acid and a base, and could therefore be split into two terms; however the measurements required are very difficult and well beyond the scope of this experiment.

in the k_{H_2O} term and write

$$k = k'_{H_2O} + k_{(H_3O^+)}[H_3O^+] + k_{(OH^-)}[OH^-]. \quad (8.3.19)$$

In the pH range 4 to 6, we may neglect the catalysis by the base, and eqn (8.3.19) becomes

$$k = k'_{H_2O} + k_{(H_3O^+)}[H_3O^+]. \quad (8.3.20)$$

Thus on plotting k against $[H_3O^+]$ under these conditions, a straight line is obtained of slope $k_{(H_3O^+)}$ and intercept k'_{H_2O}.

The equivalent experiment with a base is not as straightforward. The mutarotation becomes too rapid for accurate measurement above pH 9, and so buffer solutuions must be used in which it is necessary to make assumptions about the activity of the hydroxide ion. Evaluation of the rate constant is also complicated by ionization of the sugar, a subject not considered by early workers.

Apparatus[5]

In its simplest form a polarimeter comprises a light source, polarizer to plane polarize the light, sample cell, and analyser for determining the plane of polarization after the light has traversed the substance under investigation.

Usually both the polarizer and analyser employ a *Nicol prism*. This is a calcite crystal which produces two refracted rays which are plane-polarized in directions which are mutually at right angles. The crystal is cut so that only one of these beams emerges, the other being returned by total internal reflection in the direction of the light source. Thus the light emerging from the Nicol prism is plane polarized.

A measurement is made by rotating the analyser until a position of maximum, or minimum, light transmission is observed. The two prisms are then angled to each other by an extent α, or $\alpha + 90°$. In practice, it is almost impossible to measure the position of maximum or minimum brightness with the eye. Commercial polarimeters therefore employ some device for splitting the field into two or more portions, which become equally illuminated when the analyser is rotated by an extent α.

Figure 8.3.2 shows the layout of a typical split-field polarimeter. Monochromatic light, usually from a sodium lamp O, passes through a Nicol prism P which polarizes it. Half of the beam then strikes a specially cut quartz plate. This alters the phase of the light by half a wavelength, which is equivalent to rotating this half of the beam. The two halves of the beam then pass through the sample cell Q, which incorporates an outer jacket through which water is passed from an adjacent thermostat tank. They then strike the analysing Nicol prism. This is mounted on a

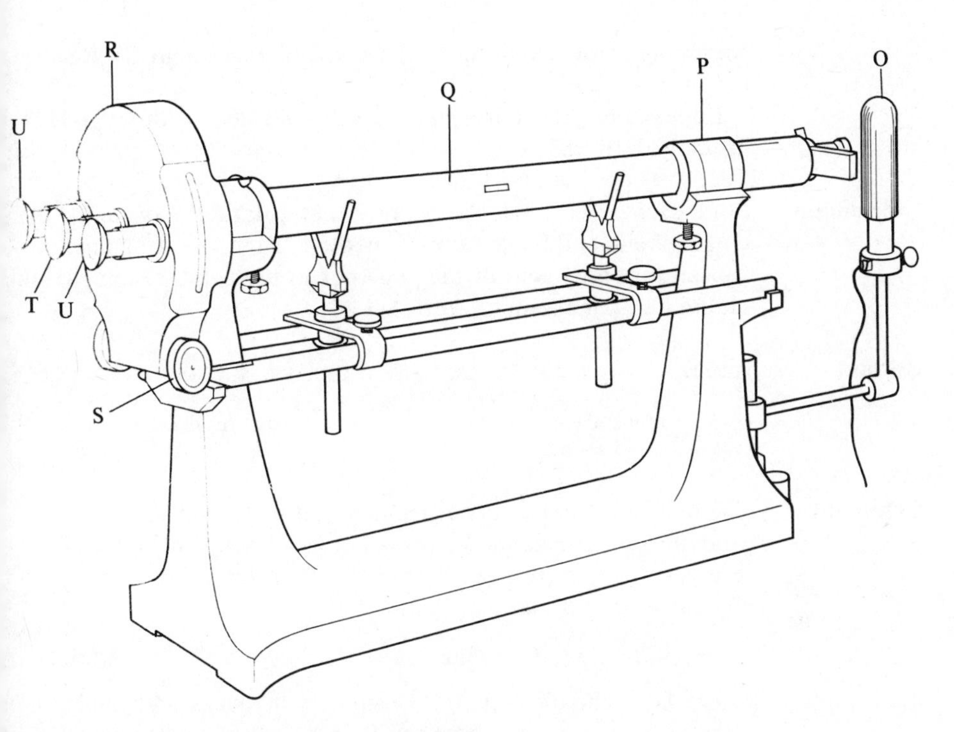

Fig. 8.3.2. A standard form of polarimeter

vertical turntable R which is rotated at S until the two halves of the field in T are equally illuminated. The angle of rotation α caused by the sample is then read off a scale by taking the mean reading of the two outer telescopes U.

Procedure

Switch on the light source and set up the polarimeter.* Wash and dry the polarimeter tube.

Allow 100 cm^3 of 0.1 M hydrochloric acid and 1 dm^3 of de-ionized water to reach thermal equilibrium in the thermostat tank.

Weigh out 10 g of dextrose. Dissolve it as quickly as possible in about 50 cm^3 of the de-ionized water in a 100 cm^3 volumetric flask. While dissolving the dextrose, keep replacing the flask in the thermostat tank to maintain the temperature of the solution as constant as possible. Add 10 cm^3 of the 0.1 M HCl to the dextrose solution and note the time. Make up to 100 cm^3 with de-ionized water. Fill the polarimeter cell with the solution, being careful to exclude air bubbles.*

On some polarimeters there are two positions of equal intensity, and the 'dull' position is much more precise than the very bright one.

Take readings of the angle of rotation of the solution at 5 min intervals until the reaction is at least 3/4 complete, i.e. until the change in rotation in 5 min is less than a quarter of that at the beginning of the experiment. This should take about 1 h for the first run.

Repeat the procedure, but with 20 and 40 cm^3 of 0.1 M HCl instead of 10 cm^3.

Calculation For each run, calculate the mutarotation constant by Guggenheim and Swinbourne's method. Then plot k against the concentration of acid in the cell (not in the solution added) to obtain k'_{H_2O} and $k_{(H_3O^+)}$, eqn (8.3.20).

Results[1,2]

Accuracy

Make a realistic estimate of the accuracy of your results, based on the scatter of the points.

Comment The next stage in this experiment would be to find values of the catalytic rate constants k_a for a series of weak acids.[2,4] The results of such experiments obey the *Brønsted catalysis law* for general acid and base catalysis:

$$\log_{10}k_a = x \log_{10}K_a + \text{const} \qquad (8.3.21)$$

where K_a is the dissociation constant of the weak acid, and x is a factor having a value between 0 and 1. This expression arises from the dependence of the catalytic activity of an acid, as measured by k_a in a given reaction, on the readiness with which proton transfer takes place, and thus on K_a.

Equation (8.3.21) is an expression of the type known as a *linear free energy relationship.*[6] The appropriateness of this term may be understood by remembering that the equilibrium constant K of a reaction is related to its standard free energy change $\Delta G^{\ominus}$ by the relation

$$\Delta G^{\ominus} = -RT \ln K.$$

A relation between the logarithms of K and k at constant temperature is thus essentially a relationship between the standard free energy change of the reaction and the free energy of activation.

The mutarotation of dextrose has also been studied by measuring the very small change in volume of the solution which occurs during the reaction.[2]

Technical notes

Apparatus

Polarimeter,$ \alpha$-D-glucose (research grade), standardized hydrochloric acid, stopclock.

References

[1] *Mutarotation of sugars in solution:* Part II. H. S. Isbell and W. Pigman. *Adv. in Carbohyd. Chem. and Biochem.* **24,** 13 (1969).
[2] *Contribution to the theory of acid and base catalysis. The mutarotation of glucose.* J. N. Brønsted and E. A. Guggenheim. *J. Amer. Chem. Soc.* **49,** 2554 (1927).
[3] *Method for obtaining the rate coefficient and final concentration of a first-order reaction.* E. S. Swinbourne *J. Chem. Soc.* 2371 (1960).
[4] *The proton in chemistry.* R. P. Bell. Chapman and Hall, London, 2nd edn (1973). ch. 8.
[5] *Techniques of chemistry.* ed. A. Weissberger. Wiley Interscience, New York (1972). Vol. 1, part IIIC, pp. 79 and 109.
[6] *Correlation analysis in organic chemistry: an introduction to linear frequency relationships.* J. Shorter. Clarendon Press, Oxford (1973), ch. 1.

8.4 Kinetics of 2-iodo-2-methylbutane hydrolysis by conductivity

If the concentration of ions in a reaction mixture changes as the reaction proceeds, then the rate of the reaction may be measured by following the change in conductivity of the solution.

In this experiment we use measurements of conductivity to study the kinetics of the hydrolysis of 2-iodo-2-methylbutane (*tert*-amyl iodide) in aqueous ethanol. The conductivity of the solution increases as the reaction proceeds because of the formation of hydroxonium and iodide ions:

$$C_5H_{11}.I + 2H_2O \rightarrow C_5H_{11}.OH + H_3O^+ + I^-.$$

The experimental results show that the reaction is first order in $C_5H_{11}I$. Its rate is independent of the concentration of hydroxonium or hydroxyl ions, and the reaction is not subject to acid and base catalysis, as is the mutarotation of glucose in the previous experiment.

Having verified the order of the reaction, we measure the effect of temperature on the reaction rate. Assuming Arrhenius behaviour, the activation energy and pre-exponential factor A are calculated. An estimate of the entropy of activation is then made from A, using the arguments of activated complex theory.

Theory

(*i*) *Rate theory*[1]

The hydrolysis of 2-iodo-2-methylbutane is a two-stage process:

$$C_5H_{11}I \xrightarrow[\text{(slow)}]{k_1} C_5H_{11}^+ + I^-,$$

$$C_5H_{11}^+ + 2H_2O \xrightarrow[\text{(fast)}]{} C_5H_{11}OH^- + H_3O^+.$$

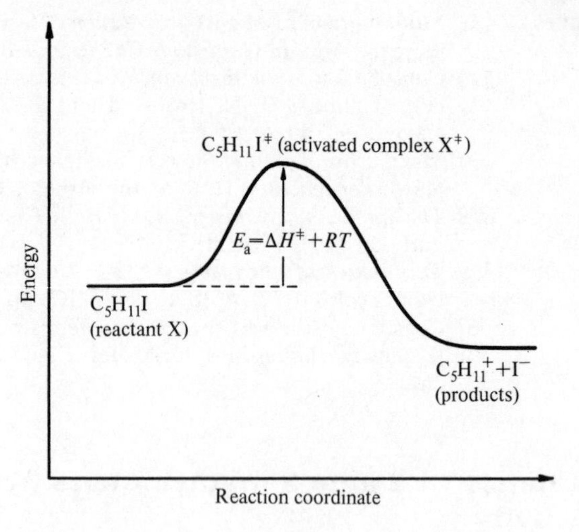

Fig. 8.4.1. Reaction profile for the hydrolysis of 2-iodo-2-methylbutane

The rate determining step is the formation of $C_5H_{11}^+$, which may be represented on a reaction profile as shown in Fig. 8.4.1. If we assume the reaction exhibits *Arrhenius behaviour* (p. 401) then

$$k_1 = A \cdot e^{-E_a/RT}, \tag{8.4.1}$$

and

$$\ln k_1 = \ln A - E_a/RT. \tag{8.4.2}$$

Thus a plot of the logarithm of the first order rate coefficient k_1 against $1/T$ will yield the activation energy E_a and the pre-exponential factor A.

It is also of interest to interpret the results in terms of simple *activated complex* (*transition state*) *theory*. The approach relies on the assumption that the activated complex is in equilibrium with the reactants.[2] We write the present reaction in the form

More detailed proofs, not relying on this assumption, yield the same results.

$$X \overset{K^{\ddagger}}{\rightleftharpoons} X^{\ddagger} \overset{\nu}{\longrightarrow} \text{products},$$

where X is the reactant, $X^{\ddagger}$ the activated complex, ν the first order rate coefficient for the reaction of $X^{\ddagger}$ to form products, and $K^{\ddagger}$ the equilibrium constant for X and $X^{\ddagger}$. Thus, since equilibrium is assumed,

$$K^{\ddagger} = [X^{\ddagger}]/[X], \tag{8.4.3}$$

and the rate of forming products is $\nu[X^{\ddagger}] = \nu K^{\ddagger}[X]$. Since the rate of forming products is also $k_1[X]$, where k_1 is the effective first order rate coefficient, then

$$k_1 = \nu K^{\ddagger}. \tag{8.4.4}$$

A more detailed application of activated complex theory shows that, to a fair approximation, ν may be set equal to $k_B T/h$, where k_B is Boltzmann's constant, and h is Planck's constant. Thus

$$k_1 = (k_B T/h) K^{\ddagger}. \tag{8.4.5}$$

We define a Gibbs free energy of activation $\Delta G^{\ddagger}$ using an equation similar to the van't Hoff expression:

$$\Delta G^{\ddagger} = -RT \ln K^{\ddagger}. \tag{8.4.6}$$

Therefore

$$k_1 = (k_B T/h) \mathrm{e}^{-\Delta G^{\ddagger}/RT}, \tag{8.4.7}$$

and substituting $\Delta G^{\ddagger} = \Delta H^{\ddagger} - T\Delta S^{\ddagger}$:

$$k_1 = (k_B T/h)\, \mathrm{e}^{\Delta S^{\ddagger}/R}\, \mathrm{e}^{-\Delta H^{\ddagger}/RT}. \tag{8.4.8}$$

Differentiation of the logarithm of this relation gives the expression

$$\frac{\mathrm{d}(\ln k_1)}{\mathrm{d}T} = \frac{\mathrm{d}}{\mathrm{d}T}(\ln k_B/h + \ln T + \Delta S^{\ddagger}/R - \Delta H^{\ddagger}/RT)$$

$$= \frac{1}{T} + \frac{\Delta H^{\ddagger}}{RT^2}, \tag{8.4.9}$$

where we have assumed $\Delta S^{\ddagger}$ and $\Delta H^{\ddagger}$ to be independent of T. However from eqn (8.4.2),

$$\frac{\mathrm{d}(\ln k_1)}{\mathrm{d}T} = \frac{E_a}{RT^2}. \tag{8.4.10}$$

Therefore from eqns (8.4.9) and (8.4.10),

$$E_a = \Delta H^{\ddagger} + RT. \tag{8.4.11}$$

The entropy of activation may be found by equating the logarithmic forms of the experimental and theoretical rate coefficients shown in eqns (8.4.2) and (8.4.9):

$$\ln A - \frac{E_a}{RT} = \ln\left(\frac{k_B T}{h}\right) + \frac{\Delta S^{\ddagger}}{R} - \frac{\Delta H^{\ddagger}}{RT}. \tag{8.4.12}$$

Substituting for $\Delta H^{\ddagger}$ from eqn (8.4.11) gives

$$\Delta S^{\ddagger}/R = \ln A - \ln(k_B T/h) - 1. \tag{8.4.13}$$

Thus $\Delta S^{\ddagger}/R$ may be calculated from the value of A obtained from the Arrhenius plot, using an average value of T.

(*ii*) *Analysis of conductivity data*

Since the hydrolysis of 2-iodo-methylbutane is first order,

$$\frac{d[X]}{dt} = -k_1[X], \tag{8.4.14}$$

where [X] is the concentration of the reactant. Integration gives:

$$[X] = [X]_0 e^{-k_1 t}, \tag{8.4.15}$$

where $[X]_0$ is the initial concentration.

Now, at time t the concentration of dissociated HI is $[X]_0 - [X]$, and the conductivity λ_t is given by:

$$\lambda_t = \lambda_0 + \lambda_{HI} = \lambda_0 + \alpha([X]_0 - [X]), \tag{8.4.16}$$

where λ_0 is the conductivity at $t = 0$, and α is a constant determined by the molar conductivity of HI and the cell constant.

The direct method of analysing the results is to take measurements of λ_t, and also of λ_∞, the conductivity at $t = \infty$, where

$$\lambda_\infty = \lambda_0 + \alpha[X]_0. \tag{8.4.17}$$

k_1 may then be found from a plot of $\ln(\lambda_\infty - \lambda_t)$ against t, since from eqns (8.4.15), (8.4.16), and (8.4.17),

$$\ln(\lambda_\infty - \lambda_t) = \ln(\alpha[X]_0) - k_1 t. \tag{8.4.18}$$

However, this method is subject to the same two problems that arose in the analysis of optical rotation data in Expt 8.3. The first is in the measurement of λ_∞. To obtain this quantity, one has either to wait a long time, or else the solution has to be taken out of the thermostat, heated to accelerate the reaction, cooled back to the right temperature and finally its conductivity has to be measured. Both procedures are tedious. The second problem is that any error in λ_∞ is repeated in every point on the graph and often leads to a curved plot.

We avoid these problems by pairing the conductivity readings, taking $\lambda_{t_1}, \lambda_{t_2}, \lambda_{t_3}, \ldots$ with $\lambda_{t_1+\Delta t}, \lambda_{t_2+\Delta t}, \lambda_{t_3+\Delta t} \ldots$ where Δt is a constant time interval, preferably of several half-lives. We then employ Guggenheim and Swinbourne's analysis[3] in the same way as in Expt. 8.3, and obtain the working equation (p. 409) modified for a quantity which increases, rather than decreases, with time:

$$\lambda_t = \lambda_{t+\Delta t} \cdot e^{k_1 \Delta t} - \text{const.} \tag{8.4.19}$$

It follows that k_1 may be determined from the slope of a graph of λ_t against $\lambda_{t+\Delta t}$.

Apparatus

The apparatus used for the measurement of conductivity, namely a conductivity meter and cell, has been described on p. 365.

Ideally the conductivity cell should be placed in a sample tube which is mounted in a constant temperature water bath and which contains a continuously-running stirrer. In practice, however, the usual procedure is to mix the reactants thoroughly in a separate tube in the tank, and then pour the mixture into a second tube containing the conductivity cell.

The solvent supplied for the experiment is 80 per cent by volume of ethanol in water.

Procedure

(*i*) *Measurement of conductivity*

Pour about 50 cm^3 of solvent into a stoppered boiling tube, and mount the tube in the water bath.

Connect the conductivity probe to the conductivity meter. Mount the probe in a second, empty tube in the water bath. Switch on the conductivity meter.

Wait at least 15 min for the solvent to reach thermal equilibrium and for the electronics in the meter to warm up.

Pipette about 0.3 cm^3 of 2-iodo-2-methylbutane into the solvent, and stir the mixture vigorously with a glass rod. Remove the conductivity probe from its tube and immerse it in the mixture. Measure the conductivity of the solution at exactly thirty-second intervals for 20 min.

(*ii*) *Influence of initial concentration*

Determine the influence of the initial concentration of the alkyl halide by repeating the procedure described in section (i) with (a) 0.2 cm^3, and (b) 0.4 cm^3 of the reactant in 50 cm^3 of solvent.

(*iii*) *Influence of temperature*

Repeat the procedure described in section (i) at three or more other temperatures in the range 18 to 35 °C. Carry out the experiments in order of increasing temperature, and allow plenty of time for the solutions to reach thermal equilibrium. At the higher temperatures, take readings every 20 sec for about 8 min.

Calculation

For each run, determine the rate coefficient k_1 by Guggenheim and Swinbourne's method.

Find whether k_1 depends on the initial concentration of the halide.

From the influence of temperature on the rate of the reaction, determine the activation energy E_a and the pre-exponential factor A.

Find the entropy of activation, $\Delta S^{\ddagger}$, and, most importantly, make a realistic estimate of the errors involved in its determination.

Results

Accuracy

One of the procedures for estimating the accuracy of this experiment should be the drawing of bounding lines on the conductivity and Arrhenius plots, to give error limits for the parameters derived from the gradients of these lines.

Comment

Measurements of the pre-exponential factor A, and hence of the entropy of activation $\Delta S^{\ddagger}$, can give information about the mechanism of a reaction, since $\Delta S^{\ddagger}$ indicates the change in the degree of disorder between the reactant and the activated complex. A value of A of $\sim(k_B T/h)$ (eqn (8.4.13)) shows that there is no significant change in the degree of disorder, whilst A factors much greater or smaller than this indicate respectively an increase or decrease. For reactions in solution, it is important to remember the changes in the ordering of the solvent consequent on forming the activated complex.

Technical notes

Apparatus

Conductivity meter,$ conductivity cell, thermostat tank, aqueous ethanol (80 per cent by volume), 2-iodo-2-methylbutane (= *tert*-amyl iodide).

References

[1] *Chemical kinetics.* K. J. Laidler. McGraw-Hill, New York, 2nd edn (1965), pp. 49–90 and 198–217.
[2] Ref. 1, p. 72.
[3] *Method for obtaining the rate coefficient and final concentration of a first–order reaction.* E. S. Swinbourne. *J. Chem. Soc.* 2371 (1960).

8.5 Kinetics of consecutive and parallel reactions by spectrophotometry

Sometimes the kinetics of a chemical process are complicated by the occurrence of consecutive and parallel reactions. An example is the hydrolysis of methyl chloroformate (X) in the presence of pyridine. This is a second-order reaction leading to the formation of the 1-methoxycarbonylpyridinium ion, MeO.CO.Py^+ (Y). In the presence of excess pyridine, the kinetics are pseudo first-order:

$$X \xrightarrow{k_1[Py]} Y, \tag{1}$$

where k_1[Py] is the pseudo first-order rate coefficient, and [Py] the concentration of pyridine in the free base form (which is *not* the same as $[Py]_{tot}$, the total concentration of pyridine added). The intermediate Y then undergoes a first-order dissociation into the products Z, which are MeOH, CO_2 and HCl:

$$Y \xrightarrow{k_2} Z. \tag{2}$$

There is also an accompanying direct first-order hydrolysis of X to Z:

$$X \xrightarrow{k_3} Z, \tag{3}$$

so that the overall scheme is

$$X \xrightarrow{k_1[Py]} Y \xrightarrow{k_2} Z, \quad X \xrightarrow{k_3} Z \tag{4}$$

In this experiment the concentration of the intermediate Y is measured with a spectrophotometer. We then find both k_1 and k_3 by measuring the time taken for [Y] to reach its maximum value, analysing the results by computer, and incorporating a known value of k_2.

Theory[1,2]

It can be seen from the overall reaction scheme that the rate of hydrolysis of X is given by the expression

$$\frac{-d[X]}{dt} = k_1[Py][X] + k_3[X] = k_{eff}[X], \tag{8.5.1}$$

where k_{eff} $(= k_1[Py] + k_3)$ is the effective first-order rate coefficient for the overall reaction. By integrating and writing $[X] = [X]_0$ at $t = 0$, we obtain the expression

$$[X] = [X]_0 \cdot e^{-k_{eff}t}. \tag{8.5.2}$$

It follows from the reaction scheme that

$$\frac{d[Y]}{dt}=k_1[Py][X]-k_2[Y]. \tag{8.5.3}$$

Equation (8.5.3) may be written in the form:

$$\frac{d}{dt}(e^{k_2t}[Y])=e^{k_2t}\,k_1[Py][X]$$
$$=e^{k_2t}\,k_1[Py][X]_0\,e^{-k_{eff}t}. \tag{8.5.4}$$

Integrating this expression we find that

$$e^{k_2t}[Y]=\frac{k_1[Py][X]_0\,e^{[k_2-k_{eff}]t}}{(k_2-k_{eff})}+\text{const.} \tag{8.5.5}$$

Since $[Y]=0$ at $t=0$, it follows that

$$[Y]=\frac{k_1[Py][X]_0(e^{-k_{eff}t}-e^{-k_2t})}{(k_2-k_{eff})}. \tag{8.5.6}$$

The mathematically inclined may be interested in working through the solution to the kinetics of this reaction when pyridine is not in excess.[3]

It can be seen that [Y] is dependent on the difference of two negative exponential terms in *t*. If we plot [Y] as a function of time, we see that it increases to a maximum and then falls again, eventually to zero. Figure 8.5.1 shows two such graphs for different relative values of k_{eff} and k_2. It is the second graph, in which $k_{eff} \ll k_2$, which applies to this experiment.

In this experiment it is not [Y] but an absorbance proportional to [Y] which is measured; the molar absorption coefficient of Y cannot readily be obtained. It is therefore convenient to proceed as follows.[4]

Let the maximum value $[Y]_m$ of [Y] occur at time t_m. At this moment $d[Y]/dt=0$, so that from eqn (8.5.3)

$$k_1[Py][X]_m=k_2[Y]_m, \tag{8.5.7}$$

where $[X]_m$ is the (non-stationary) concentration of X at $t=t_m$.

On substituting for $[X]_m$ and $[Y]_m$ from eqns (8.5.2) and

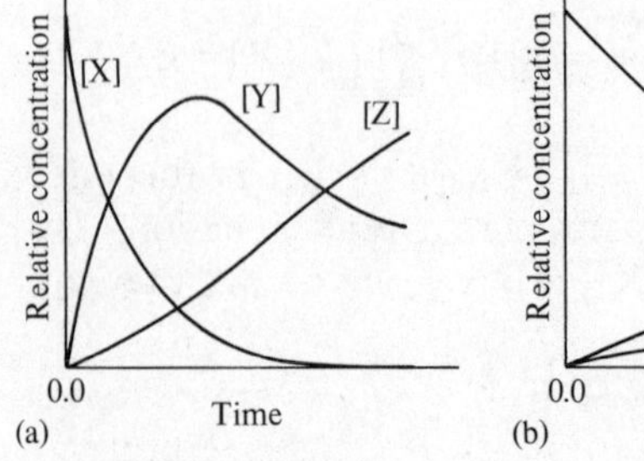

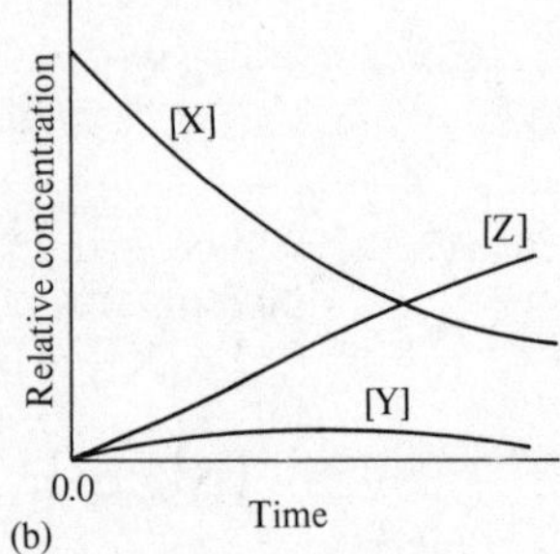

Fig. 8.5.1. Relative concentrations of the species X, Y, and Z: (a) $k_{eff} \gg k_2$; (b) $k_{eff} \ll k_2$

(8.5.6), and solving for t_m, we find that

$$t_m = \frac{\ln(k_{eff}/k_2)}{k_{eff} - k_2}. \quad (8.5.8)$$

Substituting this value into eqn (8.5.6), and simplifying, we obtain an expression for $[Y]_m$:

$$[Y]_m = [Py][X]_0\left(\frac{k_1}{k_2}\right)\left(\frac{k_2}{k_{eff}}\right)^{k_{eff}/(k_{eff}-k_2)}. \quad (8.5.9)$$

Thus, again from eqn (8.5.6), it follows that

$$[Y] = [Y]_m k_2\left(\frac{k_{eff}}{k_2}\right)^{k_{eff}/(k_{eff}-k_2)}(e^{-k_{eff}t} - e^{-k_2 t}). \quad (8.5.10)$$

The spectrophotometer used in this experiment is set at a wavelength (278 nm) at which only Y absorbs. The absorbance a recorded by the instrument is proportional to [Y], and thus

A darts player trying for a bull's-eye corrects his aim by an *iterative* process—the power and direction of each throw is corrected by an amount judged from the distance by which the previous dart missed the target. If the process is also *convergent*, the correction will become progressively smaller until the bull's-eye is hit.

$$a = a_m k_2\left(\frac{k_{eff}}{k_2}\right)^{k_{eff}/(k_{eff}-k_2)}(e^{-k_{eff}t} - e^{-k_2 t}). \quad (8.5.11)$$

The maximum absorbance a_m can be measured quite accurately. k_{eff} and k_2 may then be found by measuring a as a function of t, and fitting the results by a computer program which employs an iterative, convergent, least-squares fitting procedure as detailed in the Technical Notes section. k_1 and k_3 may be found by repeating the experiment so that k_{eff} is known at a series of different concentrations of pyridine (eqn (8.5.1)).

Apparatus

The apparatus comprises an ultraviolet spectrophotometer (p. 251) with a deuterium lamp source, connected to a chart recorder which plots optical density against time, rather than the usual graph of optical density against wavelength. The reaction takes place in a 4 cm path-length cuvette supplied with a small plastic plunger for stirring and a lid, and mounted in the spectrophotometer.

The methyl chloroformate is in the form of a 0.3 M solution in ethanonitrile (acetonitrile), and is stored in a desiccator. It is added to the other reactants in the cuvette by means of a 50 mm³ (50 μl) syringe.

Procedure

Make up the mixtures shown in Table 8.5.1 using the solutions provided.

Rinse and fill the ultraviolet cuvette to about 3/4 capacity with the first mixture. Place the cuvette in the cell holder of the spectrophotometer and close the lid. Allow 15 min for the solution to reach thermal equilibrium.

Table 8.5.1. Mixture compositions

Run	0.1 M pyridinium hydrochloride/cm^3	0.1 M hydrochloric acid/cm^3	0.1 M sodium chloride/cm^3
1	50	40	10
2	50	30	20
3	50	20	30
4	50	10	40

Set the spectrophotometer wavelength to 278 nm, switch on the power to the chart recorder with its drive switch off, and set the chart paper speed to about 4 cm min^{-1}. Lower the chart recorder pen onto the paper and mark its position as $t = 0$.

When the cell has attained thermal equilibrium, adjust the spectrophotometer so that it gives an absorbance reading of zero. Use the 'set zero' knob on the recorder to set the pen just *below* zero on the paper, to make allowance for the absorption by the ethanonitrile solvent.*

Rinse and fill the syringe with 50 mm^3 of the 0.3 M methyl chloroformate solution. Lift the spectrophotometer lid and carry out the following in steady and reasonably rapid succession:

(i) Inject the contents of the syringe into the cuvette.
(ii) Turn on the chart recorder drive.
(iii) Stir the cuvette contents with the small plastic plunger, using about four lateral movements.
(iv) Stopper the cuvette and close the spectrophotometer lid.
(v) Mark the $t = 0$ line on the spectrum.

This sequence should be completed in less than 30 s.

Monitor the absorbance of the solution for 15 min.

Repeat the procedure with the other mixtures. For run 4 use 25 mm^3 rather than 50 mm^3 of the methyl chloroformate solution and leave running for 10 min only.

At the end of the experiment rinse the syringe with ethanonitrile.

Calculation It is apparent from the procedure just described that the starting time of the experiment is not known with any precision. For this reason we introduce a parameter s which is added to the measured times t' to yield the true times t.

The computer program used to analyse the data chooses simultaneously the 'best-fitting' values of k_{eff} and s. Brief details are given in the Technical Notes section. The program requires initial estimates of the two parameters. Since we do our best to start the chart drive at the start of the reaction, the initial estimate of s is zero. The initial estimate of k_{eff} is obtained by taking account of the fact that for $k_{eff} \ll k_2$, the fall in absorbance after the maximum has been reached approaches more and more closely a first-order decay with rate constant k_{eff}. The estimate of k_{eff} may therefore be obtained by the half-life method. Measure on the chart the absorbance at $t = 180$ s. Then find the time which *elapsed* before this absorbance halved in value. Call the elapsed time $t_{1/2}$. Then the initial estimate of k_{eff} is that $k_{eff} = (\ln 2)/t_{1/2}$.

Now read off the absorbances from the first chart at 30-second intervals from 30–40 s. (Only relative values of absorbance are needed, so use the chart paper scale.) Note also the value of a_m on the same scale.

Given that k_2 is $0.035\ s^{-1}$, use the computer program to find the best-fitting values of s and k_{eff}. Note the root mean square deviation between the calculated and experimental points for each iteration, and use the values from the iteration with the lowest deviation. Plot the calculated data from this iteration against the experimental results to check the goodness of fit.

Repeat the calculation for the other runs.

For each of the four runs, calculate [Py], the concentration of pyridine in free base form, from the total concentration of pyridine added, $[\mathrm{Py}]_{tot}$, and the relations

$$[\mathrm{Py}] + [\mathrm{PyH}^+] = [\mathrm{Py}]_{tot}, \tag{8.5.12}$$

and

$$K_a \approx \frac{[\mathrm{Py}][\mathrm{H}^+]}{[\mathrm{PyH}^+]} = 10^{-5.18}\ \mathrm{mol\ dm^{-3}}, \tag{8.5.13}$$

where $[\mathrm{PyH}^+]$ is the concentration of protonated pyridine.

Then plot k_{eff} against [Py] to obtain k_1 and k_3 (eqn (8.5.1)).

Results

The maximum absorbances should occur after about 90 seconds.

Accuracy

It is possible to obtain k_1 to within 10 per cent of the true value. k_3 is much less accurately determined, but should be within the correct order of magnitude.

Comment[5]

The kinetics of consecutive reactions form the basis of the consideration of many chemical processes. The arguments can be developed to explain the concept of the rate-determining step of a reaction, and the steady-state approximation applied to a reaction intermediate. If there also exists a back-reaction from the intermediate Y to the reactants X, then a pre-equilibrium may be set up. This type of kinetics occurs in enzyme reactions and many simple gas phase reactions.

Technical notes

Apparatus

Ultraviolet spectrophotometer with chart recorder,$$ ultraviolet cuvette (4 cm path length). plastic stirrer for cuvette, syringe ($\geq$50 mm^3 (50 μl)). 0.1 M pyridinium hydrochloride, 0.1 M hydrochloric acid, 0.1 M sodium chloride, 0.3 M methyl chloroformate in ethanonitrile (acetonitrile) stored in desiccator.

Program[4]

The computer program, available from the author, carries out a least squares refinement. It determines the optimum value of k_{eff} and s to minimize the sum of squares of residual errors between the calculated and observed values of the absorbance a. The calculation is as follows.

We may deduce from eqn (8.5.11) that the absorbance a is a function A of k_{eff} and s;

$$a = A(k_{\text{eff}}, s). \tag{8.5.14}$$

Let the initial estimates be k_0 and s_0 and let the correct values be

$$k_{\text{eff}} = k_0 + \delta k_{\text{eff}}, \tag{8.5.15}$$

and

$$s = s_0 + \delta s. \tag{8.5.16}$$

The first terms of a Taylor expansion give

$$A(k_{\text{eff}}, s) = A(k_0, s_0) + \left(\frac{\partial A}{\partial s}\right)_{k_{\text{eff}}} \delta s + \left(\frac{\partial A}{\partial k_{\text{eff}}}\right)_s \delta k_{\text{eff}}. \tag{8.5.17}$$

The derivatives $(\partial A/\partial s)_{k_{\text{eff}}}$ and $(\partial A/\partial k_{\text{eff}})_s$ are measured at k_0, s_0, and for simplicity we refer to them as P_0 and Q_0 respectively. Then

$$A(k_{\text{eff}}, s) = A(k_0, s_0) + P_0\delta s + Q_0\delta k_{\text{eff}}. \tag{8.5.18}$$

The difference between a calculated value of absorbance and an observed value a_{obs} is the residual, R. Then

$$\begin{aligned} R &= A(k_{\text{eff}}, s) - a_{\text{obs}} \\ &= [A(k_0, s_0) - a_{\text{obs}}] + P_0\delta s + Q_0\delta k_{\text{eff}} \\ &= R_0 + P_0\delta s + Q_0\delta k_{\text{eff}}. \end{aligned} \tag{8.5.19}$$

By the least squares principle, the best-fitting values of δs and δk_{eff} are those which minimize ΣR^2. The conditions for this minimization must therefore be that $[\partial(\Sigma R^2)/\partial(\delta s)] = 0$ and that $[\partial(\Sigma R^2)/\partial(\delta k_{\text{eff}})] = 0$. These conditions lead to the equations

There are sometimes *two* solutions to these equations, and an absorbance which rapidly increases and slowly decreases does *not* necessarily imply a fast first and slow second reaction.[5]

$$\begin{aligned} \delta s(\Sigma P_0^2) + \delta k_{\text{eff}}(\Sigma P_0 Q_0) &= -\Sigma P_0 R_0, \\ \delta s(\Sigma P_0 Q_0) + \delta k_{\text{eff}}(\Sigma Q_0^2) &= -\Sigma Q_0 R_0. \end{aligned} \tag{8.5.20}$$

These are solved for δs and δk_{eff}, new values of k_{eff} and s are calculated using eqns (8.5.15) and (8.6.16), and the process is repeated until there is no further improvement (usually 2 or 3 iterations).

References

[1] *Kinetics and mechanism.* J. W. Moore and R. G. Pearson. John Wiley and Sons, New York, 3rd edn (1981), p. 290.
[2] *Physical chemistry.* P. W. Atkins. Oxford University Press, Oxford, 3rd edn (1986), Sect. 28.3(d).
[3] *An exact solution to a consecutive reaction sequence.* R. L. Anderson, R. S. Nohr and L. O. Spreer. *J. Chem. Educ.* **52,** 437 (1975).
[4] R. B. Moodie. Chemistry Department, Exeter University, unpublished.
[5] *Kinetics of series first-order reactions.* N. W. Alcock, D. J. Benton, and P. Moore. *Trans. Far. Soc.* **66,** 2210 (1970).

8.6 Radiochemical measurements and the kinetics of radioactive decay[1]

Radioactive substances may decay by emitting either alpha (α), beta (β) or gamma (γ) radiation, as well as other types. An alpha particle comprises two protons and two neutrons bound together in a stable unit, and is therefore a helium nucleus, a beta particle is an electron, and gamma-ray photons have neither mass nor charge. In this experiment we shall be concerned solely with the radiation of beta particles.

The experiment provides a general introduction to radiochemistry, and is in three parts. In Part A we examine the characteristics of a Geiger–Müller tube, which we use as a detector of beta radiation, and in Part B we investigate the statistics of radioactive disintegration measurements. Part C is a study of the kinetics of the radioactive decay of protactinium-234, separated from a solution of uranyl nitrate by solvent extraction.

Theory

A. Characteristics of a Geiger–Müller tube

Alpha radiation is prevented from entering the Geiger–Müller tube because it is absorbed by the entry window, and γ-rays are detected with relatively low efficiency.

One of the most common detectors of beta radiation is the Geiger–Müller tube, shown in Fig. 8.6.1, which acts as a counting device for high energy electrons. It comprises a cylindrical container (the cathode) filled with a detector gas such as argon or krypton, with an insulated central wire (the anode) to serve as a collector for electrons. Beta radiation entering the tube through the thin entry window causes ionization of the detector gas. The electrons so produced drift towards the anode, and enter the region close to the wire where the electric field strength is greater than a critical value E_c. Within this region electrons may gain sufficient energy between collisions with the gas atoms to produce ionization of these atoms. This ionization yields electrons which can themselves produce further ionization

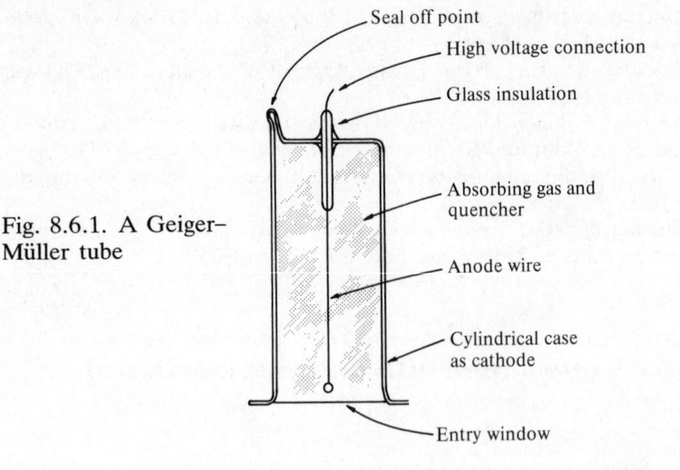

Fig. 8.6.1. A Geiger–Müller tube

by collision, so that an avalanche of ionization results, Fig. 8.6.2. The processes continue until the whole length of the anode wire is saturated with avalanches. The electrons (negative ions) produced in these avalanches are rapidly collected, in the order of 1 μs, leaving behind a sheath of positive ions surrounding the anode. This sheath increases the effective radius of the anode, decreases the field strength and thus no further avalanches can be produced.

The positive ions move slowly towards the cathode and in doing so induce a negative pulse on the anode of large amplitude (about 1 V). They reach the cathode after a time of about 100 μs. Their excess excitation energy after neutralization may be used to produce secondary electrons which can cause avalanches and start the whole process again. Thus instead of the Geiger–Müller tube delivering a single output pulse, multiple pulses will result. To prevent this, some form of quenching must be used, the two most common forms being *chemical* or *self-quenching*, and *electronic quenching*.

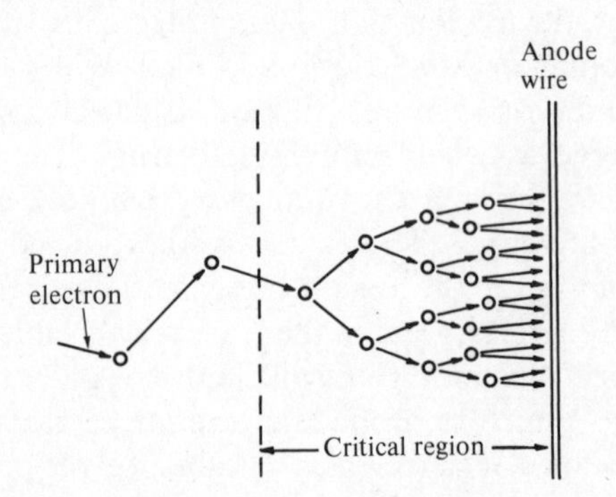

Fig. 8.6.2. Avalanche of electrons from ionization of the absorbing gas near the anode wire of a Geiger–Müller tube

Self-quenched Geiger–Müller tubes contain a polyatomic gas in addition to the detector gas; for example the addition of 10 mmHg pressure of ethanol to 100 mmHg pressure of argon in an organic quenched tube, working voltage ~1200 V, or 0.1 per cent bromine in neon in a halogen quenched tube, working voltage ~400 V. As the excitation energy of the quenching vapour molecules (~11 eV) is lower than that of argon or neon, the ions arriving at the cathode are quenching vapour ions. Following neutralization the excess energy of the quenching vapour molecules is dissipated in dissociation of the molecules rather than in the release of electrons from the cathode. The lifetime of organic quenched tubes is limited to about 10^9 counts as the decomposition products do not recombine.

With electronic quenching a negative pulse of large amplitude (several 100 V) is applied to the anode of the tube following an output pulse. This reduces the field in the counter to below the critical value for ionization by collision. The pulse remains on the anode until after all the positive ions have been collected at the cathode, thus preventing the production of further avalanches.

Although all commercial Geiger–Müller tubes are self-quenched it is common practice to use electronic quenching also. This is because following a count pulse there is a *dead time* during which the counter is insensitive to further incident radiation. If this dead time τ is known, a correction can be applied to compensate for the lost counts:

$$N_t = \frac{N_r}{1 - N_r\tau}, \tag{8.6.1}$$

where N_r is the recorded count rate and N_t the true count rate corrected for dead time losses. The duration of the negative pulse applied in the electronic quenching method sets the dead time of the counting system to an accurately known value, allowing a precise correction to be made.

The characteristics of a Geiger–Müller tube may be studied experimentally by plotting a graph of count rate against the voltage applied to the anode, Fig. 8.6.3. The pulses from the

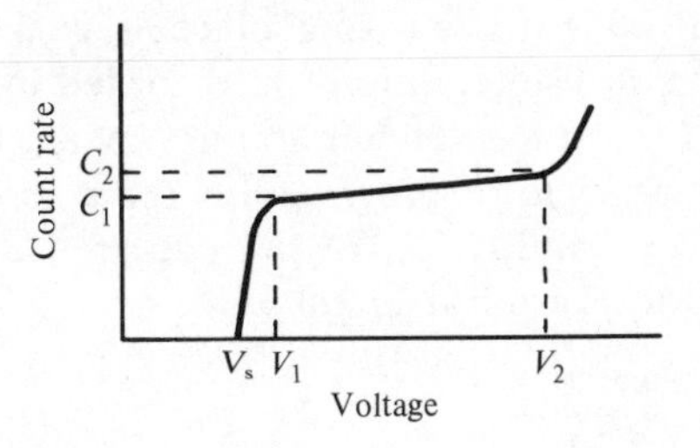

Fig. 8.6.3. Characteristic curve of a Geiger–Müller tube

Geiger–Müller tube are all practically equal in size and independent of the energy released in the initial ionizing event. As the high voltage is increased the height of all the pulses increases, and at a certain voltage V_s they exceed the threshold of the detector in the counter and are recorded. The graph of count rate against voltage rises steeply since the pulses all reach the critical height at about the same value of high voltage, known as the *counter threshold.* At the plateau voltage V_1 all the pulses are being detected, and thus although further increases in voltage increase the pulse height, there is little increase in count-rate. The slight rise which does occur is due to a small increase in the effective sensitive volume of the tube, and also to the increased probability of spurious pulses occurring when the voltage is raised, because of factors such as inefficient quenching.

The plateau region is characterized in terms of its length $V_2 - V_1$, Fig. 8.6.3, and its slope given by

$$\text{slope} = \frac{C_2 - C_1}{C_m} \times \frac{100\%}{V_2 - V_1}, \tag{8.6.2}$$

where the mean count rate C_m is $(C_1 + C_2)/2$. For a good counter the plateau length should be of the order of 200 V and the slope less than 0.1 per cent per volt. As the counter ages the counter threshold voltage, V_s, rises and the plateau becomes shorter and steeper.

Generally a Geiger–Müller tube is operated at a *working voltage* of (V_s + 100 V) for organic quenched tubes, or (V_s + 50 V) for halogen quenched tubes.

B(i) Statistics of radioactive disintegration measurements

Note that the variations in the observations of radioactive disintegrations are not experimental errors; they cannot be reduced by changes in apparatus or technique.

Suppose that we measure the radiation of a sample, and that the activity of the sample does not change perceptibly during the duration of the experiment. The radioactive disintegrations in the sample take place at random, and thus the number of counts recorded from a given sample in a fixed period of time will vary from one observation to another. However since the probability of a disintegration taking place is constant, the value of the number of disintegrations, n, recorded in equal periods of time, t, will form a *distribution* about a mean, m. The *count rate* C is n/t for a single reading, or m/t for a series of readings.

The probability, P_n, of the occurrence of a given value of n obeys the *Poisson distribution*

$$P_n = \frac{(mt)^n e^{-mt}}{n!}. \tag{8.6.3}$$

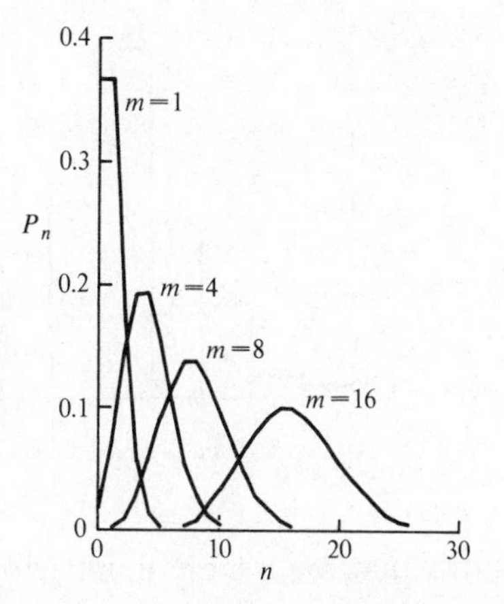

Fig. 8.6.4. Poisson distributions for the number of counts n recorded in time $t = 1$. Note the rapidly increasing symmetry of the distribution as the mean m becomes larger

The shape of the distribution varies with m, the asymmetry rapidly decreasing as m increases, Fig. 8.6.4. The distribution has a physical meaning only for integral values of n.

The *standard deviation* σ of a distribution is a measure of the spread of observations about the mean. It can be shown that in a Poisson distribution

$$\sigma = \sqrt{m}. \tag{8.6.4}$$

It follows that the standard deviation in the count rate is $n^{1/2}/t$ for a single reading, and $m^{1/2}/t = (tC)^{1/2}/t = (C/t)^{1/2}$ for a series of readings. Since this standard deviation gives an idea of the error which may occur in the estimation of the mean count rate, it is usually referred to as the *standard error* of the count rate.

Another type of distribution which commonly occurs in other types of measurement is the *Gaussian* or *Normal distribution*, shown in Fig. 7.1.2 (p. 330). For values of m greater than about 30 the Poisson distribution is almost identical in shape to the particular Normal distribution with the same mean, and the simpler deductions from the theory of the Normal distribution can be applied. In a Normal distribution m and σ are independent variables, and the width and mean value of the distribution are totally unconnected. However, since we are using a Normal distribution as an approximation to a Poisson distribution, the relationship $\sigma = \sqrt{m}$ still holds.

For a Normal distribution the standard deviation is defined as:

$$\sigma = \left(\frac{\sum (n-m)^2}{p-1}\right)^{1/2}, \tag{8.6.5}$$

where p is the number of readings. $\sum$ represents the summation of the terms which follow it, so that in this case we calculate each value of $(n-m)^2$, and then add them.

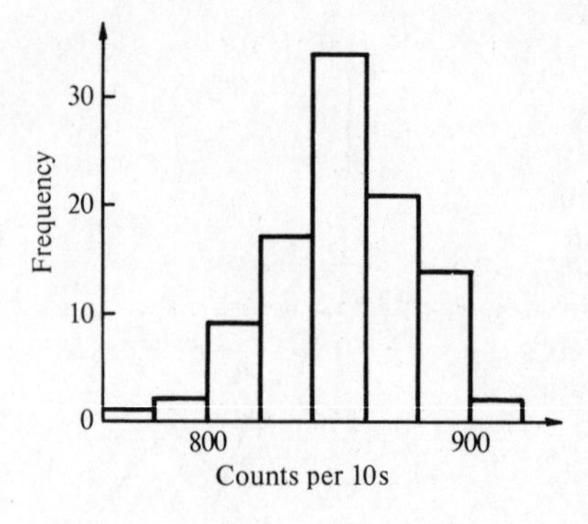

Fig. 8.6.5. Grouped count rate measurements expressed as a histogram

It follows that for a large number of readings

$$\sigma^2 \simeq \frac{\sum (n-m)^2}{p} = \frac{\sum n^2}{p} - \frac{2\sum mn}{p} + \frac{\sum m^2}{p}.$$

m^2 is constant, and thus $\sum mn = m \sum n$ and $\sum m^2 = pm^2$. Also $m = \sum n/p$ and therefore

$$\sigma^2 = \frac{\sum n^2}{p} - \left(\frac{\sum n}{p}\right)^2. \tag{8.6.6}$$

In practice it is easier to split the results up into groups, and plot them as a histogram as shown in Fig. 8.6.5. Applying eqn (8.6.6) to this grouped frequency distribution,

The analysis of a grouped distribution is best understood by following through the example given in the Results section.

$$\sigma_w^2 = \frac{\sum (fx)^2}{p} - \left(\frac{\sum fx}{p}\right)^2 \tag{8.6.7}$$

and

$$\sigma = \sigma_w \cdot w, \tag{8.6.8}$$

where f is the frequency, w is the size of the working unit (the number of intervals in each group), σ_w is the standard deviation in working units, and x is the number of working units between a given group and an arbitrary origin x'. The calculations are most manageable if this origin is taken at the centre of the group with the highest distribution. The mean m is given by

$$m = x' + \frac{\sum fx \cdot w}{p}. \tag{8.6.9}$$

Although the use of a grouped distribution has a negligible effect on the mean, it does lead to a slight over-estimate of the standard deviation σ and variance σ^2. It is therefore necessary to carry out a so-called *Sheppard adjustment*[2] on the variance (σ^2), which involves reducing it by $w^2/12$.

B(ii) Testing the distribution

In a Normal distribution, 68.3 per cent of the results lie between $m-\sigma$ and $m+\sigma$, and 95.4 per cent lie between $m-2\sigma$ and $m+2\sigma$. These properties may be used to test whether the distribution is Normal. However, if the results differ from these figures, as they almost certainly will, there is no way of judging whether the difference is significant or not.

A more informative, but nevertheless still qualitative, test of the distribution may be made by plotting the cumulative frequency against count rate on probability graph paper, as explained on p. 330 and p. 334. A Poisson distribution will yield a curved plot, whereas a Normal distribution will give a straight line.[3]

A quantitative test of the distribution is the χ^2 (*chi-square*) *test.*[2] We employ a variate which, for large samples, is distributed approximately like χ^2:

$$\text{variate } (\approx\chi^2) = \sum \frac{(f_0 - f_N)^2}{f_N}. \qquad (8.6.10)$$

f_0 are the observed frequencies and f_N are the expected values of the frequency, based, in this case, on an assumed Normal distribution.

Clearly if the observed frequencies follow a Normal distribution exactly, χ^2 will be zero. In order to calculate the significance of other values of χ^2, we must first calculate the number of *degrees of freedom* ν. This parameter is equal to the number of classes into which the results are split, less the three parameters which have to be made equal before a meaningful comparison between the actual and Normal distributions can be made. These are the sum of the frequencies in the classes, the mean of each distribution and the standard deviation. In the

Table 8.6.1. Values of χ^2 with probability P of being exceeded in random sampling

Degrees of freedom ν	P/per cent 99	5	1
4	0.30	9.5	13.3
5	0.6	11.1	15.1
6	0.9	12.6	16.8
7	1.2	14.1	18.5
8	1.7	15.5	20.1
9	2.1	16.9	21.7
10	2.6	18.3	23.2
11	3.1	19.7	24.7

Note that this method of comparison is not valid for classes with small values of f_0 or f_N—these should be avoided by combining them with neighbouring classes.

example given in Fig. 8.6.5, the readings have been split into 8 groups, and the number of degrees of freedom ν is therefore $8-3=5$. We see from Table 8.6.1 that if our variate exceeds 11.1, the two distributions are different at a *significance level P* of less than 5 per cent, which is the conventional level for comparison.

It may prove convenient to split the results up into a different number of classes, and to use the corresponding values of χ^2 listed in Table 8.6.1.

C(i) Decay series

The residual atom after radioactive decay is generally that of a different element, and is known as a *daughter* product. This daughter atom may be stable, or may itself be radioactive, particularly if the element concerned is heavier than lead. In the latter case a *radioactive series* can occur.

In this experiment we are concerned with the uranium series:

$$^{238}_{92}\text{U} \xrightarrow{\alpha} {}^{234}_{90}\text{Th} \xrightarrow{\beta} {}^{234}_{91}\text{Pa} \xrightarrow{\beta} {}^{234}_{92}\text{U} \xrightarrow{\alpha} {}^{230}_{90}\text{Th} \longrightarrow \text{etc.}$$

Half-lives:	4.5×10^9 years	24.1 days	1.18 mins	2.5×10^5 years	8×10^4 years

The superscripts refer to the nuclear mass of each species, and the subscripts are their atomic numbers.

Solutions of uranium salts, such as uranyl nitrate $UO_2(NO_3)_2$, contain the daughter products of this decay series. One of them, $^{234}_{91}\text{Pa}$ or protactinium-234, is relatively easy to separate from the others by solvent extraction.

C(ii) Decay curve

The rate of radioactive decay is unaffected by physical or chemical change, and we find that

$$-\frac{\mathrm{d}N}{\mathrm{d}t}=\lambda N, \qquad (8.6.11)$$

where N is the total number of radioactive atoms present, and $-\mathrm{d}N/\mathrm{d}t$ the decrease in this number with time. λ is a proportionality constant called the *decay constant* of the radionuclide, and is characteristic of that nuclide. If we integrate both sides of the equation between the limits $N=N_0$ at $t=0$ and $N=N$ at time t, we find that

$$N=N_0e^{-\lambda t}. \qquad (8.6.12)$$

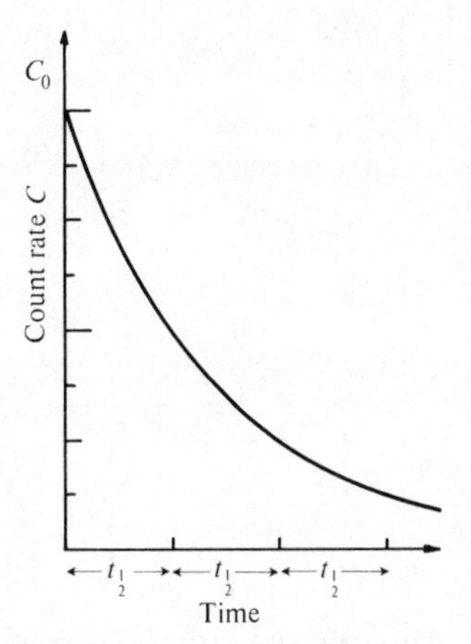

Fig. 8.6.6. First–order decay of count–rate with time. The count–rate reduces to $\frac{1}{8}$th of its initial value after 3 half–lives $t_{1/2}$

Since the count rate C is directly proportional to the number of radioactive atoms present, we also find that

$$C = C_0 e^{-\lambda t}. \tag{8.6.13}$$

Thus the count rate declines exponentially, as shown in Fig. 8.6.6.

The *half-life* $t_{1/2}$ is the time taken for half of the nuclei to disintegrate. Integration of eqn (8.6.11) between the limits $N = N_0$ at $t = 0$ and $N = N_0/2$ at $t = t_{1/2}$ yields the expression

$$\lambda t_{1/2} = \ln 2. \tag{8.6.14}$$

In this experiment we may find the half-life directly from a graph of count rate against time. For example, as shown in Fig. 8.6.6, the time taken for the count rate to reduce to 1/8th of its initial value is three half-lives.

(*iii*) *Growth curve*

After the initial removal of protactinium-234 from the sample, the concentration of this isotope in the sample rapidly increases from zero until it once again reaches its equilibrium concentration as a result of the decay of its precursors uranium-238 and thorium-234. The growth curve which results has the form shown in Fig. 8.6.7. C is the observed count rate, which rises to a value A at $t = \infty$. B is the initial count rate, due mainly to β-emitting isotopes other than protactinium-234 in the uranium series. Thus

$$C = A - (A - B)e^{-\lambda t}$$

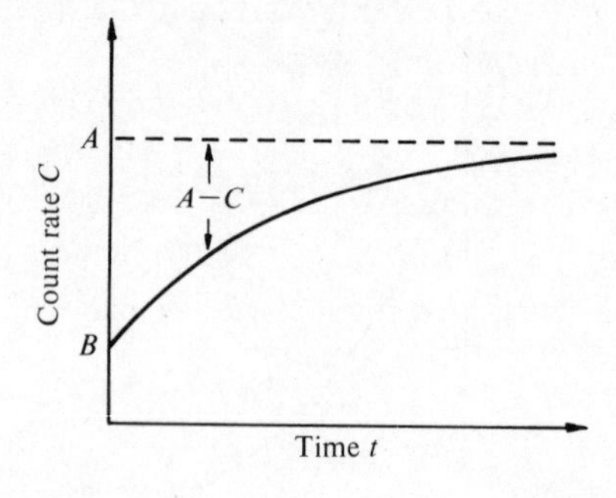

Fig. 8.6.7. Growth curve of protactinium–234

or

$$(A-C)=(A-B)e^{-\lambda t}, \tag{8.6.15}$$

and the interval on the graph represented by $(A-C)$ decays exponentially with time.

Taking logarithms we see that

$$\ln(A-C)=\ln(A-B)-\lambda t, \tag{8.6.16}$$

and that a graph of $\ln(A-C)$ against time should yield a straight line of slope $-\lambda$.

Apparatus The Geiger–Müller tube used for the detection of radiation has been described in the Theory section. It is usually housed with the sample in a lead chamber known as a *lead castle*, which masks it from the background radiation. Part C of the experiment employs a special Geiger–Müller tube designed to measure the activity of solutions.

The particular tube in use is connected to a unit containing its high voltage power supply, and a detector and amplifier for the signals from the tube. The unit also displays a count of the number of signals received. The instructions in the Procedure section assumes that there are facilities for counting for a preset time, or timing a preset number of counts. If such facilities are not available, the counting may be carried out over an interval timed by a stopwatch. The Geiger–Müller tube and associated equipment is collectively termed a *Geiger counter.*

Procedure *A. Characteristics of a Geiger–Müller tube*

Using forceps place the β-emitting source supplied for this experiment beneath the Geiger–Müller tube, as instructed.* Set the discriminator control on the counter unit to its lowest value, and the timer switch to its highest preset count or preset time, so that the counts are recorded without interruption. Check that the high voltage control is set to its lowest value, switch on the unit, and start counting.

Find out the maximum working voltage of the Geiger–Müller tube you are using.* For organic quenched tubes this will lie between 1000 and 1600 V, and for halogen quenched tubes between 400 and 600 V.

Refer to Fig. 8.6.3. Initially no counts will be recorded, as the voltage is below V_s.

Slowly increase the voltage until counts are registered at voltage V_s. Then very slowly increase the voltage until the count rate does not increase sharply with voltage, i.e. until you have reached a voltage of approximately V_1. (V_1 is typically ~20 V above V_s). If no counts occur before the maximum working voltage is reached, seek advice. Under no circumstances should the voltage be increased above the maximum working voltage.

Now set the preset time control to a suitable timing period, e.g. 100 s. If necessary, adjust the distance between the source and counter so that ~10 000 counts are recorded per 100 s, or 100 counts s^{-1}. A count rate of 500 counts s^{-1} should not be exceeded because of the length of the dead-time.

Once these adjustments have been made, reduce the high voltage until no counts are recorded, and then raise the voltage again to the point where counting just commences. Determine the count rate at this threshold voltage V_s. Then increase the voltage in suitable steps, e.g. 20 V, and determine the count rate for each voltage over a period of about 30 s. Plot a graph of count rate against voltage (Fig. 8.6.3) as you are making the measurements. Continue increasing the voltage until *either* the count rate shows a rapid increase, whereupon the voltage must be reduced at once, *or* until a value of (V_s+200 V) is reached. On no account should the tube be allowed to go into discharge, i.e. to give a very high counting rate at the top end of its characteristic curve.

When the tube is not in use, the voltage should be reduced below V_s, otherwise the counter is still detecting radiation even though the scalar may not be recording pulses. For reasons mentioned previously, this may cause the tube to deteriorate.

B. *Statistics of radioactive disintegration measurements*

Place the β-emitting source supplied for the experiment below the Geiger–Müller tube as instructed.* Set the voltage control to the correct working voltage, as described above. To keep dead time losses low, adjust the distance between the source and Geiger–Müller tube to give a count rate of about 80 counts s^{-1}.

Measure the count rate over a short interval (e.g. 10 s), and repeat until 100 readings, each over the same time period, have been obtained.

Remove the source and take a few measurements of the background count rate.

C. Kinetics of radioactive decay

It is essential that the terminals on the liquid counter are dry before connection to the scalar to prevent an electric short-circuit. Radioactive solutions should be poured into a radioactive waste container, not down the sink.

(i) Measurement of background count rate

This part of the experiment employs the special Geiger–Müller tube for solutions. Use tissues to ensure that the terminals and cell of this device are clean and dry. Place the counter in the lead castle or other appropriate holder and set a small, sealed, radioactive source on the counter. Adjust the voltage so that it is 60–100 V above the threshold or starting voltage. Remove the sealed source.

Note that some tubes become light sensitive with age and do not give correct readings unless completely shielded from daylight.

The tube may have been contaminated by previous users, and it is therefore important to measure the background count rate. Make measurements over five separate 10s periods while the tube is empty. Remove the tube from its mounting, pour 10 cm^3 distilled water into it, ensure that its terminals are absolutely dry, and return it to the castle. Measure the background count rate again.

Empty the distilled water from the counter and stand the counter in a test tube rack while the protactinium-234 solution is prepared.

(ii) Measurement of protactinium-234 decay curve

Dissolve approximately 1 g of uranyl nitrate crystals in 3 cm^3 of distilled water. Transfer to a separating funnel. Wash out the beaker with 7 cm^3 of concentrated hydrochloric acid and add the washings to the separating funnel.

Add 10 cm^3 of 4-methyl pentan-2-one and shake the mixture in the separating funnel for about 5 min. Stop shaking, remove the stopper and as quickly as possible separate the lower aqueous solution into a beaker and run the organic layer, which contains the protactinium-234, into the liquid counter. This operation should be completed in about 30 s.

Ensure that the outside and terminals of the counter are quite dry by wiping with tissues. Put the counter into the lead castle and start the stopwatch.

Measure as many 10s count rates as possible over a period of 300 s.

(*iii*) *Measurement of growth curve*

Repeat the experiment using fresh crystals of uranyl nitrate, but this time run the aqueous phase into the counter and measure the growth of $^{234}_{91}Pa$ in the solution.

At the end of the run, rinse the tube with distilled water or leave it standing in a rack with decontaminant solution in it.

Calculation

A. Characteristics of a Geiger–Müller tube

Measure the length and slope of the plateau for the Geiger–Müller tube (eqn (8.6.2)). Comment on the state of the tube.

B. Statistics of radioactive decay measurements

The following instructions detail a method of processing the measurements by electronic calculator as shown in the Results section, but a computer program may be supplied or written to carry out these calculations.*

Note that dead time corrections are based on count rate, and that care must therefore be taken in applying the corrections to total counts.

First tabulate the measurements in ascending order of magnitude. Correct each count for dead time (eqn (8.6.1)) and then subtract the average background count rate from each.

Collect the readings into between 5 and 10 convenient groups, and determine the frequency of recording count rates within each group. Plot a histogram to show the distribution of these frequencies. The histogram should be reasonably symmetrical about a mean value, as in Fig. 8.6.5.

Then calculate the standard deviation σ and the mean (eqns (8.6.7), (8.6.8), and (8.6.9)). Ascertain, by one or more of the methods described in the Theory section, whether the results follow a Normal distribution. For the χ^2 test, use the Normal distribution table provided.*

C. Kinetics of radioactive decay

Comment on the background readings obtained when the solution tube contains water as opposed to air.

Convert the 10 s readings of the decay of protactinium-234 into count rates, and correct them for dead time and background. Plot the logarithm of the corrected count rate against time, either by converting to logarithms and plotting on linear graph paper, or by plotting the unconverted count rates on 2-cycle log-linear graph paper. Determine the half-life $t_{1/2}$ of $^{234}_{91}Pa$, and its decay constant λ (eqn (8.6.14)).

Plot the growth curve results, corrected for dead time and background, to give a growth curve of the type shown in Fig. 8.6.7. Estimate the value of A. From the smoothed growth curve, determine $(A - C)$ values at various values of t. Then plot $\ln(A - C)$ against time by one of the methods described above. A straight line should be obtained, especially if the value of A has been chosen correctly. There is no need to know the value of B. Determine the half-life and decay constant as before.

Comment on the accuracy of your results, and whether you would expect the decay curve and growth curve methods to be equally accurate.

Results

A

See Fig. 8.6.3.

B

See Fig. 8.6.5 and Table 8.6.2. Take origin x' at the centre of the 840–859 group (strictly 839.5–859.5), so that $x' = 849.5$. From eqn (8.6.7), $\sigma_w^2 = (172/100) - (10/100)^2 = 1.71$. Therefore $\sigma = \sigma_w \cdot w = 1.31 \times 20 = 26.2$. Applying Sheppard adjustment, $\sigma = [26.2^2 - (20^2/12)]^{\frac{1}{2}} = 25.6$. From eqn (8.6.9), $m = 849.5 + (10 \times 20)/100 = 851.5$ counts per 10 s = 85.15 counts s^{-1}. $\sqrt{m} = 29.2 \approx \sigma$.

Table 8.6.2. Calculation of standard deviation and mean for a grouped frequency distribution

Range of group/counts per 10 s	Frequency f at which corrected counts occur in each group	x	fx	fx^2
760–779	1	−4	−4	16
780–799	2	−3	−6	18
800–819	9	−2	−18	36
820–839	17	−1	−17	17
840–859	34	0	0	0
860–879	21	1	21	21
880–899	14	2	28	56
900–919	2	3	6	18
Totals:	100		10	172

Table 8.6.3. Calculation of χ^2 for the distribution in Table 8.6.2

Group boundaries/ counts per 10 s	$f_O = f/p$	Distance of boundaries from mean in multiples of σ	Cumulative normal probability[2]	Normal probability f_N	$(f_O - f_N)^2/f_N$
759.5		−3.60	0.000		
	0.01			0.002	0.03
779.5		−2.82	0.002		
	0.02			0.019	0.00
799.5		−2.03	0.021		
	0.09			0.085	0.00
819.5		−1.25	0.106		
	0.17			0.213	0.01
839.5		−0.47	0.319		
	0.34			0.303	0.01
859.5		0.31	0.622		
	0.21			0.240	0.00
879.5		1.09	0.862		
	0.14			0.108	0.01
899.5		1.88	0.970		
	0.02			0.026	0.00
919.5		2.66	0.996		
					Sum: 0.06

χ^2 test as shown in Table 8.6.3. Cumulative Normal probability is read from table supplied for experiment. f_N is difference between cumulative probability at each boundary. χ^2 for 8 groups (5 degrees of freedom) is 11.1 at 5 per cent confidence limit. Actual variate is well within this—in fact corresponds to confidence limit of well over 99 per cent. Distribution is therefore Normal without doubt! True, skewed Poisson distributions only arise with very low total counts.[3]

C

The half-life should be accurate to 5 per cent, or a few seconds.

Comment[1] In this experiment we have studied the radioactive decay of an isotope of short half-life. The major applications of radioisotopes in chemistry, however, often employ isotopes of long half-life whose activity is therefore simply dependent on the concentration in the sample. An example is the use of radioactive isotopes in the elucidation of the pathways of biochemical reactions such as photosynthesis. When green algae are exposed to $NaH^{14}CO_3$ in the presence of sunlight, it is found that ^{14}C is incorporated surprisingly rapidly into quite complicated molecules such as sugars. However, as the time of exposure is reduced, ultimately 80 per cent of the activity is

Fig. 8.6.8. 3-phosphoglyceric acid

$$O{=}P(OH)(OH)\!-\!O{\cdot}CH_2{\cdot}CHOH{\cdot}CO_2H$$

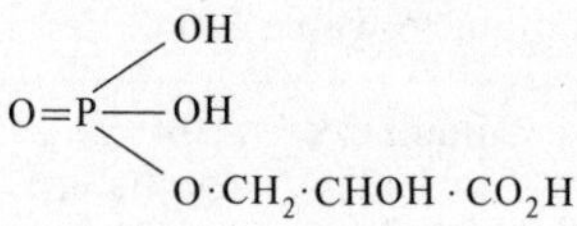

found in a single species, identified as 3-phosphoglyceric acid (Fig. 8.6.8). We may therefore deduce that this is the first compound into which $^{14}CO_2$ is fixed.

Another labelling technique is the method of *isotope dilution*. Suppose, for example, that we wish to determine the concentration of vanadium in a high tensile steel. The chemical separation procedure includes a tungstic acid precipitation to remove tungsten, and losses of vanadium occur by adsorption on this precipitate. However, the vanadium concentration can be determined accurately by isotope dilution. ^{48}V is available in almost pure form as a product of the bombardment of titanium in a cyclotron. If this radioisotope is incorporated into the production process, the final vanadium concentration can be calculated by comparing the initial and final activities.

Neutron activation analysis has been used to show that South Wales limpets contain more cobalt and tantalum relative to iron and zinc than do limpets from the Falkland Islands.

A final chemical application is in neutron activation analysis. Samples of metals or biological materials containing trace metallic elements are bombarded with neutrons. The resulting γ-ray activity is measured with a scintillation counter, which not only records the activity but also the energy of the γ-rays. The results are then displayed as a spectrum.

Technical notes

Apparatus

A and B: Geiger counter (usually in lead castle);$ power supply, scaler, timer;$ β-emitting source such as a speck of radium. B: table of cumulative frequency of a Normal distribution[2], [computer program for statistical analysis (available from author)]†. C: Geiger counter suitable for counting the beta particles emitted by a radioactive solution;$ lead castle or holder for the counter; scaler, timer, high voltage supply;$ stopwatch; rack to support liquid counter; two small beakers, spatula, 10 cm^3 measuring cylinder, separating funnel and stand; tissues; distilled water, concentrated hydrochloric acid, 4 methyl pentan-2-one (methyl isobutyl ketone, mibk), uranyl nitrate crystals.

Design note

A data logger is a very useful addition to the apparatus used in this experiment, with facilities, for example, to record 100 10 s counts unattended.

References

[1] *Handbook for the course entitled 'Radioisotopes and Radiation*

Safety for Teachers'. School of Chemical and Physical Sciences, Kingston Polytechnic, London.
[2] e.g. *A first course in mathematical statistics.* C. E. Weatherburn. Cambridge University Press, Cambridge (1968).
[3] *The statistics of radioactive decay: suggestions for a short sixth-form project.* G. P. Matthews. *School Science Review* **63,** 523 (1982).

8.7 Kinetics of ethanal photolysis by pressure measurement and gas chromatography

Ethanal (acetaldehyde) is split into radicals by ultraviolet light, a process known as *photolysis.* The radicals which are formed react further or recombine to form methane, ethane and carbon monoxide. The volume of these products is greater than the initial volume of ethanal, and the kinetics of the reaction may therefore be studied by measuring the rate of pressure change with a mercury manometer. The effect of reactant pressure on the rate of production of the various individual products may be found by analysing the products in a gas chromatograph.

In this experiment we use these methods to study the photolysis of ethanal at a temperature of 600 K, and so deduce the probable mechanism of the reaction.

Theory[1]

Deriving reaction mechanisms from experimental data requires skill and practice, and to simplify the process, we shall anticipate some of the experimental results.

(*i*) *Pressure measurement*

Our first and simplest observation is that if the photolysis is allowed to continue for a long time, the pressure of the gas approximately doubles. We therefore conclude that one mole of ethanal produces two moles of products when the reaction goes to completion. If we refer to each mole of ethanal as A, and to each mole of products indiscriminately as X, then the total number of moles of reactants and products at each state of the reaction may be written:

A	$\rightarrow 2X$	
1	0	initial
$(1-\alpha)$	2α	after degree of photolysis α
0	2	final.

It can be seen that after a fraction α of ethanal has reacted the total number of moles present will be $(1-\alpha)+2\alpha=(1+\alpha)$.

Assuming that the ethanal behaves as a perfect gas,

$$[A]=n_A/V=p_A/RT, \tag{8.7.1}$$

where n_A is the number of moles and p_A the pressure of A. The reaction takes place at constant volume and temperature, and therefore

$$[A] \propto n_A \propto p_A. \tag{8.7.2}$$

Similar considerations apply to the products X. It follows that if the initial reaction pressure is p_0, the change in pressure Δp after degree of dissociation α will be

$$\Delta p = p - p_0 = p_0(1+\alpha) - p_0 = p_0\alpha. \tag{8.7.3}$$

The relative pressure change $\Delta p/p_0$ is therefore a direct measure of the extent of the reaction. Furthermore, the initial rate of the reaction at any particular value of p_0 is equal to the rate of change of Δp, and may therefore be found by plotting Δp against t and drawing a tangent to the curve at the origin.

Suppose that the reaction is mth order with respect to the concentration of ethanal. Then since the concentration of ethanal at the start of the reaction is directly proportional to the pressure p_0,

$$\text{initial rate} = \frac{-\mathrm{d}[A]}{\mathrm{d}t} = (p_0)^m \,.\, \text{const}, \tag{8.7.4}$$

and a graph of ln (initial rate) against ln p_0 will give a straight line of gradient m.

(*ii*) *The mechanism of the photolysis of ethanal*

The photolysis of ethanal by ultraviolet light is a *chain reaction.* The first stage in the reaction, i.e. the initiation step, is the formation of methyl and aldehyde radicals:

$$CH_3CHO + h\nu \xrightarrow{k_i} CH_3^{\cdot} + CHO^{\cdot} \qquad \text{initiation.}$$

The rate of this step is equal to the product of the incident light intensity I_{abs} and the rate coefficient k_i.

The chromatographic analyses carried out in this experiment show that the products of the photolysis at around 600 K are almost entirely methane and carbon monoxide with about 1 per cent of ethane. There is no evidence of the butanedione and ethanedial (glyoxal) which are formed in the low temperature photolysis. We may therefore write the chain propagation steps as

$$CH_3^{\cdot} + CH_3CHO \rightarrow CH_4 + CH_3CO^{\cdot}$$

$$CH_3CO^{\cdot} \rightarrow CH_3^{\cdot} + CO.$$

The decomposition of the acetyl (ethanoyl) radical $CH_3CO^{\cdot}$ is fast with respect to the first step, and the chain propagation may therefore by expressed as the single process

$$CH_3^{\cdot} + CH_3CHO \xrightarrow{k_p} CH_4 + CO + CH_3^{\cdot} \qquad \text{propagation.}$$

The ethane must arise from the chain termination step

$$CH_3^{\cdot} + CH_3^{\cdot} \xrightarrow{k_t} C_2H_6 \qquad \text{termination.}$$

The rate equation for the photolysis of ethanal may therefore be written:

$$\frac{-\mathrm{d}[\mathrm{A}]}{\mathrm{d}t} = k_i I_{abs} + k_p[CH_3^{\cdot}][\mathrm{A}], \tag{8.7.5}$$

where I_{abs} is the amount of light absorbed by the ethanal.

Similarly the rate equation for the formation of methyl radicals is:

$$\frac{\mathrm{d}[CH_3^{\cdot}]}{\mathrm{d}t} = k_i I_{abs} - 2k_t[CH_3^{\cdot}]^2. \tag{8.7.6}$$

Write down the rate equations for the formation of methane and ethane.

Much more methane is produced in the reaction than ethane, which shows that a large number of propagation steps occur before termination takes place; the *chain length* of the chain reaction is long. This also implies that the rate of decomposition of ethanal is very largely dependent on the propagation steps rather than the initiation, and so to a good approximation, eqn (8.7.5) may be written

$$\frac{-\mathrm{d}[\mathrm{A}]}{\mathrm{d}t} = k_p[CH_3^{\cdot}][\mathrm{A}]. \tag{8.7.7}$$

This rate equation still contains the term $[CH_3^{\cdot}]$, which we cannot measure. To obtain the working equation for the experiment we must employ two further approximations:

(a) We apply the *steady state approximation* to the concentration of methyl radicals, so that $\mathrm{d}[CH_3^{\cdot}]/\mathrm{d}t = 0$.

(b) The amount of light absorbed by the ethanal is given by the Beer–Lambert law (p. 215):

$$I_{abs} = I_0(1 - e^{-\kappa l[\mathrm{A}]}), \tag{8.7.8}$$

where I_0 is the incident light intensity, κ the extinction coefficient, and l the path length of the light through the reaction vessel. However, κ is so low that to a good approximation the exponential term is linear with respect to [A], and therefore

$$I_{abs} = I_0(1-(1-\kappa l[A]\ldots))$$
$$\approx I_0 \kappa l[A]. \quad (8.7.9)$$

No provision has been made in our mechanism for the fate of the $CHO^{\cdot}$ radical; what suggestions can you make? What effects would there be on the reaction kinetics, and how could they be tested?

Apply these two approximations to the rate equation for the formation of methyl radicals (8.7.6). *Substitute the resulting expression into eqn* (8.7.5) *to eliminate the radical concentration term. Thus obtain an equation for the rate of decomposition of ethanal in terms of the rate coefficients, κl, and the concentration* (*and hence pressure*) *of ethanal.*

(*iii*) *Quantum efficiency*

The *quantum efficiency* or *quantum yield* Φ of a reaction may be defined most simply by the relation:

$$\Phi = \frac{\text{number of molecules that react}}{\text{number of photons absorbed}}. \quad (8.7.10)$$

The chain lengths in photochemically initiated chain reactions are often very large, and the quantum efficiencies correspondingly high.

(*iv*) *Gas chromatography*

The gas chromatograph is used to find the relative concentrations of methane and ethane. This is done by measuring the areas of the methane and ethane peaks on the chromatogram, multiplying by a relative sensitivity factor for methane and ethane, and also multiplying by range settings if they differ for the two peaks.

Your calculations in section (ii) should show that the partial pressure of methane formed is very nearly the same as the partial pressure of ethanal decomposed. Thus it follows from eqn (8.7.3) that the pressure change Δp up to the moment of letting a sample into the gas chromatograph can be taken as the partial pressure of methane. Using this information, and the relative concentrations of CH_4 and CO taken from the chromatogram, it is possible to calculate the absolute concentration of ethane.

Apparatus

The layout of the apparatus is shown in Fig. 8.7.1. The reaction vessel is about 200 cm^3 in volume and has plane silica windows. It is located in the middle of a furnace which is also provided

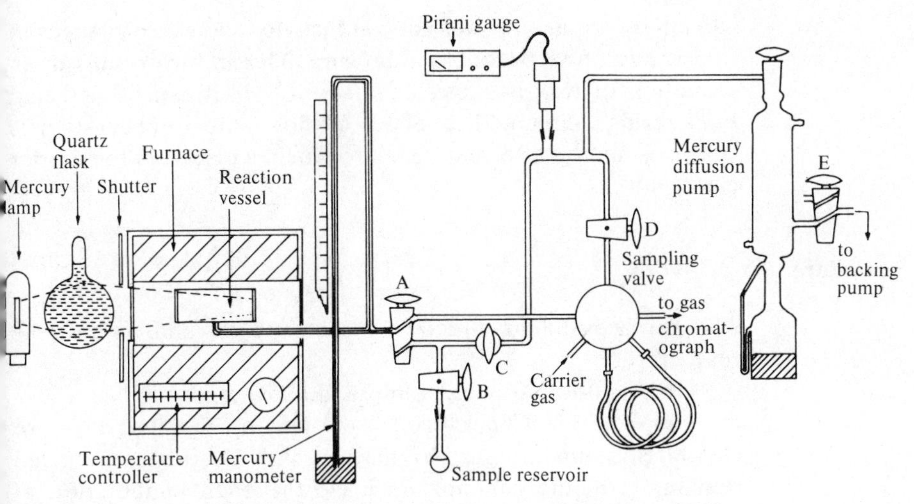

Fig. 8.7.1. Apparatus for the ultraviolet photolysis of ethanal

with silica windows. The furnace is provided with a temperature controller set at 600 K. The actual temperature of the reaction vessel is read from a mercury thermometer or a thermocouple in contact with it and should be constant to ±1 K.

The source of radiation is a high pressure mercury lamp. This type of lamp emits radiation in a group of lines around 310 nm in the near ultraviolet, close to the peak of the ethanal absorption band. It also emits radiation at around 360 nm and in the visible region, but this is not absorbed. The mercury resonance line at 254 nm is not emitted by lamps of this type, being re-absorbed by the high pressure of mercury vapour in the lamp envelope. The light from the lamp is roughly collimated and focussed onto the reaction vessel by a spherical quartz flask filled with water, which acts as a large but inexpensive lens to transmit ultraviolet radiation.

The reaction vessel is connected by a capillary tube to a capillary mercury manometer and is isolated by a two-way tap (tap A) through which ethanal can be admitted or a sample taken for analysis. The whole system is evacuated by a mercury diffusion pump and rotary backing pump, and the pressure in the line is monitored by a Pirani gauge.

The line is connected to the gas chromatograph by means of a sampling valve. Helium is used as the carrier gas to move the samples through the chromatograph column. The appearance of the sample gases at the end of the column is detected by a *katharometer.*[2] This is in the form of a heated wire which is placed in the flow of gas after the column. Sample gases in the carrier gas change its thermal conductivity and cause changes in the degree of heat loss from the wire. This in turn results in an alteration of the resistance of the wire which can be detected by a bridge circuit. The effect of flow rate on heat loss is compensated by a second detector which is placed in the carrier gas alone.

Procedure

(*i*) *Set up*

The furnace should already be switched on and at a stable temperature.

Switch on the ultraviolet lamp and allow it to stabilize. Close the lamp shutter. Check that all the taps on the apparatus are closed. Start up the gas chromatograph.* Desorb any ethanal remaining on the column and leave the oven temperature to stabilize at 40 °C.

Start the rotary vacuum pump and open tap E to the mercury diffusion pump. Then switch on the diffusion pump and Pirani gauge, and evacuate the first part of the line to about 0.03 Torr. Then evacuate the rest of the vacuum line, the reaction vessel, sample reservoir and gas chromatograph column (taps A, B, C, and D open), and also the sampling valve. Isolate the reaction vessel again (tap A), and the column (tap D and the sampling valve) once they have been evacuated.

(*ii*) *Filling of sample reservoir*

Oxygen is an extremely efficient scavenger of hydrocarbon free radicals; in the presence of even minute traces of air the kinetics are drastically altered as the methyl radicals react with oxygen to give methanal (formaldehyde) and other products instead of propagating the chain.

The sample reservoir must now be filled and any trace of air removed from it by repeated evacuation.

Close off the sample reservoir from the vacuum line (tap B). Remove the reservoir and clean and dry it carefully.

Half fill the small Dewar vessel provided with propanone (acetone). With a spatula, *slowly* add powdered solid carbon dioxide until a paste is formed.

Half fill the reservoir with ethanal using a teat pipette. Grease the joint on the reservoir with the special fluorocarbon grease provided, which is not dissolved by ethanal.

Connect the reservoir to the line and cool it with the freezing mixture. Evacuate the reservoir through taps B and C for 5 min. Isolate the reservoir again (tap B), remove the freezing mixture and allow the reservoir to warm to room temperature. Cool the reservoir again and evacuate it for a further 5 min. Finally close off the reservoir (tap B) and remove the freezing mixture.

(iii) Pressure measurements

Tap the mercury manometer gently to prevent the mercury column sticking, and take the zero reading.

Close tap C and open tap A so that the reaction vessel is connected as far as the reservoir tap (B).

Open the reservoir tap (B) slowly so as to admit about 100 mmHg pressure of ethanal vapour to the reaction vessel. Shut the two-way tap (A). Tap the manometer and measure the pressure of vapour in the reaction vessel. The difference between this and the zero reading is the initial pressure p_0.

Open the lamp shutter, noting the time. Record the manometer readings at 30 s intervals until the reaction is about $\frac{1}{4}$ complete ($\Delta p \approx \frac{1}{4} p_0$). Then close the shutter and evacuate the vessel for 5 min.

Repeat the experiment with a series of different initial pressures from 50 mmHg to the full vapour pressure of ethanal at room temperature.

(iv) Gas chromatography

The previous runs will have demonstrated that the graph of Δp against t is nearly linear up to at least $\frac{1}{10}$th of the reaction. We may therefore determine the initial rate of the reaction by sampling at this point in the reaction, determining the absolute partial pressures of the reactants, and dividing by the time from the start of the reaction.

Start the reaction again, by the procedure described in paragraph (iii), with an initial pressure of ethanal in the same range as before. Check that the tap to the sampling valve is closed (tap D). When $\frac{1}{10}$th of the reaction is completed, close the light shutter to stop the reaction. Turn the two-way tap (tap A) to pass sample into the sampling volume. Then operate the sampling valve* to allow sample into the gas chromatograph. When enough time has elapsed for all the sample to be swept into the column,* return the sampling valve to its original position and evacuate the column ready for the next sample.

The gas chromatogram will consist of three peaks, Fig. 8.7.2. The first two, with retention times typically of about 2 min, are very large peaks caused first by carbon monoxide and then methane in that order. The third, much smaller, peak, which should appear after about ten minutes, is that of ethane. The retention time for the 90 per cent of unchanged ethanal is very much longer than that of the low molar mass products and it is possible to analyse up to five successive samples of methane and

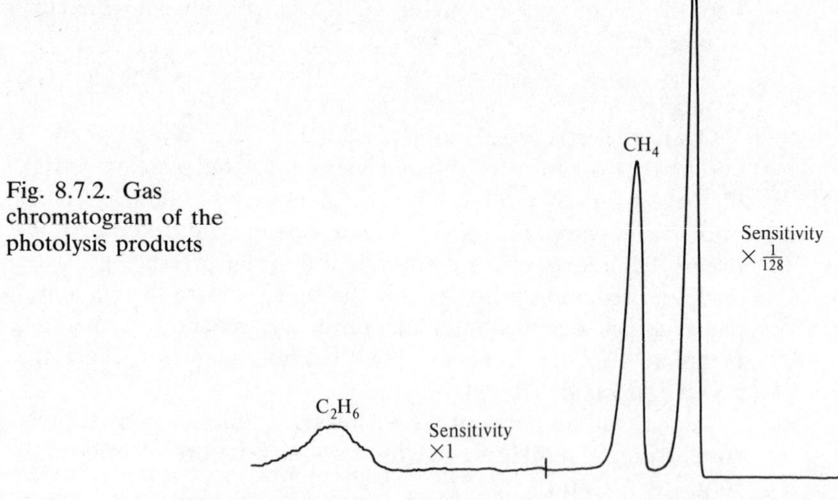

Fig. 8.7.2. Gas chromatogram of the photolysis products

ethane before the ethanal peak of the first sample reaches the end of the column—in other words, the low molar mass constituents in the subsequent samples 'leap-frog' over the slow-moving ethanal peaks. After about five samples have been analysed, raise the column temperature, whereupon the ethanal peaks of all the samples will be quickly desorbed.

Carry out a series of runs over the initial pressures as before.

Calculation

Use your pressure measurements to find *m*, the order of the reaction with respect to ethanal, eqn (8.7.4). Compare your result with the theoretical expression you have derived, and explain any discrepancies.

Measure the area of the peaks on the gas chromatogram. Use the sensitivity factor of the gas chromatograph detector* to determine the absolute concentrations of ethane at 10 per cent reaction at a series of p_0 values, as explained previously. Hence calculate the order of the reaction with respect to ethane formation and compare with your calculated value.

Estimate the overall chain length of the photolysis. Find the relationship between quantum efficiency and chain length in this reaction. What assumptions are necessary to calculate the quantum efficiency? Estimate its value.

Results

See Fig. 8.7.2.

Accuracy

If the experiment is carried out with care, the reaction orders should be accurate to within 3 per cent.

Comment This experiment demonstrates how pressure measurements and gas chromatography can be used to monitor the progress of a reaction in the gas phase. The fact that we have had to anticipate many of the experimental results underlines the difficulties which are involved in devising schemes for such reactions which adequately account for the observed behaviour.

Technical notes

Apparatus

Line as described in Apparatus section and Fig. 8.7.1;††$ ethanal; small Dewar, solid CO_2 and propanone for freezing bath.

Design notes

Photolysis lamp should be about 100 W, and have its outer glass envelope removed if it has two. Gas chromatograph column should be about 2 metres in length and packed with a non-polar porous polymer without a liquid phase (e.g. 'Poropak-Q'). Use high vacuum grease on line, and fluorocarbon grease (insoluble in ethanal) for the sample reservoir joint and the tap next to it.

References

[1] *Kinetics and mechanism*. J. W. Moore and R. G. Pearson. John Wiley and Sons, New York, 3rd edn (1981), pp. 390–401.
[2] *Gas chromatography*. D. Ambrose. Butterworths, London, 2nd edn (1971), p. 41.

8.8 Kinetics of atomic oxygen reactions by a flow method

The Rayleigh referred to is R. J. Strutt, 4th Lord Rayleigh, known as the 'airglow Rayleigh' to distinguish him from the 3rd Baron, or 'scattering Rayleigh'.

Early in this century, Rayleigh discovered that atomic oxygen is one of the products of an electric discharge through molecular oxygen. Relative concentrations of up to 10 per cent may easily be produced in this way, although in this experiment we shall use smaller values at low total pressure to simplify the reactions in the system. Atomic oxygen is not, of course, intrinsically unstable, but it is highly reactive. It is therefore not possible to study it in an ordinary static system, and a flow method is used in this experiment.[1,2]

Atomic oxygen is detected by doping the molecular oxygen with 10 per cent of nitrogen. Then not only is atomic oxygen formed in the electric discharge, but also nitric oxide. The latter reacts with the atomic oxygen to form electronically excited nitrogen dioxide, which emits light as it decays to its ground state. The greenish emission, observed by Rayleigh, is known as the *air afterglow*, and its intensity, at fixed [NO], is proportional to the atomic oxygen concentration. Thus the kinetics of oxygen atom decay can be calculated from measurements of emission intensity at measured intervals along a *flow tube* positioned downstream of the discharge.

In this experiment we derive two rate coefficients. The first is for the combination of atomic oxygen with molecular oxygen. The second is the rate coefficient for the recombination of oxygen atoms at the wall of the flow tube, and is used to calculate the fraction of wall collisions at which the atoms are lost.

Theory

(*i*) *Detection of atomic oxygen*

Before discussing the kinetics of the oxygen atom reactions, we must first consider the measurement of the relative concentration of atomic oxygen, and, in paragraph (ii), its linear flow.

The air afterglow used to detect atomic oxygen is caused by the chemiluminescent process

$$O^{\cdot} + NO + M \xrightarrow{k_1} NO_2^* + M \tag{1}$$

$$NO_2^* \xrightarrow{k_2} NO_2 + h\nu(\text{green}), \tag{2}$$

where NO_2^* represents electronically excited nitrogen dioxide. There is also a non-radiative recombination process

$$O^{\cdot} + NO + M \xrightarrow{k_3} NO_2 + M. \tag{3}$$

Even at concentrations of NO low enough not appreciably to affect the concentration of atomic oxygen, the chemiluminescence is clearly visible. NO is not, in fact, consumed because the process

$$O^{\cdot} + NO_2 \xrightarrow{k_4} NO + O_2 \tag{4}$$

is an exceptionally rapid reaction with a rate constant of 9.3×10^{-12} cm^3 molecule^{-1} s^{-1} at room temperature.[3] Thus every time an NO molecule changes to NO_2 in reaction (1) or (3), it is almost immediately regenerated by reaction (4). The effect of this is that the concentration of nitric oxide does not decay down the flow tube.

It may be shown that the intensity I of the air afterglow is proportional to the product of the NO and $O^{\cdot}$ concentrations, but independent of total pressure:

$$I = I_0[NO][O^{\cdot}], \tag{8.8.1}$$

where I_0 is the intensity at unit concentration of NO and $O^{\cdot}$. Since [NO] is constant, the intensity of the air afterglow at selected points down the tube is a relative measure of $[O^{\cdot}]$.

(ii) Measurement of linear flow velocity

In this experiment, the system rapidly settles to a stationary state after the discharge is turned on. Under these conditions, the atomic oxygen concentration measured from a point downstream from the discharge is the concentration remaining after a *contact time* t given by $t = x/v$, where x is the distance down the flow tube, and v is the linear flow velocity.

Two approximations are required to find v (as distinct from the bulk flow rate F). First, the oxygen atom concentration is sufficiently small for us to assume that its flow velocity is the same as that of the oxygen/nitrogen mixture passing from the cylinder into the discharge. Secondly, we assume that the gas flows as a 'plug' with an axial diffusion rate which is negligible compared with the flow velocity, and that the tube is wide enough for there to be no appreciable pressure drop along its length. The linear flow velocity v of the gases in the tube is then

$$v = \frac{F}{Ay}, \tag{8.8.2}$$

where A is the cross-sectional area of the tube. The total molecular concentration y (molecules per unit volume) is calculated from the total pressure in the flow line. The bulk flow rate F (molecules s^{-1}) is determined with a capillary flow meter, which responds to the different pressures p_1 and p_2 across a capillary tube mounted in the inlet line.

Poiseuille's equation for the bulk flow rate F of a compressible fluid in a uniform tube of radius r and length l has been derived on p. 351:

$$F = \frac{\pi r^4}{16\eta RTl}(p_1^2 - p_2^2). \tag{8.8.3}$$

The capillary tube used in this apparatus is simply a drawn out jet, and so is not of uniform bore. It must therefore be calibrated before the experiment. We rewrite eqn (8.8.3) in the form

$$F = \text{const.}(p_1 + p_2)(p_1 - p_2). \tag{8.8.4}$$

Then if p_1 and p_2 are very similar, so that $(p_1 + p_2) \approx 2p_1$, to a good approximation

$$F = fp_1\Delta p. \tag{8.8.5}$$

The flowmeter calibration may therefore be expressed in terms of a value of the constant f.

The volume of gas swept out per second by the vanes of the rotary pump is constant at all pressures used in the experiment. Thus the flow velocity v should remain constant and independent of the mass flow velocity F, provided that the setting of the tap between the rotary pump and flow line is not altered. In practice the connecting pipes offer increasing relative impedance to flow as the pressure drops—*why*? As a result, the flow velocity decreases somewhat with decreasing pressure.

(*iii*) *Kinetics of oxygen atom decay*

All the reactions removing oxygen atoms are first order in oxygen atom concentration $[O^\cdot]$:

$$O^\cdot + O_2 + M \xrightarrow{k_5} O_3 + M, \tag{5}$$

M is any species which takes part in the reaction by absorbing energy, but which does not make its increased energy felt in terms of its chemical behaviour.

$$O^\cdot + O_3 \xrightarrow{k_6} O_2 + O_2, \tag{6}$$

$$O^\cdot + \text{wall} \xrightarrow{k_7} \tfrac{1}{2}O_2, \tag{7}$$

where k_5, k_6, and k_7 are respective rate coefficients. Recombination reactions, second order in $[O^\cdot]$, such as

$$O^\cdot + O^\cdot + M \xrightarrow{k_8} O_2 + M, \tag{8}$$

can be neglected at the low concentrations of atomic oxygen produced in this experiment.

Reactions (5) and (8) need a third body M to stabilize the reaction products, and are kinetically third order overall. In the present instance M is molecular oxygen, since this is the major component of the reaction mixture.

It is evident from the reaction scheme that the rate of decay of atomic oxygen, $-\mathrm{d}[O^\cdot]/\mathrm{d}t$, is given by:

$$-\frac{\mathrm{d}[O^\cdot]}{\mathrm{d}t} = k_5[O^\cdot][O_2][M] + k_6[O^\cdot][O_3] + k_7[O^\cdot]. \tag{8.8.6}$$

We now apply the *steady-state approximation* to ozone, i.e. we assume that its concentration does not change with time. Then

$$\frac{\mathrm{d}[O_3]}{\mathrm{d}t} = 0 = k_5[O^\cdot][O_2][M] - k_6[O^\cdot][O_3], \tag{8.8.7}$$

and thus

$$[O_3]=\frac{k_5}{k_6}[O_2][M]. \tag{8.8.8}$$

Substituting this expression into eqn (8.8.6):

$$-\frac{d[O^{\cdot}]}{dt}=2k_5[O^{\cdot}][O_2][M]+k_7[O^{\cdot}]. \tag{8.8.9}$$

After integrating, and replacing [M] by $[O_2]$ as explained previously, we obtain

$$-\ln[O^{\cdot}]=(2k_5[O_2]^2+k_7)t+\text{const.} \tag{8.8.10}$$

This equation shows which measurements are necessary for the determination of k_5 and k_7. We fix $[O_2]$, which we assume is equal to the total gas concentration in the flow line, and require a series of measurements of $[O^{\cdot}]$ as a function of t. A plot of $-\ln[O^{\cdot}]$ against t gives a straight line of slope $(2k_5[O_2]^2+k_7)$. (In fact we can plot −ln(intensity) rather than $-\ln[O^{\cdot}]$, as shown by eqn (8.8.1).) We then carry out further measurements at different values of $[O_2]$, and thus obtain a series of slopes $(2k_5[O_2]^2+k_7)$. These are plotted against $[O_2]^2$ to obtain a straight line of slope $2k_5$ and intercept k_7.

(*iv*) *Interpretation of k_1 and k_5*

The newly formed molecule in an atom association reaction such as reaction (1) or reaction (5) is energy rich by an amount equivalent to the energy of the new bond. Unless this excited molecule is stabilized by collisions decreasing the internal energy, it can redissociate. The overall process can be represented by the (simplified) scheme

$$A+B \xrightarrow{k_a} AB^{\ddagger} \qquad \text{association}$$

$$AB^{\ddagger}+M \xrightarrow{k_s} AB+M \qquad \text{stabilization}$$

$$AB^{\ddagger} \xrightarrow{k_r} A+B \qquad \text{redissociation}$$

A steady state treatment leads to the result

$$\text{rate of reaction}=\frac{k_a k_s[A][B][M]}{k_s[M]+k_r}. \tag{8.8.11}$$

If $k_r \gg k_s[M]$ (which is the case for reactions (1) and (5) at the values of [M] used here) then the expression simplifies to

$$\text{rate of reaction}=\frac{k_a k_s}{k_r}[A][B][M], \tag{8.8.12}$$

i.e. third order, and the rate coefficient measured actually corresponds to $k_a k_s / k_r$. At a first approximation both k_a and k_s can be taken as equal to the gas collision frequency factor Z, so that the third order rate coefficients k_1 and k_5 can be used to provide estimates of the respective redissociation rate coefficients k_r from the relation

$$k_1 \text{ or } k_5 \approx Z^2/k_r. \tag{8.8.13}$$

Hence k_r may be calculated.

(*v*) *Interpretation of* k_7

If γ is the fraction of oxygen atoms which recombine on striking the wall, then the rate of wall recombination is $\gamma N\bar{c}a/4$ where N is the number of oxygen atoms per unit volume, and $\bar{c}$ their mean kinetic velocity. a, the area of the wall, is $2\pi rl$ where r is the radius of the tube, and l its length. Then the rate coefficient k_7 will be

$$k_7 = \frac{\text{rate of wall recombination per unit length}}{\text{number of oxygen atoms per unit length}} = \frac{(\gamma N\bar{c}\pi r/2)}{N\pi r^2} = \frac{\gamma\bar{c}}{2r}. \tag{8.8.14}$$

The value of $\bar{c}$ may be obtained from the kinetic theory of gases:

$$\bar{c} = (8RT/\pi M)^{1/2}. \tag{8.8.15}$$

Thus a value of γ may be calculated from the value of k_7 by using eqns (8.8.14) and (8.8.15).

Apparatus

The apparatus, shown in Fig. 8.8.1, is installed in a dark bay. A mixture of oxygen and 10 per cent of nitrogen is supplied from a cylinder with a regulator set to a pressure of about 100 mmHg above atmospheric. There is also a small control valve which is used as an on/off tap and *not* to control the pressure. The pressure is measured on a mercury manometer in the form of a short U-tube. The gas then passes through a flowmeter, in which the pressure drop across an orifice is measured by a manometer filled with dibutyl phthalate ('oil'). The flowmeter also incorporates a by-pass tap (tap A). From the flowmeter the gas passes through another tap (tap B), through a needle valve, and then into the discharge, the details of which are shown in the diagram. The excited gas then passes into the flow tube via a *light trap*—a drawn-down end of tubing to prevent the passage of stray light from the discharge.

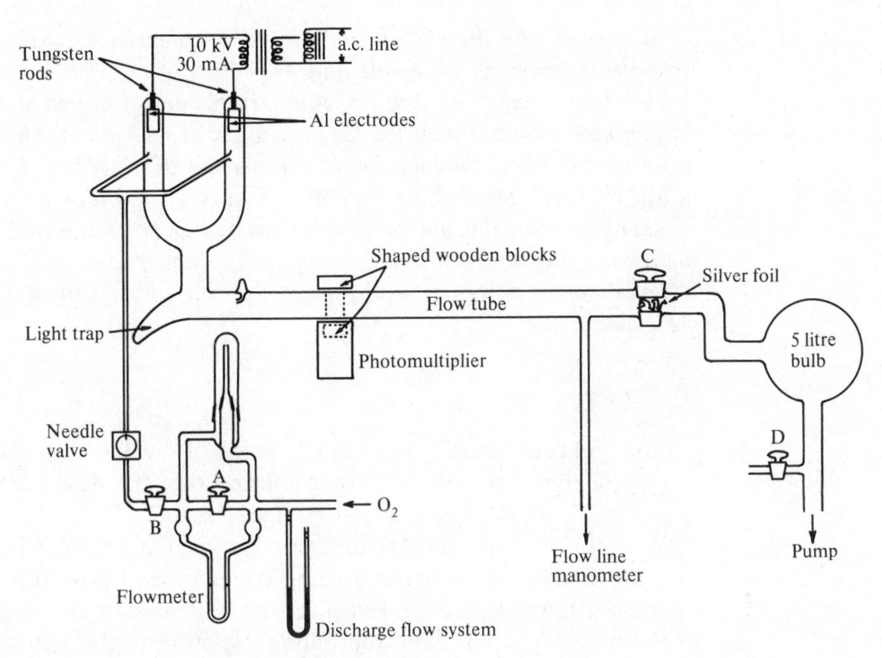

Fig. 8.8.1. Flow tube apparatus for the study of atomic oxygen reactions

A photomultiplier is mounted on the flow tube, and covered with a black cloth to minimize interference by stray light. It can be moved by sliding between positions marked along the tube. The photomultiplier is supplied by a high voltage ('EHT') power supply, and the photomultiplier anode current is measured by a digital ammeter. Alternatively a sensitive photocell may be used in similar fashion but without the need for an EHT supply.

Connected to the flow tube is an electronic or oil manometer. From the flow tube the gas passes through a large tap (tap C) which can be used to adjust the flow velocity. The channel in the tap is a convenient site for silver foil, which acts as a catalyst for the recombination of oxygen atoms to molecular oxygen, thus protecting the pump oil against oxidation. The gas then passes through a large globe which smoothes out any pressure pulses produced by the pump, and finally to the rotary vacuum pump itself. There is the usual air-leak (tap D) to prevent suck-back of pump oil when the pump is turned off.

Procedure *It is very easy to ruin this apparatus by the careless turning of taps. Taps turned too fast or too roughly may crack the glassware, or may cause mercury or oil to be blown out of the manometers. If*

air is sucked into the system, the discharge intensity will change for several hours. So be gentle and wary!

The high voltages used in this experiment are potentially dangerous—do not touch the leads associated with the discharge or photomultiplier. Photomultipliers are easily destroyed by being made to pass too high a current (>1 mA), *so under no circumstances may the photomultiplier mounting be unscrewed, or the room lights turned on, while the photomultiplier is energized.*

*Take note of any further precautions or instructions supplied for the experiment.**

(*i*) *Start up*

First check that the small control valve on the oxygen/nitrogen cylinder is turned off and that the flow tube is closed off from the vacuum pump globe (tap C).

Close the air leak (tap D) and start the vacuum pump. Check that the needle valve between the flowmeter and flow tube is closed but *finger tight only*. Evacuate the flow tube by opening it to the pump (tap C). Note the falling pressure on the flow line manometer.

Measure the ambient atmospheric pressure, p_{atm}, on a barometer.

(*ii*) *Experiment*

Open the by-pass to the flow meter (tap A) and the tap between the flow meter and its needle valve (tap B).

Turn on the cylinder supply with the spanner provided. Slowly turn on the small control valve on the cylinder regulator, making sure that the mercury in the manometer does not blow over.

Now close the by-pass tap. There should be no pressure difference on the oil manometer of the flowmeter. If there is, it means that the needle valve or flowmeter is leaking—seek assistance.

Open the flow control needle valve next to the flowmeter until the pressure difference on the flowmeter is between 10 and 100 mm oil. Remember to keep the *mercury* head between 10 and 150 mmHg.

Turn on the discharge power supply. A mauve discharge should be struck between the aluminium electrodes. Turn out the room lights and wait for a steady green air afterglow to be established—adjust the flow needle valve if necessary. Then switch off the discharge and estimate the time for the tail end of the glowing gas column to move down the tube. The flow velocity should be of the order of 100 cm s^{-1}—if it is not, adjust the tap between the flow tube and the pump. Switch on the discharge again, and leave the flow to stabilize for a few minutes.

Cover the photomultiplier with the black cloth provided. Switch on the power supply and photomultiplier. Read the oil, mercury and flow-line manometer pressures. Measure the emission intensity at the marked positions on the tube. Also measure the 'dark current' with the discharge turned off. Check that the three manometer readings have not changed significantly.

At the voltage used in this experiment, the discharge is stable at pressures of between about 0.1–10 mmHg as measured on the flow tube manometer. Carry out a total of six experiments in this pressure range. Allow the pressures to settle each time, and measure the discharge intensity at each point along the tube. Use only the needle valve between the flowmeter and flow tube to adjust the pressure. Be very careful not to blow over oil or mercury in the flowmeter—keep an eye on their levels whenever you move a tap or needle valve.

(*iii*) *Shut down*

First turn off the discharge and photomultiplier supplies at all switches. Leave the pump running. Open the bypass to the flowmeter (tap A). Close the small control valve on the oxygen/nitrogen cylinder, and shut off the cylinder with the spanner provided. If the pressure registered on the mercury manometer is higher than about 50 mmHg, reduce the pressure in the flowmeter by opening the needle valve between it and the flow tube. This is to prevent mercury being blown over when the apparatus is left unattended.

Close the needle valve between the flowmeter and flow tube, and the tap between it and the flowmeter.

Close off the flow tube from the pump (tap C). Turn off the pump and *immediately* open the air leak (tap D).

Calculation[4]

Make a careful check of the units used in the calculations for this experiment.

For each experiment, find the bulk flow rate F from eqn (8.8.5) using the value of f provided.* Convert the flow line pressures to units of molecule cm^{-3}, given that 13.0 mm oil = 1.00 Torr ≡ 1.00 mmHg which is equivalent to 3.24×10^{16} molecule cm^{-3} of gas at 298 K. Then calculate the linear flow velocity v from eqn (8.8.2), using the value of A provided.* Check that each linear flow velocity is compatible with your original visual estimate, and that this parameter is reasonably constant from experiment to experiment.

Knowing the distance between observation points on the tube, tabulate relative oxygen atom concentration (from eqn (8.8.1)) against contact time t for each experiment. Hence find k_5 and k_7 graphically, as described in the Theory section. Express them in the conventional units of cm^6 $molecule^{-2}$ s^{-1}, and s^{-1} respectively.

Given that the collision frequency factor Z is approximately 2×10^{-10} cm^3 $molecule^{-1}$ s^{-1}, find the rate coefficient for the re-dissociation of ozone from eqn (8.8.13). Find k_r for the redissociation of NO_2^* given that k_1 is 7.0×10^{-32} cm^6 $molecule^{-2}$ s^{-1} at 298 K.[5]

Calculate $\bar{c}$ from eqn (8.8.15). Hence from eqn (8.8.14) find γ, the fraction of wall collisions at which oxygen atoms are lost.

Results

Suppose that f for the capillary leak in the flowmeter is 8.55×10^{14} molecule/(s mmHg mm-oil), for p_1 in mmHg and Δp in mm oil, and that the cross-sectional area A of the flow tube is 3.80 cm^2. Say $p_{atm}=$ 759 mmHg, mercury manometer head = 140 mmHg. Then $p_1=$ 899 mmHg. $\Delta p = 32.5$ mm oil. $\therefore F = 2.50\times10^{19}$ molecule s^{-1}. Flow line pressure = 1.67 Torr or 21.8 mm oil and so $y = 5.41\times10^{16}$ molecule cm^{-3}. $\therefore v = 2.50\times10^{14}/3.80\times5.41\times10^{16} = 122$ cm s^{-1}. See Table 8.8.1. From graph, gradient = 4.35 s^{-1} = $(2k_5[O]^2+k_7)$ at $[O_2]=5.41\times10^{16}$ molecule cm^{-3}

Table 8.8.1. Specimen results

Distance/cm	Emission intensity	−ln(intensity)	Contact time/s
0	18.27	−2.91	0.0
20	8.92	−2.19	0.16
40	4.59	−1.52	0.33
60	2.14	−0.76	0.49
80	1.04	−0.04	0.66

Accuracy

Your discussion of the accuracy of your results should include the following factors: (*i*) *Limitations of the theory.* The accuracy of the theory is limited first by the invalidity of the steady-state approximation due to the existence of the reaction $O_2+O_3 \rightarrow O_2+O_2+O^{\cdot}$,[6] and secondly because of the plug flow assumption, with no corrections for the diffusion of the rectants.[4,7] Thirdly, some loss of atomic oxygen occurs in reaction (1), for which allowance has not been made. Finally, the oxygen atom concentration and the rate of atom loss have been shown to be highly sensitive to any traces of hydrogen, and hydrogen containing species such as H_2O, in the oxygen/nitrogen mixture.[6,8] (*ii*)

Accuracy of measurements. Dependent on calibration of flowmeter and pressure gauges, validity of assumption of Poiseuille behaviour (eqn (8.8.3)), linearity of photomultiplier, etc. (*iii*) *Precision of measurements*—make your own estimates.

Comment

This is an elegant experiment which illustrates the study of a reactive gaseous species in a flow tube. Provided the apparatus is in good working condition, results of the correct order may be obtained, although the complications mentioned above limit the accuracy.

The reactions which we have studied are of particular interest to kineticists, and there is a continuing search for a quantitative explanation for the relative rates of reactions such as (3) and (5), which do not simply depend on the relative strengths of the O—NO and O—O_2 bonds.[9]

Flow experiments are a powerful method of studying simple gas reactions. Recent experiments employ sensitive detection methods such as mass spectrometry, resonance fluorescence and laser magnetic resonance which allow very low radical concentrations to be measured.[7] The use of low concentrations minimizes interference by second order reactions and secondary chemistry in the flow tubes.

Technical notes

Apparatus

Line and pressure gauges as in Apparatus section and Fig. 8.8.1.,[†$] O_2/10 per cent N_2 mixture,[$] black cloth to cover photomultiplier.

Design notes

The oxygen/nitrogen cylinder main regulator should be adjusted to give around 100 mmHg pressure and then fixed. Flow meter: extra bulbs and sinters reduce likelihood of mercury and oil being blown over, extra black-waxed joints facilitate cleaning, sprung taps prevent blow-out. Flow line: 2–3 cm diameter, about 1 m long. Flow line manometer: oil manometer suitable but dibutyl phthalate very difficult to clean out of line if sucked back; Capsulon gauge with digital readout possible, but prone to drifting; pressure transducer (e.g. CEC Instrumentation) good but more expensive. Sensitive photocell can be used instead of photomultiplier.

Additional experiment

NO may be introduced by similar flowmeter system through backward pointing jet at up-stream end of flow-line. Adjust to give steep decay and measure k_1.

References

[1] *Reactions of oxygen atoms.* F. Kaufman. *Prog. in React. Kin.* **1,** 3 (1961).

[2] *The recombination of oxygen atoms in a discharge flow system.* B. A. Thrush. *J. Chem. Educ.* **41,** 429 (1964).
[3] *Chemical kinetic and photochemical data* NASA. JPL Publication 81–3, California Institute of Technology (1981).
[4] *Recombination of O atoms in the gas phase.* P. G. Dickens, R. D. Gould, J. W. Linnett, and A. Richmond. *Nature* **187,** 686 (1960) and refs. cited therein.
[5] *Evaluated kinetic data for high temperature reactions. vol. 2* D. L. Baulch, D. D. Drysdale, and D. G. Horne. Butterworths, London (1973).
[6] *Mass spectrometric studies of atom reactions. Part 4—Kinetics of* O_3 *formation in a stream of electrically discharged* O_2. A. Mathias and H. I. Schiff. *Disc. Far. Soc.* **37,** 38 (1964).
[7] *Kinetic measurements using flow tubes.* C. J. Howard. *J. Phys. Chem.* **83,** 3 (1979).
[8] *Rate constant of* $O + 2O_2 \rightarrow O_3 + O_2$. F. Kaufman and J. R. Kelso. *Disc. Far. Soc.* **37,** 26 (1964).
[9] *Theory of thermal unimolecular reactions at low pressures, Predictive possibilities of unimolecular rate theory* J. Troe. *J. Chem. Phys.* **66,** 4758 (1977), and *J. Phys Chem.* **83,** 114 (1979).

8.9 Kinetics of iodine atom recombination by flash photolysis

Flash photolysis is a method for studying very fast reactions, usually those involving atoms or radicals.[1] The technique involves the passing of a high intensity, short duration light flash through a solution or gas. Some of the molecules in the sample dissociate into atoms or radicals. The rate at which these short-lived species react may be obtained from the rate of change of absorption of a comparatively low intensity, steady light beam shone through the sample.

In the present experiment we examine the rate of recombination of iodine atoms and the dependence of the rate on total pressure. The iodine atoms are produced by the flash photolysis of molecular iodine vapour. Iodine molecules have a characteristic absorption at about 500 nm, but iodine atoms are transparent at this wavelength. Thus we can measure the rate of recombination of the atoms by exposing molecular iodine to a powerful flash of light and then monitoring the intensity of a beam of 500 nm light directed through the reaction vessel. The intensity of the transmitted light increases sharply during the flash as the concentration of the iodine molecules decreases, and then falls as the iodine atoms recombine. The rate at which the intensity falls can be used to determine the time-dependent concentration of the iodine atoms and hence the rate coefficient for the recombination. The recombination is essentially complete in 20 ms, so the flash must be short and the monitoring equipment capable of following such a rapid change.

Theory

(*i*) *Photolysis of molecular iodine*

Iodine has an absorption spectrum in the visible region (Expt 5.10) and radiation of wavelengths shorter than 499 nm causes *photolytic dissociation* according to the reaction:

$$I_2 + h\nu \rightarrow I(^2P_{3/2}) + I(^2P_{1/2}).$$

One iodine atom is in the ground state, ($^2P_{3/2}$), and the other in an electronically excited upper state, ($^2P_{1/2}$), which has an energy of 91 kJ mol^{-1} above the ground state.

Photolysis may also occur at $\lambda > 499$ nm by *predissociation*[2] to produce ground state atoms.

The flash lamps used in this experiment emit a continuum spectrum which typically has a maximum at 450 nm with extensive emission into the ultraviolet. The Pyrex glass of the reaction vessel absorbs radiation of wavelengths less than 300 nm. Therefore the flash causes dissociation by both of the mechanisms mentioned above, and yields an indefinite quantity of ground state and electronically excited atoms. The excited atoms are rapidly deactivated by collision with the undissociated I_2 in a period of $\sim 10^{-6}$ s and so the overall effect of the flash, on the millisecond time scale of this experiment, is to produce iodine atoms in their ground states.

(*ii*) *Kinetics of atom recombination*

The iodine atoms formed by flash photolysis react by recombination. The reaction requires a third body, M, to remove the excess vibrational energy of the nascent iodine molecules and thus prevent them dissociating:

$$I + I + M \xrightarrow{k_t} I_2 + M.$$

It is found that the reaction shows third-order kinetics overall and that, since $[I] \ll [M]$, the time-dependence of the iodine concentration, c, is given by

$$\frac{dc}{dt} = -kc^2, \qquad (8.9.1)$$

where $k = 2k_t[M]$. The integrated form of this expression is

$$\frac{1}{c_t} - \frac{1}{c_0} = kt, \qquad (8.9.2)$$

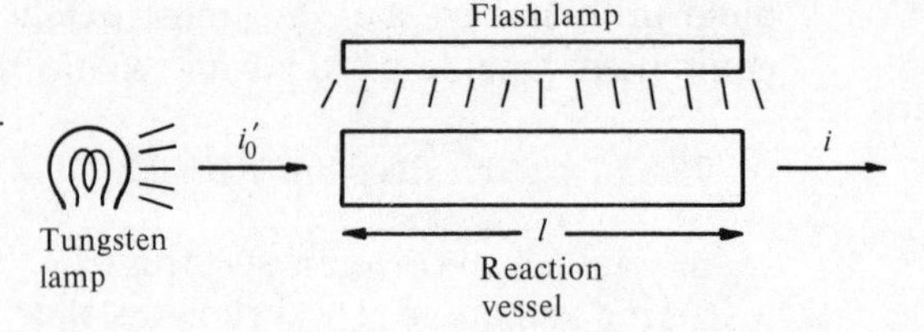

Fig. 8.9.1. Basic arrangement of the reaction vessel and lamps

where c_t is the atom concentration at time t, and c_0 is the concentration at $t=0$, i.e. immediately following the flash which, for the present experiment, may be considered instantaneous.

The concentrations are evaluated by absorption spectroscopy. Suppose the light incident on the cell at 500 nm is i_0' and that transmitted i, as shown in Fig. 8.9.1. Then by the Beer–Lambert law (p. 215),

$$\ln(i_0'/i)=\kappa cl, \tag{8.9.3}$$

where κ is the extinction coefficient of I_2 at 500 nm, and l is the length of the cell. Thus the light intensity increases from its steady value, i_s, to a value i_0, before falling back to i_s again as the iodine atoms recombine, Fig. 8.9.2.

The light intensities may be used to calculate the iodine atom concentration. From the stoichiometry of the reaction,

$$c_t=2(x_s-x_t), \tag{8.9.4}$$

where x_s is the I_2 concentration before the flash, and x_t the concentration at time t. Combining eqns (8.9.3) and (8.9.4):

$$c_t=\frac{2}{\kappa l}[\ln(i_0'/i_s)-\ln(i_0'/i_t)]. \tag{8.9.5}$$

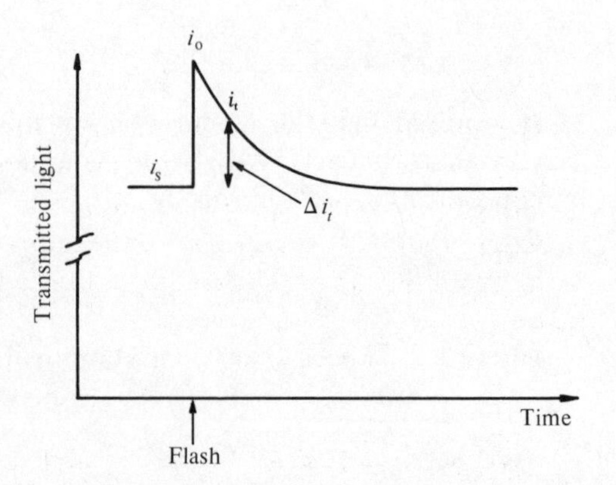

Fig. 8.9.2. Idealized trace of transmitted light intensity against time. i_s = steady (pre-flash) transmitted light intensity ($[I_2]=x_s$), i_0 = transmitted light intensity at zero time ($[I]=c_0$), i_t = transmitted light intensity at time t ($[I]=c_t$, $[I_2]=x_t$), $\Delta i_t = i_t - i_s$

If we take $\Delta i_t = (i_t - i_s)$, as shown in Fig. 8.9.2, then

$$c_t = \frac{2}{\kappa l}\ln[1 + (\Delta i_t / i_s)]$$

$$\approx \frac{2}{\kappa l}\frac{\Delta i_t}{i_s}, \tag{8.9.6}$$

because $\Delta i_t \ll i_s$. Substituting back into eqn (8.9.2),

$$\frac{i_s}{\Delta i_t} - \frac{i_s}{\Delta i_0} = \frac{2kt}{\kappa l}, \tag{8.9.7}$$

where Δi_0 is the value of Δi_t at $t = 0$. Thus a plot of $i_s / \Delta i_t$ against t has a slope of $2k/\kappa l$. The light intensities are measured by a photomultiplier and are recorded as voltages, on a digital voltmeter for i_s and on a chart recorder for Δi_t. Since the intensity is directly proportional to the voltage, V, it is valid to plot $V_s / \Delta V_t$ against t. Given the values of κ and l, k may be calculated from the slope of this line.

The influence of the third body on the recombination reaction may be divided between the effect of the added species M, and the effect of I_2 itself as a third body. k may then be shown to depend on pressure according to the relation

$$k = k_M[M] + k_{I_2}[I_2]. \tag{8.9.8}$$

The separate rate coefficients k_M and k_{I_2} may be determined by carrying out experiments at several different third body pressures. The room temperature vapour pressure is so low that, within the limits of accuracy of reading the mercury manometer, the third body pressure may be equated to the measured pressure. The slope of a plot of the measured rate coefficient k against [M] yields k_M, and the zero pressure intercept gives $k_{I_2}[I_2]$.

(*iii*) *Interpretation of rate data*

We now examine the mechanism of the reaction in more detail and try to interpret the experimental results. When the iodine atoms recombine, they initially form a vibrationally excited iodine molecule which then either dissociates to regenerate the iodine atom or is collisionally stabilized by M:[3]

$$I + I \underset{k_d}{\overset{k_r}{\rightleftharpoons}} I_2^\dagger \qquad \begin{matrix}\text{recombination}\\ \text{dissociation}\end{matrix}$$

$$I_2^\dagger + M \xrightarrow{k_s} I_2 + M \qquad \text{stabilization}$$

where k_r, k_d, and k_s are the rate coefficients for recombination, dissociation and collisional stabilization. Applying the steady-state approximation to $I_2^\dagger$ which is very short-lived,

$$\frac{d[I_2^\dagger]}{dt} = 0 = k_r[I]^2 - k_d[I_2^\dagger] - k_s[M][I_2^\dagger]. \tag{8.9.9}$$

Thus

$$[I_2^\dagger] = \frac{k_r[I]^2}{(k_d + k_s[M])}, \tag{8.9.10}$$

and the rate of regenerating I_2 is given by

$$\frac{d[I_2]}{dt} = k_s[M][I_2^\dagger] = \frac{k_r k_s[M][I]^2}{(k_d + k_s[M])}. \tag{8.9.11}$$

Because I_2 is a diatomic molecule, k_d is of the order of a vibrational frequency (10^{12}–$10^{13}\ s^{-1}$) and so $k_d \gg k_s[M]$ for all the pressures we employ. As a consequence,

$$\frac{d[I_2]}{dt} = \frac{k_r k_s}{k_d}[M][I]^2. \tag{8.9.12}$$

The rate at which the iodine atoms disappear is twice this rate, thus our measured rate coefficient $k_t = k/[M]$ (eqn 8.9.1) is given by

$$k_t = k_r k_s / k_d. \tag{8.9.13}$$

Apparatus The experimental arrangement is shown in Fig. 8.9.3. A quartz

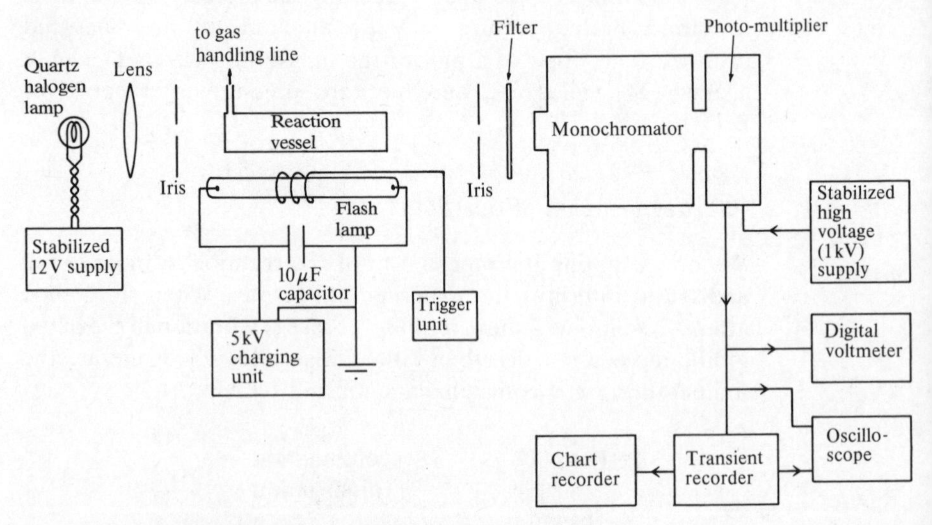

Fig. 8.9.3. Arrangement of the flash photolysis equipment

halogen lamp, connected to a well stabilized low voltage power supply and mounted in a housing with reflector, acts as the stable monitoring light source. The light passes through the Pyrex reaction vessel, and then through an iris and a green filter, both of which cut down the amount of reflected light from the flash lamp. The light is dispersed by a monochromator and then detected by a photomultiplier connected to a high voltage supply. The photomultiplier signal passes to a transient recorder, which converts it to a digital signal and then stores it. The signal is reconverted to analogue form and displayed on an oscilloscope; it can also be transferred to a chart recorder. Alternatively a storage oscilloscope can be used in place of the transient recorder and conventional oscilloscope.

The monochromator is used in second-order and so is set to 1000 nm in order to pass light at 500 nm to the photomultiplier. Any first order (1000 nm) light is removed by the green filter and would, in any case, be undetected by the photomultiplier.

The flash lamp is powered from a capacitor, which is charged and triggered by a purpose-built unit as shown. The signal from the flash also triggers the transient recorder or storage oscilloscope, so that it starts to record the signal at the correct time.

The reaction vessel is connected to a vacuum line, Fig. 8.9.4, which is used to alter the pressure of iodine and third-body gas. The iodine is stored in a reservoir which can be surrounded by coolant. The iodine may be pumped out through a by-pass line direct to a rotary vacuum pump with air-leak. The by-pass minimizes the formation of mercuric iodide in the gas line.

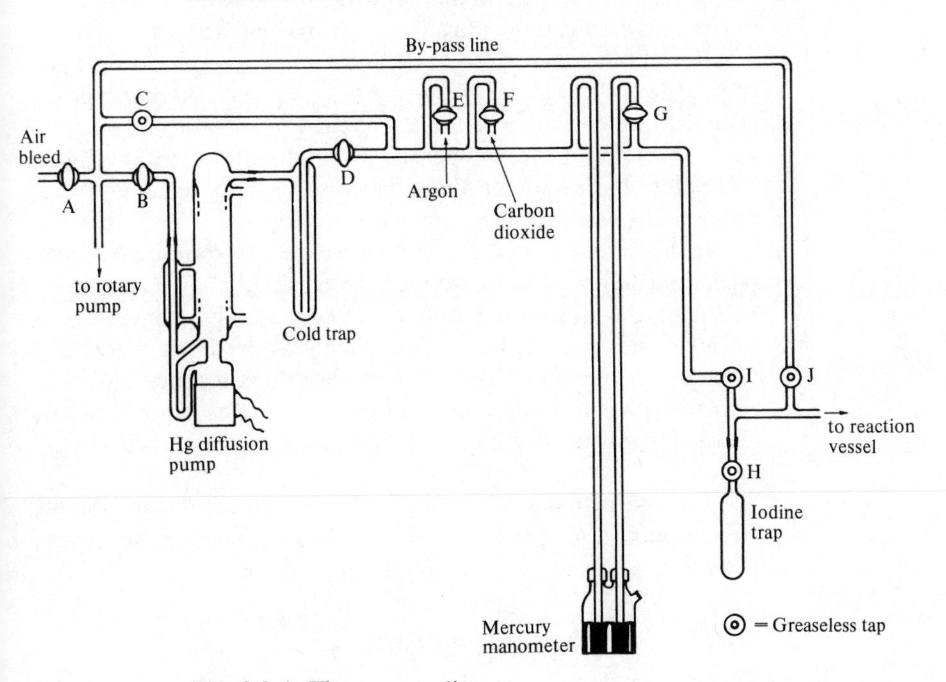

Fig. 8.9.4. The vacuum line

The third-body gases are admitted to the line from cylinders. Their pressures are measured on a mercury manometer, which also incorporates a blow-off. Tap G is closed after the line is evacuated, and the difference in mercury levels then indicates the pressure of gases in the line. The gases can be pumped away through a cold-trap, and mercury or oil diffusion pump connected to a rotary vacuum pump.

Procedure

The voltages supplied to the photomultiplier, capacitor and flash lamp are lethal. Under no circumstances tamper with the safety measures taken to prevent you touching these components.

(*i*) *Set up*

Any trace of oxygen in the reaction vessel will invalidate the results of the experiment, and the apparatus must therefore be pumped out and flushed very thoroughly.

Check that all the taps A–J are closed. Open taps H and J and turn on the rotary pump, thus roughly evacuating the reaction vessel and iodine trap. After about 5 min a reasonable vacuum should have been established. Close off the iodine trap (tap H), and open taps C and I to evacuate the line. Then after a few minutes close taps C and J, open taps B, D, and G, and turn on the diffusion pump and its water supply. Evacuate the line, reaction vessel, and manometer arm for about 15 min. Then close tap G.

Cool the iodine trap in liquid nitrogen. Open tap H and pump out the trap to remove any traces of oxygen from the iodine.

After about 5 min close tap I and remove the liquid nitrogen trap. Allow the iodine to warm up for at least 30 min, and so establish its room-temperature vapour pressure in the reaction vessel. Switch on the halogen lamp, photomultiplier and digital voltmeter. Iodine in the reaction vessel absorbs light and thus reduces the voltage from the photomultiplier. Check that the iodine has equilibrated by observing the photomultiplier voltage, which should have stopped decreasing.

Isolate the diffusion pump by closing tap D. Open tap E. Turn on the argon cylinder and bubble argon fairly vigorously through the mercury blow-off for at least 5 min. Then close the cylinder valve and open tap C to pump away the argon and any residual oxygen. Close tap C, open the cylinder valve and bubble gas for a further 5 min.

Close the cylinder valve and tap E. Open tap C fractionally and reduce the pressure of argon in the line to about 100 mmHg. Accurately record the pressure.

(*ii*) *Flash photolysis measurements*

Switch on the rest of the apparatus. Check all the settings on the oscilloscope, and transient recorder and chart recorder if used.*

Arm the storage oscilloscope or transient recorder.* Charge the capacitor to the specified voltage,* and fire the flash lamp.* The ARM light on the storage oscilloscope or transient recorder should go off, and a trace should be displayed on the oscilloscope.

If using a storage oscilloscope, carefully draw (or photograph) the trace on the screen.

If using a transient recorder, oscilloscope and chart recorder, zero the chart recorder as instructed* and switch on its paper drive. Lower the pen onto the paper. Press the PLOT button on the transient recorder. The chart recorder will then draw the trace. When the plot has finished, the oscilloscope will once again display the trace, and the chart recorder should then be stopped.

Pressures below ~50 mmHg should be avoided; a great deal of energy is liberated in the flash and a reasonable pressure of an inert gas is needed to act as a heat sink to obviate any significant change in temperature. For the same reason, k_{I_2} cannot be determined by carrying out experiments with no inert gas present.

Repeat the experiment at a range of about 5 argon pressures between 50 mmHg and atmospheric pressure. Argon should be bubbled through the blow-off between each measurement, but there is no need to evacuate the line each time.

Then carry out similar experiments with CO_2 as the third body, following through the entire Procedure section once again.

(*iii*) *Shut down*

Close off both cylinders and all the taps. Switch off the mercury diffusion pump and its water supply. Switch off the rotary vacuum pump and immediately open the air bleed (tap A).

Switch off all the power supplies to the apparatus.

Calculation

You will realize that the voltage trace obtained after each flash does not have the same shape as the light intensity curve shown in Fig. 8.9.2. Immediately following the flash, the voltage produced by the photomultiplier reaches very high values, because of scattered light from the high intensity photolysis flash falling on the photomultiplier. This section of the trace should be ignored (Fig. 8.9.5). The validity of starting the analysis at the point indicated can be checked by recording a trace with the background tungsten light turned off. Only the flash will be seen and its time-width can be determined.

Find out the relationship between the distances measured on the chart recorder and the absolute values of voltage and time.* Measure ΔV_t for at least 10 points on the unaffected portion of each trace, at measured time intervals. Plot $V_s/\Delta V_t$ against time in milliseconds for each trace.

Measure the gradient of each of these graphs. Calculate k (eqn (8.9.7)) in the conventional units of cm^3 $molecule^{-1}$ s^{-1},

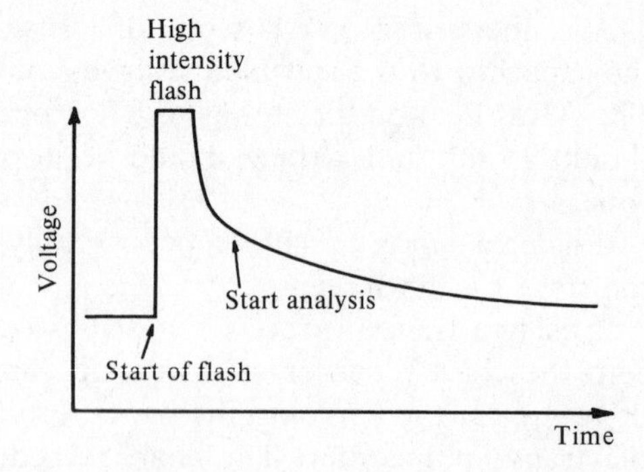

Fig. 8.9.5. Trace of voltage against time

given that $\kappa = 590$ in the awkward conventional units of $dm^3 \, mol^{-1} \, cm^{-1}$. Plot the gradients against pressure of third body gas in units of molecule cm^{-3}, given that 1.00 Torr ≡ 1.00 mmHg is equivalent to 3.24×10^{16} molecule cm^{-3} of gas at 298 K.

Hence calculate k_M and k_{I_2}, eqn (8.9.8), given that the vapour pressure of iodine at 298 K is 0.31 Torr.

(k_r/k_d) represents an equilibrium constant relating the unbound states of I_2 and the separated atoms; thus the only rate coefficient that varies on changing the third body is the stabilization rate coefficient, k_s. Determine the relative values of k_s for Ar, CO_2 and I_2 and comment on their values.

Table 8.9.1. Low pressure limit of third order rate constants k_t for recombination reactions at 300 K with Ar as third body

Reaction	k_t/cm^6 molecules^{-2} s^{-1}
H+OH	2.3×10^{-31}
CH_3+H	1.5×10^{-28}
CH_3+CH_3	1.7×10^{-26}

Table 8.9.1 shows limiting third order rate coefficients k_t for a few recombination reactions at 300 K, with the Ar as the third body, so that k_s is comparable for all the reactions. Given that k_r is approximately constant (it is usually quite close to the collision frequency) account for the observed variations. Comment on the fact that for the second and third reactions listed, the rate coefficient increases more slowly with [M] than the theory presented above suggests.

Results See Fig. 8.9.5.

Accuracy and precision

Generally accuracy is always less than precision. The precision in this experiment is limited by the storage oscilloscope, or the transient recorder with a typical precision of 8 bits or 0.4 per cent (p. xviii). However, if the signals from successive flashes under the same conditions are stored electronically in a *signal averager*, an average signal may be obtained which has an accuracy not limited by the precision, and which may therefore be better than 0.4 per cent.

Comment

The technique of flash photolysis was first reported by Norrish and Porter in 1949, and they shared a Nobel prize for their work in 1967.

Flash photolysis has been widely used, and has been particularly successful in the detection of transient intermediates.[1] The basic technique has been extended in two directions: (i) to shorter times, into the nanosecond regime with Q-switched lasers and into the picosecond regime using mode-locking, and (ii) to more energetic radiation, as in pulse radiolysis using van der Graaff electron accelerators.[1]

Technical notes

Apparatus

See Figs. 8.9.3 and 8.9.4.

Design note

The flash lamp must dissipate at least 100 J, with a decay time of less than 1 ms. Flash photolysis 'teaching' units are not powerful enough for this experiment. The heating effect of the flash may be reduced by surrounding the sample tube with a jacket containing copper sulphate solution.

Safety

Capacitor, flash lamp and charging unit must all be enclosed and operated by remote control.

References

[1] *Fast reactions*. J. N. Bradley. Clarendon Press, Oxford (1975), chap. 4.
[2] *Fundamentals of molecular spectroscopy*. C. N. Banwell. McGraw-Hill, London 2nd edn (1972), p. 222.
[3] *Kinetics and dynamics of elementary gas reactions*. I. W. M. Smith. Butterworths, London (1980), pp. 227–239.

8.10 Kinetics of the hydrogen–oxygen reaction from properties of the second explosion limit

Whereas the rates of most gas-phase reactions increase uniformly with pressure, the hydrogen–oxygen reaction exhibits the unusual behaviour shown in Fig. 8.10.1. Under conditions represented by the region to the left of the curve, the reaction is either slow or undetectable, whereas to the right of the curve,

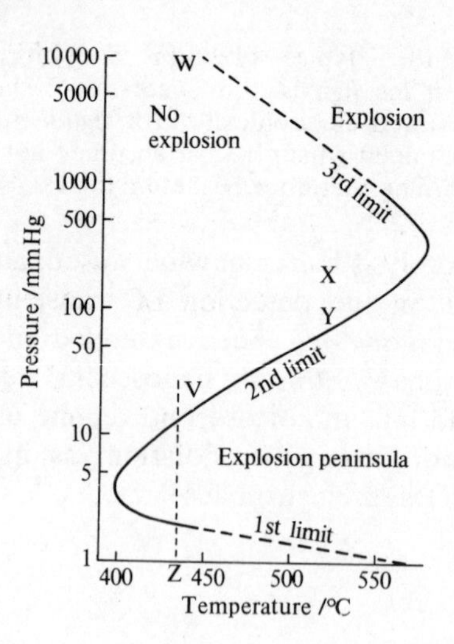

Fig. 8.10.1 Explosion limits of a stoichiometric hydrogen–oxygen mixture in a spherical KCl-coated vessel of 7.4 cm diameter. First and third limits are partly extrapolated. First limit is subject to erratic changes[3]

the mixture explodes. The three portions of the curve are termed the *first*, *second*, and *third explosion limits*, and the region between the first and second limits is generally known as the *explosion peninsula.*

The purpose of this experiment is to determine part of the second explosion limit. We do this by admitting hydrogen into a heated pyrex reaction vessel up to a pressure well above the second limit, such as that indicated by the point X on the graph. We then mix it with air, which increases the pressure still further, and finally reduce the pressure of the mixture until explosion occurs at Y. The explosion is not violent, but is detected by a 'kick' on the mercury manometer, and can also be observed as a glow in the reaction vessel.

The variation of the second explosion limit with temperature yields the activation energy of the branching step $H^{\cdot} + O_2 \rightarrow OH^{\cdot} + O^{\cdot}$. Further, by determining the composition necessary for explosion at a constant temperature, we can measure the ratio of the efficiencies of air and hydrogen as terminators of the chain reaction.

Theory[1–3]

(*i*) *Reaction scheme*

Many experiments have been carried out on the hydrogen–oxygen reaction to determine the effect of changing the pressure, altering the diameter of the vessel, the coating on the vessel walls, and so on. The conclusion which has been drawn from the results is that the (secondary) initiation process is the gas-phase thermal dissociation of hydrogen peroxide:

$$H_2O_2 + M \xrightarrow{k_1} 2OH^{\cdot} + M \qquad \text{initiation} \qquad (1)$$

where M is any suitable molecule. The hydrogen peroxide is created initially by the slow combination of H_2 and O_2 at the wall, and is both formed and destroyed by other steps of the chain reaction, so attaining a steady-state concentration.

The reaction then proceeds as follows:

$$OH^{\cdot} + H_2 \xrightarrow{k_2} H_2O + H^{\cdot} \qquad \text{propagation} \qquad (2)$$

$$H^{\cdot} + O_2 \xrightarrow{k_3} OH^{\cdot} + O^{\cdot} \qquad \text{branching} \qquad (3)$$

$$O^{\cdot} + H_2 \xrightarrow{k_4} OH^{\cdot} + H^{\cdot} \qquad \text{branching} \qquad (4)$$

$$H^{\cdot} + O_2 + M \xrightarrow{k_5} HO_2^{\cdot} + M \qquad \text{termination.} \qquad (5)$$

Reaction (5) is a chain termination step because HO_2 is fairly unreactive and can diffuse to the wall. M is any *third body* gas molecule which can take part in the reaction.

Let us suppose that we pump out the reaction vessel and heat it to some temperature above 400 °C, shown as point Z in Fig. 8.10.1. If a stoichiometric mixture of hydrogen and oxygen is slowly admitted into the vessel, no detectable reaction will occur. This is because at low pressures below the first limit, hydrogen atoms diffuse to the wall and are consequently removed.

The diffusion of $O^{\cdot}$ and $OH^{\cdot}$ can be ignored, except at low $[H_2]/[O_2]$, because of the high rates of their gas phase reactions.

As the pressure is increased to the first limit, the rate of removal of hydrogen atoms is no longer sufficient to balance steps (3) and (4), and an explosion takes place. The first explosion limit is lowered as the vessel diameter increases. The limit is also sensitive to the nature of the vessel surface and, if the surface is efficient, is lowered by the addition of inert gases. These observations confirm the occurrence of a chain reaction, with chain carriers being destroyed on the surface.

Any mixture within the area above the first limit is explosive. As the pressure is increased further, the three-body collisions in the termination step (5) begin to dominate the two-body collisions required for the propagation and branching reactions, until at the second limit (point V) no explosion takes place. The second limit is almost independent of the diameter of the vessel and the nature of its surface, but the addition of an inert gas makes the mixture less explosive. These observations support the conclusion that a chain centre is now being destroyed in the gas phase.

Just above the second explosion limit the reaction rate is very small and increases with pressure until the third explosion limit is reached at point W. As well as having some thermal characteristics, this limit is markedly dependent on the size of the reaction vessel, and must therefore depend on the presence or absence of some reaction at the vessel wall. In terms of the reaction scheme, the chains become longer because diffusion of $HO_2^{\cdot}$ to the wall is hindered, and radical regeneration takes place by the reaction

$$HO_2^{\cdot} + H_2 \xrightarrow{k_6} H_2O_2 + H^{\cdot} \qquad \text{propagation.} \qquad (6)$$

(*ii*) *The second explosion limit*

In the explosive region below the second limit, the net rate of production of radicals H, OH, and O in the branching steps exceeds the rate of removal. At the explosion limit itself, we may assume that the rate of removal of each type of radical is equal to its rate of formation.

Let us apply this condition to the concentration of H atoms which are produced by the propagation and branching steps shown in Fig. 8.10.2. In this process, three H atoms are produced from one, and since the first step is rate determining, the rate of production is therefore $2k_3[H][O_2]$. The rate of production via the initiation reaction (1) is negligible in comparison to that produced by branching steps, and reaction (6) only becomes important at the third limit. Balancing the rate of production of H atoms by their rate of removal by reaction (5), we find that

$$2k_3[O_2][H] = k_5[O_2][M][H],$$

Find an expression for the overall rate of the reaction (i.e. the rate of production of water) by applying similar arguments to the concentrations of O and OH; how does your equation represent the second explosion limit?

so that at the second explosion limit

$$2k_3 = k_5[M]. \qquad (8.10.1)$$

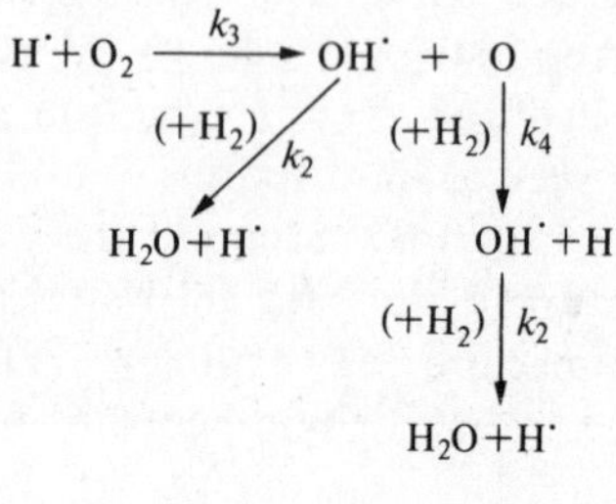

$$H^{\cdot} + O_2 \xrightarrow{k_3} OH^{\cdot} + O$$

$$OH^{\cdot} \xrightarrow[(+H_2)]{k_2} H_2O + H^{\cdot}$$

$$O \xrightarrow[(+H_2)]{k_4} OH^{\cdot} + H^{\cdot}$$

$$OH^{\cdot} \xrightarrow[(+H_2)]{k_2} H_2O + H^{\cdot}$$

Fig. 8.10.2. Reactions determining the concentration of H atoms

(*iii*) *Determination of activation energy and third-body efficiency*

The first part of this experiment is concerned with finding the activation energy of the chain branching reaction (3) from the temperature dependence of the second explosion limit. For a mixture of constant composition, the total pressure p is proportional to [M], and thus from eqn (8.10.1),

$$p = \text{const. } k_3/k_5 . \tag{8.10.2}$$

The rate coefficients of reactions (3) and (5) are related to their activation energies E_3 and E_5 by the Arrhenius equation $k = A \cdot e^{-E/RT}$, (p. 401), and therefore

$$p = \text{const. } e^{(-E_3+E_5)/RT}. \tag{8.10.3}$$

Reaction (5) has, in fact, a small negative temperature coefficient, determined by a different technique, which corresponds to $E_5 = -6.7 \text{ kJ mol}^{-1}$. To a good approximation, since E_5 is so small,

$$p = \text{const. } e^{-E_3/RT} \tag{8.10.4}$$

Strictly speaking the explosion pressures should be converted to concentrations to obtain the line activation energy E_3(line), which is therefore a factor of RT less than the activation energy calculated from the equation opposite.

and E_3 may be found from a plot of $\ln p$ against $1/T$.

In the second part of the experiment we determine the relative efficiencies of hydrogen and air as chain terminators in the form of the third-body species M in reaction (5). Expanding eqn (8.10.1) to include the partial pressures p of both gases *at the explosion limit*, we find that

$$2k_3 = k_5[\text{M}] = \text{const. } [k_5(\text{H}_2) \cdot p_{\text{H}_2} + k_5(\text{air}) \cdot p_{\text{air}}]. \tag{8.10.5}$$

This part of the experiment is carried out at constant temperature and therefore k_3 is itself constant. Thus we may rearrange eqn (8.10.5) to give

$$p_{\text{H}_2} = \frac{\text{const.} - k_5(\text{air}) \cdot p_{\text{air}}}{k_5(\text{H}_2)}. \tag{8.10.6}$$

The relative efficiency of air and hydrogen as chain terminators, $k_5(\text{air})/k_5(\text{H}_2)$, may therefore be obtained from a plot of p_{H_2} against p_{air} at the explosion limit.

Apparatus

The apparatus comprises a pyrex reaction vessel mounted in a furnace, the latter incorporating a temperature controller and thermocouple with read-out. The pyrex reaction vessel is connected to a vacuum line as shown in Fig. 8.10.3. Dried air is

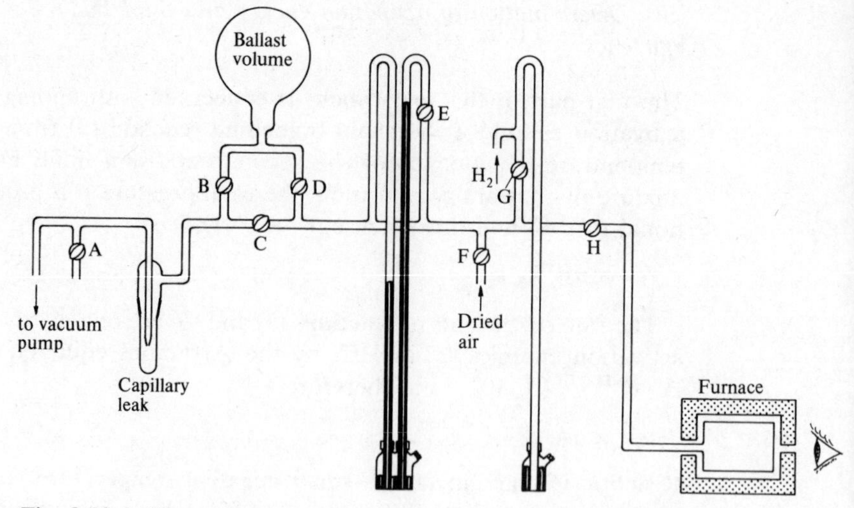

Fig. 8.10.3. The vacuum line

admitted to the line through tap F, and hydrogen from a cylinder, with fine control valve, connected next to tap G.

Procedure

This experiment is potentially hazardous. It must be carried out with a partner, and both of you should cross-check that the instructions are being followed correctly at each stage.

(*i*) *Set up*

The furnace should already be hot. Set the temperature controller to 450 °C. Evacuate the whole system (taps A, F, and G closed) for at least 15 min. Check that the hydrogen cylinder is connected next to tap G.

Close off the ballast volume and manometer side-arm (taps B, C, D, and E closed). Open the fine control valve on the hydrogen cylinder and allow the gas to bubble gently through the blow-off for a few moments.

Allow hydrogen into the line by fractionally opening tap G. By careful use of this tap and the cylinder fine control valve, adjust the hydrogen pressure to about 100 mmHg.

After a few minutes open the line to the pump (tap C) and pump off the hydrogen.

Measure the pressure of the atmosphere with a barometer.

(*ii*) *Temperature dependence of the second explosion limit*

Close the line to the pump again (tap C). Admit hydrogen into the reaction vessel to a pressure of 200 mmHg. Close off the reaction vessel (tap H) and pump off the excess hydrogen (tap C).

Close off the pump again (tap C). Allow air slowly to enter the line (tap F). When the pressure *exceeds* that of the hydrogen in the reaction vessel (*not before*), fully open the line to the reaction vessel (tap H), and allow air more quickly into the line (tap F fully open) until atmospheric pressure is reached.

Close off the air tap and the reaction vessel (taps F and H) and allow thirty seconds for the gases to mix. During the mixing period open the ballast volume to the pump (tap B). When the gases have mixed, open the reaction vessel to the line (tap H), and withdraw gas smoothly from the vessel through the ballast volume (tap D).

Hydrogen and air form water by a slow reaction under these conditions even though they do not explode; water acts as a very efficient chain terminator (reaction (5)) and will alter the explosion limit if more than thirty seconds is allowed for mixing.

Observe both the manometer and reaction vessel closely. The explosion is detected by a kick or slight arrest in the rise of the manometer mercury column, and/or a glow in the reaction vessel. It may help to open the other arm of the manometer (tap E) and observe both arms at once. The manometer reading immediately *before* the explosion should be noted, together with the furnace temperature.

Repeat the procedure with the same pressure of hydrogen at about six furnace temperatures in the range 450–510 °C.

(*iii*) *Relative efficiencies of* H_2 *and air*

Repeat the procedure described in section (ii) with a constant furnace temperature as close as possible to 510 °C, for about six hydrogen pressures in the range 100–500 mmHg. At hydrogen pressures less than 150 mmHg it is desirable to 'sandwich' the hydrogen between two amounts of air so as to avoid passing through the explosion peninsula. For an experiment requiring, say, 130 mmHg of hydrogen in the initial mixture, a suitable procedure would be to allow about 300 mmHg of air into the reaction vessel, add the required 130 mmHg of hydrogen, and then make up the mixture to atmospheric pressure with more air.

Finally, measure the ambient atmospheric pressure either with a barometer or by evacuating the manometer and measuring the length of the mercury column above the level of the reservoir.

(*iv*) *Shut down*

Turn off the hydrogen cylinder at its main valve.

Evacuate and close off the reaction vessel (tap H). Switch off the rotary vacuum pump and open its air leak (tap A).

Leave the furnace switched on at 450 °C is requested.*

Calculation Calculate the activation energy of the chain propagation step in the hydrogen/oxygen reaction by the method described previously. Calculate the relative efficiencies of hydrogen and air as chain terminators, and comment on the value obtained.

Results See Table 8.10.1. Note that the absolute results will not be reproduced in a different apparatus because (a) the surface affects the limit slightly, and (b) the temperature refers to a fixed point in the furnace, whereas the explosion is initiated at the hottest point. There factors should not affect the activation energy or the relative efficiency determination.

Table 8.10.1. Specimen results (all pressures in mmHg)

Temperature		Pressure		
mV	°C	tap side	line side	Δh
4.396	494.5	779	695	84

H_2 pressure	Air pressure	Manometer pressure			p_{H_2}	p_{air}
		tap side	line side	Δh		
100	662.9	766	684	82	10.75	71.25

Comment As well as being intrinsically interesting, explosions underline the critical dependence of the rate of branched-chain chemical reactions on temperature, pressure and the characteristics of the reaction vessel. The hydrogen/oxygen reaction has been studied extensively,[1] and serves as a prototype for the investigation of other combustion reactions.[3]

Technical notes

Apparatus

See Fig. 8.10.3 and Apparatus section.

Design notes

(i) The furnace temperature should be prevented from rising above 570 °C and melting the pyrex reaction vessel. (ii) The explosion limit is very sensitive to temperature, and therefore good temperature control is important. If the temperature distribution of the furnace is uneven, it is imperative that the thermocouple be placed in the plateau region so that minor movements of the thermocouple do not alter its e.m.f. significantly. As we are concerned only with activation energy and relative efficiency, the *actual* average temperature need not be known with great accuracy. (iii) Sometimes Pyrex reaction vessels give irreproducible results, especially if the withdrawal rate is low. This arises from rapid water formation, which changes the explosion limit for the reasons mentioned in the marginal note earlier. Undoubtedly the best results are obtained by coating the surface with KCl, although this is a messy operation.

References

[1] *Branching-chain reactions: the hydrogen–oxygen reaction.* R. R. Baldwin and R. W. Walker. In *Essays in Chemistry* **3,** 1 (1972).

[2] *Kinetics and mechanism.* J. W. Moore and R. G. Pearson. John Wiley and Sons, New York, 3rd edn (1981), p. 408.

[3] *Combustion, flames and explosions of gases.* B. Lewis and G. von Elbe. Academic Press, New York, 2nd edn (1961), ch. 2.

Index

F, figure; T, tables; italics, explanation or definition of term

AB spin system *272*
Abbé refractometer 42, 46, 190
absorbance *199*, *215*, 249
absorption
  coefficient *215*, 249
  spectra 206–69
accumulation of errors xx–xxii
accuracy *xviii*
acetaldehyde
  n.m.r. 269–78
  photolysis 441–9
acetic acid
  acidity constant 359–67
  /chloroform/water system 21–8
  density 27(T)
  dimerization 89–95
acetone iodination 396–99
acetylene line frequencies 228(T)
acknowledgements viii
acid
  base catalysis, general *409*
  dissociation constant *139*
    of polyacids *155*
acidity constant *139*, *363*
acoustic interferometer 237(F)
activated complex theory *414*
activation
  energy 369, *401*, 414, 473
  enthalpy *415*
  entropy *415*
  free energy *415*
activity *116*
  coefficient *116*, 138–9, 146–7, 151–4, 378–80
    Debye–Hückel expression 117
    ionic 116, 121, 135, 138–9, 146–7, 151–4, 378–80
    mean ionic *116*, 135(T)
  optical *406*
adiabatic
  bomb calorimeter 80–9, *81*, 84(F)
  change *81*
adsorbate *73*
adsorption *73*, 73–9
  isotherms *73*, 78
air
  afterglow *450*
  composition 178
  viscosity 357
allowed transition 194
α-transition *313*, 317
amalgam electrode 143
ammonium nitrate
  salt bridge 127
  transitions 61(F)
amphoteric compounds, titration of 156(F)
amyl iodide hydrolysis 413–8
analogue to digital convertor xviii, 158, 162
analysis
  chromatography 444–9
  gases 102(F), 175–82
  infrared 169
  ionic solutions 157
  liquid mixtures 26, 43
  mass spectrometry 170–83
  nuclear magnetic resonance 269–80
  radiochemical 439
  stratosphere 213
anharmonic oscillator 220, 254(F)
anharmonicity constant *200*, *254*
anion reversible electrode *123*
anode *128*
anti-Stokes scattering *241*
Apple microcomputer 163
approximation, orbital *280*
Archimedes principle 4
Arcton adsorption 73–9
argon
  BBMS potential 338(F)
  Lennard–Jones potential 338(F)
  neutron scattering 341(F)
  phase diagram 344(F)
  radial distribution function 341(F)
  structure of phases 340
  trajectories 348
  viscosity 350–9
  X-ray scattering 341(F)
Arrhenius
  behaviour 414
  equation *401*
  expression *363*
  pre-exponential factor *401*, 418

association
  ethanoic acid 89–95
  nitrogen dioxide 95–100
atmosphere
  colour 302
  composition 178, 213
  photochemistry 449–60
atomic
  masses *back endpaper*
  orbitals 193–8, 281
  spectra 193–206
autogenic boundary *372*
autotitrator 157(F)
avalanche of electrons 426(F)
Avogadro's constant, determination 326–36
AX_2 spin system *273*
axis of symmetry 230
azeotrope *40*

balance
  gas density 5(F)
  point 5, 132
Balmer series 195(F), 204
band
  heads 260
  hot 254(F)
  origins 260
  overtone 221, 226(T)
  parallel *224*, 231
  perpendicular *224*, 231
  spectrum 253
Barker–Bobetic–Maitland–Smith potential 338(F)
base
  catalysis 409
  salt hydrolysis 364
BASIC 159, 161
basic cell 339, 340(F)
BBC microcomputer 158–64
BBMS potential 338(F)
Beckmann
  apparatus 48–50
  thermometer 48–9
Beer–Lambert law *214*, 249
benzene/tetrachloromethane mixtures 37–46
  liquid/vapour composition 44(T)
benzoic acid, combustion enthalpy 82(F)
benzyl alcohol n.m.r. 269–80
best-fit straight line xxvi(F), 334
BET isotherm 78
β-transition *313*, *317*
binary mixture
  distillation 37–46
  liquid/vapour composition 44(T)
  temperature/composition 40(F)
  vapour pressure 40
binodal curve *23*
Birge–Sponer extrapolation *257*
bits xviii, 162

black body temperature corrections 200(T)
blacking 134, 136, 365
blank spectrum 178(F)
boiling point
  elevation 40(F)
  pressure dependence 33–7
Boltzmann
  distribution 198
  factor 401
bomb calorimeter 80–9, *81*, 84(F)
bond
  angles 284
  co-ordinate 105
  dissociation enthalpy 83, 95
  hydrogen 89, 214–9, 271
  length 284
  van der Waals 337
Born–Oppenheimer approximation *255*
Bourdon pressure gauge 15, 95–8
bouyancy method, partial molar volume 28–33
Boyle temperature *10*
branch, P, Q, R *222*, 226(F), 232
branching chains 471
bridge circuit
  capacitance 318, 365
  radio frequency 274(F)
  resistance 365
brightness *199*
bromoethane, mass spectrum 182(F)
Brønsted
  catalysis law *412*
  equation *402*
Brownian
  motion 326–36
  particles 333(F)
Brunauer–Emmett–Teller isotherm 78
bubble-plate fractionating column 38(F)
buffer solution 139
Burnett expansion method 10–13

cadmium, atomic lines 205(T)
calomel electrode 124(F)
  e.m.f. 124(T)
calorimeter
  adiabatic bomb 80–9, *81*, 84(F)
  differential scanning 59(F)
  differential thermal analysis 59(F)
  micro 58–65
  solution 107(F)
Cannon–Fenske viscometer 295(F)
capacitance 316
  cell 316
capillary
  depression *66*, 76
  flow viscometer
    gases 350(F), 355(F)
    ionic solutions 369
    polymer solutions 295(F)
  rise *66*

carbon 100
  dioxide
    bending energy 280–87
    Joule–Thomson coefficient 9–16
    reduction 100–4
    vivial coefficient 19(T)
  disulphide
    i.r. spectrum 244(F)
    normal vibrations 245(F)
    Raman spectrum 243
  monoxide, oxidation 100–4
  tetrabromide, melting and transition 55–8
  tetrachloride
    /benzene mixtures 37–46
    normal vibrations 245(F)
  Raman spectrum 245
catalysis, general acid and base *409*
cathetometer *68*, 355
cathode *128*
Ce(III)/Ce(IV) couple 145–51
cell
  amalgam 143(F)
  concentration *377*
  constant *361*
  electrochemical *120*, 120–53
  e.m.f., *see* electromotive force
  half 122
  Harned *137*, 140(F)
  modified Harned 139
  reaction
    enthalpy 142
    entropy 142
  vessel 127, *133*
  Weston standard *132*
  with liquid junction *126*, 381(F)
  without liquid junction *126*, 137
centrifugal stretching constant *221*
chain
  branching 471
  freely-jointed 292(F)
  length 443
  reaction 442, 470–2
character table, C_{3v} 231(T)
charcoal 101(F)
charge-to-mass ratio *172*
charge transfer excited state 263
chelate compound *105*
chemical
  exchange 382–6
  kinetics, *see* kinetics
  potential 47, 52, 117, 121
  quenching 397, 426
  shift *270*
chemiluminescence 450
chemisorption *73*
chi-square
  calculation 439(T)
  test 431(T)
chloroform
  density 27(T)
  /ethanoic acid/water system 21–8
chloromethane
  heat capacity 230–240
  i.r. spectrum 239(T)
  normal vibrations 231(F)
chromatography 444–9
cis-trans isomers 87–8
Clapeyron equation *34*
Clark cell 142, 143(F)
Clausius–Clapeyron equation xx, *34*
Clausius–Mosotti equation *185*
clock reaction 400–6
coexistence curve *344*
coil, random 290(F), 291, 300, 303
cold finger principle *70*
colligative properties *46*
collision
  diameter *337*
  frequency factor *458*
collisional deactivation/stabilization 463, 471
column
  bubble-plate 38(F)
  distillation 37–46
  fractionating 37–9
  Vigreux *38*
combination electrode 153(F)
combustion
  enthalpy *81*, 80–9
  explosion 469–77
  gases 198–205, 469–77
  solids 80–9
Commodore microcomputer 349
common-ion effect *115*
comparator *258*
complex
  cuprammonium 105
  ethylenediamine 105
  ferric-salicylate 248
  formation 104–12, 248–52, 376
    enthalpy 104–12
    entropy 104–12
component *22*
composition/temperature diagram 40(F)
compressibility, gas 9
compression
  factor *9*
  wave *235*
computer v
  Apple 163
  BBC 158–64
  Commodore 349
  mainframe 286, 349
  micro 158–64, 286, 349
  mini 349
  program v, 158–64, 211, 286, 310, 336, 345–9, 424, 437, 440
  Rockwell Aim 158–64
  simulation 285, 336–49

computer v (*cont.*)
  titration 151–64
concentration
  cell *377*
  profiles 388(F)
conductivity
  cell 360(F), 417
  electric *360*
  ionic solutions 363(T)
  meter 365(F), 417
  molar *360*
  reaction kinetics 413–18
  strong electrolytes 361(F), 362(F)
  weak electrolytes 361(F), 362(F)
configuration of electrons in
  atoms 193–8
  helium 197
  hydrogen 194
  iodine 256
  molecules 207, 256
  nitric oxide 207
  sodium 195
conformation
  energies 280–7
  macromolecules *289*, 289–321, 290(F), 292(F)
consecutive reactions 418–25
constant
  Avogadro 326–36
  coupling 272
  cryoscopic *46*
  decay *432*
  Faraday *121*
  Huggins 294
  mutarotation *407*
  Planck *193*
  rate, *see* rate coefficient
  relaxation 407
contact
  angle *67*
  time *451*
conversion factors *front endpapers*
cooling curve 55
coordinate bond 105
copper sulphate, activity coefficients 135(T)
Coriolis effect *233*
count rate 428
counter
  electrode 392(F)
  ion 122
  radiation 434
  threshold *428*
coupling constant *272*
critical
  point 34, 344(F)
  solution point 24(F), *25*
cryoscopic constant *46*
cuprammonium complex 105
curve fitting xxv
cyclohexane
  cryoscopy 46–52
  solvent 183, 290
daughter product *432*
dead
  space 103
  time *427*
Debye equation 302, 316
Debye–Hückel
  complete equation *117*
  constant *117*
  limiting equation *117*, 138, 403
  reaction rate 402
  theory 116–17
decay, *see* radioactive
defined constants *front endpapers*
degeneracy 196, 198
density
  chloroform 27(T)
  ethanoic acid 27(T)
  hexane 54
  toluene 54
  water 27(T)
depression
  capillary *66*, 76
  freezing point 46–52, 55–8
  transition temperature 55–8
deshielding *270*
detector
  molecular ions 174
  radiation 200, 209, 224, 246, 251, 258, 265(F), 274(F), 306(F), 426(F), 455(F), 464(F)
deuterium chloride
  i.r. spectrum 318(F)
  overtone frequencies 226(T)
dextrose, mutarotation 406–13
diatomic molecules
  energy levels 220, 253
  spectra 220–30, 253–61
dichlorotetra fluoroethane adsorption 73–9
dielectric
  cell 190(F), 316(F), 318(F)
  constant *185*
  relaxation 312–21
differential
  scanning calorimetry 58–65
  thermal analysis *59*
diffraction
  neutron 341(F)
  X-ray 341(F)
diffusion
  coefficient *378*
  first law *378*, 383
  reaction kinetics 382–6
digital voltmeter *132*
dihedral angle 285(F)
dilution law *363*
dimerization
  enthalpy 89–95
  ethanoic acid 89–95

noble gases 359
dinitrogen tetroxide, dissociation 95–100
dipole moment 183–93, *184*
measurement 189(F)
direct reading spectrometer 201(F)
discharge lamp 201
dispersion 314
forces *337*
dissipation factor *315*
dissociation
constant
water *130*
weak bases 364
weak monoprotic acids *139*, 363
weak polyacids *155*
dinitrogen tetroxide 95–100
energy *253*
enthalpy 95–100
distillation 27, 37–46
column 37–46
efficiency 39
DME *386*, 392(F)
double-focussing mass spectrometer 183
doublet *196*
dropping mercury electrode *386*, 392(F)
Dumas
bulb 92(F)
method 89–95

Einstein's expression, Brownian motion 328–30
electric
conductivity *360*
field 184–6, 313–17
potential 121
electrochemical
cell *120*, 120–53
potential *121*
electrode *120*, 120–6
amalgam 143
anion reversible *123*
blacking 134, *365*
calomel *124*(T), 129
combination 153(F)
copper 134
counter 392(F)
dropping mercury 392(F)
gas *125*
glass 126, *152*, 153(F), 380
hydrogen 125(F), 152
inert metal *125*
ion selective *126*
membrane *126*, 153(F)
metal *123*
metal, insoluble salt *123*, 134, 379
metal | metal ion *123*
pH 153(F)
platinizing *134*, 365
platinum 125
platinum | hydrogen 125(F), 152
potential *121*
determination 131–3
standard *121*, 129, 138
preparation 133–5, 140, 148
process 386
redox *125*
reference 124, 392(F)
reversible *122*
secondary standard *124*
silver 120–3
silver, silver chloride *123*, 134, 379
specific ion *126*
standard 122, 125
standardization 153
types 122–6
working *391*
electrolytes
strong *361*
weak *361*
electromagnetic radiation, spectrum 168(F)
electromotive force *120*
half-cell 122
liquid junction *379*
membrane 152–4
pH titration 155(F), 156(F)
polarography 387(F)
potentiometric titration 150(T)
solubility product 129
standard *121*, 129, 138
temperature dependence 142–5
thermodynamics 142–5
electron
impact source *174*
spin 196
transfer rate coefficient 263–5
electronic
partition function 233, 235
polarizability *186*
temperature *200*
transitions 168(F), 193–8, 248–69
e.m.f., *see* electromotive force
emission spectra 193–205
end point 146–50, 155
energy
activation 369, *401*, 414, 473
adsorption 73, 78
barrier 280, 414
free, *see* free energy
levels
anharmonic oscillator 220, 254(F)
diatomic molecules 220, 253
ethyne 223–9
harmonic oscillator 220, 253, 254(F)
helium atom 198(F)
hydrogen atom 195(F)
hydrogen chloride 222(F)
hydrogen cyanide 206
iodine molecule 254(F)
ionization 197, 204
linear polyatomics 223–9

energy (*cont.*)
  levels (*cont.*)
    n.m.r. 269–73
    rotation 206–8
    rotation-vibration 220–4
    sodium atom 196(F)
    symmetric top 230–3
enol 271, 399
ensemble *349*
enthalpy
  activation *415*
  adsorption 78
  cell reaction 142
  combustion 80–9, *81*
  complex formation 104–12
  configuration 87–8
  dimerization 89–95
  dissociation 95–100
  formation 83(F)
  freezing 47–8, 55–65
  fusion 47–8, 55–65
  melting 47–8, 55–65
  phase transition 55–65
  reduction 100–4
  vaporization 33–7
entropy
  absolute molar 104(T)
  activation *415*
  cell reaction 142
  complex formation 104–12
  freezing 47–8, 55–65
  fusion 47–8, 55–65
  melting 47–8, 55–65
  phase transition 55–65
  reduction 100
  Trouton's rule *34*
  vaporization 33–7
EPROM 162
equation, *see also* law
  Arrhenius *401*
  Boltzmann 117, 198
  Clausius–Clapeyron *34*
  Debye 302, 316
  Einstein 328–30
  Eyring 414
  Gibbs–Helmholtz *47*, 53
  Guntelberg *117*
  Ilkovic *388*
  Laplace *67*
  Lorentz *187*
  Mark–Houwink *294*
  Nernst 122, *128*, 138, 379
  of state
    perfect gas *4*, 34, 55
    virial *9*, 18
  Onsager *361*
  perfect gas *4*, 34, 55
  Poiseuille *351*, 369, 451
  Poisson 117, 428
  Poisson–Boltzmann 117
  Schrödinger 280
  Stokes 330, *368*
  Van't Hoff
    isochore *91*, 100, 216, 401
    isotherm *90*, 101, 217, 415
    virial *9*, 18
equilibrium
  constant
    acid dissociation 139, 155, 156, 363
    base salt hydrolysis 364
    cell e.m.f. 130–57
    complex formation 106
    dimerization 90
    gas dissociation 95
    gas reaction 90, 95, 100
    gas-solid reaction 100
    $K_p^{\ominus}$ 90, 95, 100
    precipitation 157
    reduction 100
    temperature dependence *91*, 100, 216
  gas-liquid 34
  liquid-solid 46–55
  oscillation 220, *254*
  separation *337*
equipartition of energy 234, 327
equivalence point *147*
errors *xviii–xix*
  accumulation xx
  in theory xix
ethanal
  n.m.r. 269–78
  photolysis 441–9
ethane, rotation barrier 280–7
ethanoic acid
  acidity constant 359–67
  /chloroform/water system 22–8
  density 27(T)
  dimerization 89–95
ethyl benzoates, hydrolysis 390(F)
ethylenediamine complex 105
ethyne i.r. frequencies 228(T)
eutectic 58–65
excess volume *30*
exchange reaction kinetics 382–6
explosion 85, 469–77
  limits 470(F), 470–3
  peninsula 470(F)
extended Hückel theory 282
extinction coefficient *215*
Eyring equation 414

face centred cubic lattice 340
Faraday constant *121*
Fe(II)/Fe(III) couple 145–51
Fenske helices 38
Fermi resonance *233*
ferric-salicylate complex 248
ferroin *148*
Fick's law of diffusion *378*, 383
field

electric 184–6, 313–17
magnetic 269–79
film, Nernst *383*
fine structure
atomic spectra 196
n.m.r. 272
first
law
diffusion *378*, 383
thermodynamics *81*
order
focussing *174*
reaction *397*, 432
flame temperature 198–200
apparatus 202(F)
flash photolysis 460–9, 464(F)
Flory–Fox formation 293
flow
capillary 294–7, 350–2
free molecular *353*
hydrodynamic 330, 368
Knudsen number *353*
laminar 350–2, 451
method 100–4, 449–60
plug 451
Poiseuillian 350–2, 451
sonic *353*
Stokes equation 330, *368*
tube 455(F)
turbulent *353*
viscous 330, 368
flowing junction *126*
flowmeter
bubble 102(F)
capillary 451, 455(F)
oil 451, 455(F)
fluorescence 261–9
spectrophotometer 265(F)
flux, ions 373, *378*, 383
focusing, first order *174*
forbidden transition 194
force
constant *220*
dispersion *337*
virial term *339*
formal redox potential *147*
FORTRAN 345
Fourier transformation 210, 279
fractionating columns 37–9
fragment
ion 172, 180
primary *172*
secondary *172*
Franck–Condon principle *255*
free
energy 33
activation *415*
cell reaction *142*
complex formation 107
e.m.f. *142*
gas equilibrium 90
Gibbs–Helmholtz equation *47*, 53
liquid-vapour equilibrium 33–4
polymers 289
solid-gas equilibrium 101
solid-liquid equilibrium 47
temperature variation 47, 53, 143
molecular flow *353*
path, mean *348*, *353*
freely jointed chain 292(F)
freezing point depression 46–52, 55–8
frequency
overtone *221*, 226(T)
radiation 168(F), 193–280, *194*, *see also* energy levels
sound 236
Freundlich isotherm 79
fugacity 125
fumaric acid, combustion enthalpy 83(F)
functional group analysis
i.r. 169
n.m.r. 269–80

gas
adsorption 73–9
analysis 102(F), 175–82, 444–9
burette 75(F)
chromatogram 448(F)
chromatography 444
composition 102, 175–82, 444–9
densities, limiting *4*
density balance 3–9
electrode *125*
equilibrium properties 3–20
heat capacity 230–40
ideal 4, 34, 55
Joule–Thomson coefficient 16–20
liquid equilibrium 34
mixtures 102, 175–82, 444–9
molar mass 3–9
pressure 5, 11, 15, 95–8, 441, 451
reactions 89–104, 441–77
solid reactions 100–4
viscosity 350–9
gauge, *see* pressure gauge
Gaussian distribution 330(F), 331(F), 347(F)
Geiger counter *434*
Geiger–Müller tube 425–428, 426(F)
characteristic curve 427(F)
threshold *428*
gel *305*
gelatin 304
general acid and base catalysis *409*
Gibbs
free energy, *see* free energy
Helmholtz equation, *47*, 53
phase rule, *see* phase rule
glass
electrode 126, *152*, 153(F), 380

glass (*cont.*)
  helical pressure gauge 95–8
  transition *313*, *317*
glucose, mutarotation 406–13
good solvent *289*
graphs xxiv–xxvi
Grotrian diagram
  helium 198(F)
  hydrogen 195(F)
  sodium 196(F)
growth
  curve 434(F)
  droplet 388(F)
Guggenheim and Swinbourne method *408*, 416
Guntelberg equation *117*
gyration, radius of *291*, 304

half-cell potential 122
half-life *433*
half-wave potential *388*
Halverstadt–Kumler procedure 192
Hamiltonian 280
Hammet σ parameter *390*, 391(T)
hard sphere model 354
harmonic oscillator 220, 253, 254(F)
Harned cell *137*, 140(F)
  modified 139
heat *81*
  capacity
    bomb calorimeter 82
    chloromethane 230–40
    gas 230–40
  enthalpy, *see* enthalpy
heating curve 88(F), 110(F)
helical pressure gauge 95–8
helices, Fenske *38*
helium
  Grotrian diagram 198(F)
  spectrum 197
Hess's law 83, 87
heterodyne beat method 188
heterogeneous reaction 382–6
hindered rotation 285–6
histogram
  count rate 430(F)
  particle velocities 347(F)
Hittorf method *376*
homogeneous catalysis 409
hot bands 254(F), *255*
Hückel theory, extended 280–7
Huggins constant 294
hydrodynamic flow 330, 368
hydrogen
  atom, spectrum 194
  bond 89, 214–19, 271
  chloride
    liquid junction potential 377–82
    overtone frequencies 221, 226(T)
    rotation-vibration transitions 222(F)
    solution activity coefficient 135(T)
    /water azeotrope 40(F)
  cyanide
    energy levels 206
    interferogram 212(F)
  electrode 125(F), 152
  oxygen explosion 469–77
hydrolysis
  anilinium hydrochloride 364
  constant *364*
  ethyl benzoates 390(F)
  iodomethylbutane 413–8
  methyl chloroformate 418–25
  salts 364

ideal
  dilute solution *47*
  gas, *see* perfect gas
  mixture *37*
  solution 47, 52
Ilkovic equation *388*
immiscible liquids 22–8
imperfect gas 9–16
inaccurate theory xix
independent migration of ions *362*
indicator 145–51
inert
  metal | gas electrode *125*
  salts, effect
    reaction rate 402–5
    solubility 114–20
inertia, moment of 221
infrared
  spectrometer 217, 224(F)
  spectrum
    acetylene 223–9, 228(T)
    carbon disulphide 244(F)
    chloromethane 239(T)
    deuterium chloride 226(F)
    diatomics 220–7
    ethyne 223–9, 228(T)
    fundamentals 221
    hydrogen bonding 215(F)
    hydrogen chloride 221(F), 222(F), 225–7
    isotope effect 225
    overtones 221
inherent viscosity 293(T)
initial rates method *400*, 400–6
interaction potential 337–9
interferogram 212(F)
  background 210(F)
  signatures 210
interferometer
  acoustic 237(F)
  Michelson 209(F)
intermolecular forces 336–59
inter-system crossing 263
intrinsic viscosity 293(T)
inversion method 358

iodination of propanone 396–9
iodine
  atom recombination 460–9
  clock reaction 400–5
  correlation of states 256(T)
  dissociation energy *253*
  flash photolysis 460–9
  molecular potential energy curves 254(F)
  molecular spectrum 253–61
iodomethylbutane hydrolysis 413–8
ion(ic)
  activity *116*, 121, 135(T), 138–9, 146–7, 151–4, 378–80
  association 376
  atmosphere *116–7*
  conductivities 363(T)
  dynamic properties 359–94
  electron multiplier *174*
  equilibrium properties 113–64
  exchange 382–6
  fragment 172
  metastable *172*
  mobility *368*, 373
  molecular *172*
  selective electrode *126*
  source 174
  strength *116*, 402
ionization
  efficiency curves 173
  energy 197, 204
  potential 197, 204
i.r., *see* infrared
irreversible waves 389
isenthalpic Joule–Thomson coefficient *17*
isochore, Van't Hoff *91*
isoelectric point *304*
isolating gauge 97
isosbestic point *248*(F)
isoteniscope *36*
isotherm
  adsorption 73, 78
    BET 78
    Freundlich 79
    Langmuir *73*
    Temkin 78
  Van't Hoff *90*, 217
isotope
  abundance 180(T)
  decay 432–4
  dilution *440*
  effect 225
iterative process *421*

Job plot *249*
Joule–Thomson
  apparatus 18(F)
  coefficient 16–20
  effect 13
junction potential 126, 377–82

$K_p^{\ominus}$ 90, 95, 100
katharometer *445*
keto-enol tautomerism 271, 399
kinetic
  salt effect *402*
  theory 354
kinetics, reaction
  consecutive 418–25
  diffusion-controlled 382–6
  exchange 382–6
  fluorescence 261–9
  gas-phase 441–77
  heterogeneous 382–6
  parallel reactions 418–25
  polarography 386–94
  radioactive decay 432–4
  solution 396–441
Kirkwood–Riseman theory 293
Knudsen
  flow *353*
  number *353*
Kohlrausch law *362*
Koopman's theorem 281

laminar flow *350*
Langmuir isotherm *73*
Laplace equation *67*
laser v
  light scattering 311
  magnetic resonance 459
  Q-switched 469
  Raman spectroscopy 240–8
law, *see also* equation
  Beer–Lambert *214*, 249
  Boyle 10
  Debye-Hückel *117*
  Fick *378*, 383
  First thermodynamic *81*
  Kohlrausch *362*
  Ostwald dilution *363*
  perfect gas *4*, 34, 55
  Raoult *37*, 40, 41
  Second thermodynamic 80
LCAO approximation *281*
lead castle *434*
least-squares method xxv, xxvi(F)
Lennard–Jones
  fluid 344(F)
  potential *337*, 338(F)
Le Roy–Bernstein extrapolation *260*
lever rule *25*
ligand *105*
light
  absorption 206–69, 461
  emission 193–205, 450
  scattering
    photometer 306(F)
    polymers 300–12
    Raman 240–8
  trap *454*

limiting
  gas densities *4*
  law *117*, 138, 403
limpets 440
line reversal 199
linear
  combination of atomic orbitals *281*
  free energy relationship 412
  polyatomic molecules, spectra 223–9
liquid
  gas equilibrium 34
  junction 377(F), 381(F)
    potential 126, 377–82
  solid equilibrium 46–55, 382
  structure 66, 341(F)
  surface tension 66–72
  vapour composition 39(F), 44(T)
  viscosity 294–7, 368–72
lithium chloride solution
  conductivity 368–72
  liquid junction potential 377–82
  viscosity 368–72
LJ, *see* Lennard–Jones
local field, electric *185*
London forces *337*
long range order *341*
longitudinal pressure wave *235*
Lorentz equation *187*
loss tangent *315*
Lyman series 195(F)

machine language 162
macromolecules, structure 288–321
magnetic
  field 269–79
  moment, proton 270
magnetogyric ratio *270*
maleic acid, combustion enthalpy 83(F)
malonic acid, titration 152, 155–6
manometer 5(F), 18(F), 35(F), 69(F), 75(F), 97(F), 236(F), 355(F), 445(F), 455(F), 465(F)
Marcus theory 264
Mark–Houwink equation *294*
mass
  action 395
  atomic, *back endpaper*
  average molar mass *294*, *300*
  charge ratio *172*
  spectrum 182(T, F)
    bromoethane 182(F)
    interpretation 179–81
  spectrometer 170–5, 171(F)
    double-focussing *183*
McCabe–Thiele plot 39
McLeod gauge 355
mean
  free path *348*, *353*
  ionic activity coefficients *116*, 135(T)
mechanics, quantum 220, 241, 253, 280–7
melting point depression 46–52, 55–8
membrane electrode *126*, 153(F)
meniscus 67
mercury
  atomic lines 205(T)
  manometer, *see* manometer
metal, insoluble salt electrode *123*
metal | metal ion electrode *123*
metastable
  energy levels *198*
  ions *172*
  peaks *173*
methoxycarbonylpyridinium ion 418
methyl
  bromide, mass spectrum 182(F)
  chloride, *see* chloromethane
  chloroformate hydrolysis 418–25
methylamine, Newman projection 285
Michelson interferometer 209(F)
microcomputer v, 158–64, 286, 349
  Apple 163
  BBC 158–64
  Commodore 349
  Rockwell Aim 158–64
microprocessor 162
microscope 331
  comparator *258*
minimum imaging *340*
mixed mode arithmetic *346*
mixtures
  azeotropic 40
  distillation 37–46
  ideal *37*
  phase diagram 24(F)
mobility, ionic *368*, 373
modified stillhead 41(F)
molality *29*
molar
  conductivity *360*
  entropy 104(T)
  mass
    elements, *back endpaper*
    freezing point depression 46–52
    gas 3–9
    mass average *294*, *300*
    number average *294*
    polymers 292–4, 300–12
    solutes 46–52
    viscosity average *294*
molarity *29*
mole fraction *29*
molecular
  collisions 348(F)
  diameter *337*, 354
  dynamics 336–49
  orbital theory 207, 280
  quantum mechanics 280–7
  structure, *see* structure
  velocity distribution 347(F)
  weight, *see* molar mass

molecules
  diatomic 220–30, 253–61
  linear 223–9
  polyatomic 223–9
  symmetric 245
  symmetric top 230–40
moment
  inertia 206, 221
  magnetic 270
monochromator 224(F), 265(F), 464(F)
Monte Carlo calculations *349*
moving boundary 372–7
multiplicity *196*
mutarotation
  coefficient *407*
  dextrose 406–13
  glucose 406–13

naphthalene, solution properties 52–8
Nernst
  equation 122, *128*, 138, 379
  film *383*
neutralization 154
neutron
  activation analysis 440
  diffraction 341(F)
Newman projection 285(F)
Nicol prism *410*
nitric oxide
  molecular orbitals 207
  rotational energies 208
nitrobenzene reduction 386–94
nitrogen tetroxide, dissociation 95–100
n.m.r., *see* nuclear magnetic resonance
noise 211, 224
Normal distribution 330(F), 331(F), 347(F)
normal modes, vibration
  chloromethane 231(F)
  ethyne 223
  linear polyatomic 223
  symmetric 245(F)
  symmetric top 230
  tetrachloromethane 245(F)
nuclear magnetic resonance 269–80
  coupling 272
  hydrogen bonding 271
  keto-enol tautomerism 271
  saturation *276*
  shielding *270*
  side-bands *277*
  spectrometer 274(F)
number average molar mass *294*

omega integral *355*, 358(T)
Onsager equation *361*
operator, Hamiltonian 280
optical
  activity *406*
  pyrometer *201*
orbital
  approximation *280*
  atomic 193–8, 281
  linear combination *281*
  molecular 207, 280
order
  long range *341*
  overall *397*
  reaction *397*
  short range *341*
  spectra 259
orientation polarization *313*
oscillator
  anharmonic 220, 254(F)
  polymer 300
  simple harmonic *220*, *253*, 254(F)
osmosis 289, 311
Ostwald
  dilution law *363*
  viscometer 295(F), 369
outer sphere redox reaction *264*
overpotential *389*
overtone *221*, 226(T)
oxidation-reduction, *see* redox
oxygen
  atom reactions 449–60
  hydrogen explosion 469–77
  liquid vapour pressure 71(T)
  surface tension 66–72

P branch *222*, 226(F) 232
pair
  distribution function *340*, 341(F)
  potential energy function *337*
    BBMS 338(F)
    Lennard–Jones 338(F)
parallel
  bands *224*, 231
  reactions 418–25
  vibrations 224, 231
parent ion 172
partial molar volume 28–33
particles, motion 325–94
partition function 233–5
Paschen series 195(F)
PEMA transitions 312–21
penetration 195
perfect gas law *4*, 34, 55
periodate 114, 359
periodic boundary condition *340*
permittivity, vacuum 184
perpendicular
  bands *224*, 231
  vibrations 224, 231
persulphate-iodide reaction 400–6
pH *151*
  electrode 153(F)
  measurement 151–64
  meter *152*
  titration 155(F), 156(F)

phase *22*
  diagram
    argon 344(F)
    binary liquid/vapour 37
    chloroform/ethanoic acid/water 22–8
    Lennard–Jones fluid 344(F)
    three-component system 22–8
  equilibria 21–79
  rule *22*
  structure 340
  transition 55, 61(F)
phenol/dioxan, i.r. spectrum 215(F)
phenyl methanol, n.m.r. 269–80
phosphorescence *261–3*
photochemical reactions 441–69
photolysis *441*
  ethanal 441–9
  flash 460–9
  iodine 461
photometer, light scattering 306(F)
photomultiplier 455(F)
photosynthesis 439
physical constants, *front endpapers*
physisorption *73*
pK_a *155*
plait point 24(F), *25*
Planck's constant *193*
planimeter 63
plate, theoretical *38*
plating of electrodes 133–5, 140
platinizing solution *136*
platinum | hydrogen electrode 125(F), 152
plug flow 451
point group 230
Poiseuille's equation *351*, 369, 451
Poiseuillian flow *350*
Poisson distribution *428*
polarimeter 411(F)
polarity
  cell *128*
  dipole *184*, 191
  electrode *128*
polarizability, molar *186*
polarization *184*, 313
polarogram 387(F)
polarographic maximum 393
polarography 386–94
  apparatus 392(F)
polyatomics, i.r. spectra 223–9
polyelectrolyte *304*
polyethylmethacrylate, transitions 312–21
polymer
  conformation 290(F)
  crystallinity 305
  light scattering 300–12
  molar mass 292–4, 300–12
  osmosis 289, 311
  structure 288–321
  unperturbed dimensions *289*
poor solvent *289*
population 196, *198*, 276
potassium
  chloride
    liquid junction potential 377–82
    salt bridge 126
  nitrate, salt bridge 127
potential
  chemical 47, 52, 117, 121
  difference, *see* e.m.f.
  energy curve
    argon 338(F)
    iodine 254(F)
  function *336*, 338(F)
  half-cell *122*
  redox *145*, 145–51
  standard electrode *121*
potentiometer *131*(F)
potentiometric
  selectivity coefficient *154*
  titration 145–51, 150(T)
Pourbaix diagrams *139*
precipitation titrations 156
precision *xviii*
predictor/corrector method 343
predissociation 461
pre-exponential factor *401*, 418
pre-saturator 125(F)
presentation of results xvii
pressure gauge
  Bourdon 15, 95–8
  glass helical 95–8
  hysteresis 15
  isolating 97, 99
  manometer 5(F), 18(F), 35(F), 69(F), 75(F), 97(F), 236(F), 355(F), 445(F), 455(F), 465(F)
  McLeod 355
  transducer 11(F)
primary
  fragment *173*
  kinetic salt effect *402*
  structure *300*
principal quantum number *194*
prism 200, 201(F)
  Nicol *410*
probability
  cumulative 331(F)
  graph paper *331*
  Normal 330(F), 429
  Poisson *428*
program, *see* computer program
propagation of errors xviii–xxiv
propanol/water system 43–5
propanone iodination 396–9
properties, colligative *46*
protactinium 425, 432
pure rotation spectroscopy 206–14
pyknometer 28
pyrometer *201*
Pythagores' theorem 339, 345

Q branch *222*, 232
quantum
  efficiency *444*
  mechanics 220, 241, 253, 280–7
  number
    angular momentum *232*
    principal *194*
    rotational *221*, 224, 232
    vibrational *220*, 224, 232, 242(F), *253*
  yield *444*
quaternary structure *300*
quenching of
  fluorescence 261–9
  Geiger–Müller tubes 426
  reactions 397
quinonoid structure 391

R branch *222*, 226(F), 232
r^2 space 345
radial distribution function *340*, 341(F)
radiation
  black body 200(T)
  detector 200, 209, 224, 246, 251, 258, 265(F), 274(F), 306(F), 426(F), 455(F), 464(F)
radioactive decay
  constant 432
  curve 433(F)
  kinetics 432–4
  series *432*
radiochemistry 425–41
radius of gyration *291*, 304
Raman *241*
  effect *241*
  spectrometer 246(F)
  spectroscopy 240–8
random
  coil 290(F), 291, 300, 303
  disintegrations 428
  velocities 349
  walk 327(F), 335
Raoult's law *37*, 40, 41
Rast's method *47*
rate
  coefficient *397*, 468(T)
  determining step 399
  effective coefficient *397*, 398
  electrode process 386–90
  equation *397*, 400
  initial *400*
  reactions, *see* kinetics
ratio-recording spectrometers 225
Rayleigh 300, *449*
  ratio *302*
  scattering 300–2
reaction
  chain 442, 470–2
  chemiluminescent 450
  clock 400–6
  consecutive 418–25
  diffusion controlled 382–6
  electron transfer 261–9
  enthalpy, *see* enthalpy
  first order *397*, 432
  fluorescence quenching 261–9
  gas 441–77
  heterogeneous 382–6
  intermediate 414, 419, 440, 442, 452, 453, 461, 471
  ionic 145–51, 261–9, 382–6, 396–406, 413–25
  kinetics, *see* kinetics
  mechanism 261–9, 382–6, 396–476
  order *397*
  outer sphere redox *264*
  parallel 418–25
  photochemical 441–69
  profile 414(F)
  rate
    ionic strength effect *402*
    kinetics, *see* kinetics
    temperature variation 401
  redox 145–51, 264
  second order *397*
redox
  electrode 148
  formal potential *147*
  potential *145*, 145–51
  reaction 145–51, 264
  titration 145–51, 150(T)
reduced quantities *343*
reduction
  enthalpy 100–4
  nitrobenzenes 386–94
  oxidation, *see* redox
reference electrode 124, 392(F)
relations, *front endpapers*
relative
  atomic masses, *back endpapers*
  permittivity *185*, 316
    cell 190(F)
relaxation *312*
  dielectric 312–21
  spin-lattice 276
  techniques 407
  time *313*
refractive index 43, 186, 190
refractometer 42, 46, 190
repeatability *xviii*
reports xvii
reproducibility *xviii*
reptation *313*
resistivity *360*
resolution 213, 227–9
resonance
  fluorescence 459
  nuclear magnetic, *see* nuclear magnetic resonance
results
  errors xviii–xxiv

results (*cont.*)
  graphs xxiv
  least-squares analysis xxv
  line-fitting xxv
  presentation xvii
reversible waves 388
Reynold's number *353*
rigid rotor *221*
ringing 275(F)
RKR analysis 261
Rockwell Aim microcomputer 158–64
rotation
  constant *221*
  energy levels 206–8
  partition function 233
  quantum number *206*, 221, 232
  specific optical *406*
  spectrum 206–14, 212(F)
  vibration spectrum 220–30, 226(F)
rotor, rigid *221*
rubber 317
Rydberg constant *194*
Rydberg–Klein–Rees analysis 261

safety xxvi
salt
  bridge *126*
  hydrolysis 364
  kinetic effective *402*
salting-in *114*
sampling interval 445
saturation
  n.m.r. *276*
  solution 114
scattering
  anti-Stokes *241*
  neutron 341(F)
  polarized light 301(F)
  Raman 241
  Stokes *241*
  X-ray 341(F)
Schrödinger equation 280
second
  law of thermodynamics 80
  order
    reaction *397*
    spectra 259
  virial coefficient *9*
    carbon dioxide 19(T)
secondary
  standard electrode *124*
  structure *300*
secular
  determinant *282*
  equations 282
seeding *50*
selection rules
  atoms 193–8
  rotational 206–8
  vibrational 220–9, 232–3
  vibrational-electronic 254–7
  vibrational-Raman 241–5
selectivity coefficient, potentiometric *154*
self-exchange rate coefficient *264*
self-quenching, Geiger–Müller tubes 426
series
  Balmer 195(F)
  Lyman 195(F)
  P,R branch 223
  Paschen 195(F)
shear viscosity coefficient *350*
Sheppard adjustment *430*
shielding 195
  constant *270*
shim coils *274*
short range order *341*
sign convention
  electrochemistry *128*
  Stockholm *128*
  thermodynamics *81*
signal averaging *469*
signal-to-noise ratio *211*, 224
significance level 432
silver
  electrode 120–3
  nitrate, ionic activity coefficient 135(T)
  silver chloride electrode *123*, 134, 379
simple harmonic oscillator *220*, *253*, 254(F)
simultaneous reactions 418–25
singlet *197*, 198(F)
sinker 30
sky, colour 302
slip flow *353*
sodium
  chloride
    flame 201
    liquid junction potential 377–82
  hydroxide
    activity coefficient 135(T)
    titrations 151–64
sol *305*
solid state transition 55, 61(F)
solubility 130
  curve 52
  ideal solution 47, 52
  ionic salt 114–20, *115*, 365
  naphthalene 52–5
  potassium periodate 114–20, *115*, 365
  product *115*, 117, 130
    concentration *115*
    true thermodynamic *117*
solution
  calorimeter 107(F)
  ideal 47, 52
solvent, polymers
  good *289*
  poor *289*
  theta *290*
sonic flow *353*
sound, speed 236–8

space shuttle 72
specific
  heat, *see* heat capacity
  ion electrode *126*, 153(F)
  optical rotation *406*
spectral
  absorptivity *199*
  emissivity *199*
spectrograph, Ebert 258(F)
spectrometer
  calibration 205(T), 258(T)
  constant deviation 200
  direct reading 201(F)
  double-beam 224(F)
  i.r. 224(F)
  n.m.r. 274(F)
  u.v./visible 251
spectrophotometer
  fluorescence 265(F)
  u.v. 251
spectrum
  absorption 206–69
  acetylene 223–30
  alkali metals 195–7
  atomic 193–206
  blank 178(F)
  chloromethane 230–40
  complex 215(F)
  deuterium chloride 226(F)
  electromagnetic radiation 168(F)
  emission 193–205
  ethyne 223–30
  excitation 261–9
  fluorescence 261–9
  helium 197
  hydrogen 194
  hydrogen bonding 215(F)
  infrared 215(F), 226(F)
  iodine 253–61
  mass 182(T, F)
  n.m.r. 269–80
  phenol-dioxan 215(F)
  pure rotation 212(F)
  rotation-vibration 220–30, 226(F)
  sodium 195
  symmetric top 230–40
  ultraviolet 248–61
speed
  molecular 347(F), 354, 454
  sound 236–8
spherical top 230–3
spin-lattice relaxation 276
spin-spin coupling *272*
spinning sidebands *277*
stability constant, of a complex *106*, *249*
standard
  cell *132*
  deviation
    calculation 438(T)
    Normal 331(F), *429*
    Poisson *429*
  electrode potential *121*, 129, 138
    silver 129
    silver, silver chloride 138
  error *429*
  state, elements 81, 121
standardization, electrode 153
statistical analysis xxiii
steady state approximation 264, 443, 452, 458
Stern–Volmer plot *264*
sticking probability *73*
stillhead 41(F)
Stockholm convention *128*
stoichiometric equation 396
Stokes
  equation 330, *368*
  scattering *241*
straight line
  graphs xxiv
  least-squares fitting xxv
structure
  atoms 170–83, 193–206
  macromolecules 288–321
  phase 340
  primary *300*
  quaternary *300*
  secondary 300
  simple molecules 170–93, 206–87
  tertiary *300*
supercooling *47*, 56
surface
  adsorption *73*, 73–9
  area 78, 454, 471
  pressure 66
  reaction kinetics 382–6, 454, 471
  tension *66*, 66–72
    cell 68(F)
suspended level 295
symbols *xiii–xv*
symmetric
  molecule 245
  top molecule 230–3
symmetry
  axis 230
  centre 240
system
  AB spin *272*
  AX_2 spin *273*
  benzene/tetrachloromethane 41–6
  ethanoic acid/chloroform/water 22–8
  propanol/water 43–5
  three-component 22–8
  two-component 41–6

Temkin isotherm 78
temperature
  coefficient
    conductivity 368–72
    e.m.f. 142–5

temperature (*cont.*)
  coefficient (*cont.*)
    viscosity 354, 369
  composition diagram 40(F)
  electronic *200*
  jump method 407
  translational *200*
term
  symbol 194–8, *196*, 256, 461
  value 220, 253
tertiary structure *300*
tetrachloromethane
  benzene mixtures 37–46
    liquid/vapour compositions 44(T)
  normal vibrations 245(F)
  Raman spectrum 245
tetrathionate 114, 400
theoretical plate *38*
theory, inaccurate xix
thermal analysis 59(F)
thermocouple 101(F)
thermodynamics
  chemical reactions 80–112
  first law *81*
  hydrogen bonding 214–19
  phase equilibria 33–65
  second law 80
  solutions 46–58
  statistical 233, 349
theta solvent *290*
thiosulphate 114, 397, 400
thorium decay 432
three component system 22–8
threshold, Geiger counter *428*
tie lines *25*, 24(F)
time
  dead *427*
  relaxation *313*
titration
  amphoteric compounds 156(F)
  pH 155(F), 156(F)
  potentiometric 145–51, 150(T)
  precipitation *156*
  redox 145–51, 150(T)
trajectories
  Brownian particles 327(F)
  Lennard–Jones fluid 348(F)
  molecular ions 170–3
transfer coefficient 389
transference numbers *372*, 377–82
transient recorder 465
transition
  allowed 194
  α *313*, *317*
  ammonium nitrate 61(F)
  β *313*, *317*
  carbon tetrabromide *55*
  electronic 168(F), 193–8, 248–69
  forbidden 194
  glass *313*, *317*
  spectroscopic 168(F)
  state theory *414*
  temperature, depression 55–8, 58–65
translational partition function *234*
transmittance *215*
transport
  number
    cell e.m.f. method 377–82
    equation *373*
    Hiltorf method 376
    moving boundary method 372–7, 381
  properties, gases 350–9
triangular co-ordinates 24(F)
triple point 344(F)
triplet *197*, 198(F)
Trouton's rule *34*
tungsten lamp temperature, corrections 200(T)
turbulent flow *353*
two-component systems 37–46
Tyndall effect 241

Ubbelohde viscometer 296(F)
ultraviolet
  spectrophotometer 251, 421
  spectrum
    Fe(III)/salicylate complex 248(F)
  iodine 253–61
units, *front endpapers*
unperturbed dimensions *289*
uranium decay 432
uranyl nitrate 432

vacuum permittivity 184
van de Graaff electron accelerator 469
Van't Hoff
  isochore *91*, 100, 216, 401
  isotherm *90*, 101, 217, 415
vaporization
  enthalpy 33–7
  entropy 33–7
vapour
  liquid composition 39(F), 44(T)
  pressure
    liquid oxygen 71(T)
    temperature measurement 71
    temperature variation 33–7
velocity
  constant, *see* rate coefficient
  distribution 347(F)
  linear flow *451*
  molecular 347(F), 354, 454
vibration
  anharmonic 220, 254
  frequency 220
  harmonic 220, 254(F)
  isotope effect 225
  normal modes *223*, 231
  overtone 221
  parallel 224, 231

partition function *234*
perpendicular 224, 231
Vigreux column *38*
virial
coefficient
carbon dioxide 19(T)
measurement 9–16
equation of state *9*, 18
term *339*
viscometer
Cannon–Fenske 295(F)
gas 350(F), 355(F)
glass capillary 294–7
Ostwald 295(F), 369
Ubbelohde 296(F)
viscosity
air 357
argon 350–9
average molar mass *294*
electrolytes 368–72
gases 350–9
inherent 293(T)
intrinsic 293(T)
nomenclature 293(T)
polymer solutions 289–99
reduced 293(T)
slip correction 353
specific 293(T)
temperature dependence 354
water 335(T)
viscous flow 330, 368
voltmeter, digital *132*

Walden
product *368*
rule *368*
water
conductivity 366
density 27(T)
hydrogen chloride azeotrope 40(F)
vaporization 37
viscosity 335(T)
wavenumber *194*
weak
acids, dissociation constants 139, 155
electrolytes, conductivity 361–3
well depth *337*
Weston standard cell *132*
Wolfsberg–Helmholtz method 286
work, maximum useful 120
working
electrode 392(F)
voltage *428*

z-average distance *303*
Z-matrix 285(T)
zero
path difference position *209*, 212(F)
point energy *254*
Zimm plot 302–5, 304(F)
zinc
amalgam 143, *145*
/cadmium/mercury lines 205(T)
sulphate activity coefficient 135(T)